COSMOLOGY

*Historical, Literary, Philosophical,
Religious, and Scientific
Perspectives*

GARLAND REFERENCE LIBRARY OF THE HUMANITIES
(VOL. 1634)

COSMOLOGY

Historical, Literary, Philosophical, Religious, and Scientific Perspectives

EDITED BY

Norriss S. Hetherington

GARLAND PUBLISHING, INC.
New York & London
1993

Library of Congress Cataloging-in-Publication Data

Cosmology : historical, literary, philosophical, religious, and
  scientific perspectives / edited by Norriss S. Hetherington.
      p. cm. — (Garland reference library of the humanities ; vol.
  1634)
      ISBN 0-8153-1085-4. — ISBN 0-8153-0934-1 (pbk.)
      1. Cosmology. I. Hetherington, Norriss S., 1942-
  II. Series.
  QB981.C822 1993
  523.1—dc20   92-46276
                                                          CIP

Printed on acid-free, 250-year-life paper
Manufactured in the United States of America

for the memory of
Vic Thoren
13 May 1935–9 March 1991
teacher, scholar, colleague, and friend

TABLE OF CONTENTS

III. Medieval Cosmology and Literature

IV. The Scientific Revolution

V. Galaxies: from Speculation to Science

VI. The Expanding Universe

VII. Particle Physics and Cosmology

VIII. Cosmology and Philosophy

IX. Cosmology and Religion

PREFACE

Orson Welles in *Citizen Kane* attempted to make a movie that was not so much a narrative of action as an examination of character through intertwining perceptions. Departing from linear storytelling, from synchronous cause and effect, he drew his audience into a labyrinth of images. Indeed, he wanted the viewer to enter the film, become a part of it, and remain there to its conclusion.

In attempting to examine the character of cosmology, in attempting to understand something about the human attempt to comprehend the surrounding cosmos, recourse to multiple images also is necessary. Cosmology, too, is a story that warrants telling many times from many perspectives. You, the reader of this book, are invited to enter cosmology through the following pages and perceptions, to become a part of the human attempt to understand our universe, and to remain—not for an hour or two, but for a lifetime.

I. Cosmology and Culture

Introduction

DIFFERENT CULTURES, DIFFERENT COSMOLOGIES

he culture of a people, the totality of their human creativity and intellect, has many components. These include aesthetics, behaviors, beliefs, biases, conventions, customs, desires, emotions, fads, fashions, forms, habits, inclinations, manners, methods, modes, mores, paradigms, patterns, practices, predispositions, preferences, styles, superstitions, tastes, traits, values, and ways. Culture, as such, is more than a little arbitrary; it is even irrational at times. Culture is something pertaining to others, something alien and exotic for us to study and wonder at in the behavior of other people.

Our own ways, of course, seem to us—but *not* to others—natural, inevitable, and beyond question. At the pinnacle of our culture is our science. Its nature and content, when we think about it at all, presumably are dictated by an underlying factual universe increasingly discoverable by experiment and observation. In such a universe an account of our cosmology—our understanding of the organization and evolution of the universe—would consist of a descriptive theory based on little more than a straightforward list of observational evidence.

Yet we are not entirely unaware that we, too, have a culture, just as other peoples have their own cultures. Correspondingly, our science, although more subject to empirical test than are many of our other intellectual creations, also must be recognized as part of our culture, and thus as an intellectual creation worthy of study. Once cosmology is thought of in this manner, we have opened one of the greatest of human achievements to analysis and to an increased appreciation. Recognition of arbitrary patterns of thought and descriptive conventions in the cultures of other peoples can furnish us with a more productive perspective from which we can view our own behavior. So, too, can a study of the cosmologies of other peoples help us recognize our own cosmology for the human intellectual creation that it is.

Replacement of initial obliviousness to familiar aspects of one's own culture by astonishment, and sometimes even by outrage leading to action, is not unheard of in human history. Awakened or heightened sensitivity often follows from exposure to a familiar object or condition in a new setting. Henry George, after visiting New York, where he was shocked by the contrast between monstrous wealth and debasing want, returned home to California, now to perceive an unjust distribution of wealth there too and to write *Progress and Poverty*. Meanwhile, Edward Bellamy, a resident of the eastern United States, was about to publish *Looking Backward*, his solution for the problems of political and social inequality, after a consciousness-raising tour of urban slums in Germany, where the extent and consequences of man's inhumanity to man were somehow more apparent to him than they had been in his more familiar haunts in New York.

To bring such examples of cultural consciousness around to the topic of astronomy and cosmology, we might consider the case of Norman Lockyer, a nineteenth-century British scientist. On vacation in Greece, he noticed that the foundations of some temples had been changed long after their original construction, apparently to realign the temples in slightly different directions. Next, Lockyer expanded his investigation to Egyptian temples, using descriptions from German and French expeditions, and found solar and stellar alignments here also. Only after all this did Lockyer turn his attention to an object in England, and realize that Stonehenge, too, had been aligned to face sunrise at the summer solstice.

There is much to be said for the enlightening and intellectually liberating advantages of travel, with its exposure to different cultures. The current interest in multicultural education, for all its potential dark side of racial divisiveness and distortion of history to serve political agendas, by focusing long-lacking and well-warranted attention on the accomplishments of often-overlooked cultures promises ultimately to increase our understanding and appreciation of our own culture. Studying the cosmologies of many cultures constitutes an especially interesting part of a genuine multicultural education, and also will heighten our sensitivity to otherwise hidden aspects of our own cosmology.

Shared patterns of thought and descriptive conventions often differ from one culture to another. A look at the cosmologies of other cultures can help us recognize that some elements of our own cosmology are arbitrary choices and values. These elements might otherwise escape our notice and pass as inevitable developments. Native American cosmologies, we learn, were not tied to geometrical predictive models derived from the mathematics of spherical geometry, nor were they tied to any physics of the heavens, such as Aristotle's physics of aethereal spheres or Newton's force of gravity. Still, as with our own cosmology, Native American cosmologies were tied to empirical observations of celestial phenomena, to theoretical models rendering the observations intelligible, and to general explanatory themata influencing other aspects of the general culture as well. Who is to say, in any absolute and eternal sense, which cosmology was wrong and which was right?

Certainly our own Western cosmology has not passed through its various historical phases without occasional blind spots, without intellectual aberrations, at least when viewed from different cultural perspectives. While ancient Chinese cosmology linked the actions of people on earth to the behavior of the heavens and consequently employed astronomers to watch the heavens for irregularities, in the cosmos of both ancient Greece and the medieval West the heavens were assumed to be changeless, with the result that observations of change often either were misinterpreted or ignored. The supernova explosion of A.D. 1054, so bright that it was even visible briefly during the day as well as at night, is virtually absent from Western records but prominent in Chinese records, the longest running sequence of astronomical records made by man, and still useful in certain aspects of modern research. Nor was Chinese cosmology, sometimes criticized for an overwhelmingly observational bias, tied to any such dogma as that of the rigid Aristotelian-Ptolemaic conception of concentric crystalline spheres, which dominated Western thought for nearly two millennia. In contrast, the Xuan Ye cosmological tradition in China considered the heavens as infinite in extent with the celestial bodies floating about at rare intervals, with the speed of the luminaries depending on their individual natures, "which shows they are not attached to anything, for if they were fastened to the body of heaven, this could not be so."

The only civilization before that of the Greeks known to have produced a scientific astronomy capable of computing lunar and planetary phenomena with precision occurred in Mesopotamia, in the fertile region between the Tigris and Euphrates rivers, now known as Iraq. Yet Babylonian scientific astronomy, as powerful and pervasive as it was, seemingly was not applied to the broader cosmological problem of understanding the universe. The general approach to the physical world in ancient Mesopotamia remained rooted in an assumption of the inherence of the divine in nature—just the opposite of what we now applaud as one of the greatest achievements of Greek philosophy: the separation of science from religion. Primarily a study of the gods, Mesopotamian cosmology is better characterized as a theological study than a scientific study.

We might be tempted to conclude that an ability, and thus an opportunity, to transform their cosmology away from myth and religion and into something like science in our Western sense was possessed by the Babylonians, but that something in their culture dictated a different path. Such speculation would have an objectionable Whiggish nuance (in the sense of "Whig" history: historians unconsciously implying through careless writing, fueled by their knowledge of what happened, that Whig politicians in England had a foreknowledge of what their country would become and a conscious plan to achieve that result). Furthermore, any implication that the Babylonians missed an opportunity to transform their cosmology away from myth and religion also would have an unattractive ethnocentric nuance. We would be judging Babylonian culture on the basis of its failure to achieve the Greeks' separation of science from religion (to whatever extent the Greeks really did achieve such a separation), not as a culture in its own right with its own values and aspirations. Still, it is tempting to wonder *What if . . . ?*

Scientific developments first appeared in the Greek world around 600 B.C., with new and distinctive approaches to the understanding of nature. This is a period of pre-Socratic science, before Socrates, from about 600 to 400 B.C. One major Presocratic tradition seems to have been to search for a basic substance or substances persisting throughout all changes. A second major tradition within Presocratic philosophy centered on Pythagoras, who believed that number was the principle of all things. For the Pythagoreans, the planets supposedly moved in a manner such that the music their motions produced was harmonious.

Any one of these studies—of Native American Cosmologies, of Chinese Cosmology, of Mesopotamian Cosmology, or of Presocratic Greek cosmology—could awaken us to an appreciation of cosmology as a human intellectual creation, not merely a collection of facts. Taken together, they must impress upon even the most resistant of us the fact that science in general, and cosmology in particular, can be done differently by different peoples in different places at different times.

The fact that different cultures have produced different sciences, including different cosmologies, suggests that science may be relative, that the validity of our modern Western cosmology may be limited geographically and temporally to modern Western civilization. Indeed, some sociologists of science have advocated a "strong program" of research in which scientific knowledge is to be understood as a product of "interests," with its categories and even its criteria of truth determined by cultural and social requirements. Seemingly there is no role in this conception of science for intellectual curiosity, nor interest in the possibility that truth may be an important objective of scientific activity. The so-called "strong program" does not deny the possibility of success in discovering the reality of the external world, but simply ignores this issue (except implicitly, to the extent that adherents of the program purport to be discovering truth while the outcome of every other intellectual endeavor is attributed to interests).

An alternate view of science sees it as cumulative, with scientists guided at least in part by the criterion of truthfulness. Cultural factors, especially in the selection of problems to pursue and the degree of social support for different sorts of studies, are not denied. There is a "Westernness" to modern cosmology. But modern Western cosmology is also seen, by some, as having a universal validity following from its grounding in empirical observations, the internal consistency of its reasoning, the comprehensiveness and explanatory power of its theories, and a progressive movement toward the ideal of discovering the reality of the external world.

In this sense of progression toward an ideal, science is very different from art. A particular work of art is not judged by its proximity to an absolute ideal of beauty. Aesthetic standards are recognized to vary from culture to culture. Nor is an early work of art superseded by later works. Scientific theories, however, are understood, by some people, to be superseded by more recent theories accepted as superior because they are based on a better set of subsequent observations, or are more comprehensive, or have greater explanatory power, or have better proofs.

There is an ongoing debate over whether every civilization has its own criteria of scientific validity, or whether scientific knowledge is universally valid. There are wide implications in the potential answers, and hence much heat in the debate. Assertions of superiority for modern Western science have been perceived, by some, as a threat to the dignity of other civilizations. Thus, those who would enhance the dignity of other civilizations may feel some need to disparage Western science. Ethnocentric arrogance, and even racism, may be charged against those who argue for the superiority of any particular civilization, though it is usually an appreciation for modern Western civilization that incites such epithets.

A cumulative and progressive understanding of science need not bar study of earlier cosmologies. But the study will not proceed far unless the cosmologies are studied as both science and cultural artifact. On the other hand, any understanding of modern Western cosmology must come to grips with its claim to universal validity (itself a cultural value) and its claim to scientific superiority over other cosmologies. Furthermore, the question of universal validity independent of particular (or other) cultural values can be a useful, heuristic probe into the character of other cosmologies, expanding our understanding of them and of their encompassing civilizations as well. Both for "arrogant ethnocentric racists" and for "politically correct multicultural relativists," a deeper understanding of all cosmologies, ancient and contemporary, Western and otherwise, including especially modern Western cosmology, will furnish both insight into cosmology and fuel for future discussions and debates over the nature of science, society, and civilization.

1. NATIVE AMERICAN COSMOLOGIES

The cosmologies of Native American peoples differ in several important respects from other, more familiar, cosmologies; and these distinctions must be made clear from the outset.

- First, as suggested by the title of this article, these cosmologies are most properly discussed in the plural. New World cosmologies do share several common themata that set them apart from the cosmologies of the Old World; nonetheless, their expressions of these shared themata differ substantially, and it would be a gross oversimplification to attempt to speak of a single Native American cosmology.

- Second, the cosmologies of the New World are not tied to a geometrical predictive model derived from the mathematics of spherical geometry, as were the cosmologies derived from Greek thought. Still less are they tied to any physics of the heavens, such as Aristotle's physical model of aethereal spheres that are moved naturally and uniformly in circles by unchanging divine beings, or the celestial physics of Newton and his successors in which the planets are material bodies that move under the same laws of motion as terrestrial bodies under the influence of a well-defined, but unexplained, force of gravity.

To the extent that in the high cultures of Mesoamerica cosmologies are tied to a predictive astronomy, these astronomies are arithmetical rather than geometrical, and they are concerned with the prediction of the dates of astronomically and astrologically significant events rather than with a kinematics describing the motions of the bodies in the heavens. These differences in astronomies reflect and contribute to the differences between Native American and European cosmological models.

Conversely, Native American cosmologies display these general characteristics of "traditional" understandings of nature:

- Conservatism.
- Resistance to change.
- A close inter-connectedness with society, myth, and ritual.

These characteristics should not be taken as defining character-istics, however, for in this regard Native American cosmologies differ only in degree from modern scientific cosmologies. Like modern cosmologies, Native American cosmologies are tied to:

- Empirical observations of celestial phenomena.
- Theoretical models that render those observations intelligible.
- General explanatory themata that guide a whole range of a culture's intellectual, political, and artistic endeavors, including those theoretical models themselves.

The two principal cosmological themata shared by many Native American cosmologies are a theme of fourness, of which the four-directionality of space seems primary, and a theme of duality, which relates the spatial pair of the below and the above, the temporal dualities of night and day and of winter and summer, and the related pair of that which is to come (the future) and that which is (the present). These themata are expressed differently as they reflect the different social and environmental circumstances of specific cultures, yet they frequently play a central role in Native American cosmologi-cal thought.

The Four Directions

The four-directionality of space seems to be the oldest and most widespread theme shared among Native American cosmologies. Olmec inscriptions from the first millennium B.C. incorporate quadruple and quintuple patternings, motifs indicating the four directions and the center of terrestrial space. Myths and rituals from a wide range of indigenous cultures, extending from Canada to Peru, incorporate the number 4, its multiples, and the four directions. Artistic representa-tions from a similar range of cultures document that the four direc-tions were crucial to the visual understanding of the cosmos for many native peoples.

A striking indication of the extent to which this theme of fourness, and particularly the four directions, permeates Native American cosmologies is found in the discussion of the number 4 by a Lakota wise man, Tyon, recorded early in this century:

> *In former times the Lakota grouped all their activities by fours. This was because they recognized four directions: the west, the north, the east, and the*

south; four divisions of time: the day, the night, the moon, and the year; four parts in everything that grows from the ground: the roots, the stem, the leaves, and the fruit; four kinds of things that breathe: those that crawl, those that fly, those that walk on four legs, and those that walk on two legs; four things above the world: the sun, the moon, the sky, and the stars; four kinds of gods: the great, the associates of the great, the gods below them, and the spiritkind; four periods of human life: babyhood, childhood, adulthood, and old age; and finally, mankind has four fingers on each hand, four toes on each foot, and the thumbs and great toes taken together form four. Since the Great Spirit caused everything to be in fours, mankind should do everything possible in fours.

Notice that the four directions come first in this Lakota discussion of fourness, reinforcing the inference that the four spatial directions are primary in the theme of fourness, as was suggested by the early appearance of the four directions in Olmec iconography.

The four directions are also related to creation and to the motions of the sun and moon in the *Walam Olum,* a chronicle of the Delaware (properly the Lenni Lenape) Indians. The account is a pictographic record obtained in the Ohio valley sometime before 1822, accompanied by a Delaware text corresponding to the pictographs. The text begins with the creation of the earth, sun, and moon:

There at the edge of all the water where the land ends,
the fog over the earth was plentiful, and this was where the Great Spirit stayed.
It began to be invisible everywhere, even at the place where the Great Spirit stayed.
He created much land here as well as land on the other side of the water.
He created the sun and the stars of night,
all these he created so they might move.

The Delaware version of the recurring cosmological theme is illustrated more clearly by the pictographs than by the text. Before the creation of the cosmos, when there was only the earth and the Great Spirit, the earth is depicted by a rectangle divided diagonally into quarters, "everywhere" by a circle marked at the center and the four directions. When land was created, the symbol of the four directions reappears; the sun, moon, and stars are depicted rising in the east, with the moon moving back and forth to either side of the sun.

This theme of a primordial fourness of space is not limited to North America or to non-literate Native Americans. Things first appear in fours in the cosmogonies of the Aztec and Maya peoples of Mesoamerica. The sixteenth-century Quiché Maya book *Popol Vuh* begins by telling of the origins of the world, of the emergence of all the sky-earth:

the fourfold siding, fourfold cornering
measuring, fourfold staking,
halving the cord, stretching the cord

in the sky, on the earth
the four sides, the four corners.

Here the poet represents the cardinal and intercardinal—dare we suggest solstitial—directions as four sides and four corners. Here again the directions come first, before the beginning of the world and, presumably, before the beginning of time. It is only later that the first people were formed to inhabit the fourth of a series of four worlds: Jaguar Quitze, Jaguar Night, Mahucutah, and True Jaguar.

A similar succession of ages is found in Aztec cosmogonic myth, but for the Aztecs, whose dominance of Mesoamerica succeeded that of the Maya, a sequence of four ages (described as four suns) has already been completed. The first four suns were named 4 Jaguar, which lasted for 676 (13×52) years; 4 Wind, which lasted 364 (7×52) years; 4 Rain, which lasted 312 (6×52) years; and 4 Water, which lasted 676 (13×52) years. Each sun ended when the people living under it were destroyed in turn by calamities appropriate to each sun's name: wild beasts, a great wind, a rain of fire, and a great flood. The present fifth sun (called 4 Movement) and the moon were created by the self-immolation of the gods Nanahuatzin and Tecuciztecatl.

After this great sacrifice, the other gods watched for the emergence of the new sun with an anxiety that reflects the importance for Native American thought of observing sunrise. The gods were uncertain where the fifth sun would rise: the myth notes (in counterclockwise order) that some thought he would appear in the north, others in the west, and others in the south, while some expected him to rise from the east. Some, seeing the predawn light all around them, thought he might emerge anywhere.

Four gods—Quetzalcóatl, Ecatl (an avatar of Quetzalcóatl as god of the winds), Tótec, and the red Tezcatlipoca (a warrior god associated through his color with the east)—and four women (presumably goddesses) are identified as guessing the place of his emergence. But even here there was uncertainty, for after emerging in the east the sun wobbled from side to side as if unsure of his proper direction.

This myth reveals two important aspects of Aztec cosmology:

- First, the present order of the universe was established when the sun found his proper rising place on the eastern horizon.

- Second, this direction became known through observation of the direction of sunrise by four major gods of the Aztec pantheon, especially by Quetzalcóatl (the plumed serpent, Venus god, and patron of learning) and by Tezcatlipoca (the smoking mirror, patron of warriors).

Astronomical observation and knowledge were, for the Aztecs, signs of sacred power and status. Thus Nezahualpilli, one of the last precolonial kings of Texcoco, observed the heavens himself and was known for his astronomical knowledge. It is not surprising, then, that the four directions were not only a guiding element of Native American cosmologies, but also that the four directions often guided and were based on native astronomical observations.

Observations and the Four Directions

This linkage of four-directional cosmology, astronomical observation, and ritual is best documented among the puebloan peoples of the southwestern United States, especially the Hopi Indians of northern Arizona. For the Hopi, as for many other Native American peoples, the four directions are not the familiar cardinal directions of the European tradition of geometric astronomy, but are directly related to observation of the rising and setting sun at the solstices.

This difference reflects a continuing problem that limits our attempt to reconstruct Native American cosmologies. Early investigators of Native American cultures and languages tended to interpret native directions in terms of the familiar European cardinal directions, north, south, east, and west, overlooking the subtle differences of native concepts. It was only after years of careful study among the Hopi that Alexander Stephen first came to understand the astronomical basis of the Hopi directions. Stephen wrote how the discovery "flashed upon him" as he contemplated a shrine where the Hopi had deposited offerings to the sun:

> The Hopi orientation bears no relation to North and South, but to the points on his horizon which mark the places of sunrise and sunset at the summer and winter solstices. He invariably begins his ceremonial circuit by pointing to (1) the place of sunset at the summer solstice, next to (2) the place of sunset at winter solstice, then to (3) the place of sunrise at winter solstice, and (4) the place of sunrise at summer solstice.

Stephen went on to describe how the old men of the village "one and all declared how glad they were that now I understood, [and] how sorry they had long been that I could not understand this *simple fact* before." For the Hopi, of course, it was a "simple fact," because they had been immersed in this four-directional cosmology since childhood.

During the Powamu initiation ceremony, when Hopi boys and girls are initiated into the first secrets of Hopi ritual, the cosmology of four directions was repeatedly presented to them through sand paintings, chants, dances, and other ritual activities. On the fourth day of the ceremony the initiates watch Chowilawu Kachina dance four times counterclockwise (i.e., following the conventional order of watching the sun's annual journey along the horizon) around a sand mosaic whose colors symbolize the four directions. The dance portrays the annual progress of the sun along the directions, relating his motion to the growth of crops. A Hopi child could no more be ignorant of the relation between the sun and the cosmological theme of the four directions than a modern child could be unaware of the notion that the earth is a sphere orbiting the sun.

And like all cosmological frameworks this theme:

- Helps the Hopi define their place in the cosmos.
- Summarizes a wide range of astronomical phenomena in an economical form.
- Makes the Hopis' observations of these phenomena intelligible.
- Helps them anticipate future phenomena.
- Provides a vocabulary to discuss both framework and phenomena.
- Provides a mnemonic symbolism to aid in the transmission of this lore to future generations.

Sacred Places at the Four Directions

Native American cosmologies, like other cosmologies, are grounded in observations of the natural world. In turn, the four-directional cosmology provides the framework within which sun-watchers observe the Sun's annual motion: the sun's set as he (the Sun) moves southward from his summer to his winter house, and his rise as he moves northward from his winter to his summer house.

Speaking in general terms, the Hopi consider all four of the places where the Sun pauses in his annual travels along the horizon to be a Sun's house, a *tawaki*. However, only two (or in the most restricted sense, only one of them) are considered the true houses of the sun. These two Sun's houses at the directions of winter solstice sunrise and summer solstice sunset—at *tátyuka* and *kwiniwi*—are marked by small shrines and play important roles in both religious ritual and astronomical observation.

The Sun's house proper is at *tátyuka*, at the southeast, where he comes out and stands directly above his house at the winter solstice. Even in recent times, a young man offers prayer sticks, or *pahos*, to the sun on the morning of the winter solstice at a shrine that precisely marks the direction of winter solstice sunrise. After watching the sun rise over these offerings, the young man runs back to the village to encourage the sun to hurry on his (the Sun's) path toward summer.

The sun's house at *kwiniwi*, at the northwest, is properly the house of Huzruing Wuhti (hard-being woman, the deity to whom all hard substances, especially shells, coral, and turquoise belong). At the summer solstice when the sun is at *kwiniwi*, he stands directly over Huzruing Wuhti's house, then descends through a hatchway in the roof into her house. As at the winter solstice, a young man deposits prayer sticks for the sun at a shrine that marks Huzruing Wuhti's house, the direction of summer solstice sunset. Here, however, the young man picks flowers in the valley as he returns slowly to the village, in order to encourage the sun to return slowly to his house, so that there will be no early frosts and the crops will have time to grow.

These sacred places are not arbitrary; they are revealed by the sun when he pauses in his annual journey. The shrine at the sun's house proper relates to the rising of the sun at the winter solstice in exactly the same way that the shrine at Huzruing Wuhti's house relates to the setting of the sun at the summer solstice. This pair of shrines marks the corresponding solar longitudes (and hence the days of the solstices when the sun arrives at his houses) to within half a day.

Although the shrines at these two directions precisely mark the place of the solstices (i.e., the sun's houses) they could not be used to determine the exact date of the solstices because they are too small to be seen from the village and because the sun moves too slowly at the solstices for observations to determine precisely when he stops and changes direction. The days of the solstices are determined by counts of days based on observations before the solstices when the sun arrives at a pair of natural landmarks near the other two directions. The San Francisco peaks (the sacred mountains where the Kachinas dwell who bring rain to the dry Hopi country) mark *tévyuna* (the direction of winter solstice sunset). The day of the coming winter solstice ceremony is fixed when the sun sets some eleven days before the solstice at Luhavwu Chochomo, a distant valley in the San Francisco peaks some 127 kilometers from the village.

In the opposite direction, *hópoko* (the direction of summer solstice sunrise) is marked by a series of minor irregularities on the distant horizon designating the times for planting various crops in specific types of fields. Most striking of these is a narrow notch called either *Kwítcala* (a narrow mountain pass) or the cosmologically suggestive *Hütcíwa* (an entrance to the beyond). This notch, some 45 kilometers from the village, marks the day to begin general corn planting, which falls some thirty days before the summer solstice.

Observations before the solstices and over these great distances can readily determine the exact day of the coming solstice. In this complementary pattern of anticipatory observations before the solstices followed by observations on the day of the solstices, we can see different interactions of observation with Hopi cosmology. The observations before the solstices fix the sacred times when the sun will arrive at his houses. The observations at the solstices of the sun's arrival at his houses validate the cosmological framework that defines the structure of sacred space and a system of astronomical coordinates by marking the location of the sun's houses from which he emerges at midwinter and into which he descends at midsummer.

Yet the four-directional cosmology is not always expressed in astronomically precise terms. The Hopi associated the Shoyóhim Kachinas with the four directions and believed that the Kachinas dwelt atop four sacred mountains standing approximately at these directions:

- *Kishyúba* (Black Mountain) to the northwest.
- *Nüvátikyauobi* (Snow Mountain, i.e., the San Francisco Peaks) to the southwest.
- *Wénima* (Zuni Mountain) to the southeast.
- *Nüvátikyauobi* (Snow Mountain, i.e., Mount Taylor) to the northeast.

Such sacred mountains are another widespread expression of the theme of the four directions, a fitting symbol since mountains reach from the terrestrial region into the celestial which orients and directs things below. For example, the Tewa of San Juan Pueblo, New Mexico, have sacred mountains at the four directions, albeit with a significantly different twist. The San Juan Tewa mark the cardinal directions:

- By four distant mountains.
- By four sacred mesas a few miles from the pueblo.
- By four shrines a short distance beyond the edges of the pueblo.
- By the four dance plazas within the pueblo itself.

Offerings are regularly made at all of these places.

Unlike the Hopi sacred mountains and directions, the Tewa direction markers are not solstitial, but instead lie unambiguously at the true cardinal directions. Despite their use of cardinal directions the Tewa, like their neighbors, also observe the sun at the solstices, and in turn, the sun's motion along the horizon is reflected in their cosmology. Like the Hopi, they commonly refer to the directions in a counter-clockwise order, which the Tewa anthropologist Alfonso Ortiz calls "sunwise." Ortiz is clearly not referring to the sun's daily path, which is clockwise; he apparently refers to the sun's annual path along the horizon, which, following Tewa custom, he sees as counter-clock-wise.

Atop the Tewa sacred mountains are bodies of water (lakes or ponds) and direction shrines which represent a *nan sipu* (earth navel), an entryway into the underworld. Similarly, the nearby mesas are said to have caves or tunnels running through them while the shrines at the village edge are associated with "objects that . . . are endowed with sacredness because they are associated with the souls and with the sacred past." Finally, at the center, at the most important of the four village dance plazas, is the primary earth navel, the "earth mother earth navel middle place." In all cases these sacred direction markers are not only associated, due to their location, with the celestial re-gions above, but they are also associated with the underworld below, a cosmological theme we will discuss later.

Mountains at the cardinal directions linking the heavens with the underworld and rain clouds with spring waters are also found in the more complex cosmologies of the high cultures of Mesoamerica. The Aztecs described mountains as great vessels or mansions filled with water; they offered sacrifices, including the sacrifice of children whose tears brought rain, to the Tlalocs, mountain gods of rain and fertility. The chief of these sacrifices was conducted at a ritual enclosure atop Mount Tlaloc, 44 kilometers due east of the Templo Mayor in the Aztec capital of Tenochtitlan (present-day Mexico City). The pyra-midal Templo Mayor was itself considered to be a sacred mountain. The equinox sun, when viewed from the nearby temple of Quetzalcóatl (who in Aztec cosmogony had correctly anticipated that the fifth sun would rise in the east) rose between twin sanctuaries to Tlaloc and Huitzilopochtli atop the Templo Mayor. Despite ongoing studies of Aztec mountain shrines, mountains marking the other directions have not yet been identified. Nonetheless, Aztec codices clearly depict the traditional four directions of space, marking them with colors, trees, birds, and other symbols.

Although Aztec direction concepts clearly favor the cardinal di-rections, Mayan accounts reflect a recognition of both. Contempo-rary Maya, both lay folk and ritual specialists, mark the corners of space by the places where the sun rises and sets at the solstices, while the cardinal directions define the sides. East then is the side where the sun rises and west the side where it sets, while north and south are secondarily defined by exclusion. We have seen the same concept in *Popol Vuh*, and the relations of space, time, mountains, and fertility are intimated in the plaint of a captured lord of the Quichés in the pre-Columbian drama, the *Rabinal Achí*:

> *Grant me thirteen times twenty days, thirteen times twenty nights, so that I may go to say farewell to the face of my mountains, to the face of my valleys, where formerly I was wont to go to the four corners, to the four sides, searching for, finding, that necessary to nourish myself, to eat.*

Far to the south of Mesoamerica the people of the Peruvian village of Misminay also divide the terrestrial space around them into four quarters, each quarter finding its place in a cruciform network of irrigation canals and reservoirs, footpaths and sacred mountains. Similarly, they quarter the celestial region above them using the solstitial directions and the changing seasonal orientation of the Milky Way, whose appearance is especially striking in the equatorial Andes.

It was not just mountains and natural markers at the four directions that provided orientation to Native Americans; houses and other structures were oriented in a fourfold spatial symbolism. Temples of the Kogi of Colombia and earth lodges of the Pawnee were both oriented to the east with posts at the intercardinal directions. The Pawnee lodge was an image of the cosmos:

> You see the head of the turtle is towards the east. That is where the gods do their thinking in the east. While in the west all things are created and you see the hind end of the turtle in the west. The four legs [*the four posts of the earth lodge*] are the four world quarter gods [*that*] uphold the heavens.

For the Pawnee these posts represent the stars that mark the intercardinal directions and limit the motions of the sun and moon.

- The female yellow star in the northwest controls the setting sun.
- The male red star in the southeast controls the rising sun.
- The female white star in the southwest is related to the moon.
- The big black meteoric star is in the northeast.

The pattern is like that of the Hopi, with a female deity at the northwest related to sunset and a male deity at the southeast related to sunrise, yet the migratory Pawnee marked the four directions by the unchanging stars, whereas the sedentary Hopi emphasized sacred mountains and other local landmarks.

Other Four-Direction Associations

A final aspect of the four-directional cosmology is that each of the four directions is associated with a broad range of sacred and profane entities as part of a grand framework for organizing disparate elements of knowledge and ritual. The most common of these associations links each direction with a specific color, although both the colors and the directions vary from place to place. Among the Hopi:

- *Kwiniwi*, the direction of summer solstice sunset, is associated with the color yellow.
- *Tévyuna*, the direction of winter solstice sunset, is associated with the color blue or green (in this context the Hopi do not distinguish between these colors).
- *Tátyuka*, the direction of winter solstice sunrise, is associated with the color red.
- *Hópoko*, the direction of summer solstice sunrise, is associated with the color white.

In much the same way, contemporary Maya locate four rain gods (*Chacs*) associated with colors at the corners marking the solstitial directions:

- The black *Chac* is at summer sunset.
- The yellow *Chac* is at winter sunset.
- The green *Chac* is at winter sunrise.
- The white *Chac* is at summer sunrise.

In Mayan codices we find the four cardinal directions associated with colors: east with red, north with white, west with black, and south with yellow. The Aztec codex *Féjévary-Mayer* uses a more complex pattern, associating east with red and west with blue, north (or, as some would argue, the above) with yellow and south (or the below) with green, and also indicating the four solstitial directions without associating them with colors. Many similar associations of colors with the four directions have been reported among colonial and contemporary tribes, ranging from the Aztecs and Maya through numerous tribes of the Southwest and California to such distant groups as the Winnebago and the Cherokee.

Ethnographers have found further associations of the directions with colors, birds, plants, animals, and mountains—all organized in seemingly arbitrary taxonomies founded on the ordering principle of the four horizontal directions and sometimes the above, the below, and the center.

At first glance, these proliferating symbolic associations seem superfluous growths on what should be a simple rational framework. Yet these associations provide a rich mnemonic device, a mnemonic repeated in art, in ritual, and in mythology. To many native peoples, then, the four directions provide an empirically derived cosmological model incorporating elements of an astronomical coordinate system, a ritually sacred space, and an entire structure of seemingly arbitrary correspondences whose combined impact assists the stable transmission of this pan-American cosmological theme.

Such variations on the theme of the four directions suggest local variations on an underlying pan-American cosmological theme. Note that we have repeatedly seen the four directions at the beginning of Native American discussions of the origins of the world. This does not seem to be an accident; I take these native accounts at face value when they make the division of space into four directions the principal source of the Native American emphasis upon fourness. In turn, this emphasis upon fourness, as well as the association of each of the four directions with a color and with other beings sharing analogous fourfold classifications, contributes to the preservation of the cosmological and astronomical theme of the four directions. This aspect of Native American cosmologies may then be considered as different articulations of the Native American theme of the four directions, just as the astronomies of Ptolemy and Copernicus are different articulations of the Greek theme of uniform circular motion.

The Above and the Below

Turning to the above and the below, we see a pair of opposed themata that are best considered as cosmological *abstractions* relating time and space, unlike the four directions which relate time and space through empirical *observations* of the sun's annual journey along the horizon. It is an almost universal concept that the sun returns from west to east during the night by way of the below; hence, when it is night in the above, it is day in the below; and when it is winter in the above, it is summer in the below.

The Hopi dealt with the sun's motion between the above and the below more specifically; they considered that the below was, in some way, the source of that which was to come. In their mythology, they saw the present world in which we live as the fourth of a succession of worlds, into which their ancestors had emerged by climbing on a reed that they had planted to grow from a third world in the below. For the Hopi, the vertical direction in which plants grow defines a quasi-temporal axis that Benjamin Whorf has called the "wellspring of the future," a concept expressed in a Hopi metaphor where "below" corresponds to before and "above" corresponds to after. In the below are made the seeds of all beings—plants, animals, and men—that live in the above; hence, in springtime the Harvest moon is already shining in the below on the fruits and vegetation that are ripe and full-grown.

When that same moon comes to shine in the above six months later, it will bring with it similar fruits and vegetation.

For the Hopi, the above and the below are not hierarchical; there is a simple duality in which things come forth from potentiality in the below into actuality in the world of everyday existence. Above this level of everyday life, the Hopi recognize a third layer in the sky. The Hopi and some of their puebloan neighbors integrate the above and the below into the symbolic associations with the four directions to form a six-directional schema.

The Hopi's simple three-layered cosmos contrasts with most Mesoamerican cosmologies, in which both the above and the below are arranged hierarchically. The Aztecs described the cosmic vision they inherited from their ancient predecessors, the Toltecs, in terms of a layered cosmos:

> *And the Toltecs knew that many are the heavens.*
> *They said there are twelve superimposed divisions.*
> *There dwells the true god and his consort.*
> *The celestial god is called the Lord of duality.*
> *And his consort is called the Lady of duality, the celestial Lady;*
> *which means*
> *he is king, he is Lord, above the twelve heavens.*
> *In the place of sovereignty, we rule;*
> *my supreme Lord so commands.*

To the Toltecs, as to their successors the Maya and the Aztecs, the axis linking the above and the below was not represented by three simple layers but was increasingly complex. In this version of the cosmos we see twelve layers in the heavens, above which is a thirteenth layer occupied by the dual god/goddess Ometeotl.

One sixteenth-century Aztec manuscript, *Codex Vaticanus 3738*, depicts the thirteen-layered heaven in greater detail. The moon travels through the first layer above the earth, and in successive layers are the stars, the sun, Venus, "smoking stars" (comets), the green heaven, the blue heaven, and storms. Subsequent layers are white, yellow, and red dwelling places of the gods, while the last is *Omeyocan* (the house of duality, the dwelling place of Ometeotl).

The below was similarly subdivided into nine layers. Immediately under the earth is the great stream that flows around the earth and into which the sun sets. Next is the true gate of the underworld, the two mountains between which the sun passes. The ninth layer is *Mictlan* (the place with no exit, the land of the dead). Between the above and the below is the terrestrial level of the cosmos, *Cemanahuac* (the land surrounded by waters).

At the time of the conquest, the Maya had much the same layered cosmology as did the Aztecs. The god of duality, the conceiver and begetter of children, reigned as the highest of thirteen gods occupying the upper levels, while nine lords of the night occupied the levels of the underworld. The sun, at his setting, entered the fangs of the earth monster to return through the underworld to the east.

Simpler versions of this cosmology are found among contemporary Maya. The villagers of Chamula distinguish three celestial layers:

- The lowest invisible layer.
- A second layer, in which are found the moon, the Virgin Mary our Mother, and the minor constellations.
- The highest layer, in which are found the sun, Christ our Father, Saint Jerome, and the greater constellations.

The underworld is the region of the dead through which the sun travels at night. There the villagers also place earthbearers, one earthbearer at the center or four earthbearers at the intercardinal directions. The earth is tilted towards the west and is surrounded by water; within the earth are limestone channels through which water flows, while the earth lords, dwelling in mountain caves, bring rains.

Though not strictly dependent on astronomical observation, the above and the below were not totally unrelated to such observations. Between the tropics, the days on which the sun passed through the zenith often formed part of the ritual calendar, and a number of observatories designed to indicate zenith passage have been found in Aztec structures. The Peruvian Incas not only noted the days when the sun passed through the zenith, but they also noted the days when the sun would pass directly through the nadir at midnight.

Away from the complex societies of Mesoamerica, we encounter increasingly simplified versions of this layered cosmos. The Cuna Indians of Panama and Colombia and the Kogi Indians of Colombia reckon four levels above and four below the earth. The Warao of Venezuela consider the world as three disks: an upper one in the sky, a lower one in the underworld, and a larger central one representing the terrestrial-aquatic world. Groups as diverse and widespread as the Iroquois and the Inca similarly divide the cosmos into three layers: the terrestrial realm, the above, and the below. The Tehuelche of Patagonia subdivide the terrestrial realm to define four levels: the celestial heavens, the sky in which birds and clouds fly, the earth, and the underworld.

Conclusion

As we compare the complex cosmologies of the Maya, the Aztecs, and their predecessors with the simpler expressions of four-directionality and of the above and the below that we find in less complex societies, we cannot avoid raising the question of how variations on these pan-American cosmological themata came to be. It has often been suggested that the simpler Native American cosmologies represent a vulgarization of cosmologies diffused from the Mesoamerican centers of high culture. Our evidence is scarce; we have no sure answers. Yet given the extent, both in time and space, of these basic cosmological themata, I would suggest another model: that the complex, hierarchical cosmologies of the Maya and the Aztecs represent elaborations of fundamental themata found throughout the Americas and thousands of years earlier among the Olmecs. In a sense, the elaboration of these themata into complex, hierarchical cosmologies both reflected and supported the more complex hierarchical ordering of their societies.

—Stephen C. McCluskey

BIBLIOGRAPHIC NOTE

The study of Native American cosmologies cannot be tidily confined within disciplinary boundaries; investigators include anthropologists, astronomers, historians of religion, historians of science, and any number of other specialists. Their studies, more than this brief essay, display how cosmology intersects with people's everyday and ritual lives. Aztec cosmology is perceived through the prism of a single structure by Johanna Broda, David Carrasco, and Eduardo Matos Moctezuma in *The Great Temple of Tenochtitlan: Center and Periphery in the Aztec World* (Berkeley: University of California Press, 1987). Astronomer-anthropologist Anthony Aveni provides a more astronomical perspective for the whole Mesoamerican region in *Skywatchers of Ancient Mexico* (Austin: University of Texas Press, 1980). Aveni's many edited volumes on archaeoastronomy also repay careful study. Tewa anthropologist Alfonso Ortiz portrays his people's cosmos in *The Tewa World: Space, Time, Being, and Becoming in a Pueblo Society* (Chicago: University of Chicago Press, 1969). A sensitive and astronomically informed ethnographic study of the intersection of astronomy and cosmology in a contemporary Peruvian village is Gary Urton's *At the Crossroads of the Earth and the Sky: An Andean Cosmology* (Austin: University of Texas Press, 1981). Von del Chamberlain examined the rich ethnohistorical record to unravel Pawnee astronomical and cosmological concepts in *When Stars Came Down to Earth: Cosmology of the Skidi Pawnee Indians of North America* (Los Altos, California: Ballena Press; and College Park, Maryland: Center for Archaeoastronomy, 1982). Current studies of Native American astronomies and cosmologies appear frequently in the *Archaeoastronomy* supplements to the *Journal for the History of Astronomy*.

2. CHINESE COSMOLOGY

Chinese views of the cosmos differed in notable respects from Western beliefs. Most significantly, for the Chinese the entire universe was a single organism. As such, it echoed conditions of mankind on earth; a bad or corrupt administration would show up as dislocations in the behavior of the celestial bodies. Thus from the earliest times, the Chinese had reason to observe the heavens with care and without preconceived ideas, while in the West such observations were colored by accepted notions of what should be found. In the third century A.D., the shape and extent of the Chinese cosmos differed markedly from the Western view. Whereas in the West people believed in a universe enclosed within a sphere with the earth fixed at the center, the Chinese considered the universe to be infinite, with the celestial bodies floating in it, driven by a celestial wind. Dating from even earlier times is the Chinese belief that the earth oscillates in space; this was the way they explained the seasons. Lastly, Chinese views of immortality and life after death differed widely from Western ideas. The Chinese believed an eternal bodily existence on earth to be possible, and they also thought that since the human soul was composed of two components, part of the soul would eventually mingle with vapors in the heavens and part would mix with the juices of the earth.

The Chinese Outlook

In the eleventh century A.D. a cosmical plan, *Explanations of the Diagram of the Supreme Pole (Tai Ji Tu Shuo)*, was discussed by the noted Daoist (Taoist) teacher Zhou Dun-Yi who otherwise left few writings to posterity. Difficulty in understanding the full import of the plan followed, but in 1173 the Neo-Confucian philosopher Zhu Xi wrote a commentary that throws light on the diagram and expresses an essential aspect of the Chinese outlook coloring the whole of their cosmology.

For our purpose, the most important part of Zhu Xi's commentary concerns the opening statement of Zhou Dun-Yi's exposition:

"That which has no Pole! And yet (itself) the Supreme Pole!" Zhu Xi shows this to be partly an identity expressed in the form of a paradox, and partly a synthetic outlook uniting the streams of Confucian and Daoist thought. From this and other writings, it becomes clear that he and other Neo-Confucians were taking a holistic view of the cosmos. They saw the entire universe as a single organism.

This view, evident in Zhu Xi's commentary, also lies behind earlier Chinese thinking and makes clear why there was always a belief that human actions on earth and the behavior of the heavens were intimately linked. The human behavior under consideration was that of the ruler and his court; but, of course, the actions of those in power were what primarily affected the lives of the Chinese people. In a country where from very early on a vast bureaucracy ran the state, a bad and corrupt administration could bring untold misery and evil. That this situation should be mirrored in the skies seemed a logical conclusion. And since malfeasance in office was something of which the emperor might not readily be aware if his ministers wished him to remain in the dark, the practice of appointing royal astronomers to watch the heavens for irregularities logically followed. Incidentally, the practice is now paying dividends, providing us with the world's longest running sequence of astronomical records, a few of which are of significance in certain aspects of modern research.

The need to keep the skies under careful and continual observation carried with it the implicit belief that the heavens could show changes. New stars or novae were recognized when they appeared. Indeed, there is at least one oracle bone from the second millennium B.C. that records such a "guest star" close to the bright red star Antares (an observation seemingly confirmed in recent times by the radio source 2C 1406). Comets and meteor showers were also recorded and understood to be celestial phenomena. This Chinese understanding is in marked contrast to perceptions of the cosmos in ancient Greece and in the medieval West, in which the heavens were assumed to be changeless. This Western view placed astronomy in a straitjacket, causing observations to be either misinterpreted or—even worse—ignored. The most glaring instance of the latter was the supernova explosion of A.D. 1054, during which the new star was visible in daylight for a short time. The event was recorded by the Chinese but was virtually ignored in Western records. The Chinese seem to have been little troubled by preconceived ideas about the heavens, and were always ready to accept new ideas.

Hemispherical Dome Cosmology
Among the physical concepts of the universe that the Chinese consid-

ered, the earliest recorded is the theory of the Gai Tian, or hemi-spherical dome, referred to in *Master Lu's Spring and Summer Annals* (*Lu Shi Chun Qiu*) of the third century B.C.

In this concept the heavens were conceived to be a dome, a hemispherical cover, lying over an inverted bowl or dome-shaped earth. The distance between the inside of the dome and the surface of the earth, between what in fact were two concentric domes, was thought to be 80,000 li (about 43,000 kilometers or 27,000 miles, although in this context a precise significance should not be given to the number). The Dipper or Great Bear constellation (Ursa Major) was in the middle of the heavens and the center of the inhabited world (China) was in the middle of the earth. The north celestial pole lay directly above the highest point of the earth's upturned dome. Though the base of the dome of the heavens was round, the base of the earth's upturned bowl was square. The distance to the center from each edge was some 10,800 kilometers (6,700 miles). Rain fall-ing on the earth flowed down to the four edges to form an ocean—the Great Trench—at its rim. The distance between the earth's edge and the sky was only some 11,000 kilometers, so the heavens came lower than the highest point of the earth.

The heavens were thought to rotate like a mill from right to left, and risings and settings of celestial bodies were only illusions, for they never passed below the base of the earth. However, celestial motion cannot be thought of as occurring along the edges where the celestial vault met the rim-ocean, and it seems likely that the technological model lying behind the workings of this Chinese concept may have been the edge-runner mill, where the grindstone is vertical and sup-ported by a horizontal axle itself pivoted vertically at its center. The vertical grindstone runs over a fixed circumference, where grinding occurs. In the cosmological analogue, however, the mill wheel would have to revolve in place instead of traveling round the circumference.

This theory appeared to have sufficient truth about it to satisfy early Chinese geometers, who had little more than what we now know as the Pythagorean theorem to help them. The great age of the theory is seemingly confirmed by its similarity to the double vault theory that existed in Babylonia during the second millennium B.C.

Celestial Sphere Cosmology

A second major cosmological concept of the early Chinese came from the Hun Tian school, which conceived of the heavens as a celestial sphere. This belief was based on what seems to be a nearly universal perception of the apparent movement of all celestial bodies around the center of the earth. This conception developed gradually among

Greek philosophers, and has come to be associated particularly with Eudoxus of Cnidus, who died in 356 B.C. [For more on Eudoxus, see chapter 5, *Plato's Cosmology*, in section II, *The Greeks' Geometrical Cosmos*.] While the theory of the heavens as a celestial sphere must have been known in China at least as early as the fourth century B.C., when star lists were being prepared by the astronomer Shi Shen and others, the earliest exponent of the celestial sphere model whose name has come down to us is Loxia Hong, who died in 104 B.C. The oldest description of the theory comes from the pen of the great astronomer and mathematician Zhang Heng, who worked during the first century A.D. In his *Spiritual Constitution of the Universe* (*Ling Xian*), of which we have fragments still, he described the basic celestial spheres with great clarity. In another book, his *Commentary on the Armillary Sphere* (*Hun Yi Zhu*), he is rather more precise, writing that: "The heavens are like a hen's egg and as round as a crossbow bullet; the earth is like the yolk of the egg, and lies alone in the center. Heaven is large and earth small. Inside the lower part of the heavens there is water. The heavens are supported by vapor (*qi*), the earth floats on the waters." Next, he discusses the number of degrees in the circumference of the heavens, and explains why only half the constellations are visible at any one time. The rotation of the heavens "goes on like that around the axle of a chariot."

The words of Zhang Heng are interesting for several reasons. His natural philosophy is of interest in connection with the beginnings of the idea of laws of nature in which the word *qi*, though just translated as *vapors*, had a subtler significance, akin to our own concepts such as *fields of force* or *matter-energy*, a subject discussed in some detail much later by Zhu Xi. Furthermore, in his full texts Zhang attributes the visualization of the celestial sphere to a much earlier time than his own, and clearly shows how the conception of a spherical earth, with antipodes, would arise naturally out of it. And he realized that space may be infinite; he was able, so to speak, to look through the immediate mechanism of the sun and stars to the unknown lying beyond.

The Hun Tian school also contributed in an important way to the growth of scientific instrumentation in China, leading to the construction of armillary rings and armillary spheres. These frameworks of metal circles depicting the celestial coordinates were to become widely used, and in China led to the development of the first astronomical measuring instruments mechanically driven to counteract automatically the apparent rotation of the heavens.

Infinite Empty Space Cosmology

A third major cosmological theory espoused by the Chinese was very much their own. Known as the Xuan Ye school, it was an extraordinarily advanced doctrine. It considered the heavens as infinite in extent with the celestial bodies floating in the heavens at rare intervals. The theory originated later than either the Gai Tian or the Hun Tian theories, and is particularly associated with Qi Meng, about whom we know little. He lived sometime between A.D. 1 and 200, in other words during the Later Han period, and may have been a contemporary of Zhang Heng.

What we know of Meng's connection with the theory of an infinite heaven comes from a passage in the *History of the Jin Dynasty (Jin Shu)*, written during the fourth century A.D. by Ge Hong. Born perhaps in A.D. 283, Ge Hong was not only the greatest alchemical writer in Chinese history, but he also was interested in astronomy. His description of the Xuan Ye school is worth quoting in some detail:

> *The books of the Xuan Ye school were all lost, but Qi Meng, one of the imperial librarians, remembered what its masters before his time had taught concerning it. They said that the heavens were entirely empty and void of substance. When we look up at them we can see they are immensely high and far away, without any bounds. It is like seeing yellow mountains sideways at a great distance, for then they all appear blue; or when we gaze down into a valley a thousand fathoms deep, it seems somber and black. But the blue of the mountains is not a true color, nor is the dark color of the valley really its own. The sun, moon and company of stars float freely in empty space, moving or standing still, and all of them are nothing but condensed vapor. The seven luminaries* [the sun, moon, and the five brighter planets] *sometimes appear and sometimes disappear, sometimes move forward and sometimes retrograde, seeming to follow each a different series of regularities. Their advances and recessions are not the same. It is because they are not rooted to any basis that their movements can vary so much; they are not in any way tied together. Among the heavenly bodies the Pole Star alone keeps its place, and the Great Bear never sinks below the horizon in the west as do the other stars. The speed of the luminaries depends on their individual natures, which shows they are not attached to anything, for if they were fastened to the body of heaven, this could not be so.*

This cosmological view is surely as enlightened as anything that ever came out of ancient Greece. Indeed, it can be claimed that it is far more advanced than the rigid Aristotelian-Ptolemaic conception of concentric crystalline spheres that dominated European thought for more than a thousand years.

Though in the past sinologists have tended to regard the Xuan Ye school as uninfluential, its views pervaded Chinese thought to a greater extent than would at first sight appear. We find it cropping up in later centuries; indeed, it persisted until the time of Matteo Ricci and other Jesuit missionaries from the West in the seventeenth century.

Because of Greek traditions—centered, as they were, on a deductive geometrical plan of the universe designed to explain the motions, both forward and backward, of the planets—totally fresh ideas were hard to formulate in the West. Even Copernicus, who took the daring step of dethroning the earth and placing the sun at the center of the universe, still made use of the ancient geometrical system of epicycles and deferents. Not until the publication in 1609 of Kepler's work, based on the precise observations of Tycho Brahe, was a thorough transformation achieved. Chinese astronomy, often reproached for its overwhelmingly observational bias, here had the advantage for at least some 1,500 years.

The Xuan Ye system has a distinctly Daoist flavor, which may account for the disappearance of the oldest writings concerning it. One senses a connection with the "great emptiness" of Lao Zi, and with the idea of heaven as piled-up *qi* in *The Book of Master Lie* (*Lie Zi*). Buddhism, however, also contributed to the Xuan Ye system, particularly its conceptions of infinite space and time and a plurality of worlds. From the middle of the third century A.D., the indigenous Xuan Ye theory must have received much support from this Indian source. A late statement of this influence appears in the thirteenth century A.D. in Deng Mu's book *Bo Ya Qin* (*The Lute of Bo Ya*). (Bo Ya was a legendary lutenist.) He wrote:

> *Heaven and earth are large, yet in the whole of empty space they are but as a small grain of rice. . . . It is as if the whole of empty space were a tree and heaven and earth were one of its fruits. Empty space is like a kingdom. Upon one tree there are many fruits, and in one kingdom many people. How unreasonable it would be to suppose that besides heaven and earth which we can see there are no other heavens and no other earths!*

To such minds the discovery of galaxies other than our own would have seemed full confirmation of their beliefs.

To substantiate the important conclusion that the Xuan Ye world picture—together with the Hun Tian spherical motions—came to form the background of Chinese astronomical thinking, it is necessary to look briefly at the subsequent history of theoretical cosmology in China. From the fifth century A.D. on, energetic attempts were made to reconcile the Gai Tian school with the Hun Tian conception. But these failed, and after A.D. 520 official histories consider the Hun Tian sphere to be the only correct view.

But running through the centuries was an additional idea, that of the "hard wind," which helped the Chinese (who could not know of the absence of all atmosphere in outer space) to imagine that stars and planets could be borne along without being attached to anything. This idea is generally thought to have been of Daoist origin, but Joseph Needham is firmly of the opinion that it goes back to the early use of bellows in metallurgy, when technicians noticed the resistance of a powerful jet of air. The third-century astronomer Yu Song may have been thinking of it when he said that the rim of the Gai Tian hemisphere floats on the primeval vapor. Certainly in the eleventh and twelfth centuries the philosophers Shao Yong and Zhu Xi constantly refer to the "hard wind" in the heavens that supports the luminaries and bears them on their way. But Zhu Xi's older contemporary Ma Yongqing expressed the general point of view well when he wrote in his *Book of the Truth-through-Indolence Master* (*Lan Zhen Zi*) in about A.D. 1115:

> ... once I was sent as government examiner to *Jinzhou* and had to climb over the highest passes; they seemed ten or twenty miles high, and I realized that it was as if I was in the heavens and looking down on the earth far below. I remembered the words of Ge Hong: "Miles above the ground the xian [the Daoist immortals] *move upon the support of the hard wind.*" So also above, far above, float the sun, moon and stars. The heavens are nothing but a condensation of air and are not like shaped substances which must have limits. The higher you go the farther away you will be. I do not believe there is any definite distance from the ground to the heavens.

Among the Confucian philosophers, Zhang Zai paid particular attention to cosmological theory. He spoke of "the great emptiness without substance" of the heavens, thus following the Xuan Ye theory. Earth, he said, consisted of pure Yin, solidly condensed at the center of the universe; the heavens, of buoyant Yang, revolved to the left at the periphery. The fixed stars were carried around endlessly by this floating, rushing wind. Then, to explain the opposite direction of the annual movements of the stars, sun, moon, and five planets, he had recourse to the interesting conception of viscous drag: these bodies were so much nearer the earth that the latter's wind impeded their forward motion. "The wind of the earth, driven by some internal force, [also] revolves continually to the left," but more slowly (because of the effect of the stationary earth), with the result that relatively—though not absolutely—the movement of the bodies of the solar system is opposite to that of the fixed stars. The degree of retardation of these bodies depended on their constitution: the moon was Yin, like the earth, and therefore most affected; the Yang sun was the least affected; and the planets were affected to an intermediate extent.

If the celestial bodies moved on the hard wind, perhaps the earth itself also moved? Not a few ancient Chinese thought that it did, though the motion envisaged was not at first one of rotation, but rather of oscillation. In the *Apocryphal Treatise on the* Shang Shu *Section of the* Historical Classic; *Investigation of the Mysterious Brightnesses* (*Shang Shu Wei Gao Ling Yao*) of the first century B.C., we read:

> *The earth has four displacements. At the winter solstice, being high and northerly, it moves to the west 10,000 miles. At the summer solstice, being low and southerly, it moves to the east 10,000 miles. At the two equinoxes it occupies a middle position (neither high nor low). The earth is constantly in motion, never stopping, but men do not know it: they are like people sitting in a huge boat with the windows closed; the boat moves but those inside feel nothing.*

This view is of considerable antiquity, going back well before the first century B.C. For us its main point of interest is that the highly anthropocentric vision of a central and immobile earth, which so dominated European thinking, was not marked in Chinese thought.

Such ideas retained their vitality until Chinese thought fused with modern science, after the coming of the Jesuit missionaries from the West. When the Jesuits came, they did not at first appreciate the achievements of Chinese cosmological ideas, but saw only that the ideas were different from Western traditions and thought. Unhappily, the Jesuits deemed the Chinese ideas inferior. Only now are we beginning to appreciate the magnitude of the Chinese achievement.

Religious and other Consequences of Chinese Cosmology

All cosmologies—or at least all early conceptions of the universe—have a religious aspect. Chinese cosmology is no exception. In China the religious aspect of cosmology centered on the question of abodes of the dead, a subject which also deeply interested other civilizations.

If we leave to one side the otherworldly realms of hells and paradises, and consider only the habitations of people's spirits, more or less disembodied, in this present world there have been three possibilities: upon the earth; under the earth in some subterranean realm; or somewhere in the starry heavens above. If ethical considerations are excluded, the essential point is that everyone went to one or another of these universal and comprehensive places. Some visualized "happy hunting grounds" as the place where the spirits went, while others thought that the spirits lived in their own tombs and barrows or dwelt in stones or trees. Or the spirits might go to some faraway place on the earth, east or west, or to some distant island. What is of considerable interest is that many peoples in different parts of the world have held similar beliefs.

One of the ideas most typical of Chinese thought, quite irrespective of which cosmological theory was favored, was that of the possible continuance of the individual person in an etherealized body upon the earth. This type of existence could last potentially through eons of time, with the deceased paying rare visits only to human habitations. More widespread in other civilizations was the idea that everybody descended after death to an underworld of deep shadows. A third idea, that of a realm of the dead situated somewhere in the heavens above, seems to have been less common. Ancient Egyptians and primitive Patagonians, though, believed that the souls of the departed went to find a home in the stars, and Daoist religious philosophy in China also maintained that the perfected immortal could in time rise up to the constellations. Such souls were part of the sky and lay within the confines of the natural world; but they formed the palaces and offices of the celestial bureaucracy, and there the distinguished immortal would find his appointed position.

These ancient ideas of universal public places were invaded gradually by another conviction—that of a judgment which would separate good and evil, rewarding the former and punishing the latter. This conviction leads to what may be called "ethical polarization," with a heaven and hell outside this world, conceived in positive terms as paradise above and a place of permanent torment below.

The idea of ethical polarization was taken up by many civilizations. An exception was the indigenous Chinese, who learned of it only when Buddhism arrived in China, not until the third and fourth centuries A.D. Afterwards, this doctrine exerted much influence, but never enough to overwhelm the ancient Daoist idea of a material immortality on earth attainable by adepts and perfected spirits, who could thus avoid the common fate of descent to the underworld.

The indigenous Chinese position was fundamentally attached to this-worldly levels. There was a conception of a dark, shadowy realm not unlike our own world, but underground somewhere in the neighborhood of the Yellow Springs (*huang quan*), the name for the Chinese *She'ol*, or Hades, which became general in the early first millennium B.C. Adherents of the Gai Tian school imagined the Yellow Springs situated beneath the lower dome, while Hun Tian supporters located it within the egg-yolk or crossbow-bullet earth, surrounded far away outside by the celestial circles.

It is important to remember that for the ancient Chinese, the conception of a human was not a person with a single unitary soul. There was a group of two sorts of souls, some Yin and some Yang, partly celestial and partly earthy, so that if they were kept together at all they would have to be in connection with the body in some form or another. This group of souls might wander the realm of the Yellow Springs, or the surface of the earth, but if there was no bodily component, then all the souls would dissipate into the world; some would go upwards and others would go downwards, to mingle with the vapor of the heavens or the juices of the earth. The body, however ethereal-ized, was the string that strung them together.

During the early days of the Shang and Zhou, the first and sec-ond millennia B.C., emphasis on mortal longevity kept growing. Since being backed by one's ancestors and surrounded by one's descendants was the greatest blessing that the heavens could confer, why should it not continue forever? Then, beginning in the fourth century B.C., during the late Zhou period, the conviction spread that there were technical means to extend life so that one could be virtually immortal, spending time not in the Yellow Springs, not out of this natural world, not in any ethical paradise or hell, but among the mountains and forests now and forever. This conviction was the basis for the conception of a "medicine of immortality," a strange part, indeed, of Chinese cosmology. The Chinese in this instance were very material-istic.

During this same period belief in a drug-plant, or herb, or some medicine of immortality strengthened. Possibly news from Babylonia, Persia, or India, maybe slightly misunderstood, reinforced Chinese worldviews. The result was a great wave of activity concerned with what is sometimes called "the cult of *xian* immortality." A *xian* was an immortal, living in or above the earth but within the world of nature, a distinctively material immortality with a lightened body. Moreover, the *xian* could be divided into two groups, the *tian xian*, or "sky immortals," and the *di xian*, or "earth immortals." Such a distinction was already present by the beginning of the first century A.D., and by the end of the Han Dynasty (A.D. 220) people were affirming the superiority of the *tian xian*. Indeed, it seems that one could choose, and according to a fascinating book, *Lives of the Divine Xian* (*Shen Xian Zhuan*), which appeared in the early fourth century A.D., some certainly preferred to be earthly immortals. In a particular case, the adept "did not intend to do away with the joys and happiness of life among men."

All this has a bearing on the later development of scientific chem-istry. Where the concept of a "drug of deathlessness" came from is unknown, but Joseph Needham is inclined to think that it originated perhaps in India, or even Babylonia. Such an herb is mentioned in the *Epic of Gilgamesh*. Conditions, however, were right in China at the time of the first century A.D. for crystallizing the idea of an elixir. And if we look for the elixir idea in other cultures, we find that although Hellenistic protochemistry did not have it—they were concerned more with gold-faking and gold-making—the Arabs undoubtedly did have it (after all, *al-iksir* is an Arabic word). From early in the eighth century A.D., protochemical writings refer to "the medicine of man and of metals."

The elixir idea came westward through many Arabic-Chinese contacts to Latin alchemists, and in time to Roger Bacon and Albertus Magnus. In about A.D.1500 it reached Paracelsus, with his wonderful slogan: "The business of alchemy is not to make gold but to prepare medicines." Modern chemistry, with Priestley and Lavoisier, may be said to have been born under the sign of chemotherapy. Traced back, its origins are found in the Daoism of ancient China.

Cosmology in China thus was associated both with astronomy and with chemistry as they developed through the history of the Old World. It contained the belief in human survival after death (or apparent death), not in some world outside space and time, but in this material universe, here and now.

Conclusion

Chinese cosmology differed from cosmology in the West primarily in being holistic. The Chinese viewed mankind and the heavens as part of a vast organism, with actions on earth echoed in the heavens. This outlook resulted in constant observations of the skies. Early on the Chinese considered the earth to move with a gentle oscillation; in this way they explained the annual changes in the seasons. They were also the first to propose, in the third century A.D., an infinite universe with stars and planets floating in it. At this time westerners still firmly believed in an enclosed universe shaped like a sphere. Chinese views of the soul and an afterlife were also different from Western views. The Chinese believed in material immortality on earth and considered the soul to be divided into two parts; at death part of the soul of most people would merge with the juices of the earth and part would mix with the vapors of the heavens.

—*Joseph Needham and Colin Ronan*

BIBLIOGRAPHIC NOTE

Joseph Needham's multivolume (fifteen already out and three more in the pipeline) history of *Science and Civilisation in China* (Cambridge: Cambridge University Press, 1954 –) is the starting point for information about science and technology in China. This large mass of material is more conveniently available in abridged form in Colin Ronan's multivolume (three out and a fourth in progress) *The Shorter Science and Civilisation in China* (Cambridge: Cambridge University Press, 1978 –). For Chinese cosmology in particular, see Needham, *vol. 3* (1959), pp. 171–461, and Ronan, *vol. 2* (1981), pp. 67–221. See also Needham, "The Cosmology of Early China," in C. Blacker and M. Loewe, eds., *Ancient Cosmologies* (London: George Allen & Unwin, 1957), pp. 87–109.

3. MESOPOTAMIAN COSMOLOGY

Gaze upon the heavens with open eyes and let thy spirit never cease to regard that most beautiful fabric of divine creation.

Firmicus Maternus, *Mathesis*, VIII, 6–7.

Before the book of Genesis, before Plato's *Timaeus* and Aristotle's *De caelo*, and so before the articulation of the single, finite, spherical, geocentric world picture and its particular variants from the Greek cultural realm, ideas about the nature and structure of heaven and earth were formulated in the ancient Near East. The approach to the physical world in ancient Mesopotamia was based on an assumption of the inherence of the divine in nature. To speak about nature was to speak about the gods. Cosmology, "the study of the world as a whole," had not yet become, as it did with the typically Western approach, an objective account of the nature and the structure of the physical universe shaped by astronomical knowledge and incorporating as the starting point for such an account theories of its origins. Nevertheless, the influence of the essentially mythological and theological accounts of cosmogony (the study of the creation of the universe, as opposed to cosmology, the study of the structure and evolution of the universe) from ancient Mesopotamia was to be felt by both the biblical and the early Greek tradition represented by Hesiod. The dual focus of Western cosmology on structure as well as etiology is the combined effect of our Jewish, Christian, and Greek heritage from antiquity, but further and considerably modified after Copernicus, Newton, Darwin, and twentieth-century discoveries in the fields of quantum and astrophysics. For us, cosmogony has become a point of departure rather than the single focus of interest. Moreover, the dominant method of inquiry for contemporary answers to the question "What is the world?" is now that of science.

Many elements of a Mesopotamian cosmology are preserved in a variety of sources, mostly literary but also iconographic, and from a wide chronological span. Whether a modern summary of these frag-

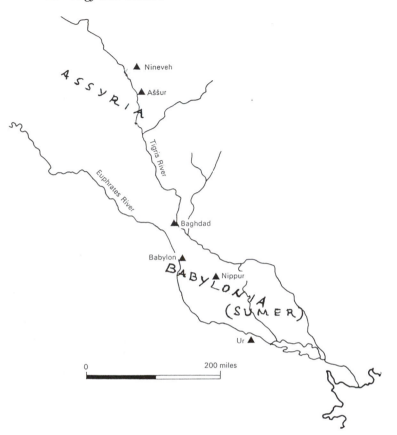

Map of ancient Sumer, Babylonia, and Assyria. Adapted from Harriet Crawford, Sumer and the Sumerians *(Cambridge: Cambridge University Press, 1991), p. 2.*

mentary remains would be congruent with the world picture of a Babylonian of some given period, or in general, will never be known— for the simple reason that no synthetic exposition of a cosmology meeting Western expectations or definitions was ever produced in Babylonia. Questions about the nature of the Babylonian inquiry into the universe can be answered in general by an identification of the kinds of sources in which various pieces of a Babylonian cosmology are found. These sources indicate that the subject was treated in Mesopotamian mythological, literary, and scholastic traditions. References to cosmogony, in a number of variant accounts, to a system of divine cosmic authority, as well as to a variety of elements of a cosmography, can be assembled from mythological, scholastic, literary, and divination (omen) texts. Considering that of all the civilizations of pre-Hellenistic antiquity, Mesopotamia was the only one to produce a scientific astronomy able to compute with great precision many planetary and lunar phenomena, it is striking that this science played only a minor role in the shaping of the Babylonian world picture. That Babylonian scientific astronomy, a product of the second half of the first millennium B.C., had no impact on the broader scope of understanding the universe, is the best proof of the continuing validity of the early non-scientific and non-astronomical ways of characterizing and imaging the cosmos. The single cosmological work in which astronomical knowledge is at all apparent is that of the late second millennium B.C. Babylonian text called *Enuma Elish* (*When Above*), whose astronomical sources are identifiable in comparatively primitive astronomical texts as well as in contemporary astrological omen texts. In general, then, the cosmos was a subject for inquiry in the context of theology ("study of the gods") much more than it was in the development of Babylonian science.

Mythology and Theology

For the physical layout of the universe and the ruling deities of the respective parts, two major traditions emerge. An early Sumerian one, based on the cosmic trio of deities AN, EN.KI, and EN.LIL, and which is not shaped by any serious empirical study of nature, particularly of the celestial phenomena, can be contrasted with a later tradition, with which it becomes interwoven to some extent. The later Babylonian tradition (that in *Enuma Elish*) is largely a politico-religious construct to legitimize the rise of the Babylonian national god Marduk to supremacy within the pantheon and, as possessor of the symbol of cosmic order, the "tablet of destinies," to authority over the entire cosmos. In this later formulation, the structuring of the universe by the young god Marduk seems to be shaped not only by the traditional mythological conception of the realms of older gods, but also by contemporary developments along the lines of objective inquiry into the physical phenomena of the heavens.

The apparently earlier cosmic mythological tradition stems from the very ancient conception of the cosmos as a polity ruled by three principal gods. AN, meaning (in Sumerian) "sky" or "heaven," ruled remote heaven, and was called Anu by the Babylonians. EN.KI, meaning "lord earth" or "lord of the below," ruled the waters around and below the earth, and was called Ea by the Babylonians. The third was EN.LÍL, meaning "lord air" or lord of the space between the great above and the great below within which exists the atmosphere and the wind. His name remained Enlil in the Babylonian pantheon.

Although Anu was conceived of as the father of the gods and the head of the pantheon in genealogical terms, the cosmic three held supreme dominion over their respective realms and together represented universal authority. The universe then was conceived of as a state comprised of the combined domains of Anu, Ea, and Enlil, and was conceptually mapped in a kind of vertical arrangement, with Anu's heaven above and Ea's waters below, divided by the airy space of Enlil.

The Sumerian word for universe was AN.KI, "above (and) below," in which KI sometimes represented simply "earth," in cosmic terms as the counterpart to heaven and in mythological terms the netherworld, which, with respect to the realm of humanity, was as below as the remote heaven of Anu was above. The Sumerian myths of creation involve the separation of a pre-existing unity (above + below) into its two component parts by Enlil, the air:

> *After heaven had been moved away from earth, after earth had been separated from heaven, after the name of man had been fixed, after Anu had carried off heaven, after Enlil had carried off earth.* [From the myth Gilgamesh, Enkidu, and the Netherworld.]

As for the original material constituents of the primordial universe, a Seleucid period (after about 300 B.C.) Babylonian text identifies Anu with fire (an identification not attested earlier), Ea with water, and Enlil with the wind, whose physical force of separation was viewed as the creative cosmic act. (Girra = Anu = fire, Primeval = Ea = water, and East Wind = Enlil = wind, according to a learned vocabulary.)

If the accounts of the creation of the cosmos as the division of heaven from earth by Enlil, the god of the "atmosphere" and wind, belong to hoary Sumerian mythological traditions, the Babylonian account of the creation and structuring of the same by their national god Marduk was a self-conscious literary composition by a learned individual (or individuals) whose goal was to legitimize the political supremacy of the city of Babylon (and the state of Babylonia) by the story of the rise to preeminence of their patron god within the pantheon as a whole. Although the date of composition of this story has not been firmly established by scholars, it is considerably later than the Sumerian cosmogonic accounts. In fact, Marduk replaces Enlil in the Babylonian story as the one who creates the physical universe, determines the nature of things (expressed as determining the "destinies"), and holds the authority of the universe and the resulting power to maintain its rightful order (symbolized by a "Tablet of Destinies"). Since Marduk was a young god, the son of Ea, the creation of physical cosmic structure comes logically after the genealogy of the gods, which proceeds to the birth of Marduk himself. But since Marduk's ancestors, the original cosmic gods Anu and Ea, are at the same time the divine essence of their respective regions (heaven above and earth below), then in some sense their very existence presupposes the existence of their realms. As a result, to Western readers the *Enuma Elish* text seems to contain two versions of cosmogony: the first accounts for the coming into being of the original gods (hence their cosmic regions) out of chaos (or, in another version, eternal time, expressed as *dūri* and *dāri*, "ever and ever"); and the second for the creation of the universe by Marduk from the defeated corpse of Tiamat, who represents the primeval salt sea.

In the genealogical section of the text, the gods are produced out of a chaos of waters, both salty and sweet. The salt sea is the primeval female, Tiamat; and her male counterpart, the sweet waters called Apsu. From them are descended Anu, the sky, and Ea, the earth, which is made as a "mirror image" of the sky. But the story develops, and by the time of the generation of young gods, mother Tiamat and father Apsu, for lack of peace and quiet, were ready to destroy their young. By means of magic, Ea fettered and slew Apsu, appropriating that watery domain for his own. Ea and his wife, Damkina, then produced the son Marduk, who would emerge as the hero to vanquish the treacherous Tiamat. When it becomes clear that neither Anu nor Ea will fight her:

> *Forth came Marduk, the wise one of the gods, your son, his heart having prompted him to go against Tiamat. He opened his mouth, saying to me: "If I, your avenger, am to vanquish Tiamat and save your lives, convene the Assembly, proclaim my destiny preeminent. [. . .] Let my word, instead of yours, decree the fates. Let nothing I create be altered.* [EnEl III, 55–63.]

The gods agree, and pronounce:

> *O Marduk, you are our avenger! We have granted you kingship over the entirety of the universe.* [EnEl IV, 13–14.]

So, in return for being made supreme over the pantheon of gods and the universe, Marduk slew the monster Tiamat, and like his mythological predecessor, Enlil, he performed his task by means of winds.

Marduk blew upon Tiamat's body with the wind and split her heart with an arrow. After surveying her dead carcass he:

> *divided her monstrous shape,* [from it] *creating something marvelous. He split her like a dried fish into two parts, half of her he set up and* [with it] *roofed the sky, spread out that which lies across it and posted guards, charging them not to let her waters escape. He went from one side of the sky to the other and examined the furthest reaches of the heavens, and made it match* [the dimensions of] *the Apsu, the abode of Nudimmud* [Ea]. [That is, he made the top part to be precisely the same size as the bottom part.] *The lord* [Marduk] *measured the form of the Apsu. He established the Esharra, an equal counterpart to the Eshgalla.* ["The great house" is here a synonym for Apsu.]

[Therefore] *Eshgalla* [Apsu], *Esharra which he had formed,* [and] *the heavens* [above] [was where] *he made Anu, Enlil, and Ea occupy their sacred dwelling places.* [EnEl IV, 137–146.]

The older symmetrical arrangement of above, below, and between becomes somewhat modified in *Enuma Elish*, as the top half of Tiamat is made the upper cosmic water—indeed, the roof of the upper heaven—and is associated with the dwelling place of Anu, while Apsu becomes the lower cosmic water, associated with the dwelling of Ea. A new region introduced between heaven and Apsu, called Esharra, was formed as a measured counterpart to the Apsu, and it became the dwelling place of Enlil. This left the lower half of Tiamat to form the earth. Marduk's creation resulted in three symmetrical cosmic regions arranged above and below the earth. Above the earth was the so-called upper heaven of Anu and the Esharra of Enlil, while the third cosmic region was the subterranean cosmic waters, formerly the Apsu of Ea. With the assignment of the three cosmic deities (Anu, Enlil, and Ea) to the respective regions of upper heaven, Esharra, and Apsu, Marduk took as his domain the earth itself. It is here that the relative position of the earth with respect to the heavens was clarified, and it is here that the notion of the temple of Marduk in the city of Babylon as the center of the cosmos is first articulated.

In sum, despite the focus of *Enuma Elish* on the birth and heroism of Marduk, and the fact that the explanation offered for the coming-to-be of the physical universe takes the form of a nature allegory in which cosmic entities (sea, earth, sky, and wind) are personified as gods, the account still offers some partial answers to the question "What is the world?" That form and order were produced from formlessness was undoubtedly of central concern; however, it was not so much a natural order as a divine one that included the natural and was in keeping with the political metaphor for the cosmos as a whole. The divine order provided a structure of authority, with Marduk at its head and the seven gods of the "destinies," the 50 Anunnaki, and 300 Igigi coming after. This divine cosmic arrangement was part of a fixed order that came into existence with creation and became a characteristic of the cosmos. The terms *cosmic designs* and *destinies* no doubt refer to that order that came into being with creation. Included within, and indeed made possible by, this fixed order were omens and portents: One text reads:

The norms had been fixed [and] *all* [their] *portents.* [EnEl VI, 78.]

Indeed, contact between humankind and the cosmos was possible through divination, since physical phenomena had the function of signs from deities. The "orderliness" of the Mesopotamian cosmos was therefore manifested in the reciprocity of heaven and earth, and further conceived of in terms of the rule of regions by designated deities. The disturbance of cosmic order following the abandonment by a deity of his designated post is the theme of a Babylonian mythological literary work, called the *Erra Epic*, in which pestilence and war break out on earth when Marduk leaves his cosmic seat to travel to the netherworld, temporarily entrusting his post to Erra, the plague god.

Cosmography

The cosmography of *Enuma Elish* set the earth in relation to two heavens above it and one cosmic region, the Apsu, below. But other configurations occur in other sources, for example, in the "three heavens" said to be consecrated to the god Lugalgirra in the magical text called *Ritual Enclosure*. Later scholastic tradition added more cosmic regions for a grand symmetry of three heavens and three earths, in which humankind is located on "upper earth." As in *Enuma Elish*, the late scheme assigns deities to dwell in each region:

> *The upper heaven is* [composed of] luludānitu-*stone of Anu. He* [Marduk] *made the 300 Igigi gods take up residence there. The middle heaven is* [of] saggilmut-*stone of the Igigi. The lord* [Marduk] *occupies the throne dais there. He sits on a throne of lapis lazuli. With* būsu-*glass and crystal, he made it shine there. The lower heaven is* [made of] *jasper of the stars. He drew the divine constellations on its surface.* [On the firm]*ness of the upper earth he settled the souls of humankind. He made Ea, his father, reside in middle earth.* [. . .] *He confined the 600 Anunnaki gods within lower* [earth]. [Ebeling, *Keilschrifttexte aus Assur religiösen Inhalts* 307: 30–37.]

Apparently from this source, Ea's domain is reassigned to the middle earth, which has come to represent a cosmic region between the living and the dead. Ea's region, termed KI.GANZIR in Sumerian (*ašar erṣeti*, "the place of the underworld," in Akkadian), is attested to in bilingual incantation literature, specifically the *Evil Demons* series from the Middle Assyrian period, around 1500 B.C. This suggests a pre-*Enuma Elish* tradition in which the plurality of earths has already emerged. Whether such a tradition was consistent with that of the present text articulating three heavens and three earths, however, is entirely unknown. Finally, in the late schema, the realm of the dead is relegated to the "lower earth," ruled over by the 600 Anunnaki gods who are locked inside. Marduk's new place of residence is with the Igigi gods in the middle heaven, where he rules the cosmos from a glittering throne. Reference is still made to Marduk's creation of the cosmos from the body of Tiamat, and the account adds that the Tigris River came from her right eye and the Euphrates from her left.

The plurality of heavens and earths, while perhaps not yet formulated in a fully symmetrical scheme, is not an innovation of first millennium Babylonia. Already, Sumerian incantations of the late second millennium B.C. make reference to seven heavens and seven earths. Beyond their number, however, their nature or extent is not qualified in this early literature. The origins of such a concept may be a good deal earlier than the date of the texts themselves, since incantations and magic seem to have been transmitted orally long before they were written down. Moreover, the seven heavens may be another derivation from the magical properties of the number 7, like the seven demons or the seven thrones, rather than evidence of an early stage in the development of a consistent cosmography.

In the context of *Enuma Elish*, the chief image, however, remains that of Tiamat as the body of the cosmos. Marduk secured bolts on either side of the "gates" of heaven (the upper part of the cosmos formed from Tiamat). Gates, also called "doors" of heaven, set up on either side, are associated frequently with the sun god. In reference to this image, prayers to the sun god, Shamash, describe him as the deity who lifts the cosmic locks. These are presumably the rising and setting points through which the sun, moon, planets, and fixed stars pass in their diurnal rotations. Indeed, the celestial bodies were placed in the upper half of Tiamat, which formed the "roof" of the sky.

Perhaps less concrete an image, but nevertheless a product of imagining the cosmos, is the motif of the "bonds of heaven and earth." The cosmic bonds were imagined as ropes or cables that tied down and controlled particularly the flow of waters (dew, rain, and clouds) from the heavens, and therefore relate to the image of the gates that locked in the waters of Tiamat. The cosmic bond is sometimes called *durmāhu*, a word for some kind of strong rope made of reeds. The term *markaṣu* also refers to a rope or cable (of a boat, for example), and in the cosmic sense, becomes a linking device which can be held, as in the description of Ishtar as the goddess "who holds the connecting link of all heaven and earth / or perhaps the netherworld." The *ṣerretu*, a lead-rope passed through the nose of an animal, becomes another synonym for this cosmological feature, which, again, can be held by a deity as a symbol of control or authority. The following passages illustrate the use of a lead-rope as a synonym for a cosmological feature and as a symbol of authority:

> *I* [Ishtar] *am in possession of the* [symbols of] *the divine offices, in my hands I hold the lead-rope of heaven.* [Ebeling, *Keilschrifttexte aus Assur* 306: 30.]

And:

Marduk made firm and took into his hands the lead-rope of the Igigi and Anunnaki, the connecting link between heaven and earth. [Craig, *Assyrian and Babylonian Religious Texts* 1 No. 31 rev. 8.]

The symbol of the "connecting link" of the cosmos—expressed variously as "great bond" or "cosmic cable" or "cosmic lead-rope"— is further present in the names given to the city of Babylon: DIM.KUR.KUR.RA (in Akkadian, *rikis mātāti*, "bond of [all] the lands"); or to the city of Nippur: DUR.AN.KI ("bond of heaven and earth"); as well as in the temples of gods, such as Marduk or Enlil; and, the implied status of these places as cosmic centers. *Enuma Elish* identifies Marduk's sanctuary at Babylon as such a cosmic center, built "opposite" and with its "horns" reaching the base of Esharra, itself located between earth and the upper heaven of Anu. This description would have us imagine the temple tower (*ziggurat*) of Marduk, called Esagila ("house that lifts its head [high]"), with its foundation penetrating to the region below earth and its pinnacle touching the heaven of Enlil. The Mesopotamian temple towers were viewed as symbolic mountains whose "peaks" reached into the heavens. From the earliest periods, the Sumerian name of the temple of Enlil in Nippur was indeed É.KUR ("mountain house"). Especially in Assyria, this name later became a generic term for a temple. Similar notions can be traced even as far back as the third millennium B.C. Evidence from the cylinder inscriptions of Gudea of Lagash concerning the building of the Eninnu, the temple of Ningirsu, reflects the same conception: namely, the cosmic centrality of the location of the temple as well as its symbolic status as a node where heaven meets earth and so, divine meets human.

The notion that the universe has a center constitutes a notable element in any systematized concept of the world as a whole. In some non-Mesopotamian contexts it serves as a diagnostic indicator of a closed (finite) rather than open (infinite) universe. The geocentric aspect of the premodern Western cosmos based largely on Plato, Aristotle, and Ptolemy is a highly articulated version of this. The idea of a world-center in ancient Mesopotamia, however, does not constitute a functional equivalent to the geocentrism of the classical and Hellenistic world pictures. That Babylon became the center of the Babylonian cosmos is more aptly paralleled by the biblical description of Jerusalem as "the navel of the earth" (Ezekiel, 38:12). Centrality of the earth or even of a specific cult center in the Mesopotamian cosmography derives from mythological and theological concerns: namely, that there should be a connection between cosmic regions metaphorically as well as physically. The physical aspect is depicted in the cables or bonds tying the universe together, and the metaphoric aspect is depicted in the view that the cult-place of the deity considered to hold ultimate authority as ruler of the cosmic polity serves as nexus. No connection may, therefore, be seen with classical geocentrism, which was argued from the point of view of a theory of motion in the universe, and the point established first by Aristotle (*De caelo*, II, *14*) that the center of the earth, the point toward which all heavy bodies fall, is at the same time the center of the universe.

The Universe

The notion of totality embedded in the oldest of Sumerian mythology was defined by the dual components above (heaven, AN) and below (earth, which sometimes meant the netherworld, KI). The more emphatic expression "the entire universe" (AN.KI.NIGIN.NA) is found in the Sumerian myth *Enmerkar and Enmushkeshdanna*. In this compound term, the Sumerian element NIGIN corresponds to the Akkadian word *lamû*, which has the basic meaning "to move in a circle" or "to encircle," and hence the semantic force of the whole encompassed by heaven and earth, possibly already connoting a round image. The basic idea, at least, of the universe defined as heaven and earth continued throughout later Mesopotamian sources.

To express universal totality in Akkadian, the language of the Babylonians and the Assyrians, words such as *kullatu* and *gimru*, both meaning "totality" in the sense of "all *x*," were occasionally used. These terms were applied to the cosmos in the sense of "all that exists," just as they were applied to express all the gods or all people or all countries. But to denote specifically the universe in a less abstract way, phrases like "the circumference of heaven and earth" and "the totality of heaven and earth" (as in the epithet of Marduk, "the one who holds the totality of heaven and earth") and "regions of heaven and earth" are used, all of which reflect the underlying mythology of the great above and below. The former evokes a spatial whole, even one of round shape, since the basic meaning of the term here translated "circumference" is "circle." A hymn to the god Ninurta describes the roof of that god's mouth as the "circumference of heaven and earth." The latter expression (constructed with the term for "regions") conveys the notion of political domains that together constitute the whole. The same term is used to connote the regions of the entire known inhabited world, or *oikoumene*. In this meaning, the word for "regions" appears frequently in royal epithets claiming universal political dominion of the world (notably *šar kiššati*, "king of the universe"), and was the term used in *Enuma Elish* when the gods proclaimed Marduk sovereign of the universe. Parallel with this usage is the phrase "four regions [of the world]," reflecting a division of the earth into four quarters, together representing the entire world.

The bipartite and essentially symmetrical cosmos persisted in later Babylonian thought. For example, in religious and scholarly texts of the first millennium B.C., reference is made to the cosmic designs, literally to "plans of the above and below." The first line of a scholastic explanatory work composed, perhaps, in the late second millennium B.C., mentions "cosmic designs, complementary elements of celestial and terrestrial parts of the universe, things of the Apsu, as many as were designed." These cosmic designs, perhaps the image of universal order, are frequently associated with what are called cosmic "destinies," as, for example, in the divine epithet "lord of cosmic destinies and designs." The two are integrally connected in virtue of the semantic force of the Akkadian term "destinies," which signifies the order or nature of things, and so did not exist before the creation of the universe itself. (It is said in *Enuma Elish* (I, 8) of the time before creation, that "no destinies [order of things] had been determined.")

The reference to the "cosmic designs" reflects another characteristic element of the Mesopotamian cosmology that seems to provide the theoretical basis for celestial divination, and later for Babylonian astrology. This is the element of complementarity between the two parts of the universe conceived in celestial and terrestrial terms. Whereas the spatial arrangement was in accord with the hierarchical status of the ruling deities—Anu on top, both spatially (in heaven) and as head of the pantheon—nevertheless, the regions above and below were mutually corresponding and had a reciprocal rather than a ruler-versus-ruled relationship. It was on the basis of this assumption of the complementarity between the above (celestial) and the below (terrestrial), rather than the domination of one over the other, that the Babylonian understanding of the mechanism of prognostication of the future from the heavens, or astrology, was made. The complementary relationship between heaven and earth could be seen as one that enabled the gods to signal impending events on earth by association with heavenly phenomena. A Babylonian textbook for celestial diviners instructs that "the signs in the sky, just like those on earth, give us signals." On such a basis, connections and associations were made between the human sphere of existence and the divine, or cosmic, sphere.

A stark contrast may be drawn between this and the cosmology that underlies Greek astrology. According to the Greek view, as represented in treatises such as Ptolemy's *Tetrabiblos* or Manilius's *Astronomia* (a view originally articulated by Plato in the *Timaeus* and developed by Aristotle in *De caelo*), the superior heavens determined change on earth because earth was physically below (sub-lunar, in the Aristotelian scheme) as well as metaphysically inferior to the celestial regions. In accepting the domination of the celestial over the terrestrial, Greek astrology developed a theory of stellar influences by which the planets physically wrought change on earth. The rationale for Babylonian astrology, on the other hand, was defined by the Babylonians' cosmological acceptance of the possibility of mutual connection between heaven and earth. Thus signs above could be associated with events below on earth. Babylonian astrology therefore never became deterministic, and did not require a physical theory of astral influence to explain the significance of celestial omens.

Influence of Early Babylonian Astronomy

In *Enuma Elish*, Tablet V, Marduk further imposes order on the heavens, establishing the constellations, the divisions of the year into the solar seasons, the lunar phases, and the "stations" of the planets. The relevant passages reflect certain elements of the contemporary astronomy of the period, roughly around the late second or early first millennium B.C. This was an astronomy characterized by schematic description of the rising and setting of stars and planets and by early attempts to quantify and calculate such astronomical phenomena as the length of lunar visibility throughout the month and the length of day throughout the year. The parts of the sky in which stars were to be observed at their heliacal risings (the first appearance in the sky after a period of invisibility due to conjunction with the sun) and the related schematic calendar of twelve 30-day months are of particular interest. They are mentioned in *Enuma Elish* V, 3–4:

> *He revealed* [the structure of] *the year demarcating its segments,* [and] *he set up three constellations for each of the twelve months.*

The reference here is to the assignation of three constellations (or stars) each to a schematic month in which they are supposed to rise heliacally each year. The text in which the specific designations of stars to months is preserved was known in antiquity as "the three stars each," but has been given the name *Astrolabe* by modern Assyriological scholarship. The *Astrolabe* text and another early Babylonian astronomical text related to it, called MUL.APIN, were certainly known to the author of *Enuma Elish*.

The arcs of the risings of the constellations mentioned belonged to different areas of the sky, called "paths" (*harrānu*). The three paths—designated by the names Anu, Enlil, and Ea—were organized with respect to the eastern horizon, where the "gates" of heaven were located and where observations of risings would naturally be made.

Anu's path was in the middle (i.e., due east for the central gates), Enlil's was to the north, and Ea's to the south. The paths correspond roughly to arcs over the horizon parallel to the equator extending between +17 degrees and -17 degrees for the path of Anu, north of +17 degrees for Enlil, and south of -17 degrees for Ea. The stars of these paths, however, do not describe fixed boundaries of declination. In fact, the circumpolar stars, which never pass through the "gates of heaven" (i.e., do not rise or set), were included in the path of Enlil, and planets, which do not have fixed points of rising and setting year after year, were also integrated within the path of Anu (except for Jupiter, which was assigned to the path of Enlil). The extreme limits (in azimuth) of the three paths correspond to the turning points of the sun in its progress back and forth across the eastern horizon, defining the winter and summer solstitial points. Before the creation of a standard celestial coordinate system, the paths of Anu, Enlil, and Ea were the chief means of locating the position of something in the sky. Even after the establishment of the ecliptical coordinates by which the Babylonian zodiac became sidereally fixed around 500 B.C., the position of objects in the heavens which were not located near the ecliptic continued to be cited with respect to the "paths," such as in the famous reference to Halley's comet in an astronomical diary of 164 B.C., in which the comet, "which previously had been seen in the east in the path of Anu in the area of Pleiades and Taurus, to the west [. . . broken text] and passed along in the path of Ea." [Stephenson and Walker, eds., *Halley's Comet in History* (1985), p. 24, lines 16–17.]

Enuma Elish also credits Marduk with the establishment of the moon as a time reckoning device:

> He [Marduk] *made the luminary* [the moon] *shine forth, entrusting* [to him] *the night. He revealed him as a phenomenon of the night to make known the days. Monthly without ceasing, he made him form shapes with* [his] *crown: "At the beginning of the month, rising over the land, you will shine with horns* [i.e., be crescent-shaped] *to signify six days, on the seventh day, a half crown. On the fifteenth day, in mid-month, you will assume the same size* [as the sun in opposition]. *Thereafter, when the sun* [sees] *you on the horizon, as is appropriate, decrease and configure yourself backwards. On the day of disappearance, you approach the path of the sun, and* [. . .] *on the 30th, you will again be the same as the sun* [but invisible]." [Enuma Elish V, 12–22.]

This lunar scheme is echoed in the commentary I.NAM.GIŠ.HUR.AN.KI.A mentioned above, essentially defining the days of the phases as 1 (*tāmartu*, "first visibility"), 7, 14/15, 21, and 27/28 (days of last visibility).

The moon is considered to be a god wearing a crown, whose aspect changes with the progress of its wearer throughout the month. Marduk's installation of the moon is therefore clearly for timekeeping purposes (i.e., "to make known the days") and to demarcate the phases of the month. The very schematic understanding of the lunar phases is paralleled in contemporary astronomical texts, particularly in *Enuma Anu Enlil*, a series of celestial omens. In this text, various phenomena related to the changing appearance of the moon are compiled as portents of future events on earth, mostly relevant to the king and to the country in general. Inserted between the lunar omens dealing with the appearance of the moon throughout the month and those exclusively concerning lunar eclipses is tablet 14, which presents a mathematical scheme for calculating the duration of the visibility of the moon on any date of the month through the year. Despite the mathematical cleverness of this second millennium scheme, which subjects two of the components of the complex problem of duration of lunar visibility on successive nights to separate but combinable mathematical functions, it stems from assumptions such as those underlying the passage quoted above from *Enuma Elish*: namely, the 30-day schematic lunar month that runs from first crescent on day 1 to invisibility on day 30. Under this assumption, the moon is not visible on day 30 or day 0, is visible for the entire night (sunset to sunrise) on day 15, and in between increases or decreases its length of visibility by a computable amount. The size of the moon (its "crown"), referred to in *Enuma Elish*, is the visual counterpart to the computational variability of the duration of visibility.

Additional sources for the Babylonians' interest in the visible cosmos are found in celestial omens, and later, after 500 B.C., in observational astronomy. Observational astronomy refers to the activity of scribe-scholars, represented by an enormous archive from Babylon of texts now called astronomical diaries, which systematically recorded night by night the positions in the sky of the moon and the five naked-eye planets, as well as reports of the weather, clouds, and any other notable phenomena observed in the sky or on the earth (such as deformed animal births or wild animals seen running through the streets of Babylon). But none of these groups of texts, omens, or diaries contains much of cosmological interest beyond references to a number of parts of the sky in which, presumably, it was thought phenomena were to be seen. These parts are, for example, the "surface of the sky," the "middle of the sky," "heights of heaven," and "base of heaven" (the horizon). There is also a term "depths of heaven," which seems to parallel "middle of the sky" in some contexts.

In the further development of astronomy in Babylonia after 500 B.C., the particular mathematical methods that were applied to various astronomical phenomena were fully independent of any physical theory of celestial motion as well as of any corresponding cosmology.

In this way, late scientific Babylonian astronomy neither influenced nor was influenced by any of the available Mesopotamian cosmologies.

Conclusion

The Mesopotamian cosmology described here reflects a collage of sources both from religious texts (of mythological and theological content) and scientific texts (of celestial divination and astronomical content). This description is not intended to portray a synchronic picture or to suggest that the Babylonians were even interested in specifically investigating the universe apart from its divine nature. Astronomy, of the relatively unsophisticated schematic type integrated within the written scholarly repertoire of at least the mid-second-millennium B.C., made an impress only on the cosmological work *Enuma Elish*. All other sources for cosmology reflect only mythological and theological considerations. *Enuma Elish* has been presented here as the major source for Babylonian cosmology. However, since variant traditions were not eradicated by that text, the status of its cosmology—whether it represented official doctrine or just one of several and not entirely incompatible systems of cosmological thought—remains unclear. But it does constitute the only system that took any account of contemporary astronomy.

The compatibility of religious and scientific elements in the Mesopotamian tradition is what we would expect in a premodern cosmology. The same is still decidedly the case until the end of the Middle Ages, when theological issues centering on the biblical six days of creation were as important a component in the development of medieval cosmology as were purely physical (or metaphysical) considerations amenable to mathematical, empirical, and philosophical methods of inquiry. The diversity of the underlying sources for a historical reconstruction of cosmology in ancient Mesopotamia, however, is not the only aspect to acknowledge. Because of the interweaving of very old myths together with theological and scholastic developments over a long historical period, the Babylonian (and Sumerian, and also Assyrian) sources present many variants sharing certain basic themes. Their contents, however, hardly constitute a consistent system that would have functioned within that culture as a single valid picture of the world. If it might be regarded as a limitation of such sources that they fail to provide clues as to why no single system was considered necessary, it must then further be acknowledged that Mesopotamian cosmologies are reflected in texts whose goals were assuredly not to construct a definitive cosmic picture to serve as the framework for inquiry about natural phenomena. Despite the diverse historical evidence for a Mesopotamian response to basic questions about the coming into existence, structure, constituent matter, and dimensions of the universe, the assumption that answers to such questions should be foundational for further empirical or speculative inquiry about phenomena of that universe seems to go back only as far as the construction of the "closed world," that is, the finite, spherical, and geocentric cosmos of the Greeks and their later inheritors.

—F. Rochberg

BIBLIOGRAPHIC NOTE

General introductions to aspects of Mesopotamian civilization, culture, and cosmology can be found in S. Dalley, *Myths from Mesopotamia: Creation, The Flood, Gilgamesh, and Others* (Oxford: Oxford University Press, 1991); S.N. Kramer, *Sumerian Mythology* (New York: Harper Torchbooks, 1961); and W.G. Lambert, "The Cosmology of Sumer and Babylon," in C. Blacker and M. Lowe, eds., *Ancient Cosmologies* (London: George Allen and Unwin, Ltd., 1975), pp. 42–66. On astronomical texts, see O. Neugebauer, *Astronomical Cuneiform Texts* (London: Lund Humphries, 1955); and T. Pinches and A. Sachs, *Late Babylonian Astronomical and Related Texts* (Providence: Brown University Press, 1955). Specific texts are available in H. Hunger and D. Pingree, *MUL.APIN: An Astronomical Compendium in Cuneiform* (Vienna: Archiv für Orientforschung, vol. 24; Horn, Austria: Ferdinand Berger & Söhne, 1989); F. Rochberg-Halton, *Aspects of Babylonian Celestial Divination: the Lunar Eclipse Tablets of Enuma Anu Enlil* (Vienna: Archiv für Orientforschung, vol. 22; Horn, Austria: Ferdinand Berger & Söhne, 1988); and A. Sachs and H. Hunger, *Astronomical Diaries and Related Texts from Babylonia. Vol. 1, Diaries from 652 B.C. to 262 B.C.*, and *Vol. 2, Diaries from 261 B.C. to 165 B.C.* (Vienna: Österreichischen Akademie der Wissenschaften, 1988 and 1989).

4. THE PRESOCRATICS

Distinctive approaches to the study of nature were formulated in the Greek world during the period from about 600 to 400 B.C., a period often labeled pre-Socratic, or Presocratic. Philosophy and science, including cosmological ideas about the universe, would change significantly if not completely at the end of this period, largely due to Socrates' influence. Pre-Socratic science of the sixth and fifth centuries B.C. would be followed by a surge of new ideas in new directions, with Hellenic philosopher-scientists of the fourth century B.C. formulating radically different approaches to science. During this second period of Greek philosophy and science, the Hellenic period, Plato and Aristotle would set major new directions for astronomy, their aspirations and achievements coming to dominate Western science for nearly two thousand years. In the following Hellenistic period, from roughly 300 B.C. to A.D. 150, scientists would develop in considerable detail the suggestions of their Hellenic predecessors. In astronomy and cosmology, Hellenistic efforts would culminate in the work of Ptolemy around A.D. 150 in Alexandria, Egypt.

This later Greek cosmology with specific, detailed questions and a large observational component was a far cry from the global questions of the Presocratics concerning how things began and of what they were made, questions and postulates which were largely untestable and unverifiable. Yet if Presocratic cosmology is now to be faulted for its dogmatic strain, the growing role of rational criticism and debate evident in references by later Presocratic philosophers to their predecessors and evident in their reactions to the ideas of their predecessors often is hailed as the beginning of science. And if not explicitly stated, still the implicit assumption that the cosmos was a rational, ordered whole accessible to human inquiry was itself an important cosmological generalization.

Sources
Any discussion of pre-Socratic philosophy might well carry a warning

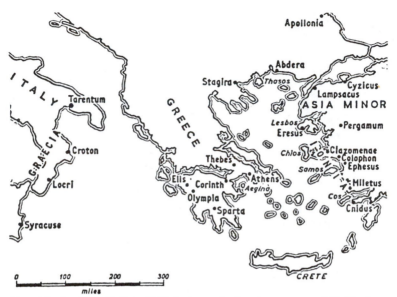

Map of the Greek World. From N. Hetherington, Ancient Astronomy and Civilization
(Tucson, Arizona: Pachart Publishing House, 1987), 42.

notice that the conclusions drawn rest on an uncertain foundation.
Indeed, not a single original writing from the period has survived.
Knowledge of pre-Socratic science comes from three sources, all
written well after the event. The sources are:

- Fragments—a few quotations from pre-Socratic works that have
  survived in books written after 400 B.C.

- Testimonia—comments in the writings, such as have survived, of
  Plato, Aristotle, and Theophrastus on pre-Socratic science. These
  books were written shortly after the pre-Socratic period.

- The doxographical tradition—consisting of summaries of the works
  of Plato, Aristotle, and Theophrastus, and also summaries of
  summaries. The primary source for the summarizers was the
  multivolume history of early philosophy by Theophrastus.

None of our sources for pre-Socratic philosophy is above question.
The fragments, the closest thing we have to the original, are perhaps
the most reliable, or—to put it more negatively and, alas, more accurately—the least unreliable of the sources available to us. But there is
no surviving original against which to check the fragmented quotations.

The testimonia have an additional uncertainty. Aristotle gives serious attention to the earlier philosophers, but his judgment and his corresponding analysis and description of earlier philosophy may well have been distorted by his own belief in the importance of the material nature of the world. Also, we may wonder whether Aristotle read into the thoughts of his Presocratic predecessors more emphasis on the problem of coming into being, on how the world was created, than was originally there. Plato, in contrast to Aristotle, offers only casual remarks on the thoughts of the Presocratics. Theophrastus offered more, but of the sixteen or eighteen books written by Theophrastus, only a single volume has survived.

The doxographical tradition consists primarily of summaries of Theophrastus's books and summaries of summaries. The summaries are especially important since almost all of Theophrastus's original writings are lost, but summaries are notoriously unreliable.

The doxographical tradition is aptly named; *doxy* means both opinion and prostitute. Knowledge of the early philosophers almost certainly has been corrupted in the process of condensation, simplification, interpretation, translation, and transmission. It may verge upon licentious behavior to base historical studies on obviously unreliable evidence. There is, however, no better evidence available.

Ionian Cosmology

One major tradition within pre-Socratic science of the sixth and fifth centuries B.C. was that established and developed by Ionian peoples, who flourished among the Aegean Islands and along the western coast of Asia Minor. A primary characteristic of their philosophical inquiries into the nature of the universe seems to have been a search for a basic substance or substances that persist throughout all changes. From the disposition of such matter the philosophers attempted to explain how the universe came to be. Neither explicitly stated nor questioned in this tradition, at least as far as we know, was the assumption that the universe was not eternal, but came into being. Much the same assumption might be found in the book of Genesis.

Potential confusion over the geographical location of the Ionians is encouraged by the fact that our modern name for that part of the Mediterranean Sea to the *west* of Greece, between Greece and Italy, is the Ionian Sea, with the islands therein called, naturally enough, the Ionian islands. But speakers of the Ionic dialect, one of the main linguistic groups of the ancient Greek people, came to reside primarily among the Aegean islands (in the Aegean Sea, also part of the Mediterranean but to the *east* of Greece, between Greece and Turkey) and on the western coast of Asia Minor (now Turkey). According to a tradition later set down by the Greek historian Herodotus, the Ionians had originally lived in southern Greece, and under pressure from invaders had fled across the Aegean Sea. Nearly identical pottery fragments from Athens and from Asia Minor as well as similarities in religious festivals and the names of months seemingly confirm some sort of migration from the Greek mainland across the Aegean. Thus it is that ancient Ionians may have come to live in and be associated with a region to the east of their original home.

Greek communities sprang up on both sides of the Aegean Sea, in Greece and in Asia Minor, with Miletus soon becoming the most prosperous of the new city-states. Now but lonely ruins, inland from the Turkish coast because the river and harbor silted up, and visited only by an occasional tourist, Miletus was—in its time—the richest city in the Greek world. Ionians from Greece had conquered earlier inhabitants, kept their women, and multiplied. Wealth followed trade, and Miletus at its height had some eighty colonies. In this expansive environment with leisure time for reflection, many activities not immediately essential for the lowest level of survival flourished, philosophy among them.

According to Aristotle, writing in his *Metaphysics* some two and a half centuries after the event, Milesian philosophers thought that matter or principles in the form of matter were the principle of all things:

> *Most of the first philosophers thought that principles in the form of matter were the only principles of all things: for the original source of all existing things, that from which a thing first comes-into-being and into which it is finally destroyed, the substance persisting but changing in its qualities, this they declare is the element and first principle of existing things, and for this reason they consider that there is no absolute coming-to-be or passing away, on the ground that such a nature is always preserved [. . .] for there must be some natural substance, either one or more than one, from which the other things come-into-being, while it is preserved. Over the number, however, and the form of this kind of principle they do not all agree; but Thales, the founder of this type of philosophy, says that it is water (and therefore declared that the earth is on water), perhaps taking this supposition from seeing the nurture of all things to be moist, and the warm itself coming-to-be from this and living by this (that from which they come-to-be being the principle of all things)—taking the supposition both from this and from the seeds of all things having a moist nature, water being the natural principle of moist things.*

The objective of Milesian science or philosophy—the two were not separate disciplines at this time—seems to have been to search for a basic substance or substances that persisted throughout all changes.

Thales

The first of the Ionian or Milesian philosophers mentioned by later writers, Thales was reputed to be the wisest of the seven wise men or sages of Greece. Asked what was difficult, he answered "to know thyself." Asked what was easy, he answered "to give advice." He was the first mathematician to demonstrate that a circle is bisected by its diameter, that the angle of a semicircle is a right angle, and that angles at the base of an isosceles triangle are equal.

Thales is said to have predicted a solar eclipse that occurred during the sixth year of a war and brought the war to an end. Either the warring parties took the eclipse as a sign to cease fighting, or they were eager for any reason to cease and found the eclipse a convenient excuse. Prediction of the solar eclipse of 584 B.C. may be more myth than historic truth, however. From a study of the periodic recurrence of solar eclipses, Thales could have predicted an eclipse before the end of 583 B.C., and then have taken credit for a different eclipse occurring slightly earlier.

Another legend involving Thales has him providing a practical justification for the study of philosophy. Reproached for his poverty, Thales determined to demonstrate to the skeptics that there were practical uses for philosophy or science. First, he studied the heavens and decided that weather conditions would produce a large olive crop the following year. He then proceeded to raise money during the winter and rent all the olive presses. The next summer growers were forced to pay Thales his price to have their olives pressed into oil. Thus did Thales demonstrate that philosophers could be rich in conventional monetary terms—if they wished to be. The philosopher's true wealth, however, is not measured in money. It is found instead in the pleasure derived from intellectual endeavor. In eschewing a myopic pursuit of wealth, Thales demonstrated again his wisdom. The tale of the olive presses, true or not, demonstrates two different motives for the practice of science. The true philosopher pursues scientific inquiry for its own sake; others demand that science pay in practical results.

Thales was not always practical. In another legend about Thales, he is said to have fallen into a well; afterwards, a witty slave girl mocked him for observing the stars rather than noticing what was at his feet.

For the basic principle of all things, Thales chose water. Aristotle guessed that Thales chose water because of its physiological importance. Or Thales might have been influenced by his interest in meteorology.

Anaximander

Thales began a tradition of searching for the basic matter or principle of things. Anaximander, probably fourteen years younger than Thales and described by Theophrastus as the successor and pupil of Thales, agreed with Thales that there was one basic principle of nature. But how could fire, destroyed by water, exist if water were the primary substance? Anaximander said that the first thing was not any specific substance, but something indefinite. A later account of Anaximander's originative substance reads:

> *Of those who say that it is one, moving and infinite, Anaximander, son of Praxiades, a Milesian, and the successor and pupil of Thales, said that the principle and element of existing things was the aperion, being the first to introduce this name of the material principle. He says that it is neither water nor any other of the so-called elements, but some other aperion nature, from which come into being all the heavens and the worlds in them.*

Anaximenes

Next in the succession of pre-Socratic philosophers seeking the basic substance unchanged through all change was Anaximenes. He was twenty-four years younger than Anaximander and his associate in Miletus. For Anaximenes, air was the originative substance and basic form of matter, undergoing change by condensation and rarefaction. According to a surviving fragmentary description:

> *Anaximenes, son of Eurystratus, of Miletus, a companion of Anaximander, also says that the underlying nature is one and infinite like him, but not undefined as Anaximander said but definite, for he identifies it as air; and it differs in its substantial nature by rarity and density. Being made finer it becomes fire, being made thicker it becomes wind, then cloud, then (when thickened still more) water, then earth, then stones; and the rest come into being from these. He, too, makes motion eternal, and says that change, also, comes about through it.*

Heraclitus

Miletus was destroyed by the Persians in 494 B.C., but the tradition of searching for a basic principle of nature had spread by then to nearby cities, where it survived and developed further. The apparent ease with which ideas seem to have travelled in these times of primitive and perilous transportation is remarkable.

In Ephesus, just north of Miletus, Heraclitus around 500 B.C. argued that fire was the basic substance. Furthermore, he continued, our surroundings contain bowls turned with their hollow sides toward us, in which bright exhalations are collected and form flames. These flames are the heavenly bodies, with the brightest and hottest the sun. Eclipses of the sun and moon occurred when their bowls turned upward, and the monthly phases of the moon occurred as its bowl was gradually turned.

Heraclitus was reputed to be exceptionally haughty and supercilious, and he may have had few friends to regret his unfortunate end. An account of his last years reads:

> Finally he became a misanthrope, withdrew from the world, and lived in the mountains feeding on grasses and plants. However, having fallen in this way into a dropsy he came down to town and asked the doctors in a riddle if they could make a drought out of rainy weather. When they did not understand he buried himself in a cow-stall, expecting that the dropsy would be evaporated off by the heat of the manure; but even so he failed to effect anything, and ended his life at the age of sixty.

Xenophanes

Farther north of Miletus, in Colophon, Xenophanes in the period 570 to 475 B.C. proposed that everything was composed of two elements, water and earth. He also speculated on the origins and arrangement of some of the celestial bodies:

> The sun comes into being each day from little pieces of fire that are collected, and the earth is infinite and enclosed neither by air nor by the heaven. There are innumerable suns and moons, and all things are made of earth.

Empedocles

From Colophon, Xenophanes wandered to Sicily, where the tradition of seeking the basic substance pops up again. Xenophanes' doubling of the original single basic element was doubled again, by Empedocles, who lived from roughly 500 to 430 B.C. in Acragas, now Agrigento, Sicily, and was the author of a physical-cosmological poem titled *On Nature*, now known only through fragments of quotations from later writers. It is probably through Xenophanes' influence that Empedocles can be placed in the Miletian tradition of philosophy that taught that principles in the form of matter were the only principles of all things.

There were four basic elements for Empedocles: air, fire, earth, and water. Different mixtures of these elements produced different substances. According to several fragments:

> Empedocles of Acragas [. . .] holds that the air that was separated off from the original mixture of the elements flowed around in a circle; and after the air, fire ran outwards and, having nowhere else to go, ran upwards under the solidified periphery around the air. There are, he says, two hemispheres revolving round the earth, one consisting entirely of fire, the other of a mixture of air with a little fire; this latter he supposes to be night. Their motion arises from the fact that the accumulation of fire in one region gives it preponderance there.
>
> The sun is not in its nature fire, but rather a reflection of fire like that which comes from water. The moon, he says, was composed of air that had been

shut in by fire; this air was solidified, like hail. The moon gets its light from the sun.

Empedocles says that the stars are made of fire, composed of the fiery element which the air originally contained but squeezed out at the first separation.

All those who generate the heavens hold that it was for this reason that the earth came together to the center. They then seek a reason for its staying there; and some say, in the manner explained, that the reason is its size and flatness; others, like Empedocles, say that the motion of the heavens, moving about it at a higher speed, prevents movement of the earth, as the water in a cup, when the cup is given a circular motion, though it is often underneath the bronze, is for this same reason prevented from moving with the downward movement which is natural to it.

Theories about the supposed basic material nature of things constituted one major tradition of pre-Socratic thought. Generations of philosophers seemingly agreed that the basic material of nature was a subject worthy of their speculations, however unsubstantiated and lacking in practical consequence these speculations were.

Pythagorean Cosmology

A second major tradition within pre-Socratic philosophy centered about Pythagoras. He was born on the island of Samos, not far north of Miletus, probably between 585 and 565 B.C., and traveled in Asia, Greece, and Egypt in search of knowledge before settling at Croton in southern Italy in about 530 B.C. From Croton, Pythagoras migrated with his companions to Metapontum in about 495 B.C. There he died, probably between 495 and 475 B.C.

Pythagoras founded a secret religious fraternity, and its rule of secrecy has not helped historians intent on prying into the community's philosophy and the lives of its individual members. Furthermore, the community was destroyed by enraged neighbors and few Pythagoreans survived to tell their secrets, even had they wished to tell.

According to Aristotle, the Pythagoreans believed that number was the principle of all things:

Contemporaneously with these philosophers, and before them, the Pythagoreans, as they are called, devoted themselves to mathematics; they were the first to advance this study, and having been brought up in it they thought its principles were the principles of all things. Since of these principles numbers are by nature the first, and in numbers they seemed to see many resemblances to the things that exist and come into being—more than in fire and earth and water (such and such a modification of numbers being justice, another being soul and reason, another being opportunity—and similarly almost all other things being numerically expressible); since, again, they saw that the attributes and the ratios of the musical scales were expressible in numbers; since, then, all other things seemed in their whole nature to be modelled after numbers, and numbers seemed to be the first things in the whole of nature, they supposed the elements of numbers to be the elements of all things, and the whole heaven to be a musical scale and a number. And all the properties of numbers and scales which they could show to agree with the attributes and parts and the whole arrangement of the heavens, they collected and fitted into their scheme; and if there was a gap anywhere, they readily made additions so as to make their whole theory coherent. E.g., as the number 10 is thought to be perfect and to comprise the whole nature of numbers, they say that the bodies which move through the heavens are ten, but as the visible bodies are only nine, to meet this they invent a tenth—the "counter-earth." We have discussed these matters more exactly elsewhere.

Aristotle's description of Pythagorean philosophy continues:

Evidently, then, these thinkers also consider that number is the principle both as matter for things and as forming their modifications and their permanent states, and hold that the elements of number are the even and the odd, and of these the former is unlimited, and the latter limited; and the 1 proceeds from both of these (for it is both even and odd), and number from the 1; and the whole heaven, as has been said, is numbers. Other members of this same school say there are ten principles, which they arrange in two columns of cognates—limit and unlimited, odd and even, one and plurality, right and left, male and female, resting and moving, straight and curved, light and darkness, good and bad, square and oblong. In this way Alcmaeon of Croton seems also to have conceived the matter, and either he got this view from them or they got it from him; [. . .] for he expressed himself similarly to them. For he says most human affairs go in pairs, meaning not definite contrarieties such as the Pythagoreans speak of, but any chance contrarieties, e.g. white and black, sweet and bitter, good and bad, great and small. He threw out indefinite suggestions about the other contrarieties, but the Pythagoreans declared both how many and which their contrarieties are. From both these schools, then, we can learn this much, that the contraries are the principles of things; and how many these principles are and which they are, we can learn from one of the two schools. But how these principles can be brought together under the causes we have named has not been clearly and articulately stated by them; they seem, however, to range the elements under the head of matter; for out of these as immanent parts they say substance is composed and molded.

The selection by the Pythagoreans of intangible number as the principle of all matter may appear implausible to modern readers. There was, however, ample reason around 500 B.C. to believe that number was important for an understanding of nature. Indeed, some of the reasons have persisted to today. Musical harmony, for example, can be explained in terms of ratios of the lengths of strings or pipes. There are definite relationships between the lengths of vibrating strings and the pitch of notes emitted by strings. A ratio of 2 to 1 is an octave, 3 to 2 is a fifth, and 4 to 3 is a fourth. Less plausible to readers in the final decade of the twentieth century A.D. may well be ancient attempts to understand qualities such as justice and goodness in terms of numerical ratios. Here the Pythagoreans seem far removed from the spirit of modern science. The search for numerical relationships among the phenomena of nature is, however, an important aspect of modern science that can be traced to Pythagorean origins.

The Pythagorean emphasis on number changed history. Their ideas about the sun also influenced subsequent thought. They disagreed with the then-common belief that the earth lay in the center of the universe. Instead, they argued that the earth moved around a central fire. Pythagorean philosophy, including the displacement of the earth from the center of the universe and its movement about another body, may well have influenced Copernicus two millennia later, in the sixteenth century A.D. Copernicus accorded to the sun a central place in his planetary system.

The Pythagoreans also were interested in the numerical harmonies of the motions of the planets. These were, in turn, related to musical harmonies. The planets supposedly moved in such a manner that the music their motions produced was harmonious.

From Aristotle's *De caelo* (*On the Heavens*) we learn:

> *Most people say that the earth lies at the center of the universe.* [. . .] *but the Italian philosophers known as Pythagoreans take the contrary view. At the center, they say, is fire, and the earth is one of the stars, creating night and day by its circular motion about the center. They further construct another earth in opposition to ours to which they give the name counter-earth. In all this they are not seeking for theories and causes to account for observed facts, but rather forcing their observations and trying to accommodate them to certain theories and opinions of their own. But there are many others who would agree that it is wrong to give the earth the central position, looking for confirmation rather to theory than to the facts of observation. Their view is that the most precious place befits the most precious thing: but fire, they say, is more precious than earth, and the limit than the intermediate, and the circumference and the center are limits. Reasoning on this basis they take the view that it is not earth that lies at the center of the sphere, but rather fire. The Pythagoreans have a further reason. They hold that the most important part of the world, which is the center, should be most strictly guarded, and name it, or rather the fire which occupies that place, the "Guard-house of Zeus," as if the word "center" were quite unequivocal, and the center of this mathematical figure were always the same with that of the thing or the natural center. But it is better to conceive of the case of the whole heaven as analogous to that of animals, in which the center of the animal and that of the body are different.*

Aristotle continues:

> *From all this it is clear that the theory that the movement of the stars produces a harmony, i.e. that the sounds they make are concordant, in spite of the grace and originality with which it has been stated, is nevertheless untrue. Some thinkers suppose that the motion of bodies of that size must produce a noise, since on our earth the motion of bodies far inferior in size and in speed of movement has that effect. Also, when the sun and the moon, they say, and all the stars, so great in number and in size, are moving with so rapid a motion, how should they not produce a sound immensely great? Starting from this argument and from the observation that their speeds, as measured by their distances, are in the same ratios as musical concordances, they assert that the sound given forth by the circular movement of the stars is a harmony. Since, however, it appears*

unaccountable that we should not hear this music, they explain this by saying that the sound is in our ears from the very moment of birth and is thus indistinguishable from its contrary silence, since sound and silence are discriminated by musical contrast. What happens to men, then, is just what happens to coppersmiths, who are so accustomed to the noise of the smithy that it makes no difference to them.

However strange the doctrine of musical harmonies may appear today, it did furnish an essential stimulus for the scientific revolution of the sixteenth and seventeenth centuries. The astronomer Johannes Kepler spent much of his working life, which ended in A.D. 1630, in a search for the numerical ratios that philosophy convinced him existed in the arrangement of the planetary system. Kepler's famous three laws, now conveniently abstracted out of a morass of number mysticism and numerous less reputable "laws," are an essential part of modern science. Newton's law of gravity followed from Kepler's work.

Parmenides' Philosophy and the Atomic Theory

In addition to Ionian materialism and Pythagorean mathematization, the thought of Parmenides holds much interest for modern philosophy. Born between 515 and 510 B.C., Parmenides became a leading member of the Eleatic school of philosophy founded by Xenophanes and centered at Elea in southern Italy. Parmenides raised disturbing questions about the unformulated assumptions of naive common sense, a characterization applicable to much of the thought of earlier philosophers. Parmenides' criticism, to the extent that it made philosophers conscious of their assumptions, was a healthy check upon potential excesses of philosophical speculation. In particular, Parmenides argued with considerable force and persuasive power that coming to be, passing away, and—indeed—any change whatsoever, was impossible. Much subsequent philosophizing was an argument between followers of Parmenides and those who resisted his conclusions.

Any significant role for Parmenides in the history of cosmology, though, is problematic. Seemingly, he neither believed in his own cosmology nor swayed others to his view, a convenient conclusion on the part of scholars who otherwise would be under considerably more pressure to reconstruct Parmenides' cosmological system. Instead, it usually is sufficient to reproduce an ancient and obfuscatory account of his ideas:

Parmenides said that there were rings wound one around the other, one formed of the rare and the other of the dense, and that there were other rings between these compounded of light and darkness. That which surrounds them all like a wall is by nature solid. Beneath it is a fiery ring. Likewise, what lies in the middle of them all is solid. Around it is again a fiery ring. The middlemost of the mixed rings is the primary cause of movement and of coming into being, for them all, and Parmenides calls it the goddess that steers all, the holder of the keys, Justice and Necessity. The air is separated off from the earth, vaporized owing to earth's stronger compression. The sun is an exhalation of fire, and so is the circle of the Milky Way. The moon is compounded of both air and fire. Aether is outermost, surrounding all. Next comes the fiery thing that we call the sky. Last comes the region of the earth.

Parmenides' criticism did stimulate the last great work of pre-Socratic philosophy, the atomic theory. Opposed to the notion of changes in a continuous material or materials making up the universe, the atomists pictured instead matter made up of small particles of the same material but differing in size and shape, with combinations of these particles, or atoms, giving rise to all substances. Each atom was ungenerated, unalterable, indivisible, and indestructible—somewhat like Parmenides' one unchanging being.

The emphasis on the matter of the universe is, perhaps, reminiscent of the Milesian cosmological tradition tracing from Thales, Anaximander, and Anaximenes. Aristotle credits the atomic theory to Leucippus, who probably was born at Miletus and was active from approximately 450 to 420 B.C. The theory was elaborated by Democritus, who was born in northern Greece around 460 B.C. Such a birth date makes Democritus younger than Socrates, but he is nonetheless considered the last of the pre-Socratic cosmological speculators.

Within atomic theory there appears to have been at least the possibility of an eternal, unchanging universe, as argued for by Parmenides, in contrast to the act of creation presumably assumed by the earlier Miletians. Any attempt now to explain why the ensuing cosmological speculations took one path rather than another constitutes an act of speculation on our part, based on the shaky premise that we even know what path was taken. With these caveats, we hesitantly put forward our own historical speculation: that some sort of overwhelming human psychological need for a creation of the universe and an explanation of that assumed creation seemingly was at work, and that from the atomic theory an atomistic cosmology was developed to satisfy this human need.

Worlds came into being, in the atomic theory, when many moving bodies of all shapes came together and produced a single whirl. As more atoms joined the whirl, its rotation decreased. The atoms separated, fine ones returning into the void but others becoming entangled and forming a spherical structure. At first the structure was moist and muddy, but dried out with the whirl of the whole structure. Eventually some of the bodies, now dry, ignited to form the substance of the heavenly bodies.

The atomic theory was rejected by Aristotle and other philosophers, who, believing that nature abhorred a vacuum, thus found the atomic theory with particles moving in a void untenable. Though revived in the first century B.C. by a Roman, Lucretius, in his poem *De rerum natura (On the Nature of Things)*, the atomic theory of the Presocratics was to have little effect on subsequent thought. The appearance of an atomic theory in the scientific revolution of the sixteenth and seventeenth centuries A.D., despite similarities with aspects of Presocratic atomistic theory, was largely independent of the earlier version. Atomistic cosmology, following from the atomic theory at the end of the pre-Socratic period, encompasses interesting features and is rightfully regarded as one of the major intellectual creations of Greek philosophy and science. However, Presocratic atomic theory seemingly had little influence on subsequent intellectual developments.

Conclusion

In Presocratic philosophy and science, nature came to be thought of as a comprehensive system ordered by law. Magical powers of gods were not invoked in explanations of nature, and the expurgation of the supernatural has been seen as marking the transition from religion to philosophy. Rational as Presocratic cosmology may have been, though, it remained in some ways little more than unfounded speculation. Presocratic philosophers, especially cosmologists, lacked the necessary methodology and experimentation to advance beyond dogmatic statements about nature to a tested and verified account of nature. As an author of *On Ancient Medicine* noted, his field of science differed from "obscure and problematic subjects, concerning which anyone who attempts to hold forth at all is forced to use a postulate, as for example about things in heaven or things under the earth: for if anyone were to speak and declare the nature of these things, it would not be clear whether to the speaker himself or to his audience whether what was said was true or not, since there is no criterion to which one should refer to obtain clear knowledge."

Controversies over the proper interpretation of Greek science have centered on the general question of the nature of scientific method and the place of observation in science. Those who see bold theorizing about the world as an essential element of science tend to rank Presocratic philosophy higher than those who find the beginnings of science ultimately in observation. Many Presocratic philosophers, impatient of empirical investigation and inclined toward abstract theorizing, began with all-encompassing cosmological questions rather than specific questions and related observations. There was, however, a role for rational criticism and debate in Presocratic science, as is evident in the development of Ionian cosmology from one generation to the next and in the development of alternative Pythagorean and atomistic cosmologies.

—*Norriss S. Hetherington*

BIBLIOGRAPHIC NOTE

For the essentials of what is known of Presocratic philosophy, begin with G.S. Kirk, J.E. Raven, and M. Schofield, *The Presocratic Philosophers: A Critical History with a Selection of Texts* (Cambridge: Cambridge University Press, 1983). G.E.R. Lloyd, *Early Greek Science: Thales to Aristotle* (New York: W.W. Norton & Co., 1970) is one of the best brief general accounts of the science of this period, while his journal article "Popper versus Kirk: A Controversy in the Interpretation of Greek Science," *British Journal for the Philosophy of Science, 18* (1967), 21–38, is an excellent introduction to an interesting and enlightening historiographic debate between Sir Karl Popper and Professor Kirk, a debate which can be studied further in Popper's and Kirk's papers cited by Lloyd. This article is reprinted in G.E.R. Lloyd, *Methods and Problems in Greek Science* (Cambridge: Cambridge University Press, 1991), preceded by an introduction assessing scholarly debate on the topic and Professor Lloyd's modifications and developments in his own position since the original publication of the article. Another discussion of the nature of Presocratic science is found in H. Cherniss, "The Characteristics and Effects of Presocratic Philosophy," *Journal for the History of Ideas, 12* (1951), 319–345. On Greek cosmology in particular, see N. Hetherington, *Ancient Astronomy and Civilization* (Tucson, Arizona: Pachart Publishing House, 1987).

II. The Greeks' Geometrical Cosmos

Introduction

TO SAVE THE PHENOMENA

ifferent cosmologies in different cultures show in the most unmistakable manner that science in general and cosmology in particular can be done differently by different peoples in different places in different times. Much the same sort of realization is also obtained from a study of Western cosmology, particularly its historical evolution. We can look back to the time before Christ and the beginning of Greek cosmology to encounter a scientific enterprise very different from that which we know today, a science with very different goals and procedures, the manifestation of very different human values and culture.

Twice in human history science has flourished in the West. Once was within Greek civilization, roughly from 400 B.C. to A.D. 150, from Athens to Asia Minor and to the delta of the Nile River, in the works of individuals such as Plato, Aristotle, Eudoxus, Euclid, Archimedes, Galen, and Ptolemy. The West's second scientific revolution has its beginning somewhat arbitrarily fixed, in A.D. 1543, with the publication of Copernicus's great work on astronomy.

Rebirth, only later followed by radical reform, characterizes this second scientific revolution, of the sixteenth and seventeenth centuries A.D. Following the so-called Dark Ages, which had seen a decline—though by no means a total cessation—of intellectual activity in Western Europe in the centuries after the fall of the Roman Empire, scholars now aspired to reclaim their Greek heritage, to reclaim the knowledge once gained and then nearly lost. They looked back to a past far greater than their contemporary civilization. In doing so, their psychological outlook must have been far different from what ours is today. All our lives we have seen continued progress in science; we look not back to the past but ahead to the future for greater knowledge.

Envisioned, initially, by scholars of the sixteenth century, was a *re-birth* of the earlier, greater culture, a *re-naissance* of Greek culture and learning. Hence the appellation *Renaissance* for this period in history. And hence also our inclusion of Copernicus in the first scientific revolution instead of in his more customary place at the beginning of the second scientific revolution, because his scientific work—both his goals and his procedures—is a continuation of Greek astronomy.

This continuity of scientific goals and procedures, and indeed of a general cosmological outlook, over nearly two millennia, from the fourth century B.C. to the sixteenth century A.D., is remarkable. It is not likely to occur again, provided the current rapid pace of scientific progress can be maintained, uninterrupted by future Dark Ages similar to the era medieval Europeans suffered through.

The Greek world had its own Dark Ages, from about 1200 to 800 B.C., after the collapse of earlier Minoan and Mycenaean civilizations. As the level of civilization rose again, this time the political base in Greece broadened and the city-state emerged as a favored form of government. New Greek cities were established throughout the Aegean region, including the coast of Asia Minor and around the Mediterranean. New trading opportunities appeared and prosperity increased.

Greek cosmology began with myths, but soon expanded to include elements of philosophy, mathematics, and astronomy. Scientific developments first appeared in the Greek world around 600 B.C., with new and distinctive approaches to the understanding of nature. This is the period of pre-Socratic (i.e., before Socrates) science and philosophy, from about 600 to 400 B.C. One major Presocratic tradition centered on a search for a basic substance or substances persisting through all changes. A second major tradition within Presocratic philosophy centered upon Pythagoras, who believed that number was the principle of all things. For the Pythagoreans, the planets were believed to move in such a way that their motions produced harmonious music.

Early cosmological thought was largely limited to the realm of the sun, moon, and planets. For most Greeks, the universe was bounded, in thought and in imagination, by an outer and unchanging sphere of the stars. Their universe, and their cosmology—the study of the structure or organization of that universe—encompassed but a tiny fraction of the universe that we observe and imagine today. Nor did the Greeks generally envision an evolving universe.

Presocratic science of the sixth and fifth centuries B.C. was followed by a surge of new ideas in new directions, with Hellenic philosopher-scientists of the fourth century B.C. formulating radically different approaches to science. During this second period of Greek philosophy and science, the Hellenic period, Plato and Aristotle set major new directions for philosophy, as did Eudoxus for astronomy. The new philosophical and astronomical ideas of the fourth century B.C. heavily influenced Greek cosmology; the aspirations and achievements of Greek cosmology would, in turn, dominate Western science for nearly two thousand years.

Plato believed that reality is to be found not in the visible world but in the world of ideas. The visible, tangible world is in a state of flux. And our senses, which perceive only the shifting surfaces of things, are easily deceived. Permanence and certainty cannot belong to perceptible things, but only to ideas.

With this philosophical predisposition, Plato naturally disparaged the role of observation in astronomy. In his dialogue the *Timaeus*, he told a little joke: birds were, in a former life, empty-headed people who were interested in the stars but foolish enough to believe that astronomy could be based on observations alone. For Plato, the correct method in astronomy and cosmology was the same as that in geometry: posing and solving problems through the effort of pure intellect. He believed that the cosmos would yield its secrets to dialectic.

The true utility of the study of astronomy and of the cosmos was, for Plato, the saving of the soul. His motive for investigating natural phenomena was ethical and moral. His conception of cosmic order was permeated with ethical overtones. His cosmological model had a moral significance.

According to an old tradition, going back to Simplicius, who wrote in the sixth century A.D. a commentary on some of Aristotle's works, it was Plato who first set for astronomers the task of *saving the phenomena*. In Plato's world of ideas, heavenly bodies moved in circular orbits with uniform speed. But planets' motions, as we observe them, are very complex; sometimes a planet seems to speed up or slow down, and occasionally even seems to reverse the direction of its motion temporarily. The task of astronomers working in the Platonic tradition was to find some combination of uniform circular motions in agreement with the observed irregular motions of the planets: to explain apparently irregular motions detected by the senses as a combination of uniform circular motions, the real motions or forms ascertained by reason and thought: to *save the phenomena*.

Much current study and discussion has focused on whether Greek astronomy was *instrumentalist* or *realist*, whether their scientific theories were intended merely as computing instruments to yield predictions or were also thought to be real representations of the world. Did their theories have the status of instruments, tools, or calculating devices used to relate observation statements (including predictions) without question concerning their truth? Did Greek astronomy consist predominantly of devices or fictions put forward purely for making calculations without any claim to any correspondence with physical reality?

Putting the question in modern terms may fail to do complete justice to Greek astronomy. Given Plato's conception of the nature of the real world, it is not impossible that what we today would consider instrumentalist he would have understood as realist.

To save the phenomena may well appear to modern readers an arbitrary and absurd task, scientifically as well as ethically and morally. Any attempt to understand Greek cosmology demands a sympathetic view of ancient concerns and values very different from our own. Most Greeks believed that the planets were divine beings, or at least that the planets were directly controlled by gods. Most Greeks also believed, from the time of Aristotle on, that celestial objects were made of a substance utterly different from the elements that make up our terrestrial world. The stars, sun, moon, and planets preserved their own natures, changeless from generation to generation. So it seemed that they must be made of a changeless, incorruptible substance, which came to be called the *aether*. The only changeless motion (appropriate to a changeless substance as well as to divine objects) was everlasting uniform motion in circles. In this cultural context, the task of saving the phenomena was a valid and intellectually defensible task. To gain an understanding of how the apparent complexity of the cosmos could be reconciled with simple philosophical and physical principles almost universally accepted was by no means an unreasonable task to undertake.

No doubt we could now with the aid of computers reproduce the observed irregular motions of the planets using, in clever combination, a near-infinite number of perfect circles of various sizes with perfect, unchanging velocities. But it would be a most cumbersome contraption. Furthermore, such a geometrical model would tell us little about the cause of the motions, at least in any acceptable modern sense. Obviously the Greeks had different values than those underlying our modern science, and consequently they accepted different tasks and different types of explanations than we demand today. Whatever we may now think of Plato, though, the fact remains that he and his pupils set for science a task accepted and pursued with varying vigor for two thousand years, from the fourth century B.C. through much of the Renaissance of the sixteenth century A.D.

Plato's role in the history of cosmology is the subject of an ongoing debate and reassessment. Historians (including this writer) have perhaps gone too far in emphasizing Plato's importance. Fads, fancies, and fashions are not unknown to historical writing, and the emphasis on Plato and the goal of saving the phenomena may be dangerously close to one of these *f* words. Recently several historians of science, including James Evans, whose chapter on Ptolemy appears in this section, have argued persuasively that the traditional history of Greek astronomy needs to be corrected. (For references to their writings, see the bibliographic note at the end of chapter 5, *Plato's Cosmology*.) These revisionist historians raise several points:

- The whole history of early Greek philosophy shows a preoccupation with physical explanations but no concern at all with quantitative model making.

- In the classical age, Greek astronomy was far more concerned with constellations and the risings and settings of stars than with planets.

- Plato himself was fundamentally inimical to empirical astronomy.

- The attribution of the principle of uniform motion to Plato was a mistake, made by late commentators, such as Sosigenes and Simplicius.

- The transition from physical allegory to geometrical model in astronomy was not far advanced before the work of Aristarchus in the third century B.C.

- Even a bit later with the work of Apollonius, to whom is credited the beginnings of what would become the Ptolemaic astronomical tradition, there was no effort at making models with quantitative and predictive capacities.

Whatever revisions of history occur, increasing our understanding of ancient astronomy and cosmology, Plato will be found to have had some sort of impact on subsequent cosmological speculation; we can debate its nature and extent.

The other major philosophical influence on cosmology came, of course, from Aristotle. Western cosmology became a combination of Platonic and Aristotelian elements, a dynamic balance varying from individual to individual, from place to place, and from time to time.

In contrast to Plato, Aristotle attempted to develop a dynamic model of the universe, seeking the cause as well as a description of the motions of the heavenly bodies. Aristotle's cosmology is intimately linked to his theory of the elements (earth, water, air, and fire, plus the heavenly aether), to his theory of forced and natural motions, and to his views on the nature of time and place. His physical principles would dominate cosmological discussions during late antiquity and the Middle Ages. *Saving the phenomena* came to mean accounting for the motions of the planets while adhering to Aristotelian principles. The overthrow of Aristotelian physics, ultimately replaced by Newtonian physics, would be, arguably, the greatest accomplishment of the Renaissance. Yet it is Plato's call to save the phenomena that characterizes the greatest achievements of Greek cosmology, and dominates Copernicus's efforts as well.

One of the first to take up the task of saving the phenomena was Eudoxus. He was of the generation between those of Plato and Aristotle, and he passed some time in Athens, where he undoubtedly knew both Plato and Aristotle. According to Simplicius (writing much later, in the sixth century A.D.), Eudoxus took up the task of saving the phenomena in response to a challenge from Plato. Eudoxus's system of revolving spheres was intended to produce the observable motions of the sun, moon, and planets. Furthermore, centered on the earth as it was, his system of revolving spheres also would satisfy Aristotle's concern for the causes of motion. We may debate whether to credit the role of Plato, but Eudoxus's influence upon Aristotle is evident. Indeed, Aristotle is our oldest authority for the details of Eudoxus's system. Aristotle even suggested some changes to make the system physically more workable.

In the following Hellenistic period, from roughly 300 B.C. to A.D. 150, scientists would develop in considerable detail the suggestions of their Hellenic predecessors. In astronomy and cosmology, Hellenistic efforts culminated in the work of Ptolemy in Alexandria, Egypt, in about A.D. 150. He developed in exquisite quantitative detail both the eccentric hypothesis and the epicycle theory. In the former, a circle carrying the sun at constant, regular speed is not centered on the earth, and thus regular motion as viewed from the center of the sun's orbit appears irregular when viewed from the earth. In the latter, a small circle moves on a large circle, and the combination of their regular motions is irregular. All this to save the phenomena!

Without rival for some 1,400 years, from its appearance in about A.D. 150 until it was supplanted by Copernicus's *De revolutionibus orbium coelestium* after 1543, Ptolemy's *Almagest* is rightfully seen as the culmination of Greek science, tremendous in its technical mastery. Yet Ptolemy failed to save the phenomena with complete success in exact detail using only combinations of regular circular motions to reproduce the apparently irregular motions of the sun, moon, and planets. Indeed, he had found himself forced to violate the accepted principle of uniform motion in order more closely to reproduce the observed motions. Major and extensive efforts to correct glaring philosophical deficiencies in Ptolemy's system would appear, but not for centuries. Intellectual activities in the Roman world were disrupted with the decline of that civilization, and there was also a shift of interest from traditional intellectual pursuits to service in the rapidly growing Christian religion. After Ptolemy, astronomy and cosmology (and Greek science, generally) went into decline.

A great revival of mathematics and astronomy began in the ninth century A.D., in Iraq and Syria, within the flourishing Islamic civilization under the Abbasid caliphs. Arabic became the new language of science and philosophy, as Greek once had been. Arabic astronomers began by mastering the details of Ptolemaic astronomy and cosmology, and soon went on to discover astronomical phenomena unknown to Ptolemy (such as the decrease of the obliquity of the ecliptic). The *Almagest* was translated into Arabic on several occasions, and Arabic astronomers began to compose their own astronomical and cosmological treatises. Arabic astronomy often showed great originality, as in the design and construction of new astronomical instruments, but it remained essentially Ptolemaic and Aristotelian at its core. Practical computation of planetary positions from theory always was carried out on the basis of Ptolemaic models, though often with improved numerical parameters. Nonetheless, several Arabic writers expressed dissatisfaction with Ptolemy—especially over his violation of the principle of uniform motion in his planetary theories. Some of the medieval doubts centering on Ptolemy's astronomy are discussed in the following chapter on Copernicus.

Notwithstanding their firsthand acquaintanceship with Ptolemy's masterpiece, Arab philosophers generally were more interested in Aristotelian physical science, and many of them may have found Ptolemy's system little more than a convenient computing device. The general Arab preference for Aristotelian physical science and the accompanying relative neglect of geometrical astronomy with its task of saving the phenomena also were characteristic of the first phases of the revival of science in the West. Not until the fourteenth and fifteenth centuries, when humanism replaced scholasticism, would circumstances be favorable for a resumption of the long-interrupted Ptolemaic effort to save the phenomena.

In Christendom the revival of astronomy and cosmology lagged several centuries behind Arab culture. A great period of reacquisition of the classics of Greek science and philosophy began in the twelfth century. At first, European scholars enjoyed access only to Arabic translations of Greek works, from which they made Latin translations. Only later were Latin translations made directly from Greek originals. The first Latin tracts on Ptolemaic astronomy and cosmology were heavily derivative, cribbed from Arabic textbooks. By the fifteenth century, though, much creative work was being done in European astronomy, even if it was still firmly set in the Ptolemaic-Aristotelian tradition. This scientific movement culminated in the middle of the sixteenth century with the great work of Copernicus—the first European astronomical treatise fully worthy of comparison with the *Almagest*.

Copernicus shared with Ptolemy the goal of saving the phenomena. And he was generally satisfied that Ptolemy had saved the phenomena. It was the manner in which Ptolemy saved the phenomena that bothered Copernicus. He was dissatisfied with Ptolemy's system not on observational grounds but for aesthetic, or philosophical, reasons. Ptolemy, struggling to reproduce the complex observed motions, had admitted into his theory of planetary motions a great deal that violated the first principle of regular or uniform motion. Also, Ptolemy was unable to deduce from his system the shape and symmetry of the universe, the commensurability of its parts. (The theory for each planet was separate from all others; relative sizes of planetary orbits were not determined by the theory.) It was, Copernicus complained, as if beautiful limbs from diverse models had been put together, resulting in a monster rather than a man. Binding the heavens together so that no part could be moved without disrupting the others was cited by Copernicus as an advantage of his system over Ptolemy's.

That Copernicus placed the sun rather than the earth at the center of his system in no way altered the underlying goal of saving the phenomena in an aesthetically stylish manner. His sun-centered system, though, would have far-reaching consequences, some unanticipated by Copernicus, leading to a radically different cosmological understanding. Because of revolutionary cosmological implications in Copernicus's seemingly simple innovation of placing the sun in the center, and also because many centuries separate Copernicus from the Greeks while but a few decades separate him from Galileo, Kepler, and Newton, Copernicus customarily is placed at the beginning of the Scientific Revolution. On the basis of his technical planetary astronomy, however, Copernicus also can be placed in the 2,000-year tradition of saving the phenomena.

Obviously it is an arbitrary decision where to place Copernicus in the context of this book. A compelling case can be made for the revolutionary nature of his cosmology, as indeed it is made in the article on Copernicus concluding this section. Admittedly, it is partly with the intention of provoking controversy, and consequent reaction and thought, that Copernicus concludes this section instead of leading off the section on the Scientific Revolution.

Shortly after Copernicus, the Greek geometrical cosmological vision disappeared from the Western world, to be replaced with one in closer accord with new values and a new sense of aesthetics. Perhaps more modern human values helped shape the newly emerging cosmology, just as the new cosmology undoubtedly has helped shape our cultural values and our civilization.

5. PLATO'S COSMOLOGY

To save the phenomena, to devise some geometrical combination of uniform circular motions that could reproduce the observed irregular motions of the sun, moon, and planets, became the primary goal of many philosopher-scientists in a tradition that ultimately stretched over nearly two thousand years of history. To Plato in the fourth century B.C. has been traditionally attributed the origin of the Greek geometrical conception of the cosmos and the articulation of the goal of saving the phenomena. Or if not to Plato himself, then perhaps to his pupil Eudoxus. Increasingly, however, scholars now question Plato's putative role and see the transition from physical allegory to geometrical models with quantitative and predictive capacities coming later in history. Nonetheless, Plato's general philosophy, at least as it was later understood, provided nurture for such visions of the universe. Historians debate the nature and extent of Plato's impact on subsequent astronomy and on cosmological speculation, but do not deny to Plato some sort of influence.

Plato believed that reality exists not in the visible world of our senses but in the world of ideas. In that world of ideas, heavenly bodies moved in circular orbits with uniform speed. From this principle followed the task of determining what combination of uniform circular motions could be devised to agree with the observed irregular motions of the planets, to discover in the world of ideas the real motions or forms ascertainable by reason and thought.

Plato's imagined cosmos and the consequent task set for philosopher-scientists was far different from the various world views and generally agreed-upon questions and types of answers acceptable to the Presocratics in Greece during the sixth and fifth centuries B.C. The discovery of the principles of nature, whether in the basic matter of the Ionians or in the mathematical ratios of the Pythagoreans, was soon followed by a surge of new ideas in new directions. Hellenic science of the fourth century B.C. formulated new approaches to science, which Hellenistic scientists from roughly 300 B.C. to A.D. 150

developed in considerable detail. During the Hellenic period, from 400 to 300 B.C., both Plato and Aristotle set major directions for science. Their aspirations and achievements were to dominate Western science for nearly two thousand years.

Plato's Life

The philosophy of an individual may be studied using the assumption or working hypothesis that philosophy is a response to the environment. Intellectual historians prefer this approach. Philosophers, in contrast, typically concentrate on the ideas themselves, not on biographical details or societal context. Plato's philosophy, including his general science and his cosmology, can be viewed as a reaction to the temporary moral values of his age, which left Plato highly dissatisfied and, consequently, searching for a new philosophy. The change in Greek science begun by Plato can be placed, and perhaps better understood, in the context of changes in Greek society.

In 479 B.C., a year after the Persians under Xerxes I had captured and burned Athens, thirty-one Greek city-states defeated the Persians in decisive land and sea battles, bringing to a successful resolution some twenty years of struggle to stop the westward expansion of the great Persian empire. In the golden age that followed, Athen's population rose to perhaps 400,000 and tribute poured in from other city-states. Soon, though, resentment at the loss of political independence grew and several Greek city-states, led by Sparta, revolted against Athenian rule, setting off the Peloponnesian Wars of 431–404 B.C. Initially, Athens had no great difficulty maintaining dominance, but eventually the fortunes of battle shifted, and Athens surrendered to Sparta in 404 B.C.

In the ensuing turmoil that enveloped Greece, people attempted to rediscover the spiritual world of the human soul and the meaning of life. Socrates led the reaction against the materialistic trends of physical science, seeking self-knowledge by first becoming aware of one's own ignorance. Though trained as a stone cutter in his father's shop in Athens, Socrates preferred to spend his time arguing in the marketplace, where he encouraged the youth of Athens to question every moral precept handed down to them. Elder citizens believed, not without justification, that Socrates was demoralizing their children. His critical questioning also extended to the government and its acts, and Socrates soon was accused of impiety.

The charge may have been intended to frighten him into fleeing Athens, but Socrates stayed and forced the issue, welcoming his trial as a forum for his ideas. A trial then consisted of two parts, with the guilt or innocence of the defendant established in the first part. If guilty, the punishment was determined in a second part. The jury found Socrates guilty, and the prosecution proposed the punishment of death. Socrates proposed board and lodging at public expense because his actions had been for the public benefit. The jury chose between the two proposed punishments, and Socrates was forced to drink a fatal cup of hemlock.

Socrates's death and the related political conditions in Athens directly influenced Plato, Socrates's pupil and close friend. Born in 427 B.C., Plato was of an age to enter public life about the time of the defeat of Athens by Sparta in 404 B.C. Furthermore, Plato's uncle was one of the thirty tyrants who ruled Athens immediately after the defeat. But the actions of the tyrants, quelling criticism by intimidation and opposition by assassination, disgusted Plato, and he chose not to join his uncle's government. A year later, the democratic faction having driven out the tyrants, Plato again considered entering politics. But then the democracy executed Socrates. At this time Plato seems to have set aside political ambition and to have begun his search for unchanging standards to hold against the shifting judgments of men.

Subsequent experiences confirmed him in this course. In 388 B.C. the dictator of Syracuse, Dionysius I, asked Plato if he did not think that Dionysius was a happy man. According to legend, Plato answered that he thought no one who was not mad would become a tyrant. Enraged, Dionysius ordered Plato sold into slavery. A friend arrived just in time with ransom money to rescue Plato. When Dionysius I died, an uncle of the young Dionysius II persuaded Plato to tutor the new tyrant. The young dictator soon banished his uncle, and also became impatient with the amount of effort demanded by his tutor, who had set out upon the ambitious task of educating Dionysius to become a philosopher-king. Later, trying to mediate the quarrel between Dionysius and his uncle, Plato was caught in the middle when negotiations broke down, and he barely escaped with his life.

Plato's philosophy and its implications for the study of astronomy and cosmology are particularly understandable as a response to the time of troubles in which Plato found himself. Reacting to the shifting moral values of his time, Plato searched for unchanging standards. The changing, visible world was itself without permanent values. So Plato turned to the world of ideas. There he could hope to find the real and unchanging standards absent in the world of experience.

Allegory of the Cave

Plato's insistence that the only real world exists in the mind is implausible, at first glance. But the concept is plausibly illustrated with a simple example. Think of a circle and draw a circle. Which is real? The circle drawn on paper is not a real circle, no matter how skilled the drafter. The drawn circle is an imperfect representation in the visible world of experience of a perfect circle. The perfect circle exists only in the mind, only in the world of thought.

Implications for the study of astronomy and cosmology of Plato's argument that reality exists not in the visible world but in the world of ideas are found in his book the *Republic*. Here Plato takes up the question of how society might be shaped to bring out the best in people, asking what justice is and how it is to be attained. Plato concluded that a philosopher-king is needed to bring the ideal state into existence. How such a person might be educated was one of Plato's concerns. In this context, the nature of reality and the study of astronomy are introduced.

In the *Republic*, as in most of Plato's books, the subject is presented in the form of a dialogue. Plato distrusted the fixed, dead words of textbooks. He believed that learning could be achieved only through discussion and shared inquiry.

In a parable to illustrate the degrees to which man may be enlightened, Plato presented his famous Allegory of the Cave. Knowledge is the power that will set people free, free from the shackles of ignorance and free to live enlightened lives in the sunlight of understanding. Plato imagined people chained from childhood in a cave, so shackled that they must sit still, looking in only one direction. A fire behind them casts shadows of objects upon the cave wall in front of them. In the absence of any other experience, the prisoners accept the shadows as reality. Furthermore, the objects casting the shadows are not real people, but statues of people.

> *Now then, compare our natural condition, as far as education and ignorance are concerned, to the following. Imagine a number of men living in a cave. There they have been confined from birth, with their legs and necks shackled. They sit still and look straight forward at the cave wall. Their chains make it impossible for them to turn their heads around. Further imagine a bright fire burning above and behind them, and a road passing between the fire and the prisoners, with a low wall built along it, like the screens which conjurors put up in front of their audience and above which they exhibit their wonders.*
> I have it.
> *Also imagine a number of persons walking behind the wall. They carry statues of men and images of other animals. The statues are of wood and stone and all kinds of materials. Together with various other articles, they overtop the wall. And, as you might expect, let some of the passers-by be talking, and others silent.*
> You are describing a strange scene, and strange prisoners.

They resemble us. Let me ask you, in the first place, whether persons so confined could have seen anything of themselves or of each other beyond the shadows cast by the fire on the wall?

Certainly not, if you suppose them to have been imprisoned all their lives with their heads unmoved.

And is not their knowledge of the things carried past them equally limited?

Unquestionably so.

And if they were able to talk with one another, do you think they would give names to the shadows they see cast on the cave wall?

Doubtless they would.

And if the cave returned an echo from the wall in front of the prisoners whenever one of the passers-by spoke, let me ask you, what could the prisoners attribute the voice to, if not to the passing shadow?

Unquestionably they would refer it to that.

Then surely such persons would believe the shadows of manufactured articles to be the only realities.

Without a doubt they would.

Now consider what would happen were the prisoners released from their fetters and their foolishness remedied in the following manner. Let us suppose that one of them has been released and forced suddenly to stand up and turn his neck around and walk with open eyes toward the light. And let us suppose that his actions are painful and that the dazzling splendor leaves him unable to see the shadows he formerly saw. What answer would you expect him to make if someone were to tell him that he was formerly watching foolish phantoms but now he is somewhat nearer reality, is turned toward things more real, sees more correctly? Above all, if he were asked what the several objects passing by were, should you not expect him to be puzzled and to regard his old visions as truer than the objects now forced upon his notice?

Yes, much truer.

And if he were further forced to stare at the light itself, would not his eyes be hurt? And would he not shrink, and turn back to the shadows which he could see distinctly? And would he not believe the shadows more clear and more real than the real objects?

Just so.

And if he were dragged violently up the rough and steep ascent from the cave into the sunlight, would he not be vexed and indignant at such treatment? And on reaching the light would he not be blinded by its radiance and unable to see any of the things he was now told were real? And would he not be ready to go through anything rather than entertain these new opinions and live in this new way?

For my part, I am quite of that opinion.

Now consider what would happen if he were to descend again and resume his old seat. Would he not find it difficult to see anything in the gloom of the cave?

Certainly he would.

And if he were forced, before his eyes had adapted to the dim light, to compete with the prisoners in describing the shadows, would they not laugh at him and say that he had gone up only to come back with his eyesight destroyed and that it was not worth while even to attempt the ascent? And if anyone endeavored to set them free and drag them out of the cave to the light, would they not ill-treat that person if they could get him in their power?

Yes, they would. They would kill him.

Now apply this imaginary case to our former statements. Compare the region which the eye reveals to the cave. Compare the light of the fire to the light of the sun. Compare the upward ascent to the contemplation of the upper world. Now you understand the mounting of the soul to the intellectual region. In the world of knowledge the essential Form of Good is the limit for our inquiries and can barely be perceived. But, when perceived, we cannot but conclude that it is in every case the source of all that is bright and beautiful—in the visible world giving birth to light and in the intellectual world dispensing truth and reason. Whosoever would act wisely, either in private or in public, must set this Form of the Good before his eyes.

The prison of the cave corresponds to the part of the world revealed by the sense of sight. Escape from the cave corresponds to the use of intelligence to reach the real world of knowledge. Would we, as students, appreciate or resist teachers who would help us to achieve freedom?

On the Study of Astronomy

The next issue in the *Republic* is how to bring man to a state of enlightenment, how to convert the soul. The answer for Plato is education, which will enable the soul to look in the right direction and learn the truth. The forms of study leading the soul away from the world of change and to the world of reality are the sciences, which deal with certain knowledge and immutable truths. Arithmetic, geometry, solid geometry, astronomy, and harmonics quicken the mind and help a pupil gain understanding of the essential form of goodness—so Plato believed. Studying these subjects for ten years would release the prisoner from the cave and enable him to look at shadows and reflections. Then there must follow an additional five years of study, devoted to dialectics.

In discussing astronomy, Plato first mentions its utilitarian benefits. Agriculture, navigation, and war are alluded to. It is not for these purposes, however, that astronomy is to be esteemed. The true utility of the regimen of study prescribed in the *Republic* is the saving of the soul. Plato's motive for investigating natural phenomena is ethical and moral. Plato's conception of cosmic order is permeated with ethical overtones. Plato's cosmological model has a moral significance.

An obvious approach to astronomy is to observe the motions of the objects in the heavens. But only a discipline dealing with unseen reality will lead the mind upward. The true motions are not to be seen with the eye. It is not by looking at the heavens that one can become truly acquainted with astronomy.

Plato's disdain for observation, his near contempt for observation, is readily apparent in his comments on the study of astronomy and has not been generally perceived as beneficial to the progress of science. Plato's denial of the value of the senses in scientific studies has contributed to a generally negative view of him as a scientist. In any emphasis on the empirical part of science, in which facts are observed, Plato fares poorly. The formation of theories is also an important aspect of science. In this context, Plato shines more favorably. If modern science is characterized as the discovery of mathematical relations between phenomena, Plato becomes a hero. In his emphasis on mathematics, Plato stands close to the Pythagoreans; the two philosophies are often viewed as one.

Plato's Universe

In the *Republic* Plato tells how to study astronomy and how not to study it. Further thoughts about the structure of the universe are presented in the *Timaeus*, Plato's dialogue on nature. It is a difficult book, seemingly more a beautiful story by a poet or mystic than a serious scientific work. Or, if it is a scientific work, it is an aberration. Yet it provides an exposition of Plato's system of nature.

The setting of the *Timaeus* is a banquet. On the previous day Socrates had discussed what kind of republic was best and what sort of men should compose such a republic. Now Timaeus, the most astronomical of the four men at the banquet, is asked to speak on the generation of the world. The object of scientific study is not a description of the world of the senses, but the discovery of the plan of the Creator of the world.

Plato frequently explained phenomena as the result of final causes. Whatever is generated is necessarily generated from some cause. The world was generated according to an idea in the mind of the artificer. The cause of the world coming into being is the idea. The world is generated to fulfill the idea.

Pervasive in the *Timaeus* and throughout much of Greek science, teleology, the use of final causes as explanation for phenomena, is at odds with modern scientific practice, which seeks instead mechanical actions or forces as explanation of phenomena observed. The teleological aspect of Plato's science is a major factor contributing to his current low esteem as a scientist.

Plato's description of the planetary system, the observable universe of the Greeks, is tantalizingly vague. He wrote:

> *To describe the evolutions in the dance of these same gods, their juxtapositions, the counterrevolutions of their circles relatively to one another, and their advances; to tell which of the gods come into line with one another at their conjunctions,*

and which in opposition, and in what order they pass in front of or behind one
another, and at what periods of time they are severally hidden from our sight
and again reappearing sent to men who cannot calculate panic, fears and signs
of things to come—to describe all this without visible models of these same would
be labor spent in vain. So this much shall suffice on this head, and here let our
account of the nature of the visible and generated gods come to an end.

This passage, when viewed with considerable hindsight, is perhaps slightly suggestive of, and may have led to, Eudoxus's system of homocentric spheres, and later the epicycle theory, in which a large circle carries a small circle, the small circle revolving about a point on the circumference of the large circle, and thus moving a planet attached to the circumference of the small circle. But it is scarcely a definitive statement from which we can trace with great confidence developments that followed.

Indeed, Plato's immediate disciples favored a metaphorical reading. Aristotle, in contrast, insisted that Plato must be held to his literal word. Plato did preface his cosmological narrative in the *Timaeus* with a political narrative, in which the story of Atlantis makes its first appearance in literature. Atlantis purportedly was a maritime empire dominating the Western Mediterranean from its island base in the Atlantic some ten thousand years earlier and threatening the independence of the Greeks. A sudden catastrophe—perhaps a flood or an earthquake—was said to have overwhelmed Atlantis, sinking it into the ocean, and demolishing all but a tiny remnant among the Greeks. The presence of the Atlantis story in the *Timaeus* suggests that Plato somehow intended to integrate his cosmology with his political theory.

Plato's Legacy

Whatever we might think Plato meant, and whatever value we might place on Plato's ideas, there remains the historical consideration of the influence of his ideas on the ideas of others. Plato's ideas were presented in books and were taught directly to pupils in a school founded by Plato in Athens in about 380 B.C. The Academy survived as a place of research and writing until A.D. 529, and can be thought of as one of the world's first universities.

Eudoxus, a pupil at Plato's Academy, did report what Plato taught, but the report has not survived. It was summarized by Eudemus in his *History of Astronomy*, also now lost. The *History* was commented upon by Sosigenes, who lived in the time of Julius Caesar and advised Caesar on the calendar reform of 45 B.C. His commentary, too, was lost, though not before Simplicius of Athens read it in the sixth century A.D. Here the string of lost but summarized works ends, because

Simplicius's work survived. In his commentary on Aristotle's *On the Heavens*, Simplicius stated:

> Plato lays down the principle that the heavenly bodies' motion is circular, uniform, and constantly regular. Thereupon he sets the mathematicians the following problem: what circular motions, uniform and perfectly regular, are to be admitted as hypotheses so that it might be possible to save the appearances presented by the planets?
>
> The curious problem of astronomers is the following: first, they provide themselves with certain hypotheses: the ancients, the contemporaries of Eudoxus and Callippus, adopted the hypothesis of "counterturning spheres"; Aristotle, who in his Metaphysics teaches the system of spheres, must be counted among them. The astronomers who followed proposed the hypothesis of eccentrics and epicycles. Starting from such hypotheses, astronomers then try to show that all the heavenly bodies have a circular and uniform motion, that the irregularities which become manifest when we observe these bodies—their now faster, now slower motion; their moving now forward, now backward; their latitude now southern, now northern; their various stops in one region of the sky; their at one time seemingly greater, and at another time seemingly smaller diameter—that all these things and all things analogous are but appearances and not realities.
>
> To save these irregularities, astronomers imagine that each star is moved by several motions at the same time—some assuming movements along eccentrics and epicycles, others appealing to spheres homocentric with the world (the so-called counterturning spheres). But just as the stops and retrograde motions of the planets are, appearances notwithstanding, not viewed as realities (they are no more real than the numerical additions and subtractions with which we meet in studying these motions), so an explanation which conforms to the facts does not imply that the hypotheses are real and exist. By reasoning about the nature of the heavenly movements, astronomers were able to show that these movements are free from all irregularity, that they are uniform, circular, and always in the same direction. But they have been unable to establish in what sense, exactly, the consequences entailed by these arrangements are merely fictive and not real at all. So they are satisfied to assert that it is possible, by means of circular and uniform movements, always in the same direction, to save the apparent movements of the wandering stars.

The goal of saving the phenomena or saving the appearances, of devising some geometrical combination of uniform circular motions that could reproduce the observed irregular motions of the sun, moon, and planets, traditionally has been attributed to Plato, largely on the strength of the above passage by Simplicius. Some scholars now question Plato's putative role. At the very least, though, his general philosophy provided nurture for such a vision of the universe. And if not to Plato himself, then perhaps to his pupil Eudoxus can be attributed articulation of the goal of saving the phenomena. At the very most, it might be said—indeed, it has been said—of Plato that the subsequent history of cosmology is largely an account, first, of attempts to solve the problem he set, and then an account of emancipation from the requirement of circular motion.

Eudoxus

Neither a very brief description of Eudoxus's system by Aristotle nor Simplicius's commentary some eight centuries later was sufficiently precise to stimulate much interest in the model of homocentric spheres. Not until the nineteenth century A.D. did Giovanni Schiaparelli, an Italian astronomer more widely known for his observations of "canali" on Mars, attempt to reconstruct Eudoxus's system. While compatible with what Aristotle and Simplicius wrote about Eudoxus, the reconstruction also depends upon the additional assumption that Eudoxus's system accounted for many of the astronomical phenomena known in Schiaparelli's time. Thus there is a danger of attributing to Eudoxus a more advanced knowledge of the planetary motions than he actually possessed and a more detailed cosmological model than he actually devised. This warning should preface any discussion of what is believed to be Eudoxus's cosmology.

A basic observational fact that any astronomical system must account for is the observed movement of the stars overhead each evening. Eudoxus seems to have accounted very simply for this phenomenon by placing the stars all on one sphere rotating with a uniform speed around the central earth with a period of twenty-four hours. This model is equivalent to a rotating earth and a fixed sphere of the stars; the observational consequences are identical. The relative motion of one star with respect to another had not yet been detected, and thus did not trouble Eudoxus, nor complicate his cosmological model.

The apparent motion of the sun presents a slightly more difficult problem:

- First, there had to be an outer sphere rotating with a period of twenty-four hours, to produce the apparent daily movement of the sun across the sky. Again, as with the stars, the outer sphere for the sun produced the apparent motion now attributed to a rotating earth.

- Eudoxus next would have needed a second sphere rotating with a period of a year, to move the sun apparently higher in the sky in summer and lower in winter, and around the heavens with a period of a year. The axis of this second sphere would be tilted relative to the first or outer sphere, to move the sun up in summer and down in winter. This second sphere would have had its axis of rotation fixed to the outer sphere, so it could be carried around with a twenty-four-hour period.

- Eudoxus seems to have ignored the changeable velocity of the sun, already discovered by his time. This omission, whatever its justification, spared him much trouble. He did, though, add a third sphere, to account for an observation now known to have been mistaken.

Eudoxus devised a similar system of spheres for the moon:

- First, as with all celestial objects, there was the outer sphere rotating once every twenty-four hours, to mimic the daily rotation of the earth.
- The moon circles the earth approximately once a month. To produce this motion, Eudoxus added a second sphere attached to the first and rotating west to east with a period of one lunar month.
- There is also a small variation in the latitude of the moon, placing it sometimes slightly above and at other times slightly below an imaginary plane containing the earth and the sun. Eudoxus added a third sphere to produce this motion.

Planetary motions presented a more difficult problem than did the motions of either the sun or the moon. The planets display retrograde motion: they appear at times to cease their movements relative to the stars, turn back temporarily, retrace small parts of their paths, and then change direction once more to resume their voyages around the heavens (figure 1). Eudoxus's task was to devise a model consisting of uniform circular motions only, yet producing the apparent retrograde motions observed:

- First, he gave each planet an outer sphere. This sphere carried the planet around the earth with a period of twenty-four hours.
- Second spheres moved the planets around the heavens, with periods of a year for Venus and for Mercury, and longer periods for the outer planets.
- To produce observed motions in latitude, Eudoxus added third spheres for each planet. So far, the planetary solutions followed the solutions for the sun and the moon.
- To produce the observed retrograde motions, Eudoxus added a fourth sphere. By a clever combination of inclinations and speeds of revolution of the third and fourth spheres (if we believe Aristotle's account), Eudoxus was able to produce in an approximate fashion the observed retrograde motions.

A system of four homocentric spheres can, in principle, produce retrograde motions (figure 2). No system limited to four homocentric

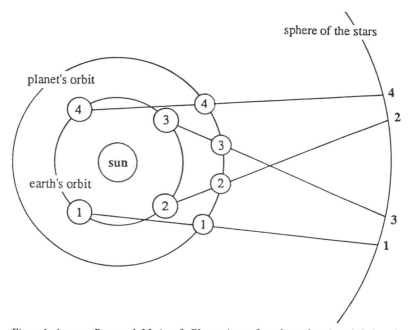

Figure 1. Apparent Retrograde Motion of a Planet. As seen from the earth at times 1, 2, 3, and 4, a planet apparently moves against the sphere of the stars from 1 to 2, turns back to 3, and then resumes its forward motion to 4. The earth, moving faster than the outer planet, overtakes and passes it.

spheres, however, can simultaneously produce with complete quantitative accuracy both the length of the retrograde motion westward and the length of the motion in latitude for each of the planets. Other astronomers apparently detected flaws in Eudoxus's system, because there occurred a series of modifications after his death in 355 B.C. Work on his spherical system was continued, first at Eudoxus's school in Cyzicus, a city on the south coast of the Sea of Marmara, now Turkey, by his pupil Polemarchus, and next by Polemarchus's pupil Callippus. Callippus later moved to Athens, where he worked with Aristotle. According to Aristotle, Callippus was satisfied with Eudoxus's combinations of spheres for Jupiter and Saturn, but added additional spheres for the other planets, the sun, and the moon. Aristotle, not content with a purely geometrical system, but also concerned with how the combination of forces caused the motions of the planets, found it necessary to add several more spheres to the system.

While Eudoxus's system was improved and brought into better agreement with observed planetary motions, there was one phenomenon for which it ultimately could not account. As Simplicius wrote in his commentary on Aristotle's *On the Heavens*, the planets move at different times closer to or farther from the earth. For the moon, apparent changes in size due to the changing distance were observed directly by the Greeks. For the planets, apparent changes in size (and therefore in actual distance) could be inferred from changes in apparent brightness. A system of homocentric spheres cannot produce

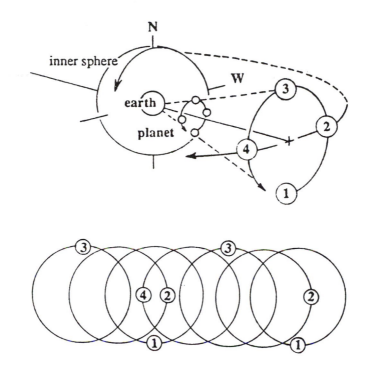

Figure 2. Retrograde Motion from Homocentric Spheres. Retrograde motion can be produced from a combination of two spheres rotating with unchanging velocities. While the geometrical system presented here is pure speculation, it may suggest an approximation of Eudoxus's planetary system, whose details have been lost to time. The two spheres have the same center; they are homocentric. The outer sphere is not shown. Its axis of rotation is vertical, in the plane of the paper, in a north-south direction. The outer sphere carries everything within it eastward. The axis of the inner sphere is horizontal and in the plane of the paper. The motion of a planet carried about by the inner sphere is up (north) and down (south) and into (west) and out of (east) the plane of the paper. The planet appears to move north and west from 1 to 2, north and east to 3, south and east to 4, and south and west back to 1. When the inner sphere is imparting an eastward motion to the planet, moving the planet from 2 to 3 to 4, the total eastward motion, including the steady eastward motion imparted by the outer sphere, will be very rapid. If the westward speed imparted by the inner sphere is greater than the steady eastward motion imparted by the outer sphere, then the planet will appear to slow down and briefly move west during the passage from 4 to 1 to 2, when the westward velocity of the inner sphere is greater than the eastward velocity of the outer sphere. Thus retrograde motion can be produced from a combination of two spheres rotating with constant velocities.

changes in the distances of objects from the center of the spheres, where the earth was assumed to reside. Possibly Eudoxus was concerned solely with movements in longitude and latitude, but the problem of distances seems to have been accepted as a legitimate problem by Greek astronomers following in his tradition.

Seemingly, no one proposed the simple expedient of centering the spheres about a point near, but not exactly coincident with, the earth. Or if anyone did, the idea quickly was rejected: Aristotelian physics required that the earth be in its natural place at the center of the universe. Furthermore, there could have been aesthetic objections to off-center celestial spheres.

Whatever the reason, Eudoxus's planetary system disappeared from history, not to be recovered for more than two thousand years. Now it is important as one of the first major attempts in Western civilization to go beyond solely philosophical reasoning and develop a scientific model in considerable detail and in agreement with observed phenomena.

The development of Eudoxus's planetary system by Callippus and Aristotle, and possibly by others after them, is also an early example of continuity in scientific work, an example illustrating the cumulative advance that is possible when a problem receives continued attention from several generations of scientists. If science now is to be an integral part of our modern civilization, supported by democratic societies in ways that encourage continuous traditions and progress, appreciation of the links between science and civilization and wisdom to preserve and advance the two together also must be democratized.

Conclusion

To save the phenomena, or the appearances, may appear to modern readers an arbitrary and absurd task. No doubt the observed irregular motions of the planets could be reproduced using, in clever combination, a near-infinite number of perfect circles of various size with perfect, unchanging velocities. But it would be a cumbersome contraption.

Furthermore, it would tell little of the cause of the motions. It would tell little unless one believed in the physical actuality in the world of the senses of the circles upon circles carrying the planets about, in addition to their reality in the real world of ideas. It also would tell little unless one accepted the final cause (i.e., the actualization of the idea in the head of the architect) as an explanation of the motion. Greek philosophers seemingly did accept such an explanation. Modern science has achieved a more elegant and informative solution to a more productively formulated problem, in the not unbiased judgment of modern scientists.

The fact remains, though, that Greek philosopher-scientists, particularly Plato and Eudoxus, set for science a task accepted and pursued with varying vigor for two thousand years, from the fourth century B.C. through much of the Renaissance of the sixteenth and seventeenth centuries A.D. Historical importance and impact are not necessarily commensurate with plausibility judged by different people in a different age.

—Norriss S. Hetherington

BIBLIOGRAPHIC NOTE

When you go to a restaurant, you eat the food, not the menu describing the food, so why settle for anything less than Plato's own words, in the *Republic*, in the *Timaeus*, and in other books. For readers in a hurry, astronomical and cosmological excerpts are conveniently collected in N. Hetherington, *Ancient Astronomy and Civilization* (Tucson, Arizona: Pachart Publishing House, 1987), though excerpts are to the original as is a hamburger from a fast food chain to a filet mignon at Maxim's. The controversy over Plato's worth as a scientist, balancing his contempt for observation against his emphasis on mathematical relations in nature, is discussed in G.E.R. Lloyd, "Plato as a Natural Scientist," *Journal for Hellenic Studies*, 88 (1968), 78–92. Lloyd further makes the important points that Plato investigates the causes of natural phenomena for ethical motives and his conception of cosmic order is permeated with ethical overtones and moral significance. Several scholars have questioned Plato's putative role in setting the problem of finding circular motions, uniform and perfectly regular, to save the appearances presented by the planets; see Bernard R. Goldstein and Allan C. Bowen, "New View of Early Greek Astronomy," *Isis*, 74 (1983), 330–340, and Wilbur R. Knorr, "Plato and Eudoxus on the Planetary Motions," *Journal for the History of Astronomy*, 21 (1990), 313–329. For a critical reappraisal of the emphasis often placed upon the methodological position of saving the phenomena, see Lloyd, "Saving the Appearances," *Classical Quarterly*, 28 (1978), 202–222; reprinted in Lloyd, *Methods and Problems in Greek Science* (Cambridge: Cambridge University Press, 1991), with an introduction assessing scholarly debate on the topic and Professor Lloyd's modifications and developments in his own position since the original publication of the article. For an interpretation of Plato's *Timaeus* combining political and cosmological elements and encompassing the legend of Atlantis, see the entry by W. Knorr on "Plato's Cosmology" in *The Encyclopedia of Cosmology: Historical, Philosophical, and Scientific Foundations of Modern Cosmology* (New York: Garland, 1993).

6. ARISTOTLE'S COSMOLOGY

A dynamic understanding of the universe was the goal of astronomers and physicists working in the Aristotelian tradition. They sought to understand the causes of motion as well as describe motion on earth and in the heavens, in contrast to followers of Plato for whom the task of saving the phenomena—of devising some geometrical combination of uniform circular motions that could reproduce the observed irregular motions of the sun, moon, and planets—had been both the beginning and the end of their aspirations.

Plato had believed that reality exists not in the visible world of our senses but in the world of ideas. Aristotle, in contrast, proposed a physical approach to the phenomena of nature. Plato's cosmos was mathematical, geometrical; Aristotle's was physical. Plato's cosmos was to be found in the mind; Aristotle's was to be found through observation.

Aristotle's Life

Just as Plato's philosophy can be understood as a response to his environment, Aristotle's very different philosophy and cosmology also may be so interpreted, though admittedly with some difficulty, inasmuch as Aristotle for some twenty years was a member of Plato's Academy.

Indeed, the standard interpretation of Aristotle's work is that he began very close to Plato's intellectual position and only gradually departed from it. Ambiguities in the dating of Aristotle's writings encourage such an analysis, because the resulting pattern then can be used to determine the chronological order of otherwise undated passages. Such developmental studies once were in vogue, but difficulties in the way of achieving solid results now are perceived as more severe than originally recognized. It is not always clear which of two different views Aristotle might have considered superior, nor thus always possible to say with certainty which view would have been the earlier and which view the later. The circularity of arguments basing a relative chronology on the supposed superiority of one version of a theory over another has become all too obvious. The assumption that Aristotle only gradually achieved intellectual independence from Plato remains plausible but not proven.

An alternative interpretation, and that to be followed here, is the view of Aristotle's philosophy as that following from the interests of a biologist. This approach to an understanding of Aristotle fits well with biographical information and makes particularly good sense of his cosmology, though of course the convenience of any particular line of analysis is no guarantee of its accuracy.

Aristotle was to experience much more favorable and pleasant encounters with the world of the senses than had Plato. While any implied linkage between personal experiences and a person's philosophy can be no more than speculation incapable of test or verification, it seems worth noting, if only for heuristic purposes, that Aristotle did not choose to reject the world of the senses for the world of ideas, as had Plato.

Aristotle was born in 384 B.C. in Macedon, a poor land of unruly people at the northern edge of the Greek peninsula. There his father had served as personal physician to the ruler, Amyntas II, and it is tempting to attribute to his physician father Aristotle's interest in the biological sciences, even though his father died when Aristotle was young. At the age of seventeen, in 367 B.C., Aristotle left Macedon and journeyed south to Athens. At Plato's Academy, where he stayed for twenty years, Aristotle would have been introduced to the problems of identifying and classifying animals and explaining their functioning.

In 347 B.C. Plato died and Aristotle entered the second major period of his life. The political situation in Greece had been changing, and the feuding Greek city-states were now ripe for conquest by Philip II. In 359 B.C. he successfully claimed the crown of Macedon, and gradually consolidated his control, with Athens emerging as his main opponent. Coupled with an anti-Macedonian mood in Athens was Aristotle's dislike of the philosophical tendencies of Plato's nephew, who had taken over direction of the Academy following the death of its founder. Aristotle left Athens, crossing the Aegean Sea to Asia Minor. Stagira, his birthplace in Macedon, would have been a more logical place to go, but Philip had destroyed it. In Asia Minor, Aristotle founded a new academy under the patronage of a local ruler, Hermias, and married Hermias's 18-year-old niece and adopted daughter. From Aristotle's later description of the ideal age for marrying as 37 for the man and 18 for the woman, we may infer that his voluntary exile was not an unhappy one. In 344 B.C. Aristotle went to the nearby island of Lesbos, where he worked with Theophrastus. Many observations in Aristotle's biological works were made in this region.

Aristotle returned to Macedon in 342 B.C. to tutor the young prince Alexander and, perhaps, also to further secret negotiations between Hermias and Philip. Many of Aristotle's political activities and experiences, in contrast to Plato's, were in successful causes, though the capture and subsequent death during torture of his father-in-law by the Persians, intent on eliciting details of the secret treaty with Philip, was a noticeable exception.

When Philip completed his conquest of Greece in 338 B.C., Aristotle returned to Athens comfortably on the side of the victors. He was in the enviable position of helping form the ideas of the heir of the leading city-state of Greece, much what Plato had thought out how to do for the education of a philosopher-king in the *Republic* but had not successfully put into practice. Alexander was appreciative of Aristotle's efforts, even to the extent of later rebuilding Stagira to please his teacher; Plato's pupil Dionysius expressed his appreciation by ordering Plato sold into slavery.

Philip planned a campaign against the mighty Persian empire, and Alexander, assuming the throne following the assassination of his father in 336 B.C., moved swiftly eastward against the Persians. While conquering the Near East, Alexander did not forget his tutor, and sent back a flood of new plant and animal specimens and Babylonian eclipse records for Aristotle to study. The political climate in Athens remained favorable for Aristotle until the death of Alexander in 323 B.C., when Aristotle again went into voluntary exile. He died a year later, in 322 B.C.

Aristotle's World View

Typical of Aristotle's writings, and also of the biological sciences, is an emphasis upon classification. Aristotle continually analyzed and classified, as if what were necessary to understand a subject was to divide it into categories. In the *Poetics*, for example, Aristotle divided poetry into tragedy and comedy, and analyzed tragedy into six factors: scenic presentment, lyrical song, diction, plot, character, and thought. And in his *Politics*, Aristotle classified the species of polity, with their normal and perverted forms. There was government of an individual: kingship and tyranny; government of a few: aristocracy and oligarchy; and government of the many: polity and democracy. In his scientific books, too, Aristotle classified and drew distinctions. He distinguished between the mathematician and the natural philosopher, between the mathematical astronomy of Plato and his own dynamic astronomy.

A major strength of Aristotle's cosmology was that it formed a complete system, a system in which every part followed logically from the rest. To understand Aristotle's cosmology, it is necessary first to understand his physics. Basic concepts in Aristotle's *Physica* (*Physics*) underlie ideas developed in *De caelo* (*On the Heavens*). Aristotle's definitions of motion and place, his conception of what constituted a cause, and his criteria for what constituted an acceptable answer to the question *"Why?"* are all essential to an understanding of his view of the physical cosmos.

To have knowledge of something, or to have grasped the *why* of it, was for Aristotle to know the cause of the phenomenon. Characteristically, Aristotle classified causes into categories: the material cause—of what the object was made; the formal cause—the shape of the object; the efficient cause—who made it; and the final or purposeful cause—the object's use or purpose.

Aristotle frequently explained a thing in terms of its purpose. For example, he explained that a man is walking because the man wants to be healthy. Thus Aristotle shared with Plato a tendency to use teleological explanations.

Aristotle's philosophy was also very animistic, often placing the source of motion within an object instead of explaining motion in terms of external forces. To the extent that Aristotle seems content in his search for causes merely to restate effects as undefined powers producing the observed effects, his scientific methodology can be strongly criticized.

The emphasis on the final cause—on the purpose of an action—helps make comprehensible Aristotle's otherwise seemingly strange definition of *motion*. Motion, for Aristotle, was not solely a change of position, but was defined more broadly as the fulfillment of potentiality. Motion thus included the growth of a plant or animal as it strove to attain its form, as well as change of position (*locomotion*, as Aristotle called it). Both types of motion, growth and change of position, involved the concepts of potentiality and purposeful action.

The Aristotelian sense of motion led to a particular sense of place, encompassing both motion and potential. Each of the four elements—earth, water, air, and fire—had its natural place. Moved away from its natural place, each element had a natural tendency to return. Thus to explain a motion, one could attribute the motion to a final cause or purpose, objects moving because they have tendencies to return to their proper places in the universe. Fire naturally moves upward toward its natural place, while earth naturally falls downward toward its natural place. If earth moves upward, it is because it has been thrown upward with an unnatural motion.

Aristotle's concepts of place and motion led him to the conclusion that void could not exist. In a void there would be no natural place for bodies to return to, and consequently the laws of natural motion would not work. Therefore there could not be a void, because if there were, Aristotle's physics would not work. He had to deny the existence of a void to maintain the logical completeness of his world view.

To argue in this fashion, of *reductio ad absurdum*, is to start with a seemingly plausible statement and then deduce consequences from it so absurd as to demand the conclusion that the original statement obviously cannot be true. Thus did Aristotle argue against the existence of a void.

Nor was projectile motion explicable in a void. Movement was assumed to require constant contact between the moved object and the mover. In a void, however, there was neither air nor, consequently, the possibility of explaining projectile motion in terms of air pushing the projectile along.

Yet another difficulty with the conception of motion in a void lay in the question of what could cause motion ever to cease if there were nothing to stop it. Modern physics, of course, does contemplate a body remaining at rest *or in a state of motion* until acted upon by another force, but the latter half of this postulate goes against common sense and observation. Friction is sufficient to eliminate perpetual motion from the real world, except for astronauts in outer space. Lacking this perspective, Aristotle thought the possibility of perpetual motion was absurd. That perpetual motion was a logical consequence of void was sufficient reason to reject the possibility of a void.

Also arguing against the possibility of a void was Aristotle's belief that the ratio of speeds of equal objects was in proportion to the thickness of the medium. If water were twice as thick as air, an object should move through water with half the speed it moved through air. But void with no thickness made such a ratio nonsense; it meant dividing by zero.

Furthermore, for falling objects, which Aristotle believed fell at different speeds depending upon their weights, a measure of the force they possessed to cut through the surrounding medium, fall in a void would produce the patently nonsensical result of bodies of different weights falling with the same speed because there was no medium to slow one more than the other. Indeed, this is exactly what modern physics says will happen, but in Aristotle's time the conclusion seemed absurd. And if the logical consequence of a void was absurd, then the very concept of void itself was absurd too.

The case of projectile motion, of a thrown object, raised a problem that would puzzle and plague generations of philosophers and scientists after Aristotle. Everything in locomotion (movement from one place to another) either moved itself or was moved by something else. And, according to Aristotle, it was impossible to move something without being in continuous contact with it. Aristotle briefly mentions projectiles in his *Physics*, attributing their continued motion either to mutual replacement or to the air being pushed by the thrower and in turn pushing the projectile. By mutual replacement, Aristotle meant the doctrine of *antiperistasis*, in which the air pushed forward by the projectile changed direction and moved around to the rear of the projectile and then, changing direction once more, gave the projectile a push forward. This explanation was not especially convincing, nor was that of the air being pushed by the thrower of the projectile and in turn pushing the projectile.

Eventually, failure to solve the problem of projectile motion would lead to the downfall of Aristotle's physics and with it the entire Aristotelian cosmology, so closely and inseparably linked were the two. Attempts to explain projectile motion led first to the concepts of impetus and momentum, and next to the concept of inertia and objects remaining in motion until some force acts to stop them. In Aristotle's time, though, and for nearly two millennia thereafter, it was necessary to explain forced or violent motion as the result of continuous contact between the moved and the mover. When the contact ceased, so supposedly would the motion.

The opposite of violent, or forced, or unnatural motion was natural motion. With it a body moved naturally toward its natural place in the universe without the interposition of any force other than the natural tendency to move to the natural place.

The one form of locomotion that could be continuous is discussed by Aristotle in the last book of his *Physics*. Locomotion was either rotatory or rectilinear, or a combination of the two, and only rotatory motion could be continuous. Furthermore, rotation was the primary locomotion because it was more simple and complete than rectilinear motion. (This value judgment regarding the relative merits of circular and rectilinear motion was to play a central role in Western cosmology for 2,000 years.)

Some of the basic considerations of motion found in the *Physics* are also present in Aristotle's *On the Heavens*. All locomotion was either straight, circular, or a combination of the two, and all bodies were either simple—that is, composed of a single element, such as fire or earth—or were compounds. The element fire and bodies composed of fire had a natural movement upward, while bodies composed of earth had a natural movement downward (more precisely, toward the center of the universe, and the earth was thus at the center). Circular movement was natural to some substance other than the four elements (earth, fire, air, and water) and more divine than these four elements (since circular motion was prior to straight movement).

Aristotle inferred that there was something beyond the region of the earth, composed of a different material, of a superior glory to our region of the earth, and also unalterable. His model is conveniently described as a two-sphere universe, with the changing region up to the sphere of the moon, the earth in the center of this sphere and surrounded by water, with air and fire at the top, beyond which are the heavenly bodies in circular motion and in a realm without change. There was a separate set of physical laws for each of the two regions, since they were composed of different types of matter.

Aristotle's universe was not infinite, he argued, because the universe moved in a circle, as we can see with our eyes if we watch the stars. Were the universe infinite, it would move through an infinite distance in a finite time, which is impossible. Thus beginning with that which he disputed, that the world was infinite, and showing that the premise led to an absurd conclusion, Aristotle by the method of *reductio ad absurdum* proved the opposite.

In another argument involving motion, Aristotle stated that bodies fall with speeds proportional to their weights. While this passage may seem scarcely relevant to cosmology, it is historically important, in addition to being incorrect. Bodies of different weights fall with the same speed, as Galileo would later argue. The problem of falling bodies, furnishing as it did an opening to Aristotle's critics, would be significant in the development of the modern laws of mechanics, even if the legend of Galileo dropping cannonballs from the leaning tower at Pisa may be only a legend.

The world was finite, and there was only one world. Aristotle demonstrated this conclusion by yet another *reductio ad absurdum* argument, this one based on the doctrine of natural place. Were there more than one world, and if each world had a center as the natural place for earth material to move to and a circumference for fire to move to, then our earth might move toward any of these centers while fire could move toward any of the circumferences, with absolute chaos the absurd result.

By an even more obscure set of arguments, Aristotle demonstrated that the heavens are unalterable. This conclusion would become another weak point in his cosmology, though not until the late sixteenth and early seventeenth centuries A.D. Then, observations of comets moving through the heavens and observations of novae, stars that flare up immensely in brightness, would find changes occurring in Aristotle's purportedly unalterable heavens.

Aristotle went on to show that the heavens rotated and that the earth was stationary in the center. The shape of the heavens was spherical, the shape best suited to its nature. The heavens also purportedly had a smooth and accurate finish.

Aristotle next argued that the motion of the heavens was regular. In support of this conclusion he produced many reasons, again often of a *reductio ad absurdum* type, beginning with a premise, deducing an absurd conclusion from the premise, and thus refuting the premise.

Then Aristotle took up the question of the composition of the stars, a question absent from Plato's kinematic astronomy, which had been concerned only with describing the motions of the stars. Nor was the question readily susceptible to observational and experimental enquiry, until the middle of the nineteenth century A.D. and the advent of the science of spectroscopy. Aristotle argued, as persuasively as usual and with as little factual basis as usual, that the stars were composed of the same element as the heavens and were fixed to circles that carried them around. His argument was all the more persuasive for an analogy with missiles and heated air, from which we may conclude that argument by analogy was no more a reliable warrant of truth in Aristotle's time than it is now.

Aristotle next asked if the stars were in motion. He answered that they did not move of their own effort, but were fixed to spheres and carried about by the spheres.

Finally in his enquiry, Aristotle arrived at the earth. Located at the center of the universe, the earth was at rest. Its shape was spherical, the shape it would acquire as its particles packed into the center. Evidence of the senses also indicated that the earth was spherical, since during eclipses of the moon the earth cast a circular shadow and different stars could be seen from different positions on the spherical surface of the earth.

Conclusion

Aristotle's use of observation in his scientific arguments is in decided contrast to Plato's rejection of the evidence of the senses, and Aristotle's work thus has been seen by some historians as a turning point in science, marking the beginning of extensive empirical enquiries. Aristotle's methodology is not identical, however, with the methods of modern science, in which observation and experiment are used differently in the development and in the verification of a theory. Aristotle formulated a problem by reviewing the ideas of other philosophers and then seeking out principles from which he could logically infer the conclusions. Observations, cited at the end, were used more to persuade his readers of the truth of his conclusions than as an aid in arriving at the conclusions. Nor did Aristotle devise critical experiments with which to test his conclusions.

Another difference between Aristotle's science and modern science lies in the degree of quantification developed in each. Aristotle's science was largely qualitative, while modern science relies heavily on mathematical quantification for its development, its verification, and its exposition.

Also, a satisfactory explanation for Aristotle usually consisted of final causes. This teleological and animistic aspect of Aristotle's science is far different in both spirit and detail from modern mechanical explanations.

While bringing to Greek science a new emphasis on the value of observation, Aristotle and his cosmology differ in significant and interesting ways from our modern outlook. But whatever the shortcomings in his cosmology when compared to our modern knowledge of the cosmos, the fact remains that for two millennia Aristotle's cosmology dominated a large part of the intellectual world. It remains an integral and important part of our intellectual heritage, of our literature, our art, our philosophy, and our very language and way of thinking.

—Norriss S. Hetherington

BIBLIOGRAPHIC NOTE

English translations of Aristotle's *Physics* and *On the Heavens* are widely available. For the standard account of Aristotle, see W. Jaeger, *Aristotle. Fundamentals of the History of His Development, second edition* (Oxford: Clarendon Press, 1948). Regarding mounting doubts concerning the developmental approach to Aristotle since Jaeger's book, see G.E.R. Lloyd, *Methods and Problems in Greek Science* (Cambridge: Cambridge University Press, 1991), particularly pp. 1–4. (This book is a collection of articles, each preceded by an introduction assessing scholarly debate on the topic and Professor Lloyd's modifications and developments in his own positions since original publication of the articles.) An alternative treatment of Aristotle as a biologist is developed in M. Greene, *A Portrait of Aristotle* (London: Faber and Faber, 1963). An attempt to link Aristotle's life, his physics, and his cosmology, including excerpts from the *Physics* and *On the Heavens*, is found in N. Hetherington, *Ancient Astronomy and Civilization* (Tucson, Arizona: Pachart Publishing House, 1987).

7. PTOLEMY

Ptolemy (Klaudios Ptolemaios, second century A.D.) is the most important of the Greek astronomical writers not only because of the sophistication and quantity of his preserved work but also because of his enormous influence on later astronomical and cosmological thought. In his cosmology Ptolemy attempted to satisfy simultaneously the demands of technical planetary astronomy and the requirements of sound physics, as he perceived them. The result was a unified world view that dominated cosmological thought throughout the entire medieval period.

From the close of Greek antiquity to the astronomical revolution of the sixteenth century, Ptolemy's astronomy and cosmology—happily wedded to Aristotle's physics—were almost universally accepted. This was the case wherever the Greek astronomical tradition was preserved—in medieval Islam as well as medieval and renaissance Christendom. Medieval astronomers and cosmologists who wrote in Greek, Arabic, Hebrew, or Latin all were heavily influenced by Ptolemy's cosmology. Each culture modified Ptolemy's views in minor ways, often to make them more consistent with religious or philosophical requirements. Nevertheless, for the great majority of the time that there has been such a thing as a science of cosmology, the reigning cosmology was Ptolemy's. Its importance in the Western tradition hardly can be overstated.

Ptolemy's Life and Works
Of Ptolemy's life we know almost nothing. His dated astronomical observations fall between A.D. 127 and 141 and were all made at Alexandria, in the Roman province of Egypt.

Ptolemy's most important work was the *Almagest*, a treatise on mathematical astronomy. Its main subject is the theory of the motions of the sun, moon, and planets. Ptolemy introduces geometrical models to explain these motions. He derives numerical values for the parameters of the models by making use of suitably selected observations. And, finally, he shows how to use the models to calculate the position of a planet for any date. In its original Greek, the *Almagest* was titled *The Mathematical Treatise of Claudius Ptolemy*. In the early Middle Ages it was translated on several occasions into Arabic. Ptolemy's treatise some time before had attracted the appellation *megiste*, the Greek adjective for the "greatest." Arabic writers transliterated this Greek word into Arabic and added the definite article *al-*. From this construction the medieval Latin title *Almagest* is descended. The very title of Ptolemy's treatise thus nicely illustrates the richness of the Western astronomical tradition, which involves ancient Greek as well as medieval Arabic and Latin writers. In a later work, the *Handy Tables*, Ptolemy grouped together the planetary tables scattered through the *Almagest* and incorporated some minor changes in theory.

Ptolemy's cosmological speculations are developed in another work, the *Planetary Hypotheses*. His main cosmological hypothesis is that the universe contains no empty space. The mechanism (deferent and epicycle) responsible for Mercury's motion immediately surrounds that of the moon. The mechanism for Venus surrounds that of Mercury, and so on, in a series of nested spheres, out to Saturn and, finally, the fixed stars. This work had a great influence on medieval cosmology, largely through Arabic intermediaries. Only a portion of it survives in Greek.

Book III of Ptolemy's treatise on music, the *Harmonics*, is devoted to the astronomical part of music theory, in keeping with Pythagorean tradition. Musical intervals are associated with astrological aspects; for example, the octave corresponds to opposition. Also, various features of the motions of the planets are given musical meaning; for example, motion in longitude corresponds to pitch. These ideas are only sketchily worked out, and the end of the treatise is missing. Partly for these reasons, Ptolemy's *Harmonics* never had the influence of his *Almagest* or his *Planetary Hypotheses*. But Book III of the *Harmonics* did encourage Kepler in his own musing on the harmony of the world, in his *Harmonice mundi*.

Ptolemy also wrote on applied astronomy, including the theory of sundials (*On the Analemma*), stereographic projection (*The Planisphere*), and astrology (*Tetrabiblos*). It is characteristic of him that he kept applied arts (making sundials) and speculative thought (astrology and cosmology) out of the *Almagest*, which was a work of pure mathematical astronomy. We also have from Ptolemy an important *Geography*, as well as works on optics and philosophy.

Ptolemaic Planetary Theory

Cosmology deals with the structure of the universe, of which our planetary system is now known to be but an infinitesimally small part. In earlier times, however, our planetary system was a large part of the known universe. Thus any discussion of ancient cosmology depends crucially on an understanding of ancient planetary theory. A planetary theory is a mathematical scheme that can be used to calculate the position of a planet, the sun, or the moon in the sky for any desired date. Here we introduce the main features of Ptolemy's planetary theory; later we will see how he incorporated them into a consistent cosmology.

The Greek planetary theory of late antiquity evolved over a period of several centuries, beginning in the third century B.C. It was brought into its substantially final form by Ptolemy in the second century A.D. Nearly all that we know of the technical details of Greek planetary theory we owe to Ptolemy's *Almagest*.

Ptolemy's work was not entirely original. He drew on the work of his predecessors, especially Apollonius of Perga and Hipparchus (third and second centuries B.C., respectively). But Ptolemy's own contribution was a large one. He not only produced a systematic textbook of mathematical astronomy, but he also introduced new features into the theories for the motion of the moon and the planets. These features brought the theories into excellent agreement with observed celestial motions. Once the *Almagest* was available, the works of Ptolemy's predecessors lost their relevance and usefulness. People ceased to read and copy them. It is for this reason that virtually nothing has come down to us from Ptolemy's predecessors on technical planetary theory.

Ptolemaic planetary theory should therefore be understood as beginning a few centuries before Ptolemy and continuing to the sixteenth century. Copernicus may be regarded as one of the last and most accomplished astronomers in the Ptolemaic tradition. In his *On the Revolutions of the Heavenly Spheres* (1543), Copernicus set forth his idea of a sun-centered planetary system. The truly new and revolutionary part of his work, the treatment of the heliocentric hypothesis, is found in Book I and constitutes the first 5% or so of the text. The rest of Copernicus's work is a kind of rewrite of the *Almagest*. Copernicus departs from Ptolemy in occasional technical details, but follows him more often than not. He relies on Ptolemy's observations (while also using some more recent ones), emulates his methods, and often adopts his numerical parameters. Thus, until the end of the sixteenth century, even astronomers who disagreed with Ptolemy on some detail were forced to couch the argument in Ptolemaic terms.

Two characteristic features of Ptolemaic planetary theory are deferents and epicycles (figure 3, page 112). The celestial body at point **P** (sun, moon, or planet) travels on a small circular epicycle, while the center **K** of the epicycle moves around a large deferent circle. The deferent may be either centered at the earth, point **O**, in which case it is said to be a concentric circle (figure 3), or it may be centered on a point **D** slightly displaced from the earth, in which case it is called an eccentric circle (figure 7, page 115).

Solar Theory

Modern readers often focus on the earth-centered nature of ancient cosmologies and planetary theories. But, for accurate astronomical prediction, *it makes no difference* whether the earth goes around the sun or the sun goes around the earth. The object that is at rest merely reflects the choice of a frame of reference. Sun-centered theories are therefore not *intrinsically* any more accurate than are earth-centered theories. The accuracy of the theory depends upon the technical details of the theory.

According to Aristotle, the motion natural to the sun is uniform circular motion. Moreover, the earth lies at the center of the cosmos. It would therefore be reasonable to suppose that the sun travels at uniform speed on a circle centered at the earth. This simple picture is fairly accurate, but it cannot be quite right, because it fails to explain the inequality in the lengths of the seasons. The equinoxes and solstices divide the zodiac into four equal 90 degree arcs. If the sun traveled at constant speed on a circle concentric with the earth, it would complete the four arcs in equal times; the seasons would all be of the same length—one fourth of the year each. But, in fact, the longest season exceeds the shortest by several days. Thus the sun appears to move more rapidly through some quadrants of the zodiac and more slowly through others. This variation in the apparent speed of the sun is called the *zodiacal anomaly*.

In the solar theory of Hipparchus and Ptolemy, the sun travels on a circle at uniform speed but the center of the circle is slightly displaced from (or eccentric to) the earth (figure 1). The earth is at the center of the sphere of the fixed stars (point **O**) and therefore is also at the center of the zodiac circle **FGHI**. Point **C** is the center of the sun's *eccentric circle* **fghi**. It is on this circle that the sun travels with uniform speed. This results in an *apparently* nonuniform motion as viewed from point **O**.

When the sun is at **f** on its eccentric, it appears to us against **F**, the spring equinoctial point (the beginning of the sign of Aries). When the sun reaches **g**, it is seen against **G**, the summer solstitial

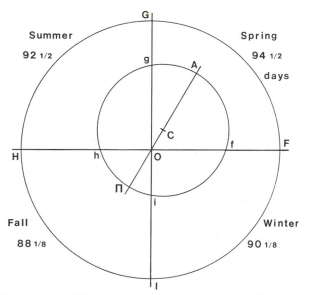

Figure 1. The solar theory of Hipparchus and Ptolemy. In this and most of the following figures, the plane of the figure is the plane of the ecliptic, as viewed from above its north pole. The earth is at **O**, *which is the center of the sphere of stars, represented by the circle* **FGHI**. *The sun travels at constant speed on its eccentric circle* **fghi**, *the center of which is at* **C**. **A** *is the apogee of the sun's eccentric circle and* Π *is the perigee.*

point (the beginning of Cancer). The eccentric circle **fghi** is divided into segments of unequal size by lines **FH** and **GI**, so we obtain seasons of unequal lengths even though the sun travels at constant speed. The season lengths shown in figure 1 are those given by Ptolemy in the *Almagest*, but were due originally to Hipparchus. The placement of **C** produces just the effect desired, making spring the longest season and fall the shortest. (The season lengths now, nearly nineteen centuries later, are slightly different: summer is the longest season and winter is the shortest.)

The line through **O** and **C** is called the *line of apsides*. It cuts the eccentric in the two apsides: the *apogee* **A** of the eccentric circle, where the sun is at its greatest distance from the earth, and the *perigee* Π, where the sun is closest to the earth in the course of the year. In Hipparchus's model, the sun travels at a constant speed, but it *appears* to travel more quickly at the perigee and more slowly at the apogee because of its varying distance from the earth.

The solar model has four numerical parameters, or elements, which must be determined from observations before the model can be used to predict future positions of the sun. These are:

(1) The length of the year, which determines the rate of the sun's motion on the circle.

(2) The longitude of the apogee (angle **FOA**).

(3) The eccentricity of the eccentric circle, which is the ratio of **OC** to the radius of circle **fghi**.

(4) The position of the sun at some one particular moment.

If these four quantities are known, the model is completely specified and can be used to predict the longitude of the sun for any date. All of these quantities can be determined from observations of equinoxes and solstices alone (i.e., by observing the *time* at which the sun reaches equinox or solstice).

Alternative Realities: Epicycles or Eccentrics?

Figure 2A summarizes the essential features of the solar theory: the sun (point **S**) travels at uniform speed on a circle centered on a point **C** that is eccentric to the earth (point **O**). Angle α increases uniformly with time. A second form of the solar theory, shown in figure 2B, is discussed by Ptolemy (and also by Theon of Smyrna, who wrote about a generation earlier than did Ptolemy). The sun (point **S**) travels clockwise on an epicycle (the small circle), while the center **K** of the epicycle travels counterclockwise around a circle concentric to the earth (point **O**). Both motions take place at the same uniform angular speed. That is, angles α and β are always equal; both go through 360 degrees in one year.

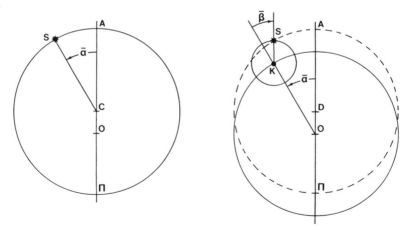

Figure 2. Two equivalent forms of the solar theory. 2A shows an eccentric circle. 2B shows a concentric circle plus epicycle. In 2A, the sun S moves uniformly around a circle centered on point C, which is eccentric to the earth O. In 2B, the sun S moves uniformly on a small epicycle whose center K moves uniformly around a deferent circle centered on the earth O.

These two versions of the solar theory are exactly equivalent. The path actually traced by the sun in the concentric-plus-epicycle model of figure 2B is a perfect circle, shown as a dashed line. The effective center of the orbit is **D**, located above **O** by a distance equal to the radius **KS** of the epicycle. Thus if the radius of the epicycle is equal to the eccentricity **OC** in figure 2A, the models of figures 2A and 2B are equivalent.

An important historical lesson follows from the mathematical equivalence of the epicycle and eccentric models. We cannot compare the "complexity" of two different planetary theories by simply counting the number of circles in each, because the number of circles is not well defined. Any valid comparison of two theories must be based on a detailed examination of the functions of the individual circles.

The equivalence of an eccentric to a concentric-plus-epicycle was proved by Apollonius of Perga in the third century B.C. This remarkable fact precipitated a debate among astronomers. Which model corresponded to physical reality? According to Theon of Smyrna, Hipparchus chose the concentric-plus-epicycle model, saying that it was more likely that the circles were arranged symmetrically with respect to the center of the cosmos (the earth). In the *Almagest*, Ptolemy chose the eccentric, saying that it is simpler because it achieves its purpose with one motion rather than two.

Both these remarks reflect the belief of Greek astronomers that their planetary models were physically real descriptions of the universe, not merely mathematical devices for saving the phenomena. Hipparchus's choice was clearly motivated by physical or cosmological principles. But this is true of Ptolemy as well, because the eccentric circle is not *mathematically* any simpler than the concentric-plus-epicycle model. The two models are mathematically equivalent. Calculations of solar positions from theory would have exactly the same degree of complexity in both models. Indeed, the calculations would be virtually identical, line by line. Ptolemy's choice of the eccentric was motivated by broader physical principles: he wanted his cosmos to be as simple as was consistent with the phenomena. This desire, explicitly enunciated by Ptolemy in several passages (*Almagest*, III, 1 and XIII, 2), stands in poignant contrast to his modern reputation for extravagant complexity. Given his physical principles, Ptolemy succeeded about as well as possible in achieving astronomical accuracy within a relatively simple cosmology.

Motions of the Planets: The Anomaly with Respect to the Sun

Each planet moves generally eastward around the zodiac, as do the sun and the moon. But each planet occasionally appears to stop and reverse its direction of motion. It then travels westward in *retrograde motion* for a few weeks or a few months (depending on the particular planet). Then it stops and again reverses direction, to resume its normal eastward motion. These reversals of direction are manifestations of the second anomaly.

The superior planets retrograde when in opposition to the sun (i.e., when they are in the diametrically opposite part of the zodiac to the sun, as viewed from the earth). The inferior planets retrograde when in conjunction with the sun. The second anomaly is therefore sometimes called the *anomaly with respect to the sun*. The connection between the sun and a planet's retrogradation is, of course, easily explained in a sun-centered model. (A superior planet appears to retrograde when the earth on its faster, smaller orbit passes between the planet and the sun; hence the planet retrogrades when in opposition to the sun.) But in the ancient, earth-centered cosmology the connection between the sun and the planets' retrogradations was simply a fact of nature.

Apollonius on the Motions of the Planets

In the earliest form of the deferent-and-epicycle planetary theory, due to Apollonius of Perga (third century B.C.), the deferent was concentric with the earth (figure 3). The planet (point **P**) moves uniformly on the epicycle while the center **K** of the epicycle moves

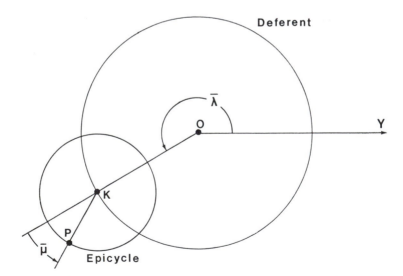

*Figure 3. Apollonius's deferent-and-epicycle theory of the planets. The planet **P** moves on a circular epicycle, while the center **K** of the epicycle moves around a deferent circle centered on the earth, **O**.*

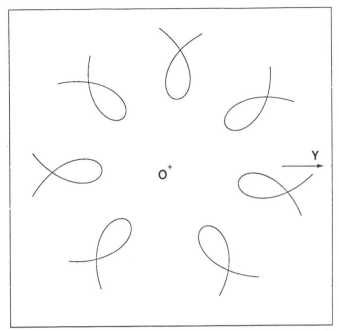

Figure 4. Retrograde loops of Mars generated by Apollonius's model (figure 3).

uniformly on the deferent circle, which is centered on point **O** (the earth). **Y** marks a reference direction fixed in space, the direction of the spring equinoctial point. The epicycle is responsible for producing retrograde motion: when the planet reaches the inner part of the epicycle, it appears to travel backward as observed from **O**. The retrograde loops produced by this model will be all of the same size and shape and uniformly spaced around the zodiac (figure 4). (Note the versatility of the epicycle. In the planetary theory it is used to produce retrograde motion. But in the solar theory it is used to produce the zodiacal anomaly, a slightly varying speed with no retrogradation.)

The Zodiacal Anomaly of the Planets

Some actual retrograde arcs of Mars (for the years A.D. 109 to 122) are shown in figure 5. Not only do the retrograde arcs vary in width by as much as a factor of two, but they are also spaced around the ecliptic in a decidedly nonuniform fashion. It is clear from this that the planets also have a *first*, or *zodiacal anomaly*, in addition to the second anomaly. Mars retrogrades when it is in opposition to the sun, but the length of the retrograde arc and the spacing between succes-

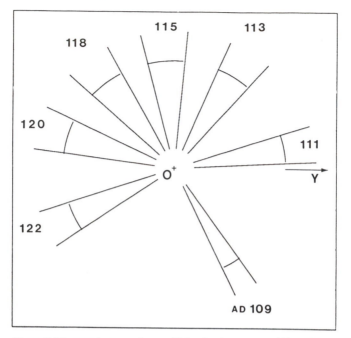

Figure 5. The actual retrograde arcs of Mars for the years A.D. 109 to 122.

sive retrograde arcs depend on just where in the zodiac the planet is when the retrogradation takes place.

The zodiacal anomaly is very striking in the case of Mars, but the simple theory of figure 3 makes no attempt to account for it. Figure 6 shows the theoretical retrograde loops generated by Apollonius's model superimposed on some actual retrograde arcs of Mars. The agreement could hardly be worse. It is clear from this that Apollonius proposed his theory only as a qualitative, physical explanation of retrograde motion, not as a quantitative model with predictive power.

Unsuccessful Efforts to Explain the Zodiacal Anomaly

By the early second century B.C., efforts had been made to explain the zodiacal anomaly (unequal spacing of the retrograde arcs) in terms of deferent-and-epicycle theory. In the solar theory, an eccentric circle accounts nicely for the anomaly with respect to the zodiac. Thus it was natural to introduce an eccentricity in the planet's deferent circle in an effort to explain the planet's zodiacal anomaly. In figure 7, the planet (point **P**) moves uniformly on its epicycle, so that the angle μ increases uniformly with time. Meanwhile, the center **K** of the epicycle moves uniformly around the deferent, so that the

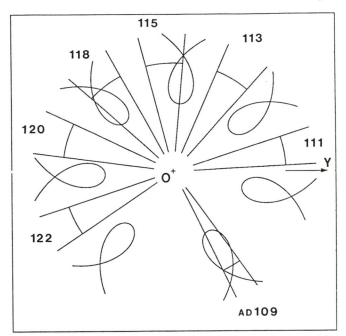

Figure 6. Comparison of theoretical retrograde loops produced by Apollonius's model with real retrograde arcs of Mars. This figure results from the superposition of figures 4 and 5.

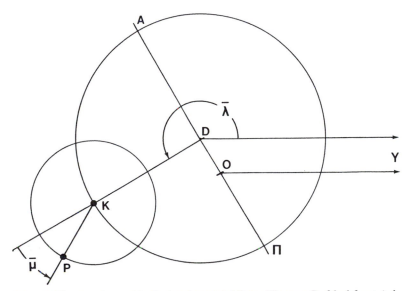

Figure 7. Planetary theory with epicycle and eccentric deferent. The center **D** of the deferent circle is displaced from the earth, **O**.

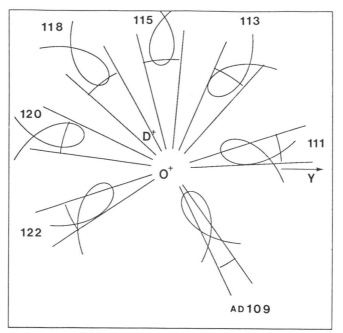

Figure 8. Comparison of the eccentric-plus-epicycle theory (figure 7) for Mars with the real retrograde arcs of Mars.

angle λ increases uniformly with time. However, the center **D** of the deferent is *eccentric* to the earth (point **O**).

The resulting retrograde loops still will be all of uniform size and shape (as in figure 4), but now the center **D** of the loop pattern will be displaced from point **O** (figure 8). The introduction of an eccentricity into the model has accounted for the nonuniform *spacing* of the retrograde arcs around the zodiac. But it has done a poor job of accounting for the *widths* of the retrograde arcs. In fact, it makes the widths even worse than did the simple theory of figure 3.

This appears to be where matters stood in the second century B.C. Ptolemy (*Almagest*, IX, 2) says that the older astronomers constructed their geometrical proofs on the assumption that each planet had a single anomaly and an unvarying retrograde arc. This would be the model of figure 3. From Ptolemy's remarks, it is also clear that Hipparchus's contemporaries or immediate predecessors had combined an epicycle with an eccentric deferent in an attempt to account for both anomalies (figure 7). Hipparchus criticized these planetary theories, pointing out their lack of agreement with the phenomena. But the inadequacy of even the eccentric-plus-epicycle model is so striking in the case of Mars that this could hardly have been news.

What was new was Hipparchus's insistence that the geometrical theory ought to work in detail—that *one could insist simultaneously on plausible physics and quantitative predictive power*. This insistence was far more important for the development of astronomy than the simple observation that the existing theories so far had failed.

Ptolemy's Theory of the Planets

The final form of the ancient deferent-and-epicycle theory was due to Ptolemy himself. Figure 9 illustrates Ptolemy's theory of Mars. (The same construction was applied to Venus, Jupiter, and Saturn. The theory of Mercury had an extra complication.) The center **C** of the deferent is again eccentric to the earth (point **O**). Ptolemy introduces a third center, **E**, which came to be called the *equant point* in medieval astronomy. **E** serves as the center of uniform motion. That is, the center **K** of the epicycle moves around the deferent in such a way that its motion appears to be uniform as seen from **E**: the angle λ increases uniformly with time. Meanwhile, the planet (point **P**) moves uniformly on the epicycle; i.e., the angle μ increases uniformly with time.

The introduction of the equant point had profound consequences for physical theory: the actual motion of **K** on the deferent is now *nonuniform*. **K** travels faster (in miles per day, say) at the perigee Π of

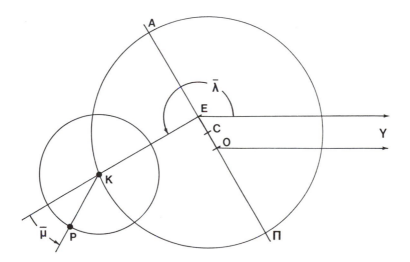

*Figure 9. Ptolemy's theory of Mars. The center **C** of the deferent is midway between the earth, **O** and the equant point **E**. The center **K** of the epicycle travels with uniform angular speed as observed from the equant point **E**. The planet **P** travels at uniform angular speed on the epicycle. **Y** marks a fixed reference direction, the direction of the spring equinoctial point.*

the deferent and more slowly at the apogee **A**. This represented a serious departure from Aristotelian physics.

What considerations led Ptolemy to such a drastic measure? Ptolemy does not tell us directly. But his remarks in book IX of the *Almagest* make it clear that he regarded the explanation of the zodiacal anomaly as the chief problem left unsolved by his predecessors. Thus Ptolemy's self-imposed task was to account for both the spacing and the widths of the retrograde arcs.

Ptolemy may have reasoned as follows. Let us examine the behavior of the eccentric-plus-epicycle model (figures 7 and 8). Here point **D** is evidently a genuine center of uniform motion, because the retrograde loops are uniformly spaced as viewed from **D**. But the theoretical loops are too small around the direction of the A.D. 118 loop and too big around the direction of the A.D. 109 loop. We must leave the center of uniform motion at **D**, in order to preserve correct spacing. But if we shift the center of the deferent circle itself to some place between **D** and **O**, we will drag the epicycle closer to the earth at apogee and thus increase the apparent size of the A.D. 118 retrograde loop. Also, the epicycle will be pushed farther from the earth at perigee, and so the apparent size of the A.D. 109 loop will be diminished. In this way we will have a fair chance of fixing the widths of the retrograde loops without destroying the correct spacing.

In the new model (figure 9), Ptolemy placed the center **C** of the deferent exactly halfway between the earth (point **O**) and the equant point **E**. The resulting retrograde loops are no longer of uniform size and shape, so it is impossible to guess exactly how things will turn out. We must work it out mathematically. Figure 10 illustrates the retrograde loops produced by Ptolemy's theory of Mars, superimposed on the planet's actual retrograde arcs for the years A.D. 109 to 122. Ptolemy's theory is a stunning success, and a dramatic improvement over all that preceded it.

Ptolemy's theory also represented the latitudes of the planets (their small departures from the plane of the ecliptic), by introducing a slight angle between the plane of the deferent and the plane of the epicycle. Somewhat different versions of the latitude theory were used in the *Almagest* and in the later *Handy Tables*. A discussion of these features would carry us too far afield. Nor will we attempt to describe the theories of the moon and of Mercury, which were rather more complicated than those for the sun and for the other planets.

The Role of the Sun

The sun plays a singular role in the earth-centered planetary theory. In what follows, we will simplify the discussion by ignoring

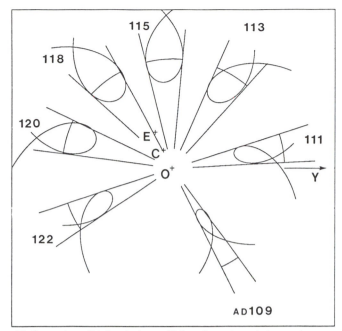

Figure 10. Comparison of Ptolemy's theory of Mars (figure 9) with the actual retrograde arcs of Mars.

the eccentricities. Figure 11A shows the sun-centered theory of a superior planet, such as Mars. The planet **P** and the earth **O** both orbit the sun **S**. Thus vectors **SO** and **SP** turn about **S**. The line of sight from the earth to **P** is in the direction of the vector **OP** = –**SO** + **SP**. But these vectors may be added in either order. Thus we may also write **OP** = **SP** + –**SO**. The geometry corresponding to the second form of the addition is shown in figure 11B. Starting from **O**, we lay out vector **OK**, equal in magnitude and direction to **SP** in figure 11A. From **K** we lay out **KP**, equal in length and opposite in direction to **SO** in figure 11A. The two vectors **OK** and **KP** in figure 11B turn at the same rates as their counterparts in figure 11A. Thus figure 11B is the *Ptolemaic theory of a superior planet.* The planet's orbit in the sun-centered model corresponds to the deferent in the earth-centered model. And the orbit of the earth corresponds to the epicycle. (In the case of an inferior planet, these correspondences are reversed.)

It results from these correspondences that the sun plays a peculiar role in the Ptolemaic theory. In the case of a superior planet (figure 12), such as Mars, the sun governs the motion in the epicycle. The radius vector **KP** in the epicycle is always parallel to the line **OS**

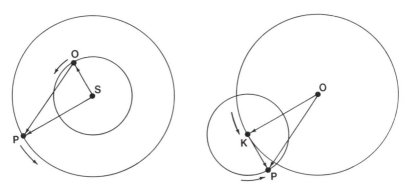

Figure 11. Sun-centered (11A) and earth-centered (11B) theories of a superior planet.

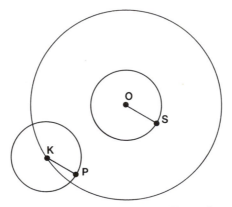

Figure 12. Relation between the sun and a superior planet. The sun, S, travels around the earth, O. The planet P travels around its epicycle, while the center K of the epicycle travels around the earth. All motions are counterclockwise. P stays in step with the sun so that KP is always parallel to OS. (In this and the following figure, the eccentricities of the circles are neglected.)

from the earth to the sun. In the case of an inferior planet (figure 13), the sun governs the motion in the deferent. The vector **OK** is in the same direction as **OS**. Thus the center **K** of the epicycle is seen in the same direction as the sun **S**. These facts are explicitly stated by Ptolemy. As he puts it in the *Planetary Hypotheses*, for each planet one motion is free and one is determined. Much later these facts were to point the way to the sun-centered system of Copernicus. But in the framework of an earth-centered cosmology, one could only offer speculative physical explanations, invoking the power of the sun, in accordance with the old Pythagorean notions.

Changing Goals of Greek Planetary Astronomy
Many modern historians of astronomy have followed the French physicist and philosopher of science Pierre Duhem in stressing a

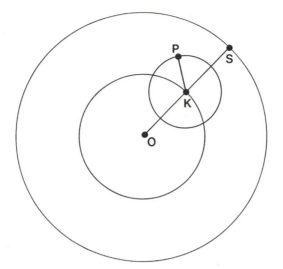

Figure 13. Relation between the sun and an inferior planet. The sun, S, travels around the earth, O. The planet P travels on its epicycle, while the center K of the epicycle travels around the earth. All motions are counterclockwise. The center K of the epicycle keeps pace with the sun, so that K always lies on OS.

Greek tradition of *saving the phenomena*, i.e. of devising a model that could account in a quantitative way for the observed motions of the planets. But in fact there was no quantitative planetary theory among the Greeks until the second century B.C.

The earliest planetary theories of the Greeks were qualitative rather than quantitative in purpose. The goal was to provide a physical explanation of the motions of the planets—a world picture. Thus it is wrong to think of Eudoxus's system of homocentric spheres as an attempt at saving the phenomena in a quantitative sense. Eudoxus's planetary theory was finally rejected not because it failed some exacting numerical test but because it seemed inadequate as a physical explanation. For example, it could not account for the fact that the planets Venus and Mars appear to be much larger, and therefore closer, at some times than at others.

In the same way, the earliest version of the deferent-and-epicycle theory aimed at no more than qualitative explanation. We know from a few remarks by Ptolemy that Apollonius of Perga (third century B.C.) proved some theorems concerning epicycle motion. Apollonius probably saw the epicycle (figure 3) as a way of accounting for retrograde motion and simultaneously explaining why the planets appear to be sometimes closer to us and sometimes farther away. But there is no evidence that anyone imagined before the second century B.C. that such a geometrical model could be endowed with quantitative, predictive powers.

As far as we know, Hipparchus was the first to show how to derive numerical values for the parameters in Apollonius's models, by performing calculations on suitably selected observations. The Babylonians had worked out quantitative methods of predicting planetary phenomena at a slightly earlier date. It is likely that the *possibility* of a quantitative, predictive theory only dawned on the Greeks after contact with Babylonian astronomy. The brilliance of the deferent-and-epicycle theory consists in its unification of quantitative, predictive power with a geometrical or physical world view. From the time of Hipparchus on, these two stringent requirements—good physics and quantitatively accurate astronomy—placed planetary theory under a new tension. In Ptolemy, the culminating figure of Greek cosmology and astronomy, we see both a mathematical astronomer and a cosmologist trying to accomplish both goals. Sometimes he was forced to make compromises.

Ptolemy as a Physical Thinker

Did Ptolemy mean his theory to be a representation of physical reality, or only a mathematical tool for calculating planetary positions? Ptolemy, like Hipparchus before him, wanted a theory that not only accurately predicted the positions of the planets but that *also provided a true image of the cosmos.* This was a difficult goal to achieve.

Ptolemy's solar theory implies an annual variation in the sun's distance from the earth. Ptolemy says that he constructed a dioptra (a sighting instrument) in order to measure the angular diameter of the sun, but was unable to detect any difference in the course of the year. This is not surprising, because the variation in the sun's angular diameter between apogee and perigee is only a minute of arc. But Ptolemy's attempt to detect a variation in the sun's apparent size shows that he fully accepted the physical consequences of the geometrical model.

The case of the moon poses some problems for the assessment of Ptolemy's aims. Hipparchus had worked out a fairly successful theory of the moon, as he had for the sun. Hipparchus's theory represented the position of the moon very well at times of full and new moon, and this was all that was necessary for the accurate prediction of eclipses. However, Hipparchus's theory showed rather large errors at quarter moon. Ptolemy modified Hipparchus's theory by introducing an extra crank mechanism, which pulled the moon in closer to the earth at the time of quarter moon (in order to increase the apparent size of the epicycle and thus correct the error in position).

Ptolemy's final theory of the moon did a very good job of predicting lunar positions. Unfortunately, it had an unwanted side effect of grossly exaggerating the monthly variation in the moon's distance. If the crank mechanism were correct, the moon's apparent diameter should nearly double between greatest and least distance. In fact, it increases by only about ten percent.

Ptolemy does not mention this obvious defect of the lunar theory. Moreover, he himself measured the variation in the angular diameter of the moon and reported reasonably accurate figures. (He needed these figures for analyses of lunar and solar eclipses.) So he was certainly aware that the lunar theory exaggerated the variation in the moon's distance. His silence about this defect is the best evidence that one could cite in favor of the view that he regarded the theory only as a machine for calculating positions and not as a physically true picture of the world. But we have already cited a good deal of evidence to the contrary. If any further proof were needed, we find it in Ptolemy's *Planetary Hypotheses.* Here Ptolemy attempts to deduce the distances of all the planets from the earth by nesting the mechanism for the moon inside that for Mercury, the mechanism for Mercury inside that for Venus, and so on, with the mechanism for Saturn surrounding them all. Here he takes quite literally the greatest and least distances deduced from the epicycle-and-deferent theory. Ptolemy's lunar theory represented a forced compromise between the requirements of physics and cosmology on the one hand and of planetary astronomy on the other. Here the difficult dual goal was realized only imperfectly.

Ptolemy's Cosmology

Ptolemy's main cosmological premise is that the cosmos contains no useless, empty space. The mechanism (deferent circle and epicycle) that produces a planet's motion fills a spherical shell. The thickness of this shell is determined by the eccentricity of the planet's deferent circle and by the radius of the planet's epicycle. The shells for all the celestial bodies are arranged one within another in the following order: surrounding the earth is the shell assigned to the moon; next come the shells for Mercury, Venus, the sun, Mars, Jupiter, and Saturn; and surrounding all is the sphere of the stars.

Figures 14 and 15 illustrate Ptolemy's cosmology as adapted by Georg Peurbach (1423–1461), an important figure in European astronomy during the Renaissance. Figure 14 shows Peurbach's system for the sun. The sun's system requires three orbs, labeled **C, D,** and **E.** The earth is at point **B,** which is the center of the cosmos. Point **A** is the center of the circle that the sun travels in the course of the year.

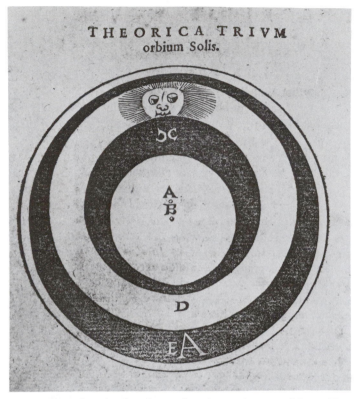

*Figure 14. Ptolemy's three-dimensional system for explaining the motion of the sun. The system requires three ethereal bodies, nested one within another. Two of the bodies, **C** and **E**, are black in the diagram. The sun is imbedded in the middle body **D**, which is white in the diagram. The earth is at **B**. This illustration is from a sixteenth-century astronomy textbook, the Paris 1553 edition of Georg Peurbach's* Theoricae novae planetarum. *Photograph courtesy of the Special Collections Division, University of Washington Libraries (negative no. UW 13653).*

Orb **C** (black in the diagram) has its inner surface centered on **B** and its outer surface centered on **A**. The sun is imbedded in orb **D** (white in the diagram). This orb turns around once in the course of the year. This is how the sun's annual motion around point **A** is effected. The outer orb **E** (black) has its inner surface centered on **A** and its outer surface centered on **B**. The two black orbs thus act as spacers for the orb carrying the sun. Furthermore, the inner hollow, bounded by the inner surface of **C**, serves as the receptacle into which the system for Venus is inserted. In a similar way, the system for Mars would be placed outside orb **E**.

Figure 15 shows Peurbach's realization of the systems for the sun and for Venus. In this figure, the three orbs for the sun are all labeled **A**. They are exactly as in the previous figure. The sun is shown as a

Figure 15. A simplified view of Ptolemy's three-dimensional system for Venus, nested inside the system for the sun. The three ethereal bodies for the sun's system are labeled A. Three ethereal bodies for Venus's system are labeled B. Venus is the asterisk riding on an epicycle set in the middle (white) body of the Venus system. From the Paris 1553 edition of Georg Peurbach's Theoricae novae planetarum. *Photograph courtesy of the Special Collections Division, University of Washington Libraries (negative no. UW 13654).*

circle with a dot in it, imbedded in the white solar orb. Three orbs for Venus are labeled **B**. Venus is the asterisk-like object located on the epicycle imbedded in the middle (white) orb of the Venus system. The epicycle is a solid sphere, which rotates inside a recess in the white orb. Points **D**, **C**, and **H** are, respectively, the earth, the center of Venus's deferent circle, and Venus's equant point. The boundary between Venus's system and the sun's system is the thin white crack between the outermost **B** orb and the innermost **A** orb. (We'll look at Ptolemy's own description of the nested-spheres cosmology near the end of this chapter.)

Peurbach simplified the picture a little by omitting some technical details. Nevertheless, Peurbach's illustrations preserve the essential features of the system described by Ptolemy in the *Planetary Hypotheses*. No better demonstration could be wished of the essential

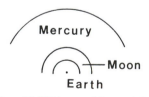

Figure 16. Thicknesses of the nested planetary spheres in Ptolemy's cosmology, drawn to scale. The scale changes by a factor of fifteen between figure 16A and 16B. That is, if figure 16A were shrunk by a factor of fifteen, it could be inserted into figure 16B. This figure is based upon the values in Table III (page 136). The gap between the shells for Venus and the sun reflects the numbers in Table III, but Ptolemy probably did not believe in the existence of an empty zone.

continuity of astronomy and cosmology from late Greek times through the Arabic and Latin Middle Ages and continuing to the Renaissance.

Ptolemy also worked out numerical values for the thicknesses of all the nested planetary systems. Ptolemy's numerical values were only slightly modified by those who followed. Figure 16 illustrates the whole of Ptolemy's cosmos to scale, based on the parameters in the *Planetary Hypotheses*. The earth is the small dot. Concentric spherical shells are assigned to the individual planets. The space between the earth and the sphere of the moon is filled by the lighter terrestrial elements, namely air and fire. Each planetary shell in figure 16 is, of course, made up of a number of orbs, as in figure 15. We will explain how Ptolemy arrived at the thicknesses of these shells, but first let us pause to examine his basic physical and cosmological premises.

Ptolemy's Physical Principles

In his physics Ptolemy was a follower of Aristotle. Among the Aristotelian doctrines accepted by Ptolemy were:

(1) Aristotle's theory of the five elements.

(2) Aristotle's theory of natural motions.

(3) The impossibility of a void place in the cosmos.

According to (1), everything below the moon is made of earth, water, air, and fire. All the changes we see before us on the earth—things coming into being and passing away—are due to combinations and recombinations of these four elements. The celestial bodies, however, are made of a fifth element, the ether, which is pure and simple and therefore incapable of change. This explains why the sun, moon, planets, and stars are changeless.

According to (2), the motion natural to the four sublunar elements is radial motion toward or away from the center of the cosmos, in keeping with their relative heaviness or lightness. Thus earth falls but fire rises. This explains why the earth we live on is found at the center of the cosmos. It explains, too, why water, air, and fire form concentric layers around the earth. The four sublunar elements can, of course, be given violent or forced motions that are contrary to their natural motions. But forced motions do not endure: a stone hurled upward soon reverts to its natural downward motion. The air, in particular, is subject to violent or turbulent motions.

In contrast, the motion natural to the ether, of which the celestial bodies are composed, is everlasting uniform circular motion. Thus the apparently complex motions of the sun, moon, and planets must be explained as combinations of simple, uniform, circular motions.

Ptolemy was conventional in his physical views; but he was not doctrinaire. Indeed, he did not hold any of these physical opinions with great tenacity. If Ptolemy was a follower in his physics, he was bold and creative in his astronomy; and he was prepared to compromise his physical principles if his astronomy seemed to require it. As we have seen in his planetary theory, Ptolemy was forced by the phenomena themselves to introduce nonuniform motions, in apparent contradiction of (2). And, although he worked out his cosmology on the assumption that there was no empty, wasted space between the spheres of neighboring planets, he added that if there *were* empty space between them the cosmos would be a bit larger—which reveals a rather cavalier attitude toward (3).

Ptolemy's *physical theology* (if we may call it that) was also conventional for his time. He never worked out its principles in great detail and his theology seems to have had no practical effect on his astronomical work. Like most astronomical writers of his time, Ptolemy believed in the divinity of the planets (*Almagest*, IX, 1). Thus, again, it is proper to attribute uniform circular motion to the planets, because disorder and nonuniformity are alien to divine beings. In the *Almagest* (I, 1) Ptolemy accepts Aristotle's unmoved mover (or in Ptolemy's words, an invisible and motionless deity) as the cause of the daily rotation of the cosmos.

Aristotle was a little inconsistent about the origins of the planetary motions. On the one hand, he was interested in improving the mechanical plausibility of Eudoxus's system for the planets. Thus Aristotle introduced the so-called counter-rolling spheres between the mechanisms of Saturn and Jupiter to prevent Saturn's motions from disturbing the motions of Jupiter. On the other hand, Aristotle believed that each individual motion (and thus each sphere in Eudoxus's model) required its own unmoved mover (*Metaphysics*, 1074a14–18). There is therefore in Aristotle an unresolved conflict between the desire for mechanical workability and the appeal to divine or other transcendental intervention.

Although Ptolemy does not follow Aristotle exactly, we see in the *Planetary Hypotheses* the same sort of unresolved conflict. Ptolemy's planetary models (figures 14, 15, and 20) are, like Aristotle's, physically realizable. That is, one could make them out of wood or other suitable materials and they would work as required—provided that the *motions* of the individual orbs could be regulated properly. But the regulation of the motions is a touchy matter in Ptolemy's cosmology, because the introduction of the equant point requires that the deferent orb revolve at a varying speed. (In figure 15, the deferent orb is the white **B** orb carrying the epicycle; in figure 20, the deferent is orb **2**.) Moreover, how could an inner planet like Venus receive its motion from an external mover when it is surrounded by the gyrating mechanisms of the outer planets? Ptolemy solves these problems by asserting that the planets move by their own wills. This appears to represent a return to the animism of Plato (*Republic*, X). Thus, for Ptolemy, neighboring planets move separately and independently of one another, just as birds do not fly through contact with other birds that would only hinder their motion. The innermost spacer orb of each planet's system (orb **3** in figure 20) plays a role similar to that of Aristotle's counter-rolling spheres: it provides a stationary receptacle into which the next innermost planet's system may be inserted. But then it is up to the inner planet to move its own orbs by the right rules. Thus the deferent orb is free to revolve with the appropriately varying speed, in accordance with the law of the equant, under the direction of the planet itself. Whatever we may think of this solution, the conflict between a mechanical view of the cosmos and the appeal to transcendental agents remains just as pointed in Ptolemy as in Aristotle. Fortunately, this conflict was of little significance for Ptolemy's technical cosmology, to which we now turn.

Fundamental Premises Concerning the Earth and the Cosmos

In the first book of the *Almagest*, Ptolemy lays out his premises concerning the earth and the cosmos:

(1) The cosmos is a sphere.

(2) The earth, too, is a sphere.

(3) The earth is at the middle of the cosmos.

(4) The earth is a mere point in relation to the cosmos.

(5) The earth is motionless.

These views were all widely accepted by Aristotle's time, if not somewhat earlier. Ptolemy's endorsement of them places him squarely in the mainstream of ancient cosmology. Of these premises, only (2) and (4) are really susceptible of proof from observation. Ptolemy, however, does argue in support of each of the five premises, drawing on Aristotelian physical principles as well as evidence from observation.

Most interesting is Ptolemy's discussion of the earth as motionless. *Certain people*, he says, claim that there is nothing to refute a supposed rotation of the earth on its axis from west to east. Here Ptolemy is plainly thinking of Heraclides of Pontos, and perhaps also of Aristarchus, although he does not mention them by name. Ptolemy adds that one could even let both the earth and the cosmos rotate about the same axis, as long as their relative speed still permitted the stars to rise and set once a day. Ptolemy freely admits that there is nothing in the celestial phenomena to contradict this view, but one must consider its physical implications. Ptolemy argues that objects not actually standing on the earth would be left behind by the rapid motion. Here, as he did so often, Ptolemy shows a clear recognition of the boundary between astronomy and physical thought. What cannot be proved from astronomical observation must be decided on physical or philosophical grounds.

Order of the Planets

In Ptolemaic planetary astronomy, the ratio of the epicycle's radius to the radius of the deferent can be determined from suitably selected observations. But the radius of the deferent in absolute units (such as earth radii) cannot be determined and is without consequence for ancient astronomy. The whole system for any individual planet can be enlarged or reduced without affecting the theory. Consequently, the theories for the several planets are completely separate from one another.

The moon is seen not only to eclipse the sun but also to occult the planets (Aristotle, *On the Heavens*, 292a4). Thus all Greek writers agree in placing the moon nearest the earth. But for the planets, there was no *astronomical* way to decide which were closer and which were farther away. The order of the planets had therefore to be decided by *physical* argument.

Several different ordering schemes were put forward in antiquity. One principle proposed at an early date was that the distances were correlated with the speeds; i.e., that the planets farthest from the earth were those with the longest tropical periods (*On the Heavens*, 291a32–b9). (A planet's tropical period is the average time required for the planet to make a complete trip around the zodiac, as viewed from the earth.) Thus most ancient writers agree in placing Saturn nearest the fixed stars, with Jupiter next below, and then Mars. The difficulty was what to do with the sun, Venus, and Mercury, since all three have a tropical period of exactly one year. Ptolemy (*Almagest*, IX, 1) says that some mathematicians placed Venus and Mercury higher than the sun because the sun had never been seen eclipsed by the planets. However, Ptolemy points out that these planets might lie a little north or south of the ecliptic at their conjunctions with the sun and therefore fail to produce an eclipse, just as the moon fails in the majority of cases to eclipse the sun at the time of new moon. In the *Planetary Hypotheses*, Ptolemy adds that the occultation of the sun by a small body might not even be perceptible, just as small, grazing eclipses of the sun by the moon are often not perceptible.

In any case, Ptolemy elects to place Venus and Mercury below the sun. In this way the sun, placed in the middle of the system, serves as a division between the planets that have limited elongations from the sun (Mercury and Venus) and those that can be at any angular distance from the sun (Mars, Jupiter, and Saturn). Ptolemy's order is therefore: moon, Mercury, Venus, sun, Mars, Jupiter, and Saturn.

This order was also favored by certain Pythagoreans, who regarded the sun as the heart and mind of the cosmos. The sun's ruling influence seemed to justify placing its orbital circle in the middle of the system—with three other bodies above it and three below it. Such views are discussed sympathetically by Theon of Smyrna and by Pliny. Ptolemy does not betray any sympathy for Pythagorean sun-mysticism, but it is possible that he was influenced by this tradition.

Ptolemy does try to justify on physical grounds his decision to place Mercury below the sun. He argues in the *Planetary Hypotheses* that the further removed a planet's astronomical hypotheses are from those of the sun, the farther the planet must lie, in real distance, from the sun. Thus Mercury must lie below the sun and close to the moon, because the rather complex theoretical system for Mercury resembles that of the moon. One might also expect the lowest planets to have the most complex motions because they are nearest to the air and their movements resemble the motion of the element adjacent to them. So again, it makes sense for the moon and Mercury, which have the most complex theories, to be lowest. This argument is Ptolemy's elaboration of an idea of Aristotle's.

Absolute Distances of the Moon and Sun in the Almagest

Although the distances of the sun and planets were irrelevant to practical astronomy, Ptolemy, like Aristarchus and Hipparchus before him, *was* interested in the absolute scale of the cosmos. The distance of the moon was the fundamental measuring stick by which the scale of the whole universe had to be judged. Moreover, the distance of the moon did have some practical significance: it was needed for a proper treatment of parallax, which affects the visibility of solar eclipses. For both these reasons, Ptolemy begins the construction of a cosmological distance scale by determining the distance of the moon.

In the *Almagest* (V, 11–13), Ptolemy attempts to find the distance of the moon by parallax methods. He compares a position of the moon observed from Alexandria with a theoretical position computed from his deferent-and-epicycle model for the moon's motion. This parallax measurement fixed the absolute scale of the moon's system. When the parallax measurement was combined with the deferent-and-epicycle theory, Ptolemy could then deduce the greatest and least distances of the moon from the center of the earth in absolute units (earth radii). (See Table I.) The numerical values, taken

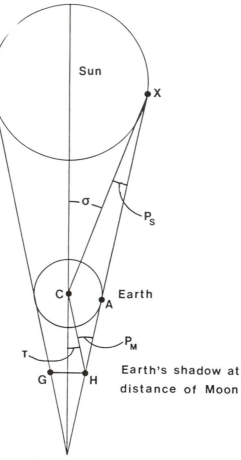

Figure 17. Ptolemy's eclipse diagram. σ is the angular radius of the sun, as seen from the earth. T is the angular radius of the earth's shadow where it falls on the moon during a lunar eclipse. P_M is the horizontal parallax of the moon and P_S is the horizontal parallax of the sun. It can be shown that $\sigma + T = P_M + P_S$.

from the *Planetary Hypotheses*, are rounded versions of the numbers in the *Almagest*.

Ptolemy's measurement of the moon's parallax is problematical. He obtained a value for the parallax that was a good deal too large, making the moon substantially too close to the earth at the time of his observation. It is likely that he "pushed" or "fudged" the parallax measurement a bit in order to make it fit with his theoretical notion of the monthly variation in the moon's distance.

To determine the absolute distance of the sun, Ptolemy used the method of the eclipse diagram (figure 17), due originally to Aristarchus of Samos. As a preliminary (*Almagest*, V, 14), Ptolemy determined the angular diameter of the moon when it was at its greatest distance from the earth, and found it to be 31' 20" (31 minutes, 20 seconds; a minute is one-sixtieth of a degree, and a second is one-sixtieth of a minute). Moreover, he took the angular diameter of the sun to be the same. Thus the angular radius of the sun (σ in figure 17) is half this, or 15' 40". For the angular radius of the earth's shadow as seen on the moon during a lunar eclipse (with the moon at greatest distance), Ptolemy found 40' 40" (T in figure 17). But in a later calculation, he said that the shadow is $2^3/5$ times as big as the moon, which would make T = 40' 44". Some of Ptolemy's predecessors had measured the angular diameter of the moon by sighting it directly with a dioptra (an instrument for measuring angles) or by timing with a water clock how long the moon took to rise. Ptolemy judged these methods to be fraught with error and difficulty. He therefore devised a clever method based on the comparison of two lunar eclipses of different degrees of totality. But Ptolemy's method is better in theory than it could ever be in practice. In any case, Ptolemy's values for T and σ did not differ much from those of his immediate predecessors.

In discussing Ptolemy's method for getting the distance of the sun, we can simplify matters a little by using modern trigonometry. In the use of the eclipse diagram we'll also follow a simpler geometrical argument than the one Ptolemy gives, but the basic method and the final results are Ptolemy's.

As a preliminary, we introduce the concept of horizontal parallax (figure 18). An observer at **A** on the surface of the earth sees a celestial body **B** (such as the moon) on his horizon. A fictitious observer at the center **C** of the earth would see **B** a little higher in the sky. The angular difference between the two lines of sight is called the *horizon-*

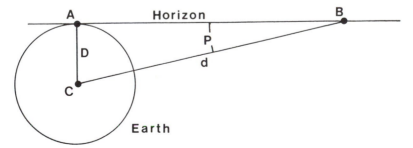

Figure 18. Horizontal parallax. Let a celestial body (the moon, for example) be located at B on the horizon of an observer who is located at A. The observer at A and a fictitious observer at the center, C, of the earth see the moon in directions that differ by an angle, P, called the horizontal parallax.

tal parallax, and it is marked **P** in the figure. The distance **d** of the object from the center of the earth is related in a simple way to the horizontal parallax: sin **P**= **D/d**, where **D** is the radius of the earth. Thus if the horizontal parallax of the object is small, the distance of the object is great. According to Ptolemy, the moon's greatest distance is $64\tfrac{1}{6}$ earth radii. (This number is rounded off in Table I to 64.) From this it follows that the moon's horizontal parallax (when at greatest distance) is $\mathbf{P}_M$ = 53' 35".

Now let's take up the eclipse diagram (figure 17), which depicts the sun, the earth, and the earth's shadow. **GH** is the path of the moon through the shadow during a lunar eclipse. σ and T are the angular radius of the sun and of the shadow, respectively. For an observer at **A**, both the edge of the sun and the edge of the shadow are on the horizon. Thus $\mathbf{P}_S$ is the horizontal parallax of the sun and $\mathbf{P}_M$ is the horizontal parallax of the moon.

In figure 17, T + σ = 180 degrees – angle **XCH**. But the three angles in triangle **XCH** must add up to 180; thus $\mathbf{P}_M + \mathbf{P}_S$ = 180 – angle **XCH**. Combining the two results, T + σ = $\mathbf{P}_M + \mathbf{P}_S$. In this equation we already have Ptolemy's results for T + σ, and $\mathbf{P}_M$. If we substitute in numerical values (15' 40", 40' 44", and 53' 35", respectively), we get $\mathbf{P}_S$ = 2' 49" for the sun's horizontal parallax. Then the distance of the sun may be found from **d** = **D**/sin $\mathbf{P}_S$. Ptolemy obtains 1,210 earth radii for the sun's distance from the center of the earth. (More accurate computation from Ptolemy's values for T, σ, and the moon's distance would give 1,218 earth radii; the difference is due largely to rounding errors in Ptolemy's calculation.) Ptolemy adopts 1,210 earth radii as the sun's average or mean distance.

Ptolemy's solar theory (figure 1) involves an eccentric circle. If the radius of the circle is taken as 60 arbitrary units, Ptolemy finds that the eccentricity **OC** is 2.5. The greatest and least distances of the sun from the center of the earth in these arbitrary units are 62.5 and 57.5, respectively. To pass over from the arbitrary units to earth radii,

Table I. *Absolute Distances of Sun and Moon*

	Least Distance	Greatest Distance
Moon	33 earth radii	64 earth radii
Sun	1,160 earth radii	1,260 earth radii

we multiply by the scale factor 1,210/60. Thus the greatest distance of the sun is 1,260 earth radii (as in Table I). The sun's least distance is 1,160 earth radii.

Scale of Cosmic Distances in the Planetary Hypotheses

For the planets, all that Ptolemy could determine astronomically was the ratio of least to greatest distance; the overall size of the deferent-and-epicycle system for each planet was unknown and unknowable. For a concrete example, let's examine the case of Mars, first defining some symbols (see figure 9):

> *R will be the radius of the deferent,* = *CA*.
> *r will be the radius of the epicycle,* = *KP*.
> *e will be the eccentricity of the deferent,* = *CO* = *CE*.

Mars will be closest to the earth (point **O**) when the center **K** of the epicycle is at the perigee ∏ of the deferent and the planet (point **P**) is also on the inside part of the epicycle. In this situation, we have the least distance of Mars = **R** − **r** − **e**. Similarly, Mars will be at its greatest distance from the earth (point **O**) when the center **K** of the epicycle is at the apogee **A** of the deferent and the planet (point **P**) is on the outside part of the epicycle. Then we have the greatest distance of Mars = **R** + **r** + **e**. In the *Almagest*, for each planet Ptolemy arbitrarily chooses **R** = 60 units. In these terms, his values for the measurable parameters of Mars are **r** = 39.5 and **e** = 6. Thus for Mars, the least and greatest distances are 14.5 and 105.5 (arbitrary units). The ratio of greatest to least distance is then about 7.3 to 1, which Ptolemy rounds to 7:1 in the *Planetary Hypotheses*.

In a similar way, we can work out the ratio of greatest to least distance for each of the remaining planets. Table II presents Ptolemy's results in the *Planetary Hypotheses*. Again, it should be emphasized that these ratios are based entirely on deferent-and-epicycle astronomy.

Table II. Astronomical Distance Ratios

	Ratio of Least to Greatest Distance
Mercury	34:88
Venus	16:104
Mars	1:7
Jupiter	23:37
Saturn	5:7

Table III. Cosmological Distance Scale

	Least Distance	Greatest Distance
Moon	33 earth radii	64 earth radii
Mercury	64 earth radii	166 earth radii
Venus	166 earth radii	1,079 earth radii
Sun	1,160 earth radii	1,260 earth radii
Mars	1,260 earth radii	8,820 earth radii
Jupiter	8,820 earth radii	14,187 earth radii
Saturn	14,187 earth radii	19,865 earth radii

To proceed further, Ptolemy had to supplement the astronomy of the *Almagest* with physical and cosmological premises. He assumes the order of the planets discussed above. Further, he assumes that the mechanisms of neighboring planets are nested one above the other, with no empty space between them.

Mercury is next above the moon. Thus Mercury's least distance must be 64 earth radii, equal to the moon's greatest distance in Table I. Then, using Table II, Mercury's greatest distance is 64 earth radii times 88/34, which works out to approximately 166 earth radii, as listed in Table III.

Then, Venus's least distance is also 166. To get the greatest distance, we again use Table II. Venus's greatest distance is 166 times 104/16, which works out to 1,079 earth radii, as listed in Table III.

Here it was possible for Ptolemy to perform a crucial check on the procedure. If the cosmological premises were correct, the sun's least distance should also be 1,079 earth radii. And the sun's least distance, found by combining the method of the eclipse diagram with the eccentric circle theory of the sun, was 1,160 earth radii (Table I). This seemed almost too good to be true!

Pure astronomy fixed the maximum distance of the moon at 64 earth radii and the minimum distance of the sun at 1,160. It turned out that the interval between these numbers was almost the right size to be filled by the mechanisms of Mercury and Venus, with no empty space left over. Of course, it did not work out quite perfectly, because the maximum distance of Venus turned out to be only 1,079 earth radii. Thus there was a gap (between 1,079 and 1,160) that Ptolemy could not account for. But he points out that if the distance of the moon is increased a little, the distance of the sun will automatically be decreased a little. This is clear from the equation we found above from the eclipse diagram: $\sigma + T = \mathbf{P_M} + \mathbf{P_S}$, with σ and T fixed by observation. If we make $\mathbf{P_M}$ (the moon's horizontal parallax) a little smaller, then $\mathbf{P_S}$ (the sun's parallax) must be made a little bigger to compensate. In this way it might be possible to fill that small gap between the shells for Venus and the sun. However, Ptolemy does not attempt to modify his numbers.

The least and greatest distances of the outer planets are easily filled in. Ptolemy puts Mars's least distance equal to the sun's greatest distance of 1,260 earth radii. Then Mars's greatest distance is 1,260 times 7/1 = 8,820 earth radii. He finds the least and greatest distances of Jupiter and Saturn in the same way. The fixed stars lie just beyond the sphere of Saturn, at 19,865 earth radii, which Ptolemy later rounds to 20,000 earth radii.

Ptolemy then converts all these distances into stades, starting from his value of 180,000 stades for the circumference of the earth.

Ptolemy concludes this portion of his discussion by reiterating his assumption that the nested spheres are contiguous, "for it is not conceivable that there be in nature a vacuum, or any meaningless and useless thing." But, he says, if there is space between the spheres, the distances cannot be any smaller than those he has set down.

Ptolemy then estimates the *sizes* of the planets and stars by combining his figures for their distances with measurements of their angular diameters. For example, assuming that the apparent angular diameter of Venus is one-tenth of the sun's disk, he arrives at an estimate of three-tenths of the earth's diameter for the diameter of Venus. The diameter of Jupiter turns out to be $4^{43}/120$ times the diameter of the earth. First magnitude stars are assigned a diameter of $4^{11}/20$ earth diameters. Ptolemy even goes on to compute the *volumes* of the planets, in comparison with the volume of the earth.

The Ptolemaic System of Nested Spheres

The system of nested planetary spheres, based on deferent-and-epicycle astronomy and supplemented by the cosmological distance scale, is often referred to as the *Ptolemaic system*. In the *Almagest* the theories of the planets are all elaborated in terms of circles, not solid spheres. The circles are all that is required for practical calculation. Even in working out the distance scale in the *Planetary Hypotheses*, Ptolemy speaks in terms of circles rather than solid spheres. But it is clear that he always regarded the spheres as physically (though not astronomically) necessary. In the second book of the *Planetary Hypotheses*, Ptolemy squarely faces the problem of reconciling deferent-and-epicycle astronomy with the solid-sphere cosmology of Aristotle and Eudoxus.

To account for the daily rotation of the whole cosmos, Ptolemy surrounds the sphere of the stars with a spherical shell of ether (figure 19). **AB** is the axis of the daily rotation. Line **CD** passes through the poles of the ecliptic. The earth lies in the middle of the diagram, at

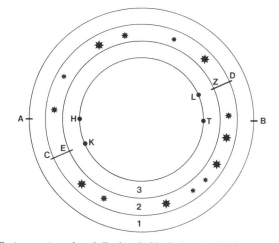

Figure 19. The intervening ether-shells, described by Ptolemy in the Planetary Hypotheses. *Surrounding the sphere of stars* 2 *is an ether shell* 1. *Ether shell* 3 *intervenes between the sphere of stars and the spherical system for Saturn. The intervening ether shells are responsible for communicating the daily rotation to the sphere of stars and to the spherical systems of the individual planets.*

the intersection of **AB** and **CD**. The exterior ether shell **1** turns once a day, from east to west, about axis **AB**. The sphere of stars **2** is pierced by axles (**CE** and **ZD**) set into the rotating ether shell. Thus the westward daily motion of sphere **1** carries the sphere of stars around with it. Meanwhile, the sphere of stars turns slowly to the east about axis **CD**: this is the precession of the fixed stars, which Ptolemy put at 1 degree in 100 years.

Inside the starry sphere is an ether shell **3**, which rotates about **EZ** in such a way that sphere **3** remains stationary with respect to the outermost sphere **1**. Thus points **H** and **T** on sphere **3** remain always directly under the corresponding points **A** and **B** of the outermost rotating ether shell. The system for Saturn may then be plugged into sphere **3** at points **K** and **L**. In other words, the daily rotation of the intervening ether shell **3** is responsible for carrying the Saturn system around once a day in exactly the same way as sphere **1** carries around the sphere of stars. The spherical system for each of the other planets is surrounded by a similar ether shell. There are seven such intervening shells (one each around the systems for the sun, moon, and the five planets) plus the outermost ether sphere (surrounding the sphere of fixed stars), for a total of eight. However, Ptolemy is not certain that the intervening shells are actually necessary. As we have seen

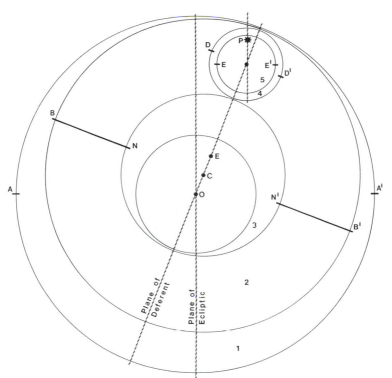

Figure 20. Solid-sphere mechanism for the superior planets, as described by Ptolemy in the Planetary Hypotheses. *If we regard this figure as representing the system for Saturn, it can be plugged into figure 19, with axis AA´ of this figure going into points K and L of figure 19.*

above, Ptolemy completely ignored the intervening shells when he worked out his scale of cosmological distances.

Ptolemy then describes the individual systems for the sun, moon, and planets in detail. The simplest case is that of the sun. Ptolemy's description is essentially the same as that given in connection with figure 14.

More complex are the systems for the planets. Figure 15, discussed above, omits a number of technical details. Let's now look a little more closely at Ptolemy's own description of his system for the planets. Figure 20 illustrates Ptolemy's system for Mars, Jupiter, or Saturn. The earth is at **O**. Surrounding the earth are three bodies **1**, **2**, and **3**, more or less as in figure 15. The exterior surface of body **1** is a sphere centered on the earth at point **O**. The inner surface of **1** is also spherical, but has its center at point **C**. Inside body **1** is a spherical shell **2**, centered on **C**. Inside body **2** is body **3**. The outer surface of body 3 is a sphere centered on C; the inner surface of 3 is a sphere centered on O, the earth.

Body 2 has a hollow that contains a spherical shell 4, which in turn contains a sphere 5. Imbedded in the sphere 5, near its surface, is the planet P itself. The spherical shell 4 and the sphere 5 are required to incorporate the planet's epicycle into the three-dimensional scheme.

Let's look now at the motions. In Ptolemy's theory of the superior planets, the planes of the deferent circles are slightly inclined to the plane of the ecliptic. These inclinations are required to explain the planets' latitudes (their observed slight departures from the plane of the ecliptic). In order to show these inclinations, Ptolemy draws his figures with the ecliptic perpendicular to the plane of the diagram. (This is different from figure 15, in which the ecliptic lies in the plane of the figure.)

Thus in figure 20, axis AA′ passes through the earth O and the poles of the ecliptic. The ecliptic plane is therefore perpendicular to AA′ and to the plane of the page. Body 2 (carrying 4 and 5 with it) rotates slowly about axis BB′, completing one rotation in a time equal to the planet's tropical period. Thus in the case of Mars, body 2 would make a rotation in about 2 years; in the case of Jupiter, 12 years; in the case of Saturn, 30 years. Axis BB′ is tilted slightly with respect to AA′ (the tilt has been exaggerated in the figure). The epicycle spheres (4 and 5) are therefore carried around a circle centered on C (the center of the deferent circle). Of course, the angular motion of the epicycles around the deferent is supposed to be uniform with respect to the equant point E, not with respect to C or O. But Ptolemy does not provide a mechanical realization of the equant in his three-dimensional theory.

Next, let's take up the epicycle spheres 4 and 5. In Ptolemy's theory of the superior planets, the plane of the deferent is tilted with respect to the ecliptic, but the plane of the epicycle is again parallel to the ecliptic plane. Ptolemy provides spherical shell 4, rotating about axis DD′, to "cancel out" the rotation about the tilted axis BB′. Thus DD′ is parallel to BB'. Sphere 4 rotates about DD′ at the same rate as sphere 2 rotates about BB′, but in the opposite direction. The net result is that sphere 4 is carried in a circular translation (i.e., without rotation) about axis BB'. Sphere 5 (carrying the planet P) rotates about axis EE′, which is parallel to AA′. The rotation of 5 is made in a time equal to the planet's synodic period.

After explaining the solid-sphere models for each of the planets, Ptolemy totals up the contents of the universe. As we have seen, each of the superior planets (Mars, Jupiter, and Saturn) requires 5 bodies, while the sun requires 3. The theory of Venus is similar to that of the superior planets; hence 5 ethereal bodies are required. It also happens that Mercury needs 7 and the moon needs 4. The fixed stars need 1 sphere of their own. Furthermore, there are 8 intervening ether shells required: 1 surrounding the sphere of the stars and 1 each around the spheres of the sun, the moon, and each of the five planets. (As explained above, these shells are supposed to impart the daily motion to the system of each planet.) The grand total therefore ought to come to 43. But for some reason, Ptolemy assigns only one ethereal body to the sun, and thus reaches a total of 41. In an alternative version of the system, Ptolemy replaces most of the complete three-dimensional orbs by rings or tambourines. In this version, Ptolemy reckons that only 29 bodies are required. However, as remarked above, he is not sure that the eight intervening shells are really necessary. One certainly is required, surrounding the whole cosmos, but perhaps the other seven can be eliminated. Thus Ptolemy conjectures that only 34 (41 − 7) bodies are necessary in the first version of his cosmology. In the second version, the elimination of seven ether shells brings the total down to 22 (29 − 7).

Ptolemy's Sources and Successors

The Ptolemaic system was a harmonious blend of several important lines of ancient thought. The basic physical principles of the system derived from the fourth century B.C., from Aristotle's physics and from Eudoxus's cosmology of three-dimensional spheres. However, neither Aristotle nor Eudoxus had provided a planetary theory with quantitative, predictive power. Quantitative planetary astronomy became possible only after the development of deferent-and-epicycle theory by Apollonius of Perga and Hipparchus (third and second centuries B.C., respectively). The final, very successful forms of the models for the motions of the moon and planets were due to Ptolemy himself. Ptolemy's problem as a cosmological thinker was to combine accepted physical principles, inherited from Aristotle and Eudoxus, with the successful planetary theory of the later astronomers. Physics and planetary theory, when supplemented with a few parallax measurements, led almost inescapably to the final scale of cosmic distances that Ptolemy worked out in the *Planetary Hypotheses*. The Ptolemaic system was almost a necessary consequence of Ptolemy's physical principles and his deferent-and-epicycle astronomy.

The incorporation of deferent-and-epicycle astronomy into solid-sphere cosmology was not original with Ptolemy. Theon of Smyrna, who was perhaps a generation older than Ptolemy, already had discussed how deferent-and-epicycle theory could be incorporated into a world view based on solid, nested spheres. Indeed, it is possible that the originator of the deferent-and-epicycle theory, Apollonius of Perga, discussed it in terms of solid spheres, although there is no direct evidence that he did. But it was apparently Ptolemy who showed how to incorporate all the technical details and who worked out a complete scale of cosmic distances on the basis of the models. In any case, the standard cosmology of the medieval period derived directly from Ptolemy: the technical astronomy of the *Almagest* supplemented by the solid spheres and the distance scale of the *Planetary Hypotheses.*

The Ptolemaic system was transmitted from the Greeks to Arabic astronomers of the early Middle Ages and thence to Latin-writing astronomers of medieval and renaissance Europe. The details of this transmission are murky, but the main outlines can be distinguished. Part of the difficulty is that the text of the *Planetary Hypotheses* has been handled roughly by history. Only the first half of the text has survived in Greek. The rest is preserved only in Arabic translation (and in a Hebrew translation made from the Arabic). The diagrams have fared even more poorly. It is unlikely that the *Planetary Hypotheses* ever circulated widely—in contrast to the *Almagest,* which did circulate widely and the text of which has been very well preserved. Often, medieval cosmologists learned the contents of the *Planetary Hypotheses* only at second or third hand. Thus we find writers who discuss the distance scale without referring to the system of nested spheres; and we find writers who discuss the nested spheres without mentioning the distance scale. We also find writers who seem not to be aware that the system originated with Ptolemy. For example, Proclus (a Greek writer of the fifth century A.D.), in his *Hypotyposis,* ascribes the principle of nested spheres not to Ptolemy but to "certain people," even though he often cited Ptolemy in other instances.

Among Arab astronomical writers of the early Middle Ages, we find echoes of the Ptolemaic system right from the start. Thābit ibn Qurrah (A.D. 827–901) was one of the central figures in the Islamic astronomical renaissance of the ninth century. He made translations of a number of the classics of Greek mathematics and astronomy. In particular, Thābit was responsible for the revision of Ishaq ibn Hunayn's Arabic translation of the *Almagest.* In another work, called *The Almagest Simplified,* Thābit gave an overview of Ptolemaic astronomy and cosmology, including Ptolemy's cosmological distance scale, borrowed from the *Planetary Hypotheses.* Thābit's numbers are the same as those in Table III, except that Thābit suppresses the gap between the spheres of Venus and the sun. He does this by making the sun's least distance from us equal to 1,079 earth radii, while the sun's greatest distance remains 1,260 earth radii; thus Thābit increases the thickness of the sun's sphere to fill the gap, even though this would result in a solar eccentricity that is far too large. Thābit has nothing to say about the astronomical consequences. The void was unsatisfactory, and he filled it in the simplest way possible. Thâbit knew Ptolemy's *Planetary Hypotheses*; indeed, he probably made or revised the Arabic translation of the *Planetary Hypotheses*. But his use of data derived from the *Planetary Hypotheses*, in a work that is purportedly an introduction to the *Almagest*, must have added to the already considerable confusion over the historical origins of the Ptolemaic system.

An elder contemporary of Thābit, al-Farghânî (ca. 800–870), wrote an *Elements of Astronomy*, an elementary survey of Ptolemaic astronomy and cosmology that achieved great popularity. Al-Farghânî seems to have been unaware of the *Planetary Hypotheses*, but he knew the general principles of the nested-sphere cosmology. He calculated the greatest and least distances of all the planets directly from the parameters in the *Almagest*. Al-Farghânî's distances were therefore a little different from those in Table III. Moreover, al-Farghânî had no gap between the spheres of Venus and the sun; nor did he fill it artificially, as Thābit did.

A later book in this same genre was Ibn al-Haytham's (ca. 965–1040) *On the Configuration of the World*, which provided a survey of astronomy, including a complete introduction to the solid-sphere cosmology (though no discussion of the distance scale). Interestingly, although he often cites the *Almagest*, Ibn al-Haytham never directly cites the *Planetary Hypotheses*, which he apparently did not know. Nevertheless, the solid-sphere cosmology he describes is essentially that of Ptolemy.

In the twelfth century, al-Farghânî's *Elements of Astronomy* was translated into Latin not once but twice, by John of Seville and by Gerard of Cremona. Ptolemy's cosmology and al-Farghânî's version of the Ptolemaic distance scale thus circulated in Latin Europe from the very beginning of the European revival of learning—but often without Ptolemy's name attached. Although European astronomers first learned their astronomy and cosmology from Latin translations of Arabic textbooks, they soon began to compose their own introductions to the science of the heavens. It is noteworthy that we find the cosmological distance scale fully present in the first complete discussion of planetary theory to be composed in the Latin West, the *Theorica planetarum* of Campanus of Novara, written around 1260.

One of the most important works for the dissemination of solid-sphere cosmology in the early Renaissance was a popular university textbook, *Thoricae novae planetarum*, by Peurbach. He composed his text for a series of lectures that he gave in 1454 for the Bürgerschule in Vienna. For the solid-sphere version of Ptolemy's cosmology, Peurbach drew upon some Arabic source in Latin translation, probably Ibn al-Haytham's *On the Configuration of the World* or a work derived from it. He called his work *New Theories of the Planets* because he meant it as a replacement for the thirteenth-century *Theorica planetarum* (sometimes attributed to Gerard of Cremona, and not to be confused with Campanus's book of the same title), a widely used but sloppy and unsatisfactory introduction to Ptolemaic planetary theory. A few manuscript copies of Peurbach's work circulated around the universities, but it was not printed until 1472, some eleven years after Peurbach's death, when it was published by Peurbach's student and friend Regiomontanus. Peurbach's work became enormously popular. It was frequently reprinted and was widely used as a university text. No fewer than fifty-six editions, including translations and commentaries, were published between 1472 and 1653. Figures 14 and 15 are from an edition published in 1553 with commentary by Erasmus Reinhold. This was ten years after Copernicus's great book, *On the Revolutions of the Heavenly Spheres*, introduced the new sun-centered cosmology. Ptolemy's cosmology remained in the standard university texts right up to the time of Copernicus, and even after.

—James Evans

BIBLIOGRAPHIC NOTE

Ptolemy's philosophical and physical premises can be studied in the opening chapters of Books I and IX of the *Almagest*, available in G.J. Toomer, trans., *Ptolemy's Almagest* (London: Duckworth, 1984). This translation is preferred to that in the *Great Books of the Western World* series. The situation for the *Planetary Hypotheses* is more complicated. It is divided into two books. Only the first part of the first book survives in Greek. However, the entire work, apart from some numerical tables that have not survived, is preserved in a medieval Arabic translation and in a medieval Hebrew translation made from the Arabic. The Teubner edition of Ptolemy's *Opera* (vol. II) contains the Greek text and a German translation of Book I, part 1 of the *Planetary Hypotheses*, as well as a German translation of the Arabic version of Book II. [*Claudii Ptolemaei Opera quae exstant omnia*, J.L. Heiberg, et al., eds. (Leipzig: Teubner, 1898–1961)]. A part of considerable cosmological interest (Book I, part 2, containing the distance scale) was inadvertently omitted from the Teubner edition. For an English translation of Book I, part 2, see Bernard Goldstein, "The Arabic Version of Ptolemy's *Planetary Hypotheses*," *Transactions of the American Philosophical Society*, new series, vol. 57, no. 4 (1967). The authoritative study of Ptolemy's mathematical astronomy is O. Neugebauer, *A History of Ancient Mathematical Astronomy* (Berlin and New York: Springer-Verlag, 1975). For a discussion of the modifications of Ptolemy's planetary theory by renaissance astronomers, see J. Evans, "The Division of the Martian Eccentricity from Hipparchos to Kepler: A History of the Approximations to Kepler Motion," *American Journal of Physics*, 56 (1988), 1009–1024. For an excellent brief history of attempts to measure the distances of planets and stars, see Albert Van Helden, *Measuring the Universe* (Chicago and London: University of Chicago Press, 1985). Several medieval Arabic and Latin astronomical handbooks are now available in English translation, providing an easy way to explore the preservation and development of Ptolemaic cosmology in the Middle Ages. The following two are especially recommended: Y.T. Langermann, *Ibn al-Haytham's "On the Configuration of the World"* (New York and London: Garland Publishing, 1990), and E.J. Aiton, "Peurbach's *Theoricae novae planetarum*: A Translation with Commentary," *Osiris*, 2nd series, *3* (1987), 5–43.

8. COPERNICUS

Nicholas Copernicus was born in 1473 at Torun, in the kingdom of Poland. In 1491 he entered the Jagiellonian University of Cracow. Founded more than a century before, in 1364, and reorganized in 1400 along the lines of the universities of Paris and Prague, Cracow was one of the first of the northern universities to be influenced by the new spirit of Humanism from Italy.

Like every student of the age, Copernicus followed the curriculum of the *trivium*—grammar, rhetoric and logic; then that of the *quadrivium*—arithmetic, geometry, music and astronomy. Instruction in astronomy was usually mediocre, consisting largely of the reading of two introductory treatises on astronomy, one of them John of Sacrobosco's *Tractatus de sphaera*. But at Cracow there was also a tradition of astronomical practice, with tables computed for the meridian of Cracow dating from 1428 and Cracovian commentaries on Sacrobosco's *de sphaera* circulating as early as 1430. Furthermore, some time before Copernicus entered the university, Albert of Brudzewo, author of a commentary on Georg Peurbach's *Theoricae novae planetarum*, founded there a school of astronomy and mathematics, and he may have given Copernicus private instruction.

In 1496 Copernicus went to Italy and the University of Bologna, where he studied with the astronomer Domenico Maria Novara, known for his assertion that the latitudes of Mediterranean cities were now 1 degree, 10 minutes higher than reported in Ptolemy's *Geography*. From this systematic increase in latitudes, Novara inferred a slow and progressive modification of the direction of the earth's axis with a period of 395,000 years. Copernicus could well have noted that if the North Pole changed its direction, the earth could not enjoy that absolute immobility that Ptolemy, following Aristotle, had imposed upon it; there is a motion of the axis of the earth in Copernicus's later *De revolutionibus*.

Although officially registered at Bologna as a student in canon law, and later in civil law, Copernicus seems already to have preferred astronomy. On 9 March 1497, he made his first known observation: of the moon approaching the star Aldebaran. Other early indications of his interest in astronomy are a lecture he is reported to have given in Rome in 1500 and his observation there of a partial eclipse of the moon.

Back home, Copernicus took possession of the office of canon in the cathedral chapter of Frauenburg (Frombork) that his uncle, Lucas Waczenrode, Bishop of Varmia (also known as Ermland), had obtained for him. Statutes of the chapter allowed a grant to any canon who desired to complete studies already begun, and in 1501 Nicolas asked for two more years, in order to learn medicine at the University of Padua. In 1503, with his medical studies uncompleted, he successfully applied to the University of Ferrara to have himself proclaimed doctor of canon law before returning to Poland.

For a few years Copernicus accompanied his uncle on his ecclesiastical and diplomatic travels, as secretary and physician, before he settled permanently in the little town of Frauenburg. There, in an isolated corner of the Bishopric of Varmia, an enclave of the Kingdom of Poland surrounded by the last fiefs of the Knights of the Teutonic Order, Copernicus participated in the administration of the chapter and in the last struggles against the knights, continued to practice a little medicine, and devoted himself to his researches in astronomy.

In 1510 Copernicus composed his *Commentariolus*, presenting in a few pages a theory of the earth in motion and the sun immobile at the center of the world. He shared this draft with a few close friends, but it was not published until the nineteenth century, under the title *Nicolai Copernici de hypothesibus motuum coelestium a se constitutis commentariolus.*

Copernicus continued to work at astronomy, and gradually compiled the book *De revolutionibus orbium coelestium libri sex*, published in 1543, the year of his death. It might never have been published, at least not during his lifetime, without the intervention of his sole disciple, Georg Joachim Rheticus, and without the firm and constant pressure of the Bishop of Chelmno, Tiedemann Giese, Copernicus's closest friend. Perhaps Copernicus deferred publication of his book out of fear of unfavorable reaction from the Catholic hierarchy. But the preface, dedicated to Pope Paul III himself, shows that Copernicus finally decided to be bold. The unfinished state of the last sections of *De revolutionibus* may also explain Copernicus's reticence.

Medieval Cosmological Heritage

In Copernicus's time astronomy was dominated by Ptolemy's *Almagest*, which could be studied either directly through Gerard of Cremona's Latin translation or indirectly through elementary textbooks, such as Peurbach's *Theoricae novae planetarum* or Regiomontanus's substantially deeper *Epytoma in Almagestum Ptolemaei (Epitome of Ptolemy's Almagest)*, published in 1496. But Ptolemy's *Planetary Hypotheses*, which described the concrete machinery of the heavens, was unavailable to medieval Europeans, who knew only simplified and anonymous discussions of Ptolemaic physical astronomy.

Three distinct components are mixed together in the *Almagest*:

- A global vision of the world, or a cosmology.

- A mathematical tool, essentially trigonometry applied to the solution of plane and spherical triangles.

- A practical astronomy, represented by a collection of geometrical models, by tables of numbers, and by rules of calculation, which permitted astronomers to find, for any given time, the positions of the planets, the sun, and the moon against the changeless background of the sphere of fixed stars.

For ancient and medieval scholars, trigonometry and practical astronomy were merely branches of mathematics, and therefore had no deep metaphysical or ontological significance. Cosmology had a higher status, because it presupposed a philosophy of nature or, at least, a physics—Aristotelian physics in the case of Ptolemy's cosmology. This physics involved fundamental propositions whose acceptance largely determined, down to the sixteenth century, the hypotheses of astronomy.

Four propositions or theses fundamental to the Aristotelian world view were:

(1) *Geocentrism.* The earth, the only body in the world that is absolutely at rest, is at the center of the universe, whose outer boundary is the rotating sphere of fixed stars.

(2) *Dichotomy of the two worlds.* The world of the four elements (earth, water, air, and fire), which reaches as far as the moon, is a world of change, of generation and corruption, and of rectilinear motion (upward for the light elements and downward for the heavy ones). In contrast, the supralunar world, reaching from the moon's orb outward to the stars, is a changeless and imperishable world, a world of a fifth element (the ether) devoid

of all lightness or heaviness, for which rectilinear motion is impossible.

(3) *Principle of plenitude.* Between the earth and the outer boundary of the world the elements as well as the ethereal spheres surround one another concentrically without break or failure of continuity. In other words, all void or empty space is excluded by nature in the interior of the cosmos.

(4) *Uniform circular motion.* The only motion proper to the ethereal bodies, made up of various circles or spheres carrying the planets, is uniform circular motion around the fixed center of the universe.

To account for facts of observation, Aristotle's principles had been somewhat modified by Ptolemy's time. In particular, variations in the apparent sizes of the planets (which, because of the second thesis, could not be explained in terms of real changes in the ethereal bodies) led astronomers to abandon Eudoxus's and Aristotle's idea of strictly concentric spheres. Postulating eccentric circles and epicycles, Hellenistic astronomers multiplied the centers of rotation and de-throned the earth from its privileged status as unique center of all circular motion (theses 1 and 4). But since the circles assigned to each planet were supposed to turn in the interior of a sphere concentric with the earth and with the sphere of the fixed stars, the earth re-mained the physical center of the world system, thus preserving, at least theoretically, thesis 3, the principle of plenitude.

In his *Planetary Hypotheses*, Ptolemy offered a physical interpreta-tion of the geometrical models of his *Almagest*. This cosmology had a great influence on medieval Europe through Arabic adaptations, par-ticularly Ibn al-Haytham's simplified interpretation, *On the Shape of the Universe.* (Known in the Latin-writing West as Alhazen, he lived from 965 to 1039.) This book helped inspire Peurbach's widely read *Theoricae novae planetarum*, finished around 1450 and published by Regiomontanus around 1472. Along with the *Epitome of the Almagest*, also by Peurbach and Regiomontanus, it in great measure defined the "restored" astronomy at the close of the fifteenth century—the resto-ration of an astronomy that had been invented by Ptolemy thirteen centuries earlier.

The principal features of Alhazen's *On the Shape of the Universe* have been well summarized by the historian Willy Hartner:

> *The universe, which is spherical in shape, consists of nine spherical shells gliding inside one another; each of these shells again is composed of a set of concentric or eccentric shells or complete spheres. There is no empty space or "void." The Earth with its water is surrounded by air, which again is surrounded by fire.*

The sphere of fire is limited by that of the moon, after which come the spheres of the six other planets and of the fixed stars. Finally, the outer limit of the universe is the "sphere of the spheres." [. . .] While the four elements of which consist the earthy, watery, airy and fiery spheres are either heavy or light, the fifth element [. . .], which is the matter filling the translunar world, is neither the one nor the other. Unlike the four terrestrial elements it has the essential quality of eternal circular motion.

Medieval Arguments over Ptolemaic Models

Controversy over the status of the multiple orbs assigned to each planet may well have influenced Copernicus's later work. The many commentators on Peurbach's *Theoricae novae planetarum*, which circulated widely as an introductory textbook on Ptolemaic mathematical astronomy, were divided over the status of the multiple orbs assigned to each planet. For some, physics required that there be real bodies filling each sphere; for others, the orbs were purely imaginary constructions useful for instruction. Albert of Brudzewo, who lectured at the University of Cracow during Copernicus's student days, held the latter interpretation.

Less evident is the source of Copernicus's objection to Ptolemy's use of an equant point. Alhazen had raised this issue in his *Doubts on Ptolemy*, composed after his better known treatise *On the Shape of the Universe*, but none of the commentators on Peurbach seems to have been aware of Alhazen's criticism. Introduced by Ptolemy in his *Almagest*, the equant point resulted in physically nonuniform motions for the planets, in contradiction of thesis 4 above. (Reasons for this departure from Aristotelian physics and the results achieved with it are discussed in the preceding article on Ptolemy in this book.) According to Alhazen, it was "impossible and in contradiction with the true principles" that an astronomer should introduce, by means of this expedient, nonuniform motions into the heavens, where it was not fictitious circles but real bodies that revolved. Alhazen denied that any celestial body (planet, orb, or sphere) made of pure elemental ether could turn irregularly around its axis, and he thus proclaimed that Ptolemy had not discovered the truth regarding the theory of the planets. This consideration also led astronomers of the so-called Maragha school—in particular, Nasir al-Din al-Tusi (1201–1274) and Ibn al-Shatir (ca. 1305–ca. 1375)—to search for alternative models that could function without the equant. Their models were non-Ptolemaic, but agreed in all essentials with both the observations and the final results of the *Almagest*, even where these were deficient. Moreover, the new models did not cast any doubt on Ptolemy's underlying cosmology.

Similar preoccupation with the equant is found in the work of Copernicus. In the opening lines of the *Commentariolus*, he objects that, in contradiction of the principle requiring that a celestial body always move uniformly, Ptolemaic astronomers had imagined "certain equant circles, because of which the planet does not appear to move always with a uniform speed, neither on its deferent orb, nor around the proper center [of the world]." Such a theory Copernicus judges "neither sufficiently achieved nor sufficiently in accord with reason." These considerations led him, he says, to seek "a more rational system of circles, from which all apparent irregularity would result, while everything would move uniformly around its proper center, as the principle of perfect motion requires." Some of the technical devices adopted by Copernicus to avoid an equant are identical with ones in Ibn al-Shatir's astronomical models; while there is no evidence that Copernicus knew directly of Ibn al-Shatir's work, some sort of indirect influence is likely.

Copernicus's Cosmological Postulates

The principal characteristics of Copernicus's cosmological system are found in the famous seven postulates of the *Commentariolus*. A careful reading of these postulates remains the best introduction to the new cosmology, which completely overturned the first three fundamental theses of the traditional Aristotelian cosmology: geocentrism, the dichotomy of the two worlds, and the principle of plenitude. The Aristotelian thesis of uniform circular motion, however, was retained by Copernicus; indeed, he even rebuked Ptolemy for discarding it.

In the *Commentariolus*, Copernicus presented the following postulates:

(1) *There is not one single center for all the celestial orbs or spheres.*

(2) *The center of the earth is not the center of the world, but only of the heavy bodies and of the lunar orb.*

(3) *All the orbs encompass the sun which is, so to speak, in the middle of them all, for the center of the world is near the sun.*

(4) *The ratio of the distance between the sun and the earth to the height of the firmament [starry sphere] is less than the ratio between the earth's radius and the distance from the sun to the earth, in such a manner that the distance from the sun to the earth is insensible in relation to the height of the firmament.*

(5) *Every motion that seems to belong to the firmament does not arise from it, but from the earth. Therefore, the earth with the elements in its vicinity accomplishes a complete rotation around its fixed poles, while the firmament, or last heaven, remains motionless.*

(6) *The motions that seem to us proper to the sun do not arise from it, but from the earth and our* [terrestrial] *orb, with which we revolve around the sun like any other planet. In consequence, the earth is carried along with several motions.*

(7) *The retrograde and direct motions which appear in the case of the planets are not caused by them, but by the earth. The motion of the earth alone is sufficient to explain a wealth of apparent irregularities in the heaven.*

Copernicus's first two postulates contradict the ancient thesis of geocentrism. His third postulate contradicts not only geocentrism with regard to the body which occupies the central position of the universe, but also the dichotomy of the two worlds, in that Copernicus abolishes the distinction between the parts of the world found below and above the moon. Copernicus's fourth postulate (implying vast regions of empty space between the solar system and the firmament) implicitly brings into question the principle of plenitude. Copernicus's last three postulates not only reject geocentrism and the dichotomy of the two worlds, but also contradict another fundamental principle of Aristotelian physics: only one single motion can properly belong to any natural body.

The only ancient physical thesis preserved by Copernicus (and which he even claims to confirm, thanks to the heliocentric doctrine) is the principle of uniform motion. In *De revolutionibus* (I, 4) he writes: "The movement of the celestial bodies is uniform and perpetually circular, or else composed of circular movements."

The new Copernican cosmology revealed for the first time a necessary connection between distances of the planets and periods of their revolutions. In the *Commentariolus* Copernicus wrote:

> The celestial orbs encircle one another in the following order. The highest is the orb of the fixed stars which, motionless, contains all things and locates them. Beneath it is Saturn's orb, then Jupiter's, and then Mars's orb. Beneath this is the orb with which we are carried in a circular motion. Then comes Venus's orb and, last of all, Mercury's. The moon's orb revolves around the earth and is carried with it, like an epicycle. It is also in the same order that the speeds of the orbs' revolutions surpass one another according to the greater or lesser revolution they accomplish. So Saturn comes back to its starting point in 30 years, Jupiter in 12 years, Mars in 2 years, and the earth in 1 yearly revolution. Venus carries out 1 revolution in 9 months, and Mercury in 3.

The order of the planets, arbitrary in Ptolemaic astronomy but following naturally and necessarily in the Copernican model, helped convince Copernicus that he had discovered the real order of the cosmos, not merely a mathematical hypothesis. Copernicus's sun-centered cosmology also contained within its structure simple explanations of several basic planetary phenomena known and accepted by Ptolemy but which had no "automatic" explanation in his earth-centered cosmology. The following planetary arrangements were arbitrary in the Ptolemaic model but necessary in the Copernican:

- Mars makes a longer retrogression (appears to stop in its movement around the heavens, retrace part of its path, and then resume its forward motion) than do Jupiter and Saturn; and Venus makes a longer retrogression than does Mercury.

  This automatically occurs in the Copernican system because the closer the planetary orbit is to the earth's orbit, the greater the apparent retrograde motion must be. (See the explanation of retrograde motion in the previous chapter in this book, on Ptolemy's cosmology.)

- Saturn, Jupiter, and Mars look larger when they are in opposition to the sun (on the opposite side of the earth from the sun) than when they are in conjunction (seen from the earth in the same direction as is the sun).

  This automatically occurs in the Copernican system because an outer planet is closest to the earth when the earth is between the planet and the sun, and is farthest from the earth when the sun is between the outer planet and the earth.

- Venus and Mercury do not altogether leave the sun.

  This automatically occurs in the Copernican system because seen from the earth, Venus and Mercury are always in roughly the same direction as is the sun.

As Copernicus noted in *De revolutionibus* (I, 10) he had found "a clear bond of harmony in the motion and magnitude of the orbs such as can be discovered in no other wise."

Copernicus had not put the earth into motion on a whim. And he had steeled himself to defend his cosmology against foreseeable reactions from philosophers and theologians, as is clear from his dedicatory letter to Pope Paul III and the first book of *De revolutionibus*, in which Copernicus argues for the physical possibility of the motion of the earth. He was not merely experimenting with hypotheses in order to save the appearances in a more satisfying fashion, as Osiander maintained in the anonymous introduction to *De revolutionibus*. He was presenting the true arrangement of the machine of the world.

Modification of the Old Physics

While overthrowing traditional doctrine, Copernicus suggests that his heliocentric system is compatible with concepts of Aristotelian physics, including the categories of natural place, of simple and composite bodies, and of natural, violent, rectilinear, and circular motions, *after these concepts are modified in the context of the new cosmology.* It is almost as if Copernicus submitted the circular motions of his celestial orbs to general conditions imposed by Aristotle.

Thus Copernicus (*De revolutionibus*, I, 8) maintains against Aristotle (*De caelo*, II, 14) that if one thinks that the earth moves, one will say that this motion is *natural* and not violent. Consequently, one need not fear that the earth and all terrestrial things will be broken up by the effects of a rotation produced according to nature.

In the same way, regarding the things that fall and those that rise, one must admit that their motions are double, that is, they are composed of rectilinear and circular motions. Copernicus agrees with Aristotle (*De caelo*, I, 2; II, 14) that a simple body must have a simple motion—but only as long as the body retains its unity and remains in its natural place. And, according to Copernicus, this can only be true in the case of circular motion, the only motion proper to bodies that are in their natural place and that therefore move always equally, for it has a cause that never ceases. As for rectilinear motion, it only belongs to bodies that are far from their natural places and it only lasts for the time required for them to rejoin these natural places, where they remain at rest. Copernicus concludes that circular motion belongs to totalities and rectilinear motion to parts, and one may say that the rectilinear coexists with the circular "as illness does with living." Thus Aristotle's division of simple motion into three genres (movement from the center, toward the center, and around the center) must be regarded merely as an "act of reason."

Copernicus is ready to overturn the Aristotelian classification of motions based on an immovable earth, but conservatively clings to Aristotelian celestial dynamics. Hence his protest in the *Commentariolus* against liberties taken by Ptolemy with the equant, and hence several circular motions in *De revolutionibus* to explain the revolution of one star. For Copernicus it seems "impossible that a *simple heavenly body* should be moved with a nonuniform motion *by a single orb* [emphasis added]. For that would have to occur either by virtue of the inconstancy of the motive force (whether this inconstancy has an extrinsic or an intrinsic cause) or else by virtue of an alteration of the body in revolution. But the mind is loath to admit these two explanations and it is quite unfitting to conceive such a case for things arranged in the best order." (*De revolutionibus*, I, 4.) The Aristotelian context of these lines is evident (cf. *De caelo*, II, 6). The expression *corpus simplex* (simple body), applied here to the orbs responsible for the transport of the planets, is borrowed from Aristotle. Moreover, the potential irregularity of this motion is completely Aristotelian in spirit, because Copernicus's *virtus movens* (motive force) is double: an exterior mover (whether the intelligence postulated by Aristotle, or the moving angel of his Christian interpreters) and an interior mover, the soul of the orb, which obeys the exterior mover. Finally, the impossibility of an alteration of the "body" in revolution can only be an allusion to the excellence of the body of a sphere—which Aristotle judged incorruptible and unchangeable. Thus, everything goes as if Copernicus had submitted the circular motions of his celestial orbs to the general conditions imposed by Aristotle.

Physical Status of the Celestial Orbs and Spheres

Regarding celestial orbs and the spheres within which they revolve, Copernicus writes in a passage in *De revolutionibus* (I, 10) examining the place occupied in the heaven by the earth and moon: "... as all the planets have a single common center ... it is necessary that the space left between the convex orb of Venus and the concave orb of

Figure 1. Copernicus's diagram in Book I of De revolutionibus, *illustrating the system of the world. From Copernicus's autograph manuscript.*

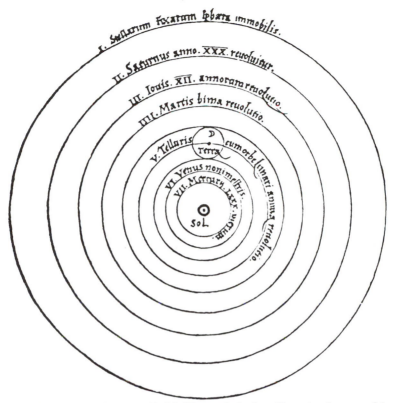

Figure 2. Copernicus's diagram in Book I of De revolutionibus, *illustrating the system of the world. From the first printed edition, published in Nuremberg in 1543.*

Mars be equally considered as an orb (or a sphere), each face of which is concentric to the orbs of Venus and Mars. And this space should receive the earth, its companion the moon, and all that is contained beneath the lunar globe." The problem here addressed and the vocabulary employed suggest that Copernicus means the concentric spheres as traditionally conceived, surrounding one another from the center of the world out to the last heaven. His diagram of the world in *De revolutionibus* (figures 1 and 2) also supports this conclusion.

But exactly what physical status does Copernicus attribute to the spheres and orbs? Does Copernicus, like Peurbach, imagine in the interior of each concentric sphere an arrangement of corporeal orbs whose motions, all uniform and circular around their own centers, would produce the apparent trajectories of the planets? (See figures 14 and 15 on pages 124 and 125 in the preceding chapter, on Ptolemy's cosmology.) Do these necessarily correspond to real mechanisms?

Do rigid material spheres really exist in the heaven? The very architecture of the heliocentric world is at issue here.

A difficulty with imagining in the interior of each concentric sphere an arrangement of corporeal orbs arises in the case of the earth surrounded by its elements. In the *Commentariolus*, in explaining how the earth's rotation axis remains parallel to itself in the course of the annual revolution, Copernicus rejected the idea that it behaves like a compass needle that points always toward the same region of the world. He preferred to attribute this effect to a carrying by a particular orb, while nevertheless declaring that he had no interest in saying what its poles were attached to. And in *De revolutionibus* he contented himself with giving a geometric description of the functions of the orb responsible for producing this so-called third movement. Had he foreseen the difficulties he would have faced in constructing a "theory of the earth" of the Peurbachian type? Copernicus's silence on this point, and more generally on the nature of the orbs that carry the planets, is difficult to interpret.

Copernicus's original diagram in *De revolutionibus* (figure 1) far more resembles the traditional geocentric representations of the ancient cosmology than it departs from them. Compare it with the diagram in a 1519 edition of Aristotle's *De caelo* (figure 3). Essential differences in Copernicus's diagram are:

- Inversion of the places occupied by the earth (along with the moon) and the sun.

- Fewer moving spheres—six in Copernicus but ten in the Alphonsine version of Ptolemy's system.

- A change in the identity of the immovable heaven. No longer is it the empyrean heaven of the theologians. For Copernicus it has become the astronomical sphere of the fixed stars.

That said, the overall structure of the universe appears identical in the two diagrams. In both we see a piling up of concentric spheres succeeding one another without break of continuity between the center and the outer spherical envelope that marks the limit of the universe.

Copernicus's schematic representation can be misleading insofar as it minimizes and even hides the actual complexity of his cosmological system. The rudimentary diagram illustrates well the simple connection between the distances of the celestial bodies and their periods of revolution, and it also allows us to understand intuitively the general economy of the system, but it neglects essential details of the doctrine. In particular, nothing in the diagram suggests a rupture

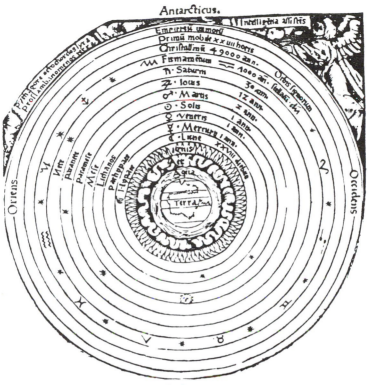

Figure 3. A typical sixteenth-century illustration of the Aristotelian-Ptolemaic cosmology. From Johann Eck's edition of Aristotle's De caelo, *published in Augsburg in 1519.*

of continuity between the world of the planets and that of the stars, an immense void space breaking with the principle of plenitude.

The rupture followed from the immobilization of the sphere of fixed stars, whose daily rotation was transferred to the earth. Copernicus thus could dispense with the otherwise immense speed of the last heaven and with the mysterious *motus raptus* (forced motion) constraining the lower heavens to accomplish each day a complete revolution around the earth.

Copernicus accompanied this considerable simplification, already imagined by Celio Calcagnini (1479–1541), with another modification. As indicated clearly in the fourth postulate of the *Commentariolus*, the annual motion of the earth requires that the distance of the fixed stars be so great that their annual parallax is even smaller than that of the sun; that is to say, immeasurably small. The medieval geocentric world, in the estimate of al-Farghânî (Alfraganus, around 800 to 870) or al-Battani (Albategnius, around 850 to 929), was about 40,000

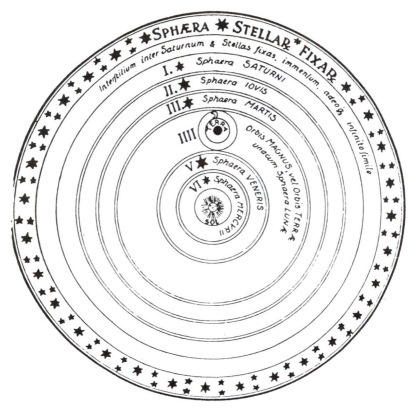

Figure 4. Copernicus's system of the world according to Michael Maestlin. From Maestlin's edition of Rheticus's Narratio prima, *published in Tübingen in 1596.*

earth radii in diameter (the inner diameter of the eighth sphere). The boundary of the heliocentric world, however, was pushed out to a quasi-infinite distance.

Michael Maestlin may have been the first to state that the space between Saturn and the sphere of the fixed stars was immense. In his 1596 diagram (figure 4), he says so explicitly, placing his words in the very gap between the sphere of Saturn and the sphere of the fixed stars. But because his diagram was not drawn to scale, it appears at first glance to differ little from Copernicus's.

Copernicus also breaks with the principle of plenitude with regard to the planetary spheres. For Ptolemy and for his Arabic and Latin popularizers as well, the planetary spheres, each defined by two envelopes concentric to the earth, were stacked up one around another, the apogee [*point in the planet's path farthest from the earth—the prefix* apo *indicating away from and* gee, *or* geo, *meaning earth*] of one planet coinciding with the perigee [*point in the planet's path closest to the earth—the prefix* peri *indicating near*] of the next higher planet. The physical principle forbidding unoccupied or void space in the interior of the universe had persuaded Ptolemy to place Mercury and Venus beneath the sun, and also permitted him to estimate the thickness of each sphere. Nothing, however, of this ancient principle survives in Copernicus. The moon, turning like an epicycle around our globe, no longer occupies its own sphere concentric to an immovable earth; rather, it now travels with the moving earth between the convex orb of Venus and the concave orb of Mars. Removed from the sequence of the planets, the moon now breaks the Ptolemaic sequence of contiguous heavens.

One of the great advantages of the heliocentric system is that it gives directly, without any "philosophical" hypotheses, the minimum and maximum distances of the planets in terms of the radius of the orb defined by the earth's motion about the sun. The resulting numerical values imply immense unoccupied spaces between the planets.

This conclusion is not evident in Copernicus's diagram in *De revolutionibus*, although he does leave a rather large interval between Venus and Mars, an interval that could not be entirely occupied by the terrestrial orb.

Copernicus gives only relative distances of the planets, without bothering to group them together in a general table. However, using Copernican parameters, modern calculations of the absolute distances tell us that:

- In placing the aphelion [*heli*, or *helio*, meaning sun] of Saturn at 11,073 terrestrial radii from the center of the great orb, Copernicus reduced the overall dimensions of the planetary system by nearly half!

- In assigning to the sphere of Saturn [the interval between its perihelion and its aphelion] a thickness of 1,192 terrestrial radii and to the sphere of Mars a thickness of 333 terrestrial radii (as compared to Alfraganus's parameters of 5,705 and 7,656 terrestrial radii, in the Ptolemaic cosmology), Copernicus also modified the internal structure of the system.

Contrary to what his diagram suggests, the planetary spheres were not contiguous for Copernicus. Furthermore, the intervals separating the spheres were greater than the thicknesses of the spheres themselves. In his *Compendiosa enarratio hypothesium Nicolai Copernici*, a treatise completed in 1594 but never published, J. Praetorius (1537–1616) points out this rupture of contiguity. In his diagram of the

Copernican system, he separates the spheres from one another by double lines and writes in the relative distances of the planets. The same separation of the spheres by double lines was introduced, no doubt intentionally, by Maestlin in his 1596 diagram (figure 4).

But it was left for Kepler to set forth the essential cosmological facts and to illustrate them in the most remarkable manner. In chapter 14 of his *Mysterium cosmographicum* of 1596, Kepler writes that in Copernicus "no orb is in contact with another, but there exist immense spaces between the diverse systems," and he illustrates it with a plate (figure 5) "showing the true size of the celestial orbs and the intervals [between them] according to Copernicus's values." This plate is exceptional for at least two reasons:

- It is the first known published representation of the scale of the complete planetary system (except for the interval between Saturn and the fixed stars, whose immense size cannot be shown to scale in the diagram, and which Kepler notes as "similar to the infinite").

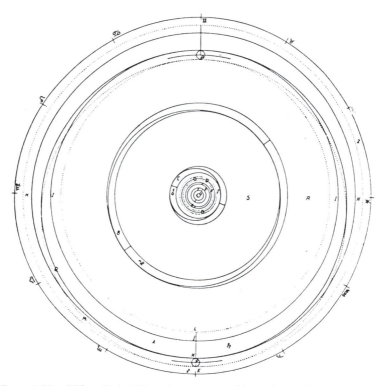

Figure 5. Plate IV from Kepler's Mysterium cosmographicum *showing the true proportions of the celestial orbs and the intervals between them according to Copernicus's numerical values. Published in 1596 in Tübingen.*

- It presents a universe in which the "void" greatly surpasses the "full" (in marked contrast to Copernicus's representation).

Postulating such enormous void spaces between the orbs logically leads to the question of what there is in the way of substance in these large intervals and, ultimately, to questions about the matter of the orbs themselves.

Copernicus seems to assume a classical structure for the heaven of the fixed stars (figures 1 and 2). He does not abandon the outer sphere and scatter the stars at different distances from the earth, though this was now a logical possibility, inasmuch as the stars no longer needed to be carried by an outer sphere around a nonrotating earth once every twenty-four hours. Nor does Copernicus, when presenting his catalog of stars, avail himself of this opportunity to state that the least bright stars seem small because they are farther from the center of the world than the more luminous and larger stars. Instead, Copernicus defines the starry heaven as the fixed reference frame for the planetary motions and the immovable place of the universe. Its concave envelope is a sphere of immense radius, infinite as far as our senses are concerned.

The only indication that Copernicus may have had doubts about the figure of the heaven, and perhaps even about the existence of the outer sphere, is his comment that one cannot know the limit of the starry heaven. More likely, though, he was referring to the distance of the sphere from the earth.

In Maestlin's diagram (figure 4), in contrast, the stars are placed at slightly unequal distances from the center. They are placed with greater variation in Thomas Digges's celebrated diagram of 1576 (page 172), which, for the first time, proclaimed an infinite universe.

Copernicus's Planetary Theory

Copernicus's astronomy offered two great advantages. First, it eliminated Ptolemy's large epicycle, replacing it with the orb of the earth's motion. Second, it gave the inferior planets (Mercury and Venus) and the superior planets (Mars, Jupiter and Saturn) the same cosmological status, thus eliminating the need (in the Ptolemaic system) to reverse the role of the deferent and the epicycle to reproduce the different behaviors of the inferior planets (which have limited elongations, or angular distances, from the sun) and the superior planets (which can have any elongation).

Another advantage, on which Copernicus and Rheticus were particularly insistent—whether through sincerity or diplomacy—was the suppression of the Ptolemaic equant, a violation of the principle of uniform circular motion. Thus Copernicus introduced new technical details into the models for the motions of the planets and the moon. Since all astronomers agreed that the moon revolves around the earth, a theory of its motion was independent of the cosmological choice between geocentrism and heliocentrism. So Copernicus had no need to modify Ptolemy's lunar model in order to adapt it to heliocentrism. But he did introduce changes, both to suppress the equant and to give a better accounting of the appearances.

In the fourteenth century Ibn al-Shatir and Levi ben Gershon had both remarked that the Ptolemaic model of the moon, while satisfactory for determining longitudes, implied large variations in the apparent diameter of the moon, variations much larger than were observed. To remedy this defect in Ptolemaic theory, al-Shatir constructed a new model based on epicycles, while Levi ben Gerson chose a system based purely on eccentrics.

Regiomontanus was probably the first in the West to point to this significant defect in Ptolemy's lunar model. In his 1496 *Epitome of Ptolemy's Almagest* (V, 22), Regiomontanus noted that according to Ptolemy's lunar theory, the moon should appear four times larger (in area) at quarter moon (if it were entirely illuminated) than at full moon.

Interestingly, the model used by al-Shatir to fix this problem is identical, other than in its numerical parameters, to the lunar model later adopted by Copernicus. We do not know for certain whether Copernicus knew of Ibn al-Shatir's model, but it seems likely that he did.

Copernicus's changes in the theories of the planets were less successful. For Mars, Jupiter, and Saturn, Copernicus eliminated the first and large epicycle (which the motion of the earth rendered superfluous), but then adopted a system of epicycles attached to an eccentric. For Venus and Mercury, he preferred eccentrics on eccentrics (his way of replacing the Ptolemaic equant-and-eccentric). Copernicus's system for the motion of Mercury resembles that for Venus, except in the addition of a small epicycle along the diameter about which the planet oscillates.

Here elimination of the equant was accompanied by the introduction of a rectilinear, to-and-fro motion in the cosmos, where only circular motion ought to reign. It seems, then, that Copernicus violated the principle of uniform circular motion in spite of his suppression of the equant. But he argued that the rectilinear to-and-fro motion along the diameter of this new little epicycle could be generated starting from two circular motions. This mechanism, called *libration*, was not invented by Copernicus; the demonstration he gives is identical to that devised by Nasir al-Din al-Tusi to account for variations in the distance of the lunar epicycle from the earth.

Copernicus's new combinations of circles devised to save the appearances offered no simplification for practitioners of astronomy—mainly astrologers who desired to cast horoscopes or produce almanacs with the least possible trouble. The use of the Copernican tables is just as complicated as that of their Ptolemaic counterparts.

Another weakness of *De revolutionibus* that must have discouraged many practitioners was the absence of tables of retrogradation. Ptolemy devotes seven long chapters in book XII of the *Almagest* to this problem, but Copernicus offers only two very short chapters in book V of *De revolutionibus*. It is not a matter of Copernicus profiting from the clarification of the phenomena of the stations and retrogradations of the planets following from the motion of the earth, and thus being able to give a new and simple technique, based directly on his theory of longitudes, for calculating the apparent retrograde motions. Far from it. Worse yet for his reputation as a mathematician, Copernicus vainly tries to produce a theory of the retrogradations following Ptolemy, going back to the lemmas and the theorem of Apollonius and trying to apply them to the case of a terrestrial observer now in motion. (In doing so, he stumbles onto a system of the type later advocated by Tycho Brahe.) Once Copernicus generalizes the problem, his demonstration becomes very abbreviated. The last chapter of book V concludes with the following paragraph:

> *Nevertheless, since the variable motion of the planet with respect to the place of observation involves a considerable difficulty and a doubt* [!] *on the subject of the stations, and since the theorem of Apollonius does not relieve us of this, I wonder whether it would not be better to investigate the stations simply, starting from the closest position of the planet to the earth, just as we investigate the conjunction of a planet in opposition by starting from the known values of their movements with respect to the line of the mean motion of the sun and by combining them. But we leave it to each who wishes to examine this question.*

Copernicus was no more able than Ptolemy to construct a theory of the stations and retrogradations that depended directly on the theory of longitudes. Comparison of the above quotation from the first edition of *De revolutionibus* with comparable sections in the autograph manuscript and in Rheticus's *Narratio prima* suggest that this sudden turn at the end of book V, which must have been disappointing for the astronomers among its readers, was due to the intervention of the young Rheticus. The intervention came too late to enable Copernicus before his death to develop a practical solution for the convenient calculation of stations and retrogradations.

The passage is also interesting for illuminating the true obstacle to a correct utilization of heliocentrism in this particular case: the absence of essential kinematic concepts. Among astronomers, the concept of speed was poorly defined and that of acceleration totally absent.

Initial Reception of Copernicus's Cosmology

In his unsigned and unauthorized foreword, "To the reader concerning the hypotheses of this work," the Protestant theologian Andreas Osiander expressed an opinion inevitably taken by many readers of *De revolutionibus* as Copernicus's own. With words similar to those he had spoken two years earlier, in April 1541, when suggesting to Rheticus and Copernicus that they preface their new heliocentric doctrine with a notice designed to appease Aristotelians and theologians, Osiander now characterized Copernicus's heliocentric theory as a convenient mathematical hypothesis not pretending to describe the real motions of the system of the world.

Copernicus does seem to have feared hostile reactions, but neither he nor Rheticus (whose *Narratio prima* appeared for the first time, although without its author's name, in 1540) chose to follow Osiander's suggestion. Indeed, it is clear that Copernicus and his friends intended to present the heliocentric cosmology at face value, and definitely not as a mere mathematical hypothesis.

To deal directly with theologians, Protestant as well as Catholic, whose objections could be foreseen, two texts had been prepared. The first, titled *Hyperaspistes* and due to Bishop Tiedemann Giese, an old friend and colleague of Copernicus, was to defend the orthodoxy of the new doctrine in regard to the Bible. Unfortunately, only the title of this work has been preserved. The second text, long believed lost, was discovered and published by R. Hooykaas in 1984. It is nothing less than a defense, by Rheticus himself, of the compatibility of heliocentrism with sacred Scripture.

Copernicus chose a different tactic. He wrote boldly in his dedication of *De revolutionibus* to Pope Paul III that he considered the inevitable attacks of the *mataiologoi* (blabbermouths) a fraudulent use of Scripture in a domain that had nothing to do with it.

Copernicus was not mistaken in foreseeing that a hard kernel of opposition would be formed by theologians (who would also use the physics of Aristotle to support their argument) and by philosophers (who would also invoke Scripture). The early Protestant rejection by Luther, Melanchthon, and perhaps also Calvin, is well known; the absurd doctrine of the motion of the earth was rejected in the triple name of good sense, received philosophy, and Revelation. To these famous reactions can be added criticisms from some Catholic theologians. By 1546, a year after the opening of the Council of Trent, the Dominican G.M. Tolosani had completed the composition of a refutation of heliocentrism based on arguments simultaneously physical (Aristotle and his commentator St. Thomas Aquinas) and scriptural. The project had been begun by B. Spina, the Master of the Sacred Palace, prior to his death in 1546. Thus even during the lifetime of Pope Paul III, an important figure in the Roman church wished to assert himself against the "heterodox" system of Copernicus.

Writing that "mathematics are written for mathematicians," Copernicus asked professional astronomers to judge his book. Few of them held unshakably to the Ptolemaic-Alphonsine system, but just as few of them before 1600 accepted Copernican heliocentrism with all its cosmological consequences. For many astronomers a middle way was possible. *De revolutionibus* is a complex work, full of teachings which need not necessarily be all accepted or all rejected *en bloc*. Astronomers could retain what they judged positive in Copernicus and useful for the "restoration" of astronomy (such as Copernicus's rehabilitation of the axiom of uniform circular movement and his new planetary models) while refusing to modify the cosmological status of the earth. That the earth is at rest in the center of the universe was for many astronomers a truth of physics confirmed by the authority of Scripture—a demonstrated verity concerning which astronomical science had no authority to make pronouncements and, consequently, no proofs to oppose to it.

Erasmus Reinhold accepted part, but by no means all, of Copernican astronomy. He composed and calculated the *Tabulae Prutenicae*, published in 1551, based on the planetary theories of *De revolutionibus*. The *Prutenic Tables* rapidly dethroned the famous *Alphonsine Tables* based on Ptolemaic models, although, as we know today, their precision was not greatly superior. A member of Melanchthon's circle, Reinhold paradoxically helped to confirm Copernicus's high reputation among both practitioners of astronomy and non-specialists. Annotations in Reinhold's personal copy of *De revolutionibus* show that, like P. Wittich in the 1570s, he was much more interested in details of the geometric constructions of books III to VI than in the cosmological chapters of book I. Moreover, though he never published them, we know that Reinhold elaborated "new hypotheses" for astronomy differing as much from Copernicus's as from Ptolemy's and that he tended toward the "geo-heliocentric" compromise, which was to seduce other astronomers as well. Thus Reinhold was by no means a heliocentrist.

Although his *Prutenic Tables* contributed greatly to the diffusion of the Copernican system, Reinhold's contemporaries must nevertheless have been struck by his eloquent silence on the foundations of heliocentrism. Some may even have seen here a renewed invitation to take Copernicus's fundamental hypothesis for a mathematical fiction.

One astronomer who proclaimed loudly and strongly his cosmological convictions—even to the point of drowning out the voices of some of his competitors—was Tycho Brahe. As early as 1574, while acknowledging Copernicus's rejection of the Ptolemaic equant, Brahe also criticized Copernicus for having sinned against physical principles. The alleged sins were:

- Placing the sun, motionless, at the center of the universe.
- Making the earth move with its elements and the moon in a triple motion around the sun.
- Immobilizing the eighth sphere.

Three years later Brahe envisaged a circumsolar orbit for the comet of 1577, with Mercury and Venus moving around the sun in a scheme that recalls Martianus Capella's, from the late third or early fourth century. And in 1588 Brahe published the outlines of his own system of the world, whose distinguishing characteristic (compared with other geo-heliocentric systems conceived around the same time) is that the orbit of Mars cuts that of the sun in two places.

Brahe rejected Ptolemaic astronomy because the mathematical advantages of the Copernican hypothesis saved "the first principles of the art." Furthermore, the Copernican hypothesis banished from the sky numerous and disproportionate epicycles now recognized as superfluous, but which had occupied an immense space in Ptolemy's cosmology. In his models for the librations, however, Copernicus introduced against the same "first principles of the art" rectilinear movements in the heaven. Also, Brahe maintained physical objections against Copernican astronomy based on:

- The concept of natural place.
- The concept of simple and composite bodies.
- The concept of natural, violent, rectilinear, and circular motions.
- The effect that the rotation of the earth would have on bodies in free fall.
- The effect that the rotation of the earth would have on projectiles fired in opposite directions.

Finally, Brahe did not hesitate to invoke arguments drawn from the Bible that conferred the status of dogma on the immobility of the earth.

With astronomical instruments of unmatched precision, Brahe employed systematic observations to criticize numerous Copernican parameters and also major cosmological implications of heliocentrism. In particular, he believed that he had found solid arguments against the annual motion of the earth and against the enormous dimensions of the world implied by the motion of the earth.

Misled, as were all of his predecessors, by the diffusion of stellar images on the human retina, Brahe assigned apparent diameters of two minutes of arc to stars of the first magnitude (even three minutes to some of them) and a bit more than one minute of arc to third magnitude stars. A minute of arc was also Brahe's estimate for the upper limit of the annual parallax of the stars. This estimate followed from the fact that Brahe could not measure any perspective displacement of the stars over intervals of six months, during which period the earth ought to be at two diametrically opposed points of the great Copernican orb. He concluded that if the apparent diameter of a third magnitude star were greater than the apparent diameter of the terrestrial orbit as viewed from that star, then the actual diameter of the star must be greater than the actual diameter of the terrestrial orbit (2,284 terrestrial radii!).

Such a universe, in which the dimensions of the objects were greater than the distances that separated them, seemed most implausible to Brahe. And what to make of the 7,850,000 terrestrial radii that, in the heliocentric world, would have to separate the stars from the center of the world? That would amount to postulating that the distance between Saturn and the sphere of the fixed stars is equal to 700 times the distance between the sun and Saturn, an absolutely immense space which would be completely void of any body. Such consequences drew from Brahe but a single response, expressed in a single word: *absurd!*

Neither C. Rothmann, *mathematicus* to Wilhelm IV, the Landgrave of Hesse, nor Michael Maestlin, nor Kepler—three astronomers whose reading of *Narratio prima* and *De revolutionibus* had made them declared Copernicans before 1600—were impressed by Brahe's arguments. They rejected the supposed authority of Scripture in the domain of astronomical argument. Brahe's other arguments might have been expected to carry more weight with them. Astronomical observations conducted by the most competent of specialists had brought no confirmation of the annual motion of the earth. Furthermore, only the reigning physics possessed a theoretical apparatus capable of saving the phenomena of daily terrestrial experience. In the face of this physics, the few pages of *De revolutionibus* in which Copernicus tried to get the better of Aristotle possessed no great weight. Yet the combined authority of Brahe and Aristotle failed to disturb Rothmann, Maestlin, and Kepler, who gave absolute precedence to the architectonic element in their appraisals of planetary systems.

In the *Narratio prima*, carefully read by the first Copernicans, Rheticus numbered among the advantages of the new system the fact that "each orb moves uniformly with the movement assigned to it by nature," and that "the orbs of larger size complete their revolution more slowly, as is proper, while those that complete them most rapidly are the orbs nearest the sun, which may be said to be the principle of motion and of light." These last words were of the sort to seduce a Kepler, but they have no counterpart in *De revolutionibus*. Rheticus also tells us that it was "the rule which requires that the order and motions of the celestial orbs constitute an absolutely perfect system" that had, above all else, determined that his master be in favor of heliocentrism.

By all the evidence, this is also the consideration that made Copernicans of Rothmann, of Maestlin, and of Maestlin's brilliant disciple, Kepler. In the preface that accompanies the 1596 re-edition of Rheticus's *Narratio prima*, Maestlin explains with great clarity the reasons why, according to him, one is lead necessarily to become a Copernican. He recalls the defects of the usual hypotheses:

- The uncertain number of the spheres.
- The doubtful order of the planets.
- The dubious distances.
- The improbable speed of the daily revolution, which must be attributed to everything in the heaven except the minuscule earth.

And Maestlin exposes anew the uncertainty in the "physical" argumentation of Aristotle and underscores its major defect: in reasoning from the part (the center of the earth) to the whole (the center of the universe), Aristotle can do no better than to propose the accidental coincidence of these two centers in the world. In contrast, the "astronomical reasons" of Copernicus, which proceed from the whole to the parts, are the only ones that permit us to establish which bodies in the world are at rest and which bodies are in movement, as well as the order and the periods with which the bodies move.

Once one understood that in the Copernican system any uncertainty over the disposition and order of the spheres was impossible, one could not remain Ptolemaic—an argument later used by Galileo—nor become Tychonic. Maestlin, who made use of the freedom (denied to Kepler) to criticize Brahe publicly, denounced the "correction of hypotheses" attempted by Brahe. In the Tychonic system the centers of motion and the motive forces are separated and dissociated, and "nothing is united to anything by rapport or any proportion whatever among the sizes, the motions, and the order." According to Maestlin, no astronomer sensitive to the harmony that ought to reign among the mobile elements of the world could subscribe to such a patchwork, especially since the eminent mathematician Kepler had discovered a new and very strong reason for supporting heliocentrism. Instead of proceeding backwards like the other astronomers (Copernicus included), that is, instead of formulating hypotheses concerning the motions, sizes, and distances by starting from observations alone, Kepler in his *Mysterium cosmographicum* had taken the road of the "a priori." In other words, he had first of all assigned to the world a certain geometrical structure, which would correspond to the plan that God had followed at the time of the Creation, and then he had deduced from it the values and the dimensions to which the observations must be compared. Thus, thanks to his "invention," Kepler had demonstrably established the solid base of the Copernican order of the world, and the door was finally open to an authentic restoration of astronomy.

Perhaps the most difficult question posed by Copernicanism was whether the heliocentric world was finite or infinite. Was it unique, plural, or even indefinitely multiplied? These questions fall outside the domain of competence of the science of celestial motions as such, as Copernicus himself recognized when he abandoned to the philosophers of nature speculations on the finiteness or infinity of the world. Consequently, responses of Copernicans on this point depended less on their understanding of *De revolutionibus* than on their own metaphysical or theological convictions concerning the relation between God and the world. Among heliocentrists asserting the finite size of the world—though still accepting its considerable enlargement—were Maestlin, Rothmann, and Kepler. The immense space, void of all bodies, that they accepted between Saturn and the sphere of the fixed stars seemed to them reconcilable both with the omnipotence of God the creator and with the perfection of the divine work. Despite the new disproportion among its parts, the world remained in their eyes a cosmos.

With Thomas Digges the heliocentric world was for the first time publicly declared infinite. The heaven of the fixed stars is defined in Digges's *A perfit description of the Cælestiall Orbes, according to*

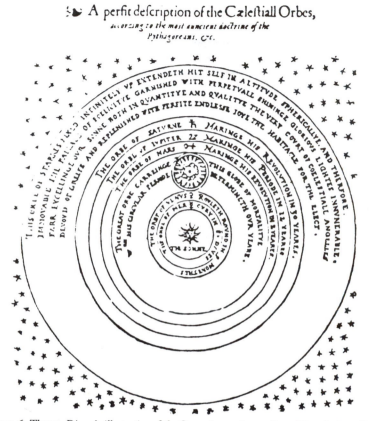

Figure 6. Thomas Digges's illustration of the Copernican universe. From Digges's A perfit description of the Cælestiall Orbes, *according to the most auncient doctrine of the Pythagoreans revived by Copernicus,* published in London in 1576.

the most auncient doctrine of the Pythagoreans revived by Copernicus as extending spherically to infinity beyond its concave surface (figure 6). Digges's diagram, however, does not represent a simple, logical extension of Copernican doctrine. His heaven of the fixed stars is also, and less clearly, a kind of "theological" heaven fused with the traditional empyrean, since the stars are placed there among the angels and God's elect. For Digges, the world was infinite but still unique.

With the philosopher Giordano Bruno, all was changed. The multiplication of worlds, with each star becoming a sun circled by its own planets; the homogenization of the celestial matter, with the stars no longer differing from the earth in their elemental composition; the infinitization of the universe, furnished with innumerable worlds, none of which can claim a fixed center—all this and many other features of Bruno's cosmology take us very far from the unique world of Copernicus, no matter how greatly enlarged. We are, indeed, so far from our starting point that neither Kepler nor, later, Galileo could see the infinity of the universe and the plurality of worlds (which Bruno, for good measure, imagined as inhabited) as necessary or even probable consequences of the Copernican hypothesis. However, with these bold cosmological speculations (which, by his own admission, would have been impossible without the "divine" inspiration that pushed Copernicus to put the earth into motion around the sun), Bruno incontestably developed an aspect of heliocentrism that was latent in the new world view. And thus, for us moderns, Bruno is an emblematic figure of the "Copernican revolution." In the year 1600 Bruno was burned for his heresies, which were many. Copernicanism was by no means his only heterodox opinion; nor was it his chief crime, although it certainly did not help his case.

The Copernican doctrine was officially condemned as heretical by the Roman Church in 1616. This event marks a new part of the story, presented elsewhere in this book, in chapters on Galileo and on Galileo and the Inquisition.

—*Michel-Pierre Lerner and Jean-Pierre Verdet*
Translated by James Evans

Bibliographic Note

For Copernicus's main work, the best English translation (based on the Nuremberg edition of 1543) is A.M. Duncan, *On the Revolutions of the Heavenly Spheres* (London: Newton Abbot, 1976). A very useful companion to Duncan's translation is N.M. Swerdlow and O. Neugebauer, *Mathematical Astronomy in Copernicus's De revolutionibus* (Berlin: Springer Verlag, 1984). This book offers the most thorough analysis of the new planetary theory, a biography of Copernicus, and a general presentation of his astronomy. Rheticus's *Narratio prima*, a helpful pedagogic presentation of the heliocentric system, is available in a French translation, with extended notes and other documentary pieces, in H. Hugonnard-Roche and J.-P. Verdet's edition, published by the Polish National Academy of Science in 1982 as *volume 20* of *Studia Copernicana*. We are not completely satisfied with the English translation in Edward Rosen, *Three Copernican Treatises. The* Commentariolus *of Copernicus. The* Letter against Werner. *The* Narratio prima *of Rheticus. Second edition, revised, with an Annotated Copernicus Bibliography, 1939–1958* (New York: Dover Publications, 1959). T. Kuhn, *The Copernican Revolution. Planetary Astronomy in the Development of Western Thought* (Cambridge, Massachusetts: Harvard University Press, 1957); A. Koyré, *From the Closed World to the Infinite Universe* (Baltimore: Johns Hopkins University Press, 1957); and *The Astronomical Revolution: Copernicus, Kepler, Borelli*, translated by R.E.W. Madison (Ithaca: Cornell University Press, 1973) provide convenient introductions to Copernican cosmology.

III. Medieval Cosmology and Literature

Introduction
The Aristotelian World View

Introduction

THE ARISTOTELIAN WORLD VIEW

osmology, culture, and civilization have always been connected, in all times and in all places, in greater or lesser degree. Perhaps never before nor since in human history, though, has a cosmological vision so pervasively penetrated and suffused culture and civilization as did the Aristotelian world view.

Aristotle's world view was a dominating theme of ancient Greek cosmology, of late medieval thought in Western Europe, and of the early Renaissance as well. For nearly two millennia, albeit with interruptions, Aristotle's cosmology was the cosmology of a large part of the Western intellectual world. During the late Middle Ages it was the cosmology of Dante and the cosmology of Chaucer. It remains an integral and important part of our intellectual heritage, our literature, our art, our philosophy, and our very language and way of thinking.

A major strength of Aristotle's world view was that it furnished a complete physical description of the universe and the forces acting within the universe. It was an intellectual system in which every aspect followed logically from the rest. Aristotle's physics and cosmology were closely interwoven; no part could be separated from the rest; no part could be changed without changing the whole.

Conversely, even the slightest flaw in Aristotle's physics could not be considered in isolation, but was necessarily a potentially fatal flaw in the entire intellectual edifice. Though dominant for two millennia, Aristotle's cosmology, the cosmology of the medieval world, would by the end of the seventeenth century be rendered totally ineffective through the work of Copernicus, Galileo, Tycho Brahe, Kepler, and Newton. Never before, nor since, has so pervasive a world view been so thoroughly overturned and replaced in all aspects of human thought, in the physics of poets as well as the poetry of physics.

Aristotle, himself, and the development of his world view can be placed in a cultural context. Born some forty-three years after Plato, in 384 B.C., Aristotle experienced much more favorable and pleasant encounters with the world of the senses than had Plato. Plato's philosophy can perhaps be understood as a response to his environment, as a reaction to the temporary moral values of his age, which left Plato highly dissatisfied and, consequently, searching for a new philosophy. Thus the change in Greek science begun by Plato can be placed, and perhaps better understood, in the context of changes in Greek society. Similarly, Aristotle's very different outlook on the world also may be interpreted as a response to his environment, though admittedly with some difficulty, since Aristotle was a member of Plato's Academy for twenty years, from about 367 B.C. when Aristotle entered the Academy as a youth of 17 until Plato's death in 347 B.C. From Athens, Aristotle went to Asia Minor, where he founded a new academy under the patronage of a local ruler and married the ruler's 18-year-old niece. From Aristotle's later description of the ideal age for marrying as 37 years for the man and 18 for the woman, we may surmise that he found at least this part of the world of the senses more hospitable than Plato had found the world of the senses.

In any event, setting aside such speculations on the causes of a philosopher's trend of thought, it is a fact that Aristotle proposed a physical approach to the phenomena of nature in marked contrast with Plato's emphasis on the world of thought. Astronomers and physicists, working in what quickly became the Aristotelian tradition, attempted to develop dynamic models of the universe, seeking in addition to a description the cause of the motions of the planets and stars.

Nor was the immediate world of the senses neglected. Aristotle's physics encompassed all motion, on the earth as well as in the heavens. Yet if Aristotelian physics ruled the entire universe, that cosmos was clearly divided into two, into what has since been characterized as a two-sphere universe, each part composed of different types of matter, each with its separate set of physical laws. The region of change, of corruption, extended up to the moon, beyond which were the heavenly bodies in circular motion in a realm without change.

As remarkable and all-encompassing as was Aristotle's world view, it did not descend directly through Greek, Roman, Medieval, and Renaissance times without various vicissitudes. Greek science almost disappeared from Western Europe from approximately A.D. 500 to 1100. What little cosmology was known then in the West was due primarily to a partial Latin translation of Plato's *Timaeus*. During the late Middle Ages, however, a "new" cosmology entered Western

Europe, primarily from Spain between 1150 and approximately 1280, in translations of the works of Aristotle and Ptolemy and in Arabic treatises and commentaries on Aristotle.

Using this inheritance as a foundation and frequently building upon their own ideas and interpretations, scholastic authors produced a large body of cosmological literature near the end of the late Middle Ages, between approximately A.D. 1250 and 1500. In these texts we can come to understand how natural philosophers viewed the operation and structure of the cosmos in terms of the science, metaphysics, and theology of their day.

Philosophers and scientists were scarcely the sole adherents of the Aristotelian world view. Poets, too, reacted to this all-pervasive aspect of their culture and reflected it in their poems. Few better representations and interpretations of the late Medieval period are to be found than in the writings of Dante and Chaucer.

Both Dante and Chaucer were active in affairs of their times. Dante served as one of the Priors of Florence, ambassador first to San Gimignano and later to the Pope in Rome, before a political revolution ended with his permanent banishment upon pain of death. Chaucer, after his marriage to Philippa, sister to John of Gaunt's third wife, undertook various commissions on behalf of the king. Some of his missions took him to France and Italy, where he became acquainted with Dante's writings and may well have met Petrarch and Boccaccio.

Both Dante and Chaucer also were well versed in Aristotelian astronomical and cosmological matters, as indeed were nearly all educated persons of this era. It is hardly a distinguishing feature of a man of the fourteenth century that he followed the cosmology of Aristotle. Aristotelian physics and metaphysics shaped Dante's poetry to a very great extent and influenced Chaucer's, which also made liberal use of astrology and the technicalities of planetary astronomy.

There was then no clear boundary between cosmology and the twin sciences of planetary astronomy and astrology. That neither subject would today be considered truly cosmological is a consequence of our changing perspective on the physical extent of the planetary system and on the interrelatedness of that system and human affairs. That few poets today are as well versed in cosmology or employ it as extensively in their work as did Dante and Chaucer is a consequence of our changing culture, particularly a weakening of the interrelatedness perceived to hold between science and human affairs.

9. MEDIEVAL COSMOLOGY

Numerous "cosmologies," written and unwritten, were conceived during the Middle Ages. Although their ideas were largely unrecorded, peasants and uneducated people generally must have held some views, however primitive and crude, about the structure and operation of the world. Among the literate, authors and poets such as Dante and Chaucer [see the following two essays] were free to envision the physical world in accordance with the dictates of their literary needs. Even occasional authors who wrote on science and natural philosophy—for example, Hildegarde of Bingen (1098–1179) and Robert Grosseteste (ca. 1168–1253)—devised interesting, albeit idiosyncratic, cosmologies that were generally without influence. Whatever the intrinsic interest and significance of these cosmologies, they are not our major concern here.

The cosmology we will discuss here is that which was taught in the medieval universities between 1250 and 1600. It was ultimately based on certain works of the Greek philosopher Aristotle (ca. 384–322 B.C.)—primarily *On the Heavens*, *Physics*, and *Metaphysics*—and, to a lesser extent, on the works of the Greek astronomer, Ptolemy (ca. 100–ca. 170), namely his astronomical works *Almagest* and *Hypotheses of the Planets*, or rather material from the *Hypotheses* in a work by the Arabic scientist Alhazen (Ibn al-Haytham; 965–ca. 1040), and his astrological work *Tetrabiblos*.

During the early Middle Ages, from approximately 500 to 1100 or 1150, the relevant works of Aristotle and Ptolemy were unknown in Western Europe, where Greek science had virtually disappeared. The cosmology that was available was based on Chalcidius's partial Latin translation of the *Timaeus* of Plato and on a series of encyclopedic treatises associated with the names of Pliny the Elder (ca. A.D. 23–79), Seneca (4 B.C.–A.D. 65), Macrobius (fl. fifth century A.D.), Martianus Capella (ca. 365–440), Cassiodorus (ca. 480–ca. 575), Isidore of Seville (ca. 560–636), and Venerable Bede (672–735). Even if to these we add certain works of the Church Fathers, especially those of St. Basil (ca.

330–379) and St. Augustine (354–430), the overall cosmology embedded in all of these treatises was meager and insubstantial by comparison with what was to come during the late Middle Ages, from approximately 1150 or 1200 to 1500.

Sources of Medieval Cosmology

In the present context, the term *sources* is used in two distinct ways. The first concerns the inherited literature that profoundly shaped the content of late medieval cosmology. The "new" cosmology that entered Western Europe (primarily from Spain) in the late Middle Ages between 1150 and approximately 1280 did so by way of Latin translations from Arabic and, to a much lesser extent, Greek. Not only did the new cosmology include translations from the works of Aristotle and Ptolemy mentioned above, but also from numerous Arabic treatises and commentaries associated with such names as Averroes (Ibn Rushd, 1126–1198), Avicenna (Ibn Sina, 980–1037), Alhazen (Ibn al-Haytham, 965–ca. 1039), al-Farabi (ca. 870–950), Thābit ibn Qurrah (836–901), and the Jewish author, Maimonides (1135–1204). Although Greco-Arabic scientific literature formed the most significant component in the development of late medieval cosmology, there was also the commentary literature of the Church Fathers on the six days of creation (known as hexaemeral literature) and the small cluster of Latin works that came from late antiquity and the early Middle Ages.

The second way in which the term *sources* is used concerns the body of writings that late medieval natural philosophers fashioned. Using their inheritance as a foundation and frequently building upon their own ideas and interpretations, scholastic authors produced a large body of cosmological literature during the late Middle Ages and Renaissance, that is, between approximately 1250 and 1700 (although we shall confine ourselves here to the late Middle Ages, the upper limit of which may be conveniently taken as 1500). Only by a study of these texts can we come to understand how natural philosophers viewed the operation and structure of the cosmos in terms of the science, the metaphysics, and the theology of their day.

Questiones:
The Basic Form of Medieval Cosmological Literature

What were medieval texts like? For cosmology, they were primarily secular in content, but with a significant intermixture of Christian theological ideas, drawn largely from the creation account in Genesis. In both categories, the texts were overwhelmingly commentaries and questions (*questiones*) on relevant authoritative texts. In a commentary, a section of text was cited and then explained with occasional elaborations by the commentator; upon completion, the next section or unit of text was expounded, and so on. By the fourteenth century, this procedure was largely superseded by the question (*questio*) method. During the Middle Ages and even into the seventeenth century, scholastic natural philosophers centered their teaching and analysis of authoritative texts around the *questio*. Natural philosophy, which constituted the core of the curriculum and of which cosmology was a major part, was thus taught by the analysis of a series of questions posed by a master and eventually determined—that is, resolved—by him. Collections of these *questiones* on the different works of Aristotle, and other works as well, were written down by the masters and published under university auspices.

Although occasional variations in the arrangement of the constituent elements of the *questio* occurred, scholastics tended to present their arguments in a rather standard format that remained remarkably constant over the centuries.

- First came the enunciation of the problem or question, usually beginning with a phrase such as "let us inquire whether" or simply "whether" (*utrum*). For example, "whether the earth is spherical," or "whether the earth moves," or "whether it is possible that several worlds exist."

- This was followed by one or more—often five or six—solutions supporting either the negative or affirmative position. If arguments for the affirmative side appeared first, the reader could confidently assume that the author would ultimately adopt the negative position; or conversely, if the negative side appeared first, it could be assumed that the author would subsequently adopt and defend the affirmative side. The initial opinions, which would ultimately be rejected, were called the "principal arguments" (*rationes principales*).

- Immediately following the description of the principal arguments, the author might further clarify and qualify his understanding of the question or explain particular terms in it.

- With the necessary qualifications completed, the author was ready to present his own opinions, usually by way of one or more detailed conclusions or propositions.

- In order to anticipate objections, the master might choose to raise doubts about his own conclusions and subsequently resolve them.

- To conclude the question, he would respond sequentially to each of the "principal arguments" enunciated at the outset.

Over the centuries, hundreds of questions—perhaps as many as 400 or 450—formed the basis of medieval cosmology. From whence did they come? From the secular side, the most important for cosmology was Aristotle's treatise *On the Heavens* (known in Latin as *De caelo*). It was probably the only specifically cosmological treatise available in the Middle Ages. The numerous treatises titled *Questions on De caelo* are therefore our most important source for medieval views on the cosmos.

Although *De caelo* was Aristotle's primary contribution to cosmology, we have already mentioned that several of his other works—*Physics*, *Metaphysics*, and the *Meteorology*—included topics relevant to the subject. Thus in his *Physics* Aristotle discussed the concepts of vacuum and place in the fourth book, continuity and contiguity in the fifth book, and the prime mover in the eighth book. And in the twelfth book of the *Metaphysics* he discussed the number of celestial spheres and the substances that move them. While the *Meteorologica* was essentially "a study of the combinations and mutual influences of the four elements," it has some bearing on cosmology because Aristotle considered its subject matter to be natural phenomena that "take place in the region nearest to the motion of the stars," such as "the milky way, and comets, and the movements of meteors."

Sections and topics in other secular treatises also provided occasions for cosmological discussions. One of the most significant was the *Treatise on the Sphere* (*Tractatus de sphaera*) by John of Sacrobosco (died 1244 or 1256). In the first of four chapters he included brief descriptions of heaven and earth. Because the treatise was a university undergraduate text for some four centuries, it was widely known. It elicited a small number of commentaries, of which those by Michael Scot (died ca. 1235) and Robertus Anglicus (fl. ca. 1271) in the thirteenth century, Pierre d'Ailly (1350–1420) in the fourteenth century, and Christopher Clavius (1537–1612) in the late sixteenth century, are noteworthy. Medieval encyclopedias by Vincent of Beauvais (ca. 1190–ca. 1264), Bartholomew the Englishman (fl. 1220–1250), and others, also contained cosmological discussions.

The theological side of medieval cosmology derives from the large genre of theological commentaries on the *Sentences* of Peter Lombard (died ca. 1160), which served as the basic textbook in theology for some five centuries. In the second of its four books, theological students were confronted with problems about the creation. It was largely from these questions that a Christian component was introduced into medieval cosmology.

Ironically, technical astronomical treatises included relatively little on cosmology. Only occasionally did they find reason to discuss cosmology. This is perhaps attributable to a long-standing division of labor between astronomers and natural philosophers. The task of the former was to predict and determine the positions of the planets and stars. To achieve this, they were expected to use geometry and arithmetic and to invoke a variety of mechanisms, some real and some imaginary. By contrast, natural philosophers, or more specifically cosmologists, were little interested in the positions of the planets, but sought rather to describe the nature of the heavens and the causes of its various motions. Their task was to explain the nature of the celestial substance, that is, to determine whether it is incorruptible and indivisible, whether it is equally perfect throughout its extent or differentially so, and whether its properties are similar to matter in the terrestrial region, and so on.

Outline of Medieval Cosmology: The World as a Whole

In his *Questions on De caelo*, Albert of Saxony (ca. 1316–1390) was one of the few, and perhaps the only one, who made explicit the fundamental divisions of cosmology that Aristotle had only implied in his *De caelo*. Albert discerned a tripartite division wherein the first part treated topics concerned with the world as a whole. The second and third parts were based on Aristotle's division of the world into two rigidly distinct domains:

- A celestial region of stars, planets, and orbs that was composed of a special ether, or fifth element, possessed of extraordinary qualities.
- A radically different, sublunar realm immediately below the celestial region, ranging from the concave surface of the lunar orb to the center of the earth and embracing the four elements—fire, air, water, and earth—and the bodies compounded of them.

Of Albert's tripartite division of cosmology, which makes eminent sense within the context of Aristotelian natural philosophy, we shall only consider the first two parts, namely the world as a whole and the celestial region, and thus ignore the arrangements and properties of the four elements in the sublunar region. However, Albert did treat the earth as relevant to the celestial region in so far as it functioned as its center. In this we shall follow Albert.

Did the World Have a Beginning?

With the introduction of the natural books of Aristotle in the thirteenth century, medieval scholastics confronted a serious dilemma. From Holy Scripture (Genesis 1.1-2.1, John 1.2–3 and 17.5) they learned that the world was created and brought into being by supernatural means and by that same method would eventually suffer destruction and pass away. From Aristotle's *De caelo* (*On the Heavens*) came a powerful contrary message: the world could not have had a beginning and could never come to an end. Left as stated, Aristotle's strong and reasoned conviction that the world was eternal, without beginning or end, was in direct conflict with one of the most powerful messages of Christianity: the divine creation of our unique world, which was destined to endure for only a finite time. Indeed, although no explicit statement in Jewish, Christian, or Muslim Scripture declares that creation was out of nothing (*ex nihilo*), the latter opinion became commonplace in medieval Latin Christendom.

During the thirteenth century, the problem of the eternity of the world was widely debated. In the well-known Condemnation of 1277, when 219 propositions were condemned by the Bishop of Paris, 27 of them involved the eternity of the world. Some, like St. Bonaventure (1221–1274), were convinced that the temporal creation of the world was demonstrable, and they sought to reduce the concept of an eternal world to absurdity, largely by showing the impossibility of infinites differing with respect to size. But others demonstrated that the concept of an eternally existent entity was not self-contradictory and that it was perfectly intelligible for one infinite to be larger or smaller than another. Yet others argued that infinites were not comparable with respect to "equal," "greater than," or "less than."

There was a considerable body of scholastic opinion that upheld the possibility that not only was the world created but that it might also be eternal. In medieval parlance, they held with Thomas Aquinas (ca. 1225–1274) that the possibility of an eternal world was a *problema neutrum*, whereby logical argument and evidence favored neither a temporal beginning of the world nor an absence of a beginning. At least one scholastic went even further, perhaps as far as a Christian could go: Marsilius of Inghen (died 1396) argued for the greater probability of an eternal world. Nevertheless, however much reason may have led Marsilius and his fellow Christian scholastics to consider the plausibility, and even probability, of a beginningless, eternal world, they believed, as a matter of faith, that the world was created supernaturally from nothing.

Will the World End?

Aristotle had also argued that if the world had no beginning, it could, of necessity, have no end and was therefore indestructible. Or, to put it another way, something created anew could not thereafter endure forever. Many denied Aristotle's claim and argued that since God created the world, He also possessed the power to confer upon it the ability to endure forever. Scholastics were thus prepared to break with Aristotle and hold that a created world—that is something that had a beginning—could indeed exist through an eternal future. But this opinion had to be reconciled with the Christian concept of a Day of Judgment. On that day, would God substantially alter the world as we know it so that we might rightly say that it will cease to exist? Or would the world of Judgment Day differ only accidentally from the present world so that we could rightly say that the world God created will endure forever through an eternal future? Although most would opt for the first opinion, there were those who chose the second.

Creation, Finitude, Shape, and Perfection of the World

Following Saint Augustine, most scholars during the Middle Ages assumed that the world was created simultaneously rather than in six natural days. All things were created in the order described in Genesis but in an instant, so quickly that "before" and "after" are indistinguishable. Augustine explains, however (in *The Literal Meaning of Genesis*, book 4, chapter 33), that God chose to narrate the creation day by day because "those who cannot understand the meaning of the text 'He created all things together' [*Ecclesiasticus* 18.1] cannot arrive at the meaning of Scripture unless the narrative proceeds slowly step by step."

Of the six days of creation, only the first four are relevant for cosmology. Of these, the first, second, and fourth are the most important.

What specific entities were created on each of these days? On the first day, heaven (*caelum*), earth (*terra*), and light (*lux*) were created. On the second day, the firmament (*firmamentum*) that divided the waters above from those below and which God called "heaven" (*caelum*) was created. On the third day, God turned His attention to the earth, where He gathered the seas together in one place, exposing dry land on which He then placed plants and trees capable of reproducing themselves. And finally on the fourth day, He made the physical light of the heavens by creating all the celestial bodies, assigning the sun to provide the light of day and the moon to provide the light of night.

Within these brief passages commentators were obliged to resolve some basic dilemmas, obscurities, and seeming inconsistencies. How, for example, does the heaven (*caelum*), or firmament, created on the first day, differ from the heaven (*caelum*) created on the second day? How does the light created on the first day compare to the light created on the fourth day? How could plants come forth on the third day when the sun, whose warmth and light are required, was not created until the fourth day? What are the waters above and below the firmament? Do they differ?

And then there were problems that arose as a consequence of the need to reconcile the Christian creation account with contemporary physics and cosmology, which, in the period we are discussing, was overwhelmingly Aristotelian. How do prime matter and the four elements relate to creation? How do mixed, or compound, bodies formed from those elements fit into the creation account? What aspect of creation embraces Aristotle's celestial ether? And so on.

However scholastic authors may have disagreed on these perplexing questions, they were agreed that the world was a finite sphere. The finitude of the world was a cardinal principle of medieval natural philosophy and cosmology. Evidence for this fundamental tenet was not provided directly, but was furnished in the manner of Aristotle: by demonstrating that no infinite body could exist and that the world was therefore necessarily finite. Sphericity was readily acceptable because the sphere was considered the largest and most capacious of isoperimetric bodies and because it had neither beginning nor end. Observational arguments were also common, as, for example, that the planets seem to move with circular motions and that from any part of the earth or water, the heaven always seems to be equidistant from the earth.

In creating this finite, spherical world, did God make it perfect? That is, did God create a world that was so perfect that it was incapable of further improvement? Medieval opinion overwhelmingly rejected this interpretation. The world taken as a whole was considered more perfect than any of its parts and was thus perceived as only relatively perfect. As the only absolutely perfect existent thing, God chose not to make an absolutely perfect world, but one that is capable of improvement. It was thus regularly assumed that God could make better and better worlds without end.

In response to the natural query as to why an omnipotent God with absolute infinite power chose not to make a perfect world in the first instance, scholastics believed that God conferred on our world all the perfection that it needed and placed in it all the species that He thought it required. God was under no compulsion to create every possible species of being. Nor indeed did He have to produce the best version of what He did create. In accordance with the medieval doctrine of God's absolute power, most scholastics would have conceded that, by His omnipotence and inscrutable will, God could do as He pleased. There was no necessity to create every possible existent or to make everything the best that it could be.

Other Worlds or Space Beyond our World?

In the Condemnation of 1277, one of the 219 condemned propositions denounced as an excommunicable offense was the claim that God could not make other worlds. Although all were convinced that God did not make other worlds, it was always assumed that by His absolute and infinite power, He could make as many as He pleased. Aristotle had argued the opposite: our world was of necessity unique and it was impossible for body, place, vacuum, time, or motion to exist beyond our world, just as it was impossible for more than one center and one circumference to exist. As an inference from this conclusion, Aristotle would presumably have denied even the possibility that God could create another world. Thus it was impossible for other worlds to exist beyond our own, or for any kind of space or vacuum.

Scholastic natural philosophers argued that, although God probably did not create other worlds, He certainly could do so if He wished. To substantiate this claim, they showed, by a variety of arguments, that simultaneously existent identical and distinct worlds—and this was the type most frequently discussed—could exist. Each of these identical, though distinct, worlds could exist as a self-contained, wholly independent closed system with its own center and circumference, contrary to Aristotle's argument that only one center and one circumference could exist and therefore only one spherical world. Also challenged was Aristotle's fundamental idea that each of the four elements had one absolutely determined natural place, a concept that was viable only if there was one possible world with a single center and circumference. With the possibility of simultaneously existent worlds, the element fire, for example, would be equally distributed in each world and could have no single natural place. What Aristotle had deemed impossible was shown to be possible and intelligible.

Aristotle was quite aware of the opinion that "outside the heaven" some people assumed an infinite void and place in which an infinite body or infinite worlds might exist. He was equally aware that we have a strong intuitive sense that something must lie beyond. Nevertheless, he denied the possibility of extra-cosmic space. On theological grounds, however, Thomas Bradwardine (ca. 1290–1349) assumed the existence of an infinite, extensionless void space—often called "imaginary" space—beyond our world, which he identified with God's infinite, omnipresent immensity.

Scholastic ideas about space and God form an integral part of the history of spatial conceptions between the late sixteenth and eighteenth centuries, the period of the Scientific Revolution. From the assumption that infinite space is God's immensity, scholastics derived most of the same properties of space as would be conferred upon it during the course of the Scientific Revolution.

Outline of Medieval Cosmology: The Celestial Region

As the "noblest" part of the world, the celestial region—from the moon to the sphere of the fixed stars and beyond—was thought to be composed of a celestial ether, a fifth element, that was incredibly subtle and rare. Following Aristotle, medieval natural philosophers assumed that this ether was in a state of perfection and therefore incorruptible. As support for this claim, Aristotle and his medieval followers insisted that no changes in the celestial region had ever been observed or recorded. Aristotle had facilitated the near unanimous acceptance of incorruptibility by denying celestial locations to shooting stars, comets, and similar phenomena, placing them instead below the moon in the upper atmosphere of the terrestrial region.

Although planets, stars, and orbs were composed of ether, medieval authors did not consider them of equal nobility. Some authors simply assumed that nobility increased with distance from the earth. Others denied this, and assumed instead that the sun was the most noble celestial body. Some of this group then adopted the principle of ascending nobility for the other planets.

Number and Order of the Planets

All medieval authors were agreed on the existence of seven planets and the fixed stars. These celestial bodies were usually located in ascending order, from farthest away to nearest the earth. In the absence of any measurable parallax, the order of the celestial bodies was made on nonastronomical grounds. Since the planets moved against the background of the stars, the latter were assumed to be farthest from the earth, and equidistant from it. The superior planets (Saturn, Jupiter and Mars) were ranked by their sidereal periods (time to circle the heavens once, as measured against the background of the stars) in that order. Because Venus and Mercury always remained in the vicinity of the sun, the order of these three planets (considering the sun as also a planet) was not determinable except on the basis of arbitrary, nonastronomical reasons. Most followed Ptolemy's arrangement, and located the sun, Venus, and Mercury in that order, followed by the moon, universally assumed to be the closest planet to the earth.

The Celestial Orbs

Although it appeared to the naked eye that the planets were self-moved, like fish in water or birds in the air, natural philosophers denied appearances and assumed, with Aristotle and Ptolemy, that each planet was embedded in a single ethereal sphere and carried around by it. Similarly, the stars, which always remained at fixed intervals from each other, were located on a single sphere.

However, not only was each planet assigned a single sphere, but *each motion* of the planet—daily motion, sidereal motion, motion in latitude, etc.—was also assigned its own orb. Thus although Ptolemy and Aristotle assumed that each planet was attached to a single sphere, both employed a plurality of spheres to account for the resultant motion of each planet. By assigning one orb for each motion of a planet, both Aristotle and Ptolemy were compelled to assign multiple orbs to each planet. All told, Aristotle assigned as many as fifty-five, while Ptolemy may have assigned as many as forty-one.

But there was a fundamental difference between the kind of orbs they assigned. Aristotle's spheres were all concentric with respect to the earth, whereas Ptolemy's were basically eccentric and epicyclic. Shortly after these rival cosmologies entered Western Europe, sometime between 1160 and 1250, it became evident that Aristotle's concentric orbs could not account for observed variations in the distances of the planets and that Ptolemy's eccentrics and epicycles had been devised to cope with that problem.

Medieval natural philosophers accepted a compromise that retained both concentricity and eccentricity. It was a compromise that had already been made by Ptolemy in the latter's *Hypotheses of the Planets* and that hinged on a distinction between the concept of a "total orb" (*orbis totalis*) and a "partial orb" (*orbis partialis*), to use medieval terminology. The "total orb" was a concentric orb whose center is the center of the earth, whereas a "partial orb" was an eccentric orb, that is, an orb whose center is a geometric point lying outside the center of the world. The concentric total orb, the concave and convex surfaces of which have the earth's center as their center, is composed of at least three partial orbs, one of which, the eccentric deferent, carries an epicycle in which a planet is embedded. Thus were the concentric orbs of Aristotle fused with the eccentric orbs of Ptolemy (figure 1).

The number of celestial orbs could thus be determined in two different ways: either as the totality of concentric, or "total" orbs; or as the totality of eccentric, or "partial" orbs. Aristotelian natural philosophers were overwhelmingly interested in the former and rarely in the latter. In fact, the most popular question in medieval cosmology concerned the number of spheres: "whether there are eight or nine, or more or less," the response to which clearly requires a count of total, or concentric orbs and not partial, or eccentric orbs. The answers to this popular question varied anywhere from eight to eleven.

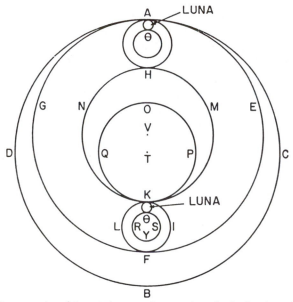

Figure 1. Representation of the moon's concentric, eccentric, and epicyclic orbs, as described in Roger Bacon's Opus tertium. *In medieval natural philosophy, there were two types of orbs: "total orbs" (concentric planetary orbits centered on the center of the earth) and "partial orbs" (eccentric planetary orbits whose centers were on geometric points lying outside the center of the world). In this figure, the total orb is the space lying between OQKP and ADBC, both circles centered on the center of the earth, T. This total orb is made up of three parts, the spaces between the circles ADBC and AGFE, between AGFE and HNKM, and between HNKM and OQKP. One of these three partial orbs, that consisting of the space between the circles AGFE and HNKM, both centered on V, is the eccentric deferent. It carries an epicycle in which a planet (in this case, the moon—luna) is embedded.*

Eight concentric orbs represent the seven planets (including the sun and moon) and the sphere of the fixed stars. However, as many as three different motions were assigned to the sphere of the fixed stars (the daily motion, precession, and trepidation). On the principle that a single sphere must be assigned to each distinct celestial motion, two additional spheres plus an immobile Empyrean sphere were usually added for a total of eleven. No celestial bodies were embedded in these three supra-firmamental orbs, which were invisible and transparent (figure 2).

Hard or Soft Celestial Orbs?

During the late Middle Ages, natural philosophers, in their *Questions on De caelo*, devoted no single question to the hardness or softness of the celestial orbs. They showed little direct interest in the problem. Only occasional remarks provide clues as to an author's choice. Prior to the thirteenth century, the heaven and orbs were

Figure 2. The movable celestial spheres ranged in order from the lunar orb to the "first movable heaven" (primum mobile). Encompassing the whole is the immobile Empyrean heaven, "dwelling place of God and all the elect." From Peter Apian's 1524 Cosmographicus liber. *Courtesy of the Lilly Library, Indiana University, Bloomington, Indiana.*

usually conceived as naturally fluid. Because of their presumed unalterable and incorruptible natures, fluid orbs could be in contact at every point and still retain their perfect, spherical shapes.

The widespread assumption of a fluid heaven and orbs in the thirteenth century was yielding to an assumption of their hardness in the fourteenth century. Already in the thirteenth century, Richard of Middleton (fl. second half of the thirteenth century) described both theories, but refused to choose between them. Shortly after, Aegidius Romanus (Giles of Rome, born before 1247–1316) proposed a combinatory theory, in which the overall heaven was hard, but the eccentric deferents were fluid. As the fourteenth century progressed, Themon Judaeus (born ca. 1330, died after 1371) and Henry of Hesse (1325–1397) explicitly opted for hard orbs and Nicole Oresme (ca.

1320–1382) did so implicitly. Overt defenders of fluid orbs or heavens had yet to turn up in the late fourteenth and fifteenth centuries. In the transition from fluid to hard orbs, Job 37.18 served as significant Biblical support. In this passage, Elihu asks Job whether, like God, he could fabricate the heavens as if they were made of molten metal, thus implying that God had made the heaven hard like metal.

By the late sixteenth century, it is evident that the concept of hard orbs was widely accepted, because Tycho Brahe (1546–1601) declared it the major opinion he had to destroy and replace with a fluid theory of celestial matter. Many scholastic authors followed Tycho. Thus did the fluid theory of the early Middle Ages triumph over the hard orbs of the late Middle Ages.

Theological Orbs

Christian theology intruded into—or better yet, was superposed on—medieval Aristotelian cosmology in the context of the fixed stars and the spheres that lay beyond. Whichever of the three astronomical motions (daily motion, sidereal motion, and motion in latitude) may have been assigned to the eighth and ninth spheres, these spheres also functioned as theological heavens representing descriptions in Genesis. Although the Biblical firmament was assigned a variety of meanings, one of the most widely accepted was its identification with the eighth sphere of the fixed stars.

But what of the waters above the firmament? Biblical exegesis demanded that they be assumed real, although their precise nature was debatable. Relatively early in the history of Christianity, these waters were conceived of as *crystalline*, a term that was sometimes thought of as applying to fluid waters and sometimes to waters that were congealed and hard like a crystal. Thus for St. Jerome (ca. 347–419?) and Bede (672–735), the waters above the firmament were conceived as crystal-like, which signified hardness, whereas for Saints Basil, Gregory of Nyssa (died 394?), and Ambrose (ca. 339–397) they were fluid. Whether fluid or hard, however, the crystalline orb was usually located above the firmament and was identified with the ninth sphere, and occasionally with both the ninth and tenth spheres.

All the orbs discussed thus far were conceived to be in perpetual motion. The Empyrean sphere, a purely theological construction, was assumed to be immobile, and to be the place and ultimate container of the universe. Indeed, it was conceived as "the dwelling place of God and the elect." As was fitting for such a privileged and holy place, only the purest light filled its world-encompassing dimensions. If the number of mobile orbs is n, the Empyrean orb was always numbered $n + 1$. In the most popular ten orb mobile system, it was the eleventh and final orb.

Cause of Celestial Motion

We have seen that the orbs carried the planets, which were themselves immobile. But what moved the orbs? It was usually assumed that each orb was assigned an immaterial, spiritual intelligence distinct and separate from the orb it moved. Intelligences, which were frequently, though by no means always, identified with angels, were conceived of as voluntary agents divinely endowed with inexhaustible powers to move their respective orbs with uniform circular motion.

Not all scholastics appealed to intelligences. Some—a small minority—held that God created the orbs with an intrinsic power for self-motion. John Buridan (ca. 1295–ca. 1358) even invoked a divinely impressed force, or impetus, that enabled the orbs to move forever through a celestial ether that was assumed by all to offer no resistance to their motion.

Dimensions of the Cosmos

By modern standards, the size of the medieval cosmos was insignificantly small. For those who lived in the Middle Ages, however, it was judged large—indeed enormous. The celestial spheres were thought to be nested one within the other. For the measurement of distances, it was assumed that the convex surface of one planetary sphere was exactly equal to the distance of the concave surface of the next upper celestial sphere. Indeed, they formed one continuous surface. Thus the convex (outer) surface of the moon's sphere was assumed identical with the concave (inner) surface of the sphere of Mercury. And the convex surface of Mercury's sphere was identical with, or exactly the same distance from the earth, as the concave surface of the sphere of Venus. Generally, the convex surface of a sphere was assumed to be exactly the same distance from the earth as the concave surface of the very next planetary sphere. Based on tables in Campanus of Novara's (ca. 1220–1296) thirteenth-century *Theory of the Planets* (*Theorica planetarum*), the following data (Tables I, II, III, and IV) provide a widely accepted set of values for the dimensions of the medieval cosmos.

From the center of the earth to the convex surface of Saturn, an approximate distance of 73 million miles was assumed. The last visible component of the universe was the eighth sphere of the fixed stars. Following the pattern we have already described, the concave surface of the sphere of the fixed stars was to be assumed continuous with the convex surface of the sphere of Saturn. From this, we may

Table I. Distances of Concave and Convex Planetary Surfaces From the Center of the Earth (in miles)

Planet	Concave Surface		Convex Surface	
Moon	107936	20/33	209198	13/33
Mercury	209198	13/33	579320	28/33
Venus	579320	560/660	3892866	550/660
Sun	3892866	550/660	4268629	110/660
Mars	4268629	110/660	32352075	420/660
Jupiter	32352075	420/660	52544702	280/660
Saturn	52544702	280/660	73387747	100/660

Table II. Thicknesses of Planetary Spheres (in miles)

Moon	101261	26/33
Mercury	370122	5/11
Venus	3313545	650/660
Sun	375762	220/660
Mars	28083446	310/660
Jupiter	20192626	520/660
Saturn	20843044	480/660

Table III. Diameters of the Planets (in miles)

Moon	1896	26/33
Mercury	230	26/33
Venus	2884	560/660
Sun	35700	
Mars	7572	480/660
Jupiter	29641	540/660
Saturn	29209	60/660

Table IV. Circumferences of the Planets (in miles)

Moon	5961	11/33
Mercury	725	11/33
Venus	9095	63/660
Sun	112200	
Mars	23800	
Jupiter	93160	
Saturn	91800	

infer that Campanus of Novara and almost everyone else would have assumed that the fixed stars were at least 73 million miles from the center of the earth. Despite their acceptance of the division of the fixed stars into six magnitudes graded according to size (from the largest in the first magnitude to the smallest in the sixth), medieval astronomers and natural philosophers seem to have accepted the idea that all stars of whatever magnitude were fixed in their sphere at an equal distance from the earth.

Some Properties of Celestial Bodies

The stars and planets were made of the same transparent ether as the orbs that carried them. The stars and planets were visible only because each celestial body was a region of highly concentrated ether that was capable of receiving light and becoming luminous (or, for those who assumed the celestial bodies to be opaque, of reflecting light from the sun). Although some people thought that stars and planets, like the sun, were self-luminous, and therefore received no light from the sun, the prevailing opinion assumed that all stars and planets did indeed receive their light from the sun. Some people further assumed that the planets might also be weakly self-luminous.

It was always assumed that the stars and planets were spherically shaped. Although many disagreed with Aristotle and assumed that the moon had a proper rotatory motion, they denied this property to the other planets.

In what was virtually a self-evident truth, derived ultimately from Aristotle, it was assumed throughout the Middle Ages that celestial bodies exerted a vital and even controlling influence over material things in the terrestrial region below the lunar sphere. Three instrumentalities were distinguished:

- Celestial motions (*motus*), which transmitted heat and light to the earthly region.
- Celestial light (*lumen*), which produced daylight and heat, and by its absence night and coldness.
- Celestial influences (*influentiae*), which were invoked to explain many otherwise inexplicable phenomena, as, for example, the formation of metals within the depths of the earth, magnetism, and the tides caused by the moon.

Of the seven planets, the sun was considered the largest and most important. It was also the noblest, even though it was not farthest removed from the earth. Counting upward from the moon or downward from Saturn, the sun occupied the fourth, or middle, position and was perceived as a "wise king in the middle of his kingdom," or as "the heart in the middle of the body." Not only was the sun the primary source of light and heat in the cosmos, but its movement through the ecliptic produced the change of seasons. It was the basis for life on earth.

The moon was more often thought to be transparent, rather than opaque. From the reception of sunlight, it became luminescent. But because the luminescence was unevenly distributed—perhaps the result of differences in rarity and density—lunar spots appeared where the luminescence was weakest.

The remaining planets also were thought to be transparent rather than opaque. They were capable of receiving light from the sun into the depths of their spherical volumes, from which the light was diffused to terrestrial observers. A few thought that the planets were self-illuminated and independent of the sun's light.

Most of the other properties assigned to the planets were derived from astrology, especially from Ptolemy's *Tetrabiblos*. In Aristotelian physical theory, wherever opposite qualities, such as, for example, hot-cold and dry-wet, were operative, change was inevitable. Since the heavens were assumed incorruptible and unchangeable, the existence of such qualities within the celestial ether usually was denied, or they were said to exist only virtually (*virtualiter*) rather than really (*formaliter*). Thus Saturn was deemed cold and dry, by which it was usually meant that Saturn was not itself cold and dry but, under appropriate circumstances, was capable of producing the effects of coldness and dryness in sublunar bodies. The same may be said of the properties of the other planets: the wetness and coldness of the moon, or the hotness and dryness of Mars, and so on.

The Earth as the Center of the World

Although the needs of technical astronomy required that the earth be removed from the center of the world by use of eccentric orbs, most natural philosophers wrote about the earth as if it lay immobile in the center of the world. An exception was John Buridan, who attributed real, but minute, motions to the earth. Buridan assumed slight, though continuous, movements of the earth as it sought to bring its center of gravity (which was always changing because of incessant variations in the density of its different parts as a consequence of geological changes) into coincidence with the geometric center of the universe.

Indeed, Buridan, along with Nicole Oresme, boldly contemplated a more radical motion of the earth when they considered whether the daily rotation of the heavens might not actually derive from a daily axial rotation of the earth. For different reasons, they both eventually rejected this alternative, but they were agreed that the earth's daily rotation, along with the assumed immobility of the heaven, saved the astronomical phenomena just as well as the traditional opinion did.

In the course of their discussions, one or the other or both argued for the earth's axial rotation by appealing to the relativity of motion as illustrated by:

- The movement of ships.
- Compound motion involving rectilinear and circular components.
- Economy of motion, with the earth requiring a much smaller velocity than the huge heavens would to complete one daily rotation.

Buridan and Oresme further assumed that air and the birds in the air share the daily rotation and, on the assumption that a state of rest is nobler than motion, it would be more appropriate for the ignoble earth to rotate than the nobler heavens. Nicholas Copernicus (1473–1543) found these arguments worthy of inclusion in defense of his heliocentric system where the earth is assigned both a daily rotation and an annual motion around the sun.

Conclusion

Medieval cosmology remained viable until the cumulative impact of the efforts of Copernicus, Tycho Brahe, Galileo, Kepler, and Newton rendered it totally ineffective by the end of the seventeenth century.

—Edward Grant

BIBLIOGRAPHIC NOTE

My forthcoming *Planets, Stars, and Orbs: The Medieval Cosmos, 1200–1687* (London and New York: Cambridge University Press) provides an overview of medieval cosmology. More specific topics are covered in "Celestial Perfection from the Middle Ages to the Late Seventeenth Century," in Margaret J. Osler and Paul L. Farber, editors, *Religion, Science, and Worldview, Essays in Honor of Richard S. Westfall* (London and New York: Cambridge University Press, 1985), pp. 137–162; and "Celestial Orbs in the Latin Middle Ages," *Isis,* 78 (1987), 153–173. See also E.J. Aiton, "Celestial Spheres and Circles," *History of Science,* 19 (1981), 75–114; Steven J. Dick, *Plurality of Worlds. The Origins of the Extraterrestrial Life Debate from Democritus to Kant* (Cambridge: Cambridge University Press, 1982); Pierre Duhem, *Medieval Cosmology: Theories of Infinity, Place, Time, Void, and the Plurality of Worlds,* edited and translated by Roger Ariew (Chicago: The University of Chicago Press, 1985); and Nicholas Steneck, *Science and Creation in the Middle Ages: Henry of Langenstein (d. 1397) on Genesis* (Notre Dame, Indiana; and London: University of Notre Dame Press, 1976).

10. DANTE'S MORAL COSMOLOGY

Poets as well as philosophers reacted to the Aristotelian world view, an all-pervasive aspect of their culture, and reflected it in their poems. No better representation and interpretation of the late Medieval period is to be found than *The Divine Comedy*. Dante was well versed in Aristotelian astronomical and cosmological matters, as indeed were nearly all educated persons of this era. This knowledge helped shaped Dante's poetry.

Dante's Life

Much of what we know about Dante Alighieri is derived from what he recounts in his own works. He was born in 1265 into a Florentine family of modest means (his father may have been a moneychanger) allied with the Guelf party. During Florence's civil conflict, dating from the early thirteenth century, the Guelfs (loyal to the pope) had finally expelled the Ghibellines (loyal to the emperor) from the city in 1267. Dante himself was involved in the Battle of Campaldino in 1289, in which Ghibellines exiled in Arezzo attempted to return to Florence. In order to become eligible to hold public office, Dante joined the guild of doctors and pharmacists, and in June 1300 was elected as one of the Priors of Florence, the city's executive council, for a two-month term. He was ambassador to San Gimignano in May 1300 and to Pope Boniface VIII in Rome in October 1301. It was during this period that the ruling Guelf party split into two rival factions, the Whites and the Blacks. While Dante was away from home, the Black Guelfs conspired to invite Charles of Valois (the military general commissioned by the pope) to invade Florence. In Siena in January 1302, Dante received news of a sentence of exile issued against him and four other important White leaders. The charge was barratry, the abuse of civic office for personal gain. In March he was permanently banished from Florence on pain of death and all his property became forfeit. Dante was thereafter compelled to roam from city to city, depending on the hospitality of various lords, notably Cangrande della Scala in Verona. Dante continued to condemn publicly the current government in Florence, in letters in which he describes himself as "Florentine by birth but not by custom." He eventually became disillusioned with the leaders of the White faction and broke with his companions in exile in 1304 to form a "party of one." His hopes for a new political order were disappointed when the invading Emperor Henry VII of Luxembourg, whom Dante called "Lamb of God who taketh away the sins of the world," died during a siege of Florence in 1313. Dante never returned to his native city, and refused the humiliating terms of an amnesty extended in 1315, declaring: "What? Can I not everywhere behold the sun and the stars? Can I not under any sky meditate on the most precious truths?" *The Divine Comedy*, which he finished in Ravenna shortly before his death in 1321, narrates a journey through the three realms of the afterlife—Hell, Purgatory, and Heaven—which the poet claims he actually experienced in the spring of 1300 just before the events leading to his exile.

Dante's literary career began with his association with a group of Tuscan poets who wrote in what he would later label the "dolce stil novo." The "sweet new style" transformed the medium of popular love poetry into a highly refined art strongly influenced by contemporary philosophical trends. These lyrics celebrate "angelicized" ladies in superlatives touching upon the whole order of the universe. The Florentine poet Guido Cavalcanti, whom Dante called his "first friend" in this new group of poets, was a leading White Guelf, an Averroist, and perhaps a heretic. Ironically, Cavalcanti would later die of malaria contracted during an exile imposed during Dante's tenure as prior. In his first long work, *La vita nuova* (*The New Life*), Dante pieced together a number of his early lyric poems with portions of prose narrating his first encounter with Beatrice when they were both about nine years old. The poet concludes that Beatrice's brief life on earth was a miraculous event, and finally proposes to speak no more about her until he can say of her what has never been said about any woman. *The Divine Comedy*, in which Beatrice returns to guide Dante through the stars, evidently fulfills that old promise. To console himself over Beatrice's death in 1290, Dante says that he began to frequent the "schools of the religious and the disputations of the philosophers" (presumably the Franciscans of Santa Croce and the Dominicans of Santa Maria Novella) and found a new love in Lady Philosophy. This education began with Boethius's *Consolation of Philosophy* and Cicero's *On Friendship*, and must have gone on to include other classical Latin authors (Ovid, Horace, Lucan, Statius, Seneca, and, above all, Virgil), Plato's *Timaeus* and its commentators, the works of Aristotle and his Arab commentators (such as Averroes), and Saint Augustine, Albert the Great, Thomas Aquinas, and other ecclesiastical writers. In his unfinished philosophical treatise, *Il convivio* (*The Banquet*), Dante proposed to disseminate some of this book learning in the form of vernacular commentary on a group of his own poems. There are also three works in Latin: *De vulgari eloquentia* (*On the Eloquence of the Vernacular*), a political treatise titled *De monarchia*, and *Quaestio de aqua et terra*—a lecture Dante apparently gave in 1320 addressing a cosmological problem provoked by Aristotelian physics.

Dante's Cosmos

The Divine Comedy, written by Dante in the early 1300s, describes the universe from top to bottom, inside and out. The journey narrated in the poem begins with Dante lost in an anonymous dark wood somewhere on the surface of the globe. Accompanied by the pagan poet Virgil, Dante descends into the bowels of the earth to the dead center of the universe. From there the wayfarer and his guide reascend through an underground passage to the shores of an ocean island in the unexplored southern hemisphere and climb its steep mountain peak, at whose summit they discover the terrestrial paradise (figure 1). Here Virgil disappears from the story and Beatrice, the lady of Dante's early love poems, appears to lead him beyond the element of fire to the heaven of the moon, and through each of the subsequent six planetary spheres. In the *stellatum*, or starry heaven, she has him turn to glance back down at the earth through the whole system of celestial wheels which he has traversed. Passing beyond the largest body of the universe, the crystalline sphere or *primum mobile*, Dante enters the Empyrean. This heaven of the blessed and of the angels is flooded with light reflected from the "divine mind" which contains the entire physical world.

The cosmos of *The Divine Comedy* is based on a simplified Aristotelian model. The earth is quiet at the center, surrounded by whirling spheres made of solid but transparent material and nested one inside the other. The first seven planetary spheres move from west to east independently. Slowest and closest to the earth is the sphere of the moon; then come Mercury, Venus, the sun, Mars, Jupiter, and Saturn. Encasing these planetary spheres, the immense sphere of the fixed stars has an almost imperceptible motion west to east of one degree in every hundred years (apparent in the precession of the equinoxes). Added to this Aristotelian cosmos is a starless ninth sphere, which imparts to all the inferior heavens the simple daily impetus from east to west. This invisible "first moved," the *primum mobile* or crystalline sphere, is the boundary of space in the finite medieval

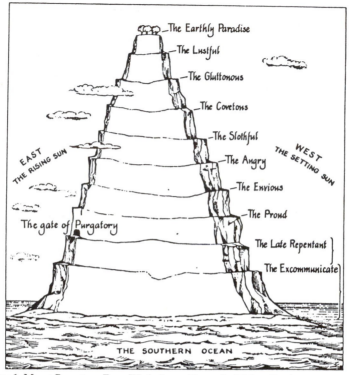

The Earthly Paradise
The Lustful
The Gluttonous
The Covetous
The Slothful
The Angry
The Envious
The Proud
The Late Repentant
The Excommunicate
The gate of Purgatory
EAST THE RISING SUN
WEST THE SETTING SUN
THE SOUTHERN OCEAN

Figure 1. Mount Purgatory. From W. Anderson, Dante the Maker *(London: Routledge and Kegan Paul, 1980), p. 257.*

universe. Yet beyond this outermost physical sphere is the spiritual universe that contains it, the mind of God or Empyrean heaven, which is infinite (figure 2).

Attempts to develop a realistic model of how the universe was actually constructed commanded great interest during the Middle Ages. Although Arabs and Christians depended upon Ptolemy's mathematical astronomy to predict planetary positions with the greatest accuracy, they found its ponderous apparatus of epicycles and eccentrics repugnant. Religious sensibilities demanded certain principles of harmony between the natural and the supernatural worlds, which geometrical hypotheses designed merely to "save the appearances" could not satisfy. The *primum mobile*, for example, which has no appearance at all since it is entirely transparent, became philosophically necessary. As Alpetragius, an Arab influenced by Neoplatonism, explained, the many must be derived from the One and thus an outermost sphere of supreme simplicity must actually exist.

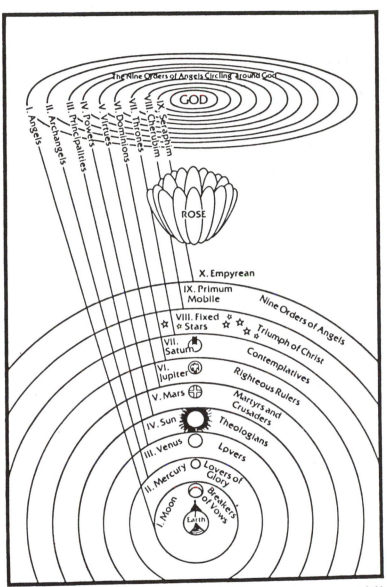

Figure 2. The cosmos of Dante's Divine Comedy, *with the earth at the center surrounded by the planetary spheres. From M. Musa,* Dante's Paradise *(Bloomington: Indiana University Press, 1984), p. 25.*

Dante's task, as a poet, is to represent these various astronomical and philosophical notions as they would appear to someone journeying through them. At times we almost glimpse the great armillary sphere underlying the medieval universe, when, for example, Dante indicates a point on the sun's path where "four circles join three crosses" (*Paradiso*, I, 39) or mentions the epicycle of Venus when his characters are inside the planet itself (*Paradiso*, VIII, 3). Unlike other dream-visions in medieval literature, this fiction claims to be an auto-biographical account of an actual journey through the three realms of the afterlife, made by a historical individual over a set number of days elapsed during a specific year. Hell's cavern comprises a certain number of miles beneath the earth's surface and the mountain of Purgatory is situated in precise relation to other known geographical locations. The heavens appear differently from the opposite hemisphere, and the tiny earth can be seen from the height of the fixed stars. Whether the fictional character makes the journey through the stars in his body or only in his mind, it is the poet's role to bring back a sensible image of the way the universe really is, to "figure Paradise," as Dante says (*Paradiso*, XXIII, 61).

Dante made literary use of astronomical facts throughout his works. In *La vita nuova*, he chronicles his first encounter with Beatrice by revolutions of the sun, her age by the fraction of a degree of the movement of the *stellatum*, and the date of her death according to a spectrum of calendars—Hebrew, Arab, and Christian. His series of poems known as the "stony rhymes" are situated at the winter solstice of 1296, as is indicated by the rising of the Gemini at sunset, the appearance of Saturn in the sign of Cancer visible all night long, and the occultation of Venus in conjunction with the sun. In *Il convivio*, Dante dates by the synodic revolutions of Venus the beginning of his consolation through philosophy after the death of Beatrice. He also discusses different opinions regarding the layout of the heavens and the movements of the planets, and employs a long comparison between the ten heavens and the classification of the sciences. There are approximately a hundred passages relating to astronomy in *The Divine Comedy*. Each of the three canticles—*Inferno*, *Purgatorio*, and *Paradiso*—closes on the same word: *stelle*. As constant reminders of the scope of the pilgrimage, the stars reappear after the exit from Hell, indicate the ascent to heaven at the summit of Purgatory, and return as the dim trace of the "Love that moves the sun and other stars" glimpsed in the final vision of Paradise.

It is essentially a Platonic notion that the physical structure of the universe (whose boldest lineaments are apparent in the passage of celestial bodies over the vault of the sky) is but a sensible copy of the supernatural pattern that informs it. In Dante's words, "the order among all things is what makes the universe similar to God" (*Paradiso*, I, 103–105). The inevitable difference between model and copy is often expressed in terms of the imperfect impression a seal makes in wax (*Paradiso*, XIII, 67–69). Yet the sensible image remains the only indication we have of the reality that lies beyond it. The concentric form of the planetary spheres, for example, is actually a reflection of the angelic hierarchies that rotate around God. (The circles of Hell that spiral down to Satan at its central pit are a grotesque parody of the same.) As the journey progresses, the deceptiveness of appearances in the physical world is exposed. In the geocentric universe, the reprobate who, as Dante says, "pierces the world like a worm," might seem to dominate the world from its dead center. From the perspective of Purgatory and the Garden of Eden located in the uncharted seas on the other side of the earth, however, it becomes clear that Satan is in fact upside down. From the perspective of Paradise, moreover, the privileged state of rest (thought to be located in the immovable center of the Aristotelian cosmos) is transferred to the circumference or container of the Christian universe: the Empyrean heaven. It is only when we pass beyond the limit of the physical world that the geocentric system is revealed, not as the true structure of the universe, but as its imperfect copy, a visible metaphor of the unseen cosmos. Dante represents the spiritual form of the universe as another system of hierarchical spheres around a fixed point, just as it appears in the corporeal world, but with one radical difference: the spiritual universe is theocentric. We might furthermore say that this structure is (at least metaphorically) heliocentric. The angels who provide the motive force of the planets that go around our earth are actually revolving around God ("Sun of the angels") who, like the sun, illuminates all things from the center. Dante's solar symbolism in his depiction of the true form of the intelligible universe derives from Hermetic and Neoplatonic texts, some of which Copernicus would later use to justify his revolutionary challenge to the geocentric world system.

Formation of the Earth

The extraordinary unifying vision in *The Divine Comedy*, which provides a physical and a moral topography of the universe, might best be appreciated by comparing it with what is thought to be Dante's last work. *Quaestio de aqua et terra* is a straightforward scientific effort to solve a classical cosmological problem: how a portion of the heaviest element, earth, came to rise above the lighter element of water. This treatise proposes that the influence of certain stars exerted on the northern part of the terrestrial globe pulled an area of dry land up out of the surrounding ocean. This primeval topographical shift was necessary to provide the initial mixture of elements that form the material basis of all things. In this solution the cause and effect of the cosmic change, however providential, are purely physical. In *The Divine Comedy*, by contrast, the same phenomenon is explained by a supernatural event: the sin of Lucifer and the other rebel angels. This primordial catastrophe precipitated the moral drama of the human race and at the same time formed the physical stage on which it was to unfold. Satan's fall from the summit of creation into the geometrical center of the cosmos was a fall into matter, into the substrate of the elements, to the lowest point in the universe from which he could fall no further. As Virgil explains in the last canto of *Inferno*, the dry land originally exposed in the southern hemisphere recoiled in fear, ducked under the ocean, and came around to form the familiar inhabitable continent in the northern half of the globe. According to this account, then, the most important characteristic of the present geography of our world is that it is post-lapsarian. It is not the perfect product of the first creation, but the deformed result of a fall.

This cataclysmic event in the first moment of history also generated the other two major landmarks of Dante's geography: Hell and Purgatory. Hell is in the shape of a hollow cone, its "apex" at the

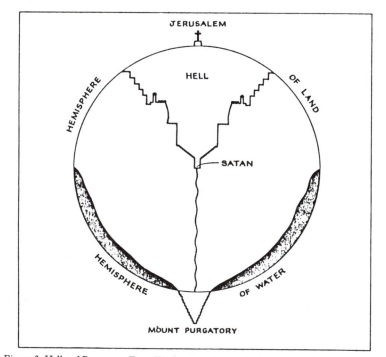

Figure 3. Hell and Purgatory. From W. Anderson, Dante the Maker *(London: Routledge and Kegan Paul, 1980), p. 254.*

center of the earth and its wide base near the surface (figure 3). Inscribed along the interior walls is a path spiraling down toward the tip, forming the various circles of sin and correspondent punishment. Purgatory is a mountain, also in the shape of a cone, terraced as is Hell with a path that winds up to its summit. The geometrical shapes of Hell and Purgatory—one a hollow void, the other a solid mass—are complementary. The cavity of the underworld roughly corresponds to the impression the mountain of Purgatory would make if it were inverted and plunged into the earth. These two inversely proportional realms have the same origin. In addition to transferring a land mass from one hemisphere to the other, Satan's fall also displaced from the interior of the globe a quantity of earth that surfaced to form the island mountain of Purgatory. If the land that formed Purgatory was taken from the cavern that forms Hell, this means that Purgatory is a mountain of outrageous scale, as well as vertiginous steepness. It is as high as the earth is thick to its core (some 3,000 miles). The third of Purgatory's ten terraces is already above the earth's atmosphere, out of reach of its weather, with its summit reaching to just beneath the elemental sphere of fire, far closer to the moon than to the surface of the earth.

The "emperor of the painful realm," who seems to command the terrestrial world from his seat at the center, is frozen where he landed headlong when he precipitated out of heaven. His head and bust poke out into Hell's pit and his legs dangle upwards toward Purgatory. To pass into the other hemisphere, Virgil carries Dante down Satan's hairy shanks until a certain point where he turns completely around and heads "up" toward the devil's feet. This point is not only the geometric zero point of the universe but also, for Dante, the center onto which all weights converge. From here the pilgrim and his guide follow a natural cave (*natural burella*) and emerge onto the shores of the mountain of Purgatory on Easter morning. This is evidently the same lone mountain in the hemisphere of water that the explorer Ulysses (who burns deep in Dante's Hell) glimpsed just before his last shipwreck. At its summit is the Garden of Eden. (Even Christopher Columbus, troubled as to what land he had in fact come upon, proposed that he may in fact have discovered the lost terrestrial paradise on the opposite side of the globe.) The process of sin's purgation thus leads to the place where the human race began and where the first people sinned. The Roman poets accompanying Dante, Virgil and Statius recognize this garden as their dreamed-of Golden Age. In Dante's peculiar Christian geography, the Golden Age becomes not a time but a place (or better, a condition) always available. Eden is located directly opposite Jerusalem. A single diametrical axis can thus be drawn from Mount Zion, through the body of Satan, to the apex of Purgatory on the other side of the earth (figure 4).

These other symmetrical relations are discovered along the journey. In the *Inferno* the descent along the subterranean ledges is easy, just as the climb in the *Purgatorio* up the steep terraces is arduous. The mouth of Hell is wide, since the lighter sins are most readily committed, but the bottom terraces of Purgatory are the most difficult to climb, since the first stages of penitence are the hardest. The place where sin is punished is a vacuous space, but the place where sin is expiated has a positive value. The concave threads of hell spiral down "to the left," whereas the convex path of Purgatory winds up "to the right." Yet because of the inverse placement of Hell with respect to Purgatory, the trajectory of Dante's pilgrimage is always in the same rectilinear direction—toward God.

There is no mention of the sun in the trek through Hell, where time is reckoned by the unseen movements of the moon and the stars. In Purgatory the pilgrim not only sees the sun and the stars but also becomes acutely aware of their motions. Gone are the familiar stars of the North Pole, and now apparent are constellations "never seen

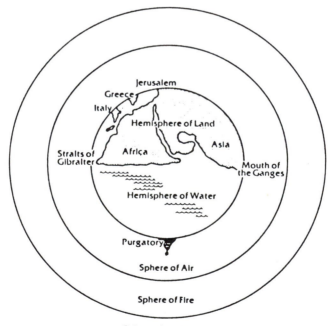

Figure 4. The earth, with Purgatory opposite Jerusalem and surrounded by air and fire. From M. Musa, Dante's Paradise (Bloomington: Indiana University Press, 1984), p. 14.

except by the first people." Dante first becomes aware of being on the other side of the earth when he is astonished to feel the sun striking his left shoulder in the early morning as he faces east. Time, of the utmost importance in the process of purgation, is meticulously tracked by references to the position of the sun ("the sphere that is always playing like a child"), and no progress up the mountain is possible after sundown. Dante passes three and a half days in Purgatory. The hours of these days are reckoned not only by the angle of the sun in Purgatory's sky, but by indications of what time it would have been elsewhere on earth. In Canto 15, for example, we are told that it was vespers "over there" (in Purgatory) when it was midnight "here" (in Italy, where the poet is composing his verses and where he could reasonably expect his reader to be reading them). In Canto 27 we are told that first rays were appearing where the Creator spilled his blood (Jerusalem), that it was midnight in Spain and noon over the Ganges, and that the day was ebbing on the mountain of Purgatory. This preoccupation with the passage of time is, however, abandoned when Beatrice and Dante leave the surface of the earth at noon and begin to move with the eternal spheres.

Configuration of the Heavens

In the *Paradiso* the narrative progresses according to the order of the heavens as the protagonist ascends through each successive celestial sphere to encounter the souls of the blessed. The stratification of Paradise in this manner is really little more than a narrative convenience since all saints are supposed to be equally happy, and the souls do not (as Plato thought) actually inhabit the heavenly spheres, but have "condescended" to appear there for the benefit of Dante's pilgrimage. The *Paradiso* is supposed to concern the place where the third category of souls dwells after death—the Empyrean heaven, or Paradise proper. Yet the action of the first 29 cantos of the *Paradiso* takes place in the astronomical heavens, from the moon to the *primum mobile*.

The mode of ascent has something to do with the reflection of light. Beatrice, who at the end of the *Purgatorio* has taken over from Virgil as Dante's guide, looks fixedly into the sun like an eagle. Just as a second (refracted) ray rebounds upward from the ray of incidence, so Dante, too, by Beatrice's gesture, is able to stare at the sun. For the rest of the trip through the stars, Beatrice will fix her gaze on the "eternal wheels," Dante will look at their reflection in her eyes, and both of them will be instantly transported into the next heaven. From the sphere of fire they pass into the body of the moon (the "eternal pearl"), where Beatrice delivers a lecture on the moon spots and discredits Dante's own earlier theory (in *Il convivio*) by suggesting an experiment involving a lamp and three mirrors (*Paradiso*, II, 94–105). The souls that meet Dante here are inconstant (as is the moon) to vows made on earth, and they appear as watery reflections. Thus each planet shares some quality with the spirits it hosts. Illustrious leaders (such as the Emperor Justinian) are found in Mercury, so often eclipsed by its proximity to the sun; and famous "lovers" are, of course, to be found in Venus. The sun, typically compared to the light of wisdom, contains wise men, theologians, and philosophers. In ruddy Mars, Dante meets martyrs of the faith including his own ancestor, Cacciaguida, who died on a crusade to the Holy Land. In Jupiter—temperate, serene, pure white—he meets the spirits of the just, who together form a series of letters that spell out "Love justice you who rule the earth." In Saturn, that cold and most distant of planets, he meets the contemplative spirits.

From there, "you could not draw your hand out of a tongue of flame and thrust it back" as quickly as Dante says he saw the constellation that follows Taurus and found himself within it (*Paradiso*, XXII, 108–111). It is thus within his birth sign, Gemini, that Dante wheels with the sphere of the fixed stars. From this immense height he looks down through all the planets and sees their true shapes, colors, sizes, velocities, and distances from one another, and smiles at the humble appearance of "this globe," "this little threshing-floor that makes us so ferocious." Six hours later, Dante turns to look again and sees that he has wheeled ninety degrees in the company of the heavens. Of the globe he can see westward beyond the straights of Gibraltar to the mad course Ulysses took over the trackless ocean, but westward only as far as the coast of Phoenicia, because the westward moving sun (two signs away from Dante) has now left that part of the globe in darkness.

Frontier of the Physical Universe

From "Leda's lovely nest" (Gemini) Dante is spirited into the "swiftest heaven." It is in this "watery sphere" (the waters over which the spirit of God moved in the first act of creation) that time has its roots. In the unmarked, outermost sphere of the *primum mobile* the pilgrim arrives at a limit, the threshold between the natural and the supernatural worlds. In the previous canto he had glanced down again at the speck of the earth surrounded by the celestial wheels; now when he looks up he perceives a point of light—as extraordinarily tiny as it is terrifically intense—surrounded by rotating circles of sparks. "From this point," says Beatrice (paraphrasing Aristotle), "depend the heavens and all of nature." The point represents God (the intersection of every "where" and "when"), and the nine fiery circles revolving around it represent the nine hierarchies of angels, each spinning more fervently the closer it is to the hub. The pilgrim recognizes in this representation the paradigm of the physical cosmos, but wonders why the exemplum and the exemplar vary inversely. In the sensible world, the relation of speed to distance is just the opposite: one sees the wheels "more divine" (in other words, faster) the more remote they are from the center. In the visible universe, it is the nearest and smallest sphere (of the moon) which is the slowest, whereas the larger ones travel at ever more fantastic speeds to complete their immense revolutions in a single day. These two structures are, however, related. Beatrice explains that the material sphere that "sweeps all the universe along with it" (the *primum mobile*) corresponds to the smallest and most fervent circle of angels around God. They "know and love the most." By the same token, the angels farthest from the center in the spiritual representation are those that move the neighboring sphere of the moon in our world.

This passage has attracted the attention of modern physicists. Mark Peterson has proposed that Dante's configuration, in which the physical universe "depends" from the spiritual universe, involves non-Euclidean geometry and amounts to what contemporary topologists call the "3-sphere." Peterson suggests that the proportion between speed and size in the planetary and the angelic spheres constitutes a "fourth dimension," and J.J. Callahan has pointed out parallels with Einstein's theory of a finite but unbounded universe.

The visible structure of Dante's universe is deceptive in that the journey toward heaven seems to be a journey away from a center and toward a circumference. With the addition of the spiritual dimension, the revelation of the proportionate degrees of ardor with which the angels turn these celestial spheres, it becomes clear that the pilgrim is really moving away from the periphery and toward the center. The geocentric universe has a theocentric soul. In fact, in the spiritual model in which each angelic sphere corresponds to a physical sphere, the earth is left without a counterpart: it has no place at all. In this way an important medieval dissatisfaction with the geocentric system (the apparent nobility of the earth's place at the center) is resolved.

The sphere of the *primum mobile* receives the impetus for its great revolution from the penetration of a single point of light, seen as surrounded by wheels of sparks representing angels. When we finally pass beyond the physical universe, it appears that the same single ray of light not only penetrates but actually bounces off the surface of the *primum mobile*. The reflected light forms a pool in the shape of an open white rose. This rose is the Empyrean, where the souls of the blessed dwell with the angels. Dante's tenth heaven of pure intellectual light is a concept of Neoplatonic origin, later adopted by Church fathers without much regard for its relation to the structure and apparent motions of the visible cosmos. Scholastic philosophers had made use of the Empyrean heaven to solve the problem of "place" in the Aristotelian cosmos. Aristotle had said that the eighth and last sphere was in no place; yet according to his own definition of movement, it appears impossible for the largest sphere to have *local* motion if it has no *locus*. The Empyrean thus serves as the place of the world. As Dante puts it, the largest body in the universe has no other "where" than the divine mind. All spatial relations within the Empyrean itself, however, are purely metaphorical. The celestial rose is depicted in simple, pastoral terms. God is, as it were, reflected in the amphitheater of the blessed "as a hillside is reflected in water, almost as if to see itself adorned with blossoms and grass"; and the angels ministering above and below are compared to bees dipping in and among the petals of a flower.

The journey through the stars to arrive in the Empyrean is by no means the extent of Dante's examination of the physical cosmos. There is also an extraordinary amount of astronomical metaphor— precise and technical examples taken from astronomy to describe what is in essence an indescribable experience. One example is the long exordium of Canto 13, in which the reader is asked to collect in his mind bright stars from specific constellations spread out over the entire vault of the sky and combine them in a new sign that will serve as a shadow of the "true constellation" he saw in the heaven of the sun. In Canto 29, the "two children of Latona" (the sun and the moon) are imagined in a moment of perfect cosmic balance in order to express the time that Beatrice takes to pause and smile. In Canto 30, the approach to the last, immaterial heaven is described in terms of a coming dawn: "Perhaps 6,000 miles away the noon hour is burning," Dante says, "and the middle of the sky, deep to us, is whitening so that the stars fade one by one until the most beautiful one disappears." In the final image of the poem, Dante compares himself to a geometer struggling to square the circle, as he tries to find a principle that could reconcile the Trinity with "our effigy." When, in a flash, his mind obtains what it wants, the pilgrim feels his "desire and his will" being revolved together, like the celestial wheels turning on their axes, by the "Love that moves the sun and other stars."

—*Alison Cornish*

BIBLIOGRAPHIC NOTE

One of the best and most accessible books on the medieval cosmos and much more is C.S. Lewis, *The Discarded Image. An Introduction to Medieval and Renaissance Literature* (Cambridge: Cambridge University Press, 1961). The most complete treatment of Dante's astronomy in English remains M.A. Orr, *Dante and the Early Astronomers* (London: Gall and Inglis, 1913). Also fundamental is the work of Edward Moore, *Studies in Dante*, Third and Fourth Series (Oxford: Oxford University Press, 1903 and 1917). Patrick Boyde's recent volume, *Dante, Philomythes and Philosopher* (Cambridge: Cambridge University Press, 1981) covers virtually every aspect of natural science in Dante's works. Jeffrey Russell provides a good explanation of the land formation caused by Satan's fall in his *Lucifer: The Devil in the Middle Ages* (Ithaca: Cornell University Press, 1984). For the opinion of contemporary physicists on the modernity of Dante's universe, see J.J. Callahan, "The Curvature of Space in a Finite Universe," in O. Gingerich, ed., *Cosmology + 1* (San Francisco: W.H. Freeman and Co., 1976), pp. 20–30; Mark Peterson, "Dante and the 3-sphere," *American Journal of Physics*, 47:12 (1979) 1031–1035; and Mark Peterson, "Dante's Physics," in G. Di Scipio and A. Scaglione, eds., *The "Divine Comedy" and the Encyclopedia of Arts and Sciences* (Philadelphia: John Benjamins Publishing Company, 1989). John Freccero, *Dante: The Poetics of Conversion* (Cambridge: Harvard University Press, 1986) contains essential interpretations of Dante's cosmological symbolism. The Penguin edition of *The Divine Comedy*, translated by Dorothy Sayers and Barbara Reynolds, 3 vols. (London: Penguin Books, 1949–1962) has very useful introductory essays, notes, and illustrations. A better modern verse translation, with original text on the facing page, is *The Divine Comedy*, translated by Allen Mandelbaum, with original drawings by Barry Moser (Berkeley, Los Angeles, and London: University of California Press, 1980–1982). The most accurate prose translation with extensive commentary, including primary sources in original and translation, is in *The Divine Comedy*, translated by Charles S. Singleton, 6 vols., Bollingen Series, 80 (Princeton: Princeton University Press, 1970–1975). *Il convivio*, Dante's philosophical treatise containing much of his scientific learning, has also been recently retranslated by Christopher Ryan, *The Banquet*, Stanford French and Italian Studies 61 (Saratoga, California: ANMA Libri, 1989).

11. CHAUCER

Geoffrey Chaucer, the premier English poet of the Middle Ages, was the son of John Chaucer, a London vintner. He was brought up in a cosmopolitan London atmosphere, and eventually entered the higher reaches of society. In the 1360s, after his marriage to Philippa, sister to John of Gaunt's third wife, Chaucer undertook various commissions on behalf of the king, some of them taking him to France and Italy. In Italy he became acquainted with the writings of Dante and Boccaccio, and he may well have met Petrarch and Boccaccio. It is possible that he studied law at the Inner Temple in London. He was certainly well read, and far better versed in astronomical and cosmological matters than most university graduates, but there is no evidence that he ever attended university at either Oxford or Cambridge.

In 1374 Chaucer received the royal appointment of chief customs officer in the port of London, and he stayed there for the next twelve years, the busiest of his life. His fortunes rose with those of his new patron, Richard II. In 1386 he was knight of the shire for Kent, where he lived for most of his remaining years, and for a time he was Master of the King's Works. With Richard's fall came a decline in Chaucer's material fortunes. He survived Richard's downfall in 1399, but died in the following year.

Chaucer's poetry developed steadily from a period of French influence (say, from the late 1360s), into a period when Italian influence began to take over. *The House of Fame* (approximately 1383) is a product of this "middle" period and shows much awareness of Aristotelian cosmology, while his great work *Troilus and Criseyde* (1385) owes much to Italian writers. In the last period of his life he gave most of his poetic attention to his best-known collection of poetry, *The Canterbury Tales*. The component parts of this unfinished collection were written at various times (largely 1385–1394), and not in the order in which they were finally presented. It seems likely that the order he wished them to follow was one based on the order of the signs of the zodiac.

It is hardly a distinguishing feature of a man of the fourteenth century that he followed the cosmology of Aristotle. Dante did so, for example, and allowed his knowledge to shape his poetry to a great extent. In Chaucer's case, Aristotelian physics and metaphysics provided only a small part of a scientific world view that permeated a considerable part of his work. In his poetry Chaucer made use of astrology no less than the technicalities of planetary astronomy. That neither subject would today be considered truly cosmological is a consequence of our changing perspective on the physical extent of the planetary system and the interrelatedness of that system and human affairs. In Chaucer's time there was no clear boundary between cosmology and the twin sciences of astronomy and astrology.

Chaucer's prose works include a translation of Boethius and two scientific works (one on the astrolabe, the other on the equatorium), both of them notable for the fact that they were written in English. The treatise on the astrolabe was written by Chaucer around 1393 for his ten-year-old son Lewis. It was drawn largely from a work that was mistakenly ascribed to Masha'allah (who flourished around 762 to 815) but in reality is a thirteenth-century blend of material from four or five different sources. The text on the equatorium, now known arbitrarily as his *Equatorie of the Planetis*, describes an instrument for determining the ecliptic positions of the Sun, Moon, and planets, minimizing the use of astronomical tables, and so speeding up what was otherwise a long and tedious process. Equatoria fall into two main classes. One type simulates by material discs (brass, parchment, etc.) the movements of the circles of Ptolemaic theory. This is intuitively easy to use, granted a knowledge of Ptolemaic theory. The other type uses a series of scales for simplifying the subsidiary stages in standard planetary calculation, but not necessarily by a straightforward simulation. Richard of Wallingford's albion (1326–27) was the most sophisticated example of this second type. Chaucer's equatorium falls into the first class but shows some originality. The manuscript describing it, now belonging to Peterhouse, Cambridge, seems to be in Chaucer's own hand—one of perhaps only two documents of which this can be said.

Chaucer's knowledge of the science of his time was extensive. Thus his canon's yeoman (in *The Canterbury Tales*) practices alchemy, a learned clerk in *The Franklin's Tale* practices geomancy, the wife of the cock Chauntecleer in *The Nun's Priest's Tale* is knowledgeable about the psychology of dreams, many characters introduce points of medicine and the life sciences generally, and in *The Squire's Tale* there are many allusions to optical theory. (Optics had an affinity with experimental sciences described as "natural magic," which must be sharply distinguished from the doctrinally dangerous spiritual magic, which invoked demonic assistance.) Above all else, Chaucer shows an affection for astronomy and astrology. Both of these he practiced in a rigorously mathematical way, in keeping with skills that we can assume he developed as head of customs.

Medieval Western scientific knowledge was derived largely from Greek sources obtained in Latin translation, often through Arabic and other intermediaries. The main routes of transmission were through Italy and Spain. Following the recovery of most of the text of Aristotle by the early thirteenth century, and for about a century and a half thereafter, there was a period of Aristotelian "scholasticism." The sciences were studied in the context of the philosophical writings not only of Aristotle but also of his commentators, Islamic, Jewish, and Christian. Scholastic writers generally accepted Aristotle's account of natural and forced motion, introducing many new arguments into their accounts of how projectile motion takes place, new ideas of spatial and temporal continuity, and new forms of mathematical proportionality, which were needed for the analysis of constant and accelerated motions. Great conceptual advances were made by Thomas Bradwardine, one of a group of scholars at Merton College, Oxford, in the early fourteenth century, of whose world Chaucer shows himself to have been aware.

Those parts of the medieval university curriculum around which Chaucer's world view was chiefly organized were the four mathematically founded subjects of the Quadrivium (arithmetic, geometry, astronomy, and music). The universities taught the Quadrivium in a rather literary way, typified by the highly influential astronomical textbook of John of Sacrobosco. As more Greek texts, together with original Muslim contributions to the sciences, became available through Arabic intermediaries, the level of European expertise in astronomy rose rapidly. Chaucer was heir to half a century of vigorous Oxford astronomical practice, based in particular on the so-called Alfonsine tables, named after King Alfonso X of Leon and Castile, under whose patronage they had been produced in the last quarter of the thirteenth century.

If we are to distinguish by the word *cosmology* a branch of medieval study that was distinct from astronomy, then we presumably have in mind problems of the sort raised by Plato in his *Timaeus*, and then more systematically by Aristotle in his *Physics*, *On the Heavens*, *Metaphysics*, and other writings. Aristotle, for example, had from an early date taken over from the pre-Socratics a concept of nature as having a source of motion within itself, and his later doctrine of a *primum mobile* had the additional attraction to later Christians (not to mention Jews and Muslims) of a parallel with the actions of God. Chaucer makes use of this idea when he refers to "first moeving" in his work on the astrolabe. The idea that this outermost heaven moved the planets and everything below by a force transmitted down from it pervades Chaucer's writing.

Chaucer learned basic Aristotelian cosmology in many ways: it permeates his *Boece* and is behind Cicero's *Dream of Scipio* and Macrobius's commentary on it. Through these intermediaries it provides part of the frame for Chaucer's *The Parliament of Fowls*. He no doubt picked up material from Martianus Capella and from Sacrobosco's *On the Sphere*; he might even have read some of Aristotle's works himself. Chaucer often speaks of the intrinsic natures ("kyndes") of objects and human individuals as determining their characters and patterns of behavior. The eagle that gives a lesson in physics in *The House of Fame* explains how speech or sound "of pure kynde" is inclined to move naturally upward and how everything, indeed, has an inclination to return to its natural place.

Although Chaucer's explicit references to Aristotle are few, he shared the indebtedness of all medieval scholarship to this man, who was often simply known as "the Philosopher"—as in *The Parson's Tale* and *The Legend of Good Women*. The clerk of Oxenford in the general prologue to *The Canterbury Tales* would rather have had twenty volumes of Aristotle and his philosophy than rich robes, fiddle, or psaltery. Although Plato's name is mentioned with Aristotle's in *The House of Fame*, the cosmological survey there offered is thoroughly Aristotelian. Several doctrines in *Boece* are ascribed to the philosopher, as are materials Chaucer took from Martianus Capella and Macrobius.

If Chaucer's acceptance of Aristotelian cosmology is more or less obvious, the same is not true of his awareness of the technicalities of the astronomy and astrology of his time. How these subjects affected his entire outlook on the work is something we can only deduce from a careful analysis of the numerous astronomical references made in what seem at first sight to be no more than casual asides. In reality, they were carefully calculated, and it can be shown that they point to an intricate use of calculated analogy that relates many of Chaucer's poems to the precise places of the planets in the heavens at specific dates and times in his lifetime—as calculated by him using Alfonsine astronomical tables for London. (More than fifty dates can be derived from his various writings.)

Here Chaucer was the heir to two traditions that owed much to Claudius Ptolemaeus, the Alexandrian astronomer of the second century who wrote the so-called *Almagest*, a synthesis of all astronomy, and the astrological work known as *Tetrabiblos*. Although the parameters of the Ptolemaic theory had been revised by astronomers in the Islamic and Iberian worlds, to the writer of the Middle Ages Ptolemy (Chaucer's "Ptholomee") was the prime source of astronomical knowledge. There is in the *Almagest* a catalogue of over a thousand stars, and there is reason to think that Chaucer made use of an appropriately modified version of it for allegories in *The Legend of Good Women*, *The Nun's Priest's Tale*, *The Squire's Tale*, and *The Parliament of Fowls*. The clerk Nicholas in *The Miller's Tale* possessed an "Almageste" and an astrolabe. The astronomical system underlying Chaucer's treatise on the equatorium, *Equatorie of the Planetis*, was, of course, in character thoroughly Ptolemaic. It was, as explained, aimed at the calculation of planetary longitudes and latitudes rather than at explaining planetary movement on the basis of physics. Aristotelian cosmology was regarded to a great extent as a different subject.

The geometrical models involved in the Ptolemaic planetary system were naturally much easier to appreciate than the derived techniques of calculation, which proceeded in many stages from the determination of mean motions to true places by the incorporation of certain so-called equations. The vocabulary of the Franklin on this subject, in an account of procedures used by the clerk of Orleans, is nevertheless accurate and correctly used. Moreover, it seems likely that some of the astronomical tables owned by Chaucer himself are still extant, bound together with the unique manuscript copy of the *Equatorie of the Planetis*.

The place of astronomy in time reckoning, with astronomical calendars used for periods of days and years and for the feasts of the Church, was taught in the Quadrivium under the name of *computus*. Numerous astronomical instruments apart from the astrolabe were used for recording hours and minutes. The "chilyndre" of *The Shipman's Tale* was a portable cylindrical sundial, and dials of many other sorts were in use, including various sorts of horary quadrant with a wider astronomical value. These are hardly cosmological symbols, but the mechanical clock was indeed just that, in its earliest manifestations. The earliest simple clocks sounded a bell only, but when they sported a dial, this was often nothing less than a moving representation of the entire planetary universe, based on the astrolabe. There are clear signs of a cosmic metaphor based on either the clock or the astrolabe in *The Squire's Tale*.

Astrological notions pervade many of Chaucer's works. The clerk Nicholas in *The Miller's Tale* "had lerned art" (i.e., had studied the liberal arts at Oxford), but he also had turned to astrology. This was not formally a part of the Quadrivium, but it was certainly studied in universities, not least as a part of medicine. He could have learned much basic astrology from Chaucer's work on the astrolabe, an instrument Nicholas owned and one that in a sense encapsulated the workings of the universe. As related in the general prologue, the Doctour of Phisik was grounded in astronomy and could calculate ascendents for use in natural magic on his patients' behalf. Chaucer makes much use of the subject, but he leaves us in some doubt as to his sources. Apart from Ptolemy's *Tetrabiblos* (known in the West as *Quadripartitum*), other important ancient sources of medieval doctrine were by such persons as Dorotheus, Firmicus Maternus, and Rhetorius. Chaucer mentions the astrologer Alkabucius on a trivial point in *A Treatise on the Astrolabe*. Contrary to common belief, astrology was doctrinally safe within the orbit of the Christian Church during the Middle Ages. Although several leading clerics preached against it—most of them only after learning it at first hand themselves—it was generally thought to lead into spiritual danger only when associated with "black" arts such as nigromancy. Astrology was not entirely concerned with the fortunes of the human individual: its branches considered meteorological questions, the rise and fall of religions and sects, epidemics, plagues, the fortunes of nations, and matters of history generally. There is scarcely a branch of astrological doctrine, as practiced in his century, with which Chaucer does not show some familiarity.

As an example of Chaucer's plain use of astrological allegory, we might take *The Complaint of Mars*, which falls into three parts: the Proem, sung by a bird on St. Valentine's day; the Story, an astrological allegory of the love of Mars and Venus; and the Complaint by Mars, whose affair with Venus was disturbed by Phebus. The astrological meaning is thinly veiled and conforms to the movements of the planets Mars and Venus and of the Sun (Sonne) between 14 February and 12 April 1385. At the outset, Venus is at her exaltation. Mars enters Taurus, her domicile, on 12 March, and by 17 March the two are abed. Phebus, the Sun, knocks on the palace gates on 27 March—an event that would have been perfectly intelligible to those aware of the astrological doctrine of rays and combustion. The rays of the Sun and Mars touched on 3 April; Venus left for Gemini five days later. A second stage in the combustive process began on 12 April, a date actually named in the poem. There is much use of technical astrological vocabulary. Taken in conjunction with the Story, the

Complaint may be seen as raising questions concerning the pre-determination of human action and fortune by the stars.

Astrology was not the only means used to attempt to divine the future of man and cosmos. Geomancy provided a related alternative, that is, divination on the basis of patterns of dots. These patterns are formed according to definite rules on the basis of random figures drawn, for instance, in sand. Geomancy is possibly, but not certainly, of Arab origin. The patterns had names, and there is a reference to two of them—Puella and Rubeus—in *The Knight's Tale*. Also in *The Canterbury Tales* the parson preaches against nigromancy, the conjuring up of spirits, and exorcism, and some versions of the manuscripts have "geomancie" in place of "nigromancie," although the two forms of divination were thought to be accompanied by very different degrees of spiritual danger. There are several places in his writings where Chaucer shows a deep philosophical interest in the general problem of reconciling freedom of the human will, required by the Christian faith, with the seeming inevitability of the future as assumed by the various sciences, including astrology and related subjects.

Chaucer's cosmological outlook, in its simple Aristotelian forms, stemmed from the formal teaching of the universities and is not particularly remarkable. Overlooking its moral dimension, the medieval world view was one created by an environment that included—at least in the upper reaches of society—numerous reminders of the other intellectual systems mentioned here. Among these reminders were the clock, with its astrolabe dial; church windows which commonly showed astronomical symbolism; astrological practices in medicine; and the ecclesiastical calendar, which linked astronomy and even astrology to ecclesiastical life. Subjected as Chaucer was to such influences, it is not surprising that his poetry was suffused with astrological and astronomical coloring. It is not this that makes it remarkable: the same could have been said of other poetry of the time. What distinguishes Chaucer from all other poets is that he imposed complex astronomical structures on his poetry with such minuteness and computational skill, and yet he managed to leave these structures largely concealed. His poetry is frequently written on a variety of allegorical planes, but he presumably thought that by adding this cosmic allegory—which is less architectonic but far more intricate than Dante's—he was adding an important dimension of truth to his writing.

—*J.D. North*

BIBLIOGRAPHIC NOTE

For Chaucer's own writings, the best single-volume edition is now that of L.D. Benson, ed., *The Riverside Chaucer* (Boston: Houghton Mifflin, 1987). The most complete discussion of astronomical themes in Chaucer is J.D. North, *Chaucer's Universe* (Oxford: Oxford University Press, 1988 and 1990). The second printing includes some additional material. On calendars, see S. Eisner, *The Kalendarium of Nicholas of Lynn* (Athens, Georgia: The University of Georgia Press, 1980). A concise general account of Chaucer's view of science and its context is found in M. Manzalaoui, "Chaucer and Science," in D. Brewer, ed., *Geoffrey Chaucer: Writers and their Background* (London: G. Bell, 1974), pp. 224–261. There is useful material in C. Wood, *Chaucer and the Country of the Stars* (Princeton: Princeton University Press, 1970), and also in W. Curry, *Chaucer and the Mediaeval Sciences*, rev. edition (London: Allen and Unwin, 1960). This book was first published in 1926 and is now somewhat dated. For astronomical styles in post-Chaucerian English writing, see J.C. Eade, *The Forgotten Sky* (Oxford: Oxford University Press, 1984).

IV. The Scientific Revolution

Introduction
   A New Physics and a New Cosmology

Introduction

A NEW PHYSICS AND A NEW COSMOLOGY

lacing the sun instead of the earth at the center of the universe was, in one sense, merely a reversal of places for these two objects in schemes designed to save the phenomena and in mathematical models used to calculate the positions of the celestial bodies. Copernicus's 1543 *De revolutionibus orbium coelestium* (*The Revolution of the Celestial Orbs*) manifested a continuation of scientific goals and procedures, and indeed of a general cosmological outlook, over nearly two millennia, from the fourth century B.C. to the sixteenth century A.D. Copernicus joined with Plato, Eudoxus, Ptolemy, and many others in accepting as a reasonable and important problem that of determining what combination of uniform circular motions could be devised to agree with the observed irregular motions of the planets. Copernicus was intent on saving the phenomena, on explaining apparently irregular motions in the heavens with some combination of uniform circular motions. Thus we have placed Copernicus in this book at the end of section II, *The Greeks' Geometrical Cosmos*, instead of in his more customary place at the beginning of the Scientific Revolution of the sixteenth and seventeenth centuries A.D.

But now we are at the beginning of that revolution, and find it necessary to speak again of Copernicus and, more importantly, of some of the immense intellectual consequences that followed from his work. By moving the earth from the center of the universe, Copernicus set in motion streams of thought that would eventually, in the hands of others, demolish the Aristotelian world view that had so dominated Western thought during the Middle Ages.

While Copernicus himself continued to accept the Aristotelian boundary of the universe (the sphere of the fixed stars), now that the apparent nightly motion of the stars across the heavens was transferred from the rotation of their sphere to the rotation of the earth, and the stars themselves no longer were thought to move, there no longer was a necessity to attach them to an outer, rotating sphere.

Copernicus did not realize that his innovation had made obsolete the idea of the sphere of the stars, but thinkers after him did. In 1576 the English astronomer Thomas Digges described the great orb of the Heavens as "garnished with lights innumerable and reaching up in *Sphaericall altitude* without end." (This was not necessarily entirely a manifestation of a simple logical extension of Copernican doctrine, though. While extending spherically to infinity beyond its concave surface, Digges's heaven of the fixed stars is also a kind of "theological" heaven fused with the traditional Empyrean, with stars placed among the angels and God's elect. Also, the historian of astronomy Colin Ronan recently has suggested that Digges possessed a telescope and that his observation of many more stars than can be seen with the human eye alone could have helped inspire his diagram.)

Digges pictured the stars as extending infinitely upwards, beyond their former limiting sphere. But he continued to accord the earth and the sun a unique position at the center of the universe, leaving it for others to recognize that the concept of a center loses all meaning in an infinite universe. And once the earth came to be thought of as but one of many planets circling many suns, thoughts would be extended even further, as far as the possibility of life on other worlds.

More is said about these matters in section VIII, *Cosmology and Philosophy*, particularly in chapter 27, *Plurality of Worlds* (and also in the introduction, *The Copernican Revolution*, to section IX, *Cosmology and Religion*). Demanding our immediate attention here is the impact of Copernican astronomy on Aristotelian physics in particular, and through the subsequent demise of that physics on the Scientific Revolution in general.

During much of the 1950s and 1960s, historians generally understood the Scientific Revolution to have been an episode of profound intellectual transformation comparable in magnitude to the rise of Christianity. Moreover, this formation of our modern mentality was placed firmly in the sixteenth and seventeenth centuries. And it was thought to have occurred primarily in the classical fields of physics, astronomy, and cosmology. The Scientific Revolution did encompass biological issues as well. In the same year as Copernicus's *De revolutionibus*, the Flemish anatomist and physician Andreas Vesalius published his *De humani corporis fabrica* (*The Fabric of the Human Body*), a book marking an equally revolutionary new understanding of human anatomy. And in 1628 the English physician and biologist William Harvey published his *Exercitatio anatomica de motu cordis et sanguinis in animalibus* (*The Motion of the Heart and Blood*), reporting his discovery of the circulation of the blood and overthrowing still-prevailing opinions of the great Roman physician Galen in the second century A.D. But however revolutionary were developments in the biological sciences, the essence of the Scientific Revolution was most often found in fundamental changes in the conception of nature and in the procedures of scientific inquiry occurring from Copernicus to Newton. An animistic Aristotelian world view was destroyed, and replaced by a mechanistic Newtonian world view, the mechanical philosophy.

Such were the organizing principles accepted by many historians; such was the dominant interpretation of early modern science. The certainty was psychologically satisfying. But recently all has been thrown into doubt by a new generation of historians of science. The old picture no longer commands universal assent. Now we must find pleasure not in the contentment of seeming understanding, but in the give and take of a lively historiographic debate.

Fascinating as current debates over the nature of the Scientific Revolution are, our immediate task is to follow developments in cosmology, whether central to the Scientific Revolution as previously supposed or now relegated to a lesser role. Reassessment of the place of the earth in the universe and the resulting change from a geocentric to a heliocentric world view set in motion by Copernicus ultimately required the creation of a new physics and led, with considerable contributions from Galileo and Kepler, to Newton's revolutionary new understanding of the nature of our universe. Even before Copernicus, though, problems already had arisen in connection with Aristotelian physics, as is evident in medieval commentaries on the works of Aristotle by Jean Buridan and Nicole Oresme. Many of their arguments would be repeated in later attempts to answer logical implausibilities inherent in the combination of Aristotelian physics and Copernican astronomy.

Buridan (1295–1358) taught natural philosophy at the University of Paris, and his teachings were of considerable influence in northern European universities for centuries thereafter. The university curriculum in Buridan's time was based largely on study of Aristotle's works in logic and in natural philosophy (science), and Buridan wrote several commentaries on books by Aristotle. The standard, accepted procedure in the fourteenth-century scholastic world was to list all the questions and then to present all possible opinions. Incorrect opinions were presented first; one could tell from the order which would ultimately be accepted as correct. Nor was the spirit of the presentation modern, or scientific, in the sense of a genuine inquiry aimed at developing a new understanding of nature. Everyone knew what the questions were and what the answers would be; the aim of the exercise was a skillful presentation of known information, not the discovery of new information. Buridan is a noteworthy exception to this generalization.

In setting the earth rotating, Copernicus would raise the problem of why a stone thrown straight up is not left behind by the motion of the earth, but instead is seen to fall straight down. The concept of a force impressed into the stone from the rotating earth and carrying the stone along with the earth would answer this potential objection to the Copernican heliocentric system, and it owed much to Buridan's earlier discussion. Buridan had sought to answer how a projectile is moved. Rejecting Aristotelian explanations of motion, Buridan proposed an alternative, that in the stone or other projectile there was impressed something which was the motive force of that projectile. The motor in moving a moving body impressed in it a certain impetus or a certain motive force. It was by this impetus that a stone was moved after the projector ceased to move. The movement of the stone continually became slower, until finally the impetus was so diminished or corrupted that the gravity of the stone won out and moved the stone down to its natural place.

Aristotelian physics could lead to truths necessary to philosophy but contradictory to dogmas of the Christian faith, and in 1270 the Bishop of Paris condemned several propositions derived from the teachings of Aristotle, including the eternity of the world and the necessary control of terrestrial events by celestial bodies. In 1277 Pope John XXI directed the bishop to investigate intellectual controversies at the university, an investigation that resulted within three weeks in the condemnation of over two hundred propositions. Excommunication was the penalty for holding even one of the damned errors.

In response, a nominalist thesis was developed, with Buridan's help. While conceding the divine omnipotence that Christian doctrine maintained, it also freed natural philosophy from religious authority. Put simply, science was understood as a working hypothesis in agreement with observed phenomena. But the truth of any particular working hypothesis could not be insisted upon because God could have made the world in some different manner that nonetheless had the same set of observational consequences. Therefore tentative, not necessary, scientific theories could pose no challenge to religious authority. Buridan came to conclusions, but he did not insist that these conclusions in any way were binding or limiting upon God's power to have created the world in any other manner.

The tradition of commentaries on Aristotelian physics was continued by Nicole Oresme (1320–1382), who studied with Buridan at the University of Paris in the 1340s and subsequently translated Aristotelian texts into French and wrote commentaries on them. Regarding the possible rotation of the earth, particularly the arguments and observations commonly raised against that possible hypothesis, Oresme wrote:

> But it seems to me, subject to correction, that one could support well and give luster to the opinion that the earth, and not the heavens, is moved with a daily movement. Firstly, I wish to state that one could not demonstrate the contrary by any experience. One experience is: We see with our senses the sun and moon and many stars rise and set from day to day, and some stars turn around the arctic pole. This could not be except by the movement of the heavens. Thus the heaven is moved with a diurnal movement. Another experience is: If the earth is so moved, it makes a complete turn in a single natural day. Therefore, we and the trees and the houses are moved toward the east very swiftly, and so it should seem to us that the air and the wind blow continuously and very strongly from the east. But the contrary appears by experience. The third experience is that which Ptolemy advances: If a person were on a ship moved rapidly eastward and an arrow were shot directly upward, it ought not to fall on the ship but a good distance westward away from the ship. Similarly, if the earth is moved so very swiftly in turning from west to east, and it has been posited that one throws a stone directly above, then it ought to fall, not on the place from which it left, but rather a good distance to the west. But in fact the contrary is apparent.

Having raised observations commonly cited against the possible rotation of the earth, Oresme proceeded to demolish them:

> I make the supposition that local motion can be sensibly perceived only insofar as one may perceive one body to be differently disposed with respect to another. If a person is in one ship that is moved very carefully, without pitching or rolling, and this person sees nothing except another ship that is moved in every respect in the same manner, I say that it will seem to this person that neither ship is moving. And if the first ship is moved and the second is at rest, it will seem to the person on the first ship as if it is the second that is moved. Similarly, if the earth and not the heavens were moved with a diurnal movement, it would seem to us that the earth was at rest and the heavens were moved. From this is evident the response to the first experience, since one could say that the sun and the stars appear thus to set and rise and the heavens to turn as the result of the movement of the earth. To the second experience, not only is the earth so moved, but with it the water and the air in such a way that the water and the lower air are moved differently than they are by winds and other causes. If the air were enclosed in a moving ship, it would seem to the person situated in this air that it was not moved. To the third experience, concerning the arrow or stone projected upward, one would say that the arrow is trajected upwards and also moved eastward very swiftly with the air through which it passes, it all being moved with a diurnal movement. For this reason the arrow returns to the place on the earth from which it left.

There were other arguments against the rotation of the earth, to which Oresme also replied:

> It is said that if the heavens could not make a rotation from day to day, all astronomy would be false. I answer this is not so because all aspects, conjunctions,

oppositions, constellations, figures, and influences of the heavens would be completely just as they are. The tables of the movements and all other books would be just as true as they are.

And there were objections from Scripture:

This seems to be against the Holy Scripture which says "The sun riseth, and goeth down, and returneth to his place: and there rising again, maketh his round by the south, and turneth again to the north: the spirit goeth forward surveying all places round about, and returneth to his circuits." And so it is written of the earth that God made it immobile: "For God created the orb of the earth, which will not be moved." Also, the Scriptures say that the sun was halted in the time of Joshua and that it turned back in the time of King Ezechias.

To these objections Oresme answered:

Concerning the Holy Scripture which says that the sun revolves, one would say of it that it is in this part conforming to the manner of common human speech. Also appropriate to our question, we read that God covers the heavens with clouds, and yet in reality the heavens cover the clouds. Thus one would say that according to appearances the heavens and not the earth are moved with a diurnal motion, while in actuality the contrary is true. According to appearances in the time of Joshua the sun was arrested and in the time of Ezechias it returned, but actually the earth was arrested in the time of Joshua and advanced or speeded up its movement in the time of Ezechias. It would make no difference as to effect whichever opinion was followed.

There were esthetic reasons to place the observed rotation in the earth rather than in the heavens:

All philosophers say that something done by several or large scale operations which can be done by less or smaller operations is done for nought. As Aristotle says, God and Nature do not do anything for nought. But if it is so that the heavens are moved with a diurnal movement, it becomes necessary to posit in the principal bodies of the world and in the heavens two contrary kinds of movement, one of an east-to-west kind and the other of the opposite kind. With this it becomes necessary to posit an excessively great speed. Consider the height or distance of the heaven, its magnitude, and that of its circuit, for if such a circuit is completed in one day, one could not imagine nor conceive of how the swiftness of the heaven is so marvelously and excessively great. It is unthinkable and inestimable. Since all the effects which we see can be accomplished and all the appearances saved by substituting for this a small operation, the diurnal movement of the earth, which is very small in comparison with the heavens, and without making the operations so diverse and outrageously great, it follows that if the heaven rather than the earth is moved, God and Nature would have made and ordained things for nought, but this is not fitting.

Oresme summed up his arguments, focusing on the concept of relative motion:

It is apparent then how one cannot demonstrate by any experience whatever that the heavens are moved with daily movement, because regardless of whether

> *it has been posited that the heavens and not the earth are so moved or that the earth and not the heavens is moved, if an observer is in the heavens and he sees the earth clearly, the earth would seem to be moved; and if the observer were on the earth, the heavens would seem to be moved. The sight is not deceived in this, because it senses or sees nothing except that there is movement. But if it is relative to any such body, this judgment is made by the senses from inside that body, and such senses are often deceived in such cases, just as was said concerning the person who is in the moving ship.*

Oresme could not leave things here; he was too close to asserting truths necessary to philosophy that could be contradictory to dogmas of the Christian faith. He thus ended by presenting his work in the context of a nominalist thesis that we cannot insist upon the truth of any particular working hypothesis because God could have made the world in some different manner that nonetheless has the same set of observational consequences:

> *Yet nevertheless everyone holds and I believe that the heavens and not the earth are so moved, for "God created the orb of the earth, which will not be moved," notwithstanding the arguments to the contrary. This is because they are persuasions which do not make the conclusions evident. But having considered everything which has been said, one could by this believe that the earth and not the heavens is so moved, and there is no evidence to the contrary. Nevertheless, this seems as much or more against natural reason as are all or several articles of our faith. Thus, that which I have said by way of diversion in this manner can be valuable to refute and check those who would impugn our faith by argument.*

This was a convenient stance to take when religious matters were taken seriously and heretical opinions could get one in serious trouble with powerful ecclesiastical authorities. We should not jump to the conclusion, however, that Oresme necessarily held a modern view of religious authority and was playing a cynical game. He may well have had a very different set of intellectual values many centuries ago and he may well have been sincere in his statements. In contrast to Copernicus, who came to believe in the reality of the rotation of the earth, Oresme seemingly treated the possibility of rotation as a topic for discussion and debate but ultimately unknowable with certainty other than through faith. In doing so, he freed science from religious oversight and in turn was himself free to develop arguments that later would be invaluable to Copernicus and others concerned with the reality of the cosmos.

Medieval Aristotelian commentaries would contribute to the acceptance of Copernican astronomy, with its inherent need for a new physics to replace Aristotle's. Even more important were Galileo's telescopic discoveries and his new physical theory. His observation in 1610 of craters and mountains on the moon showed that it was not a smooth sphere, as Aristotle had maintained, but uneven and rough, like the earth. The discovery of four satellites of Jupiter further helped break down the Aristotelian distinction between the imperfect earth and supposedly perfect heavenly bodies, and also helped refute the objection that a moving earth would leave behind its satellite moon. Later, Galileo would produce a terrestrial physics compatible with motions of the earth.

Somewhat ironically, Galileo, the ultimate revolutionary in physics, was as conservative as Copernicus when it came to the astronomy of circular orbits. It remained for Johannes Kepler to break the two-thousand-year domination of the circle and set planetary astronomy on a sound technical foundation of elliptical orbits. Also finally banished from the universe were Aristotelian crystalline spheres, by the observations of Tycho Brahe. Kepler's elliptical orbits embodied Brahe's conclusion. The central problem of cosmology in the seventeenth century—what holds the planets in their orbits around the sun once the physical support of the crystalline spheres is removed—remained for Isaac Newton to answer.

The Scientific Revolution was, arguably, the most important event in Western history. The example set by Newton in discovering the natural laws that govern the physical world soon was felt in other intellectual domains, including economics and politics. Adam Smith's *Wealth of Nations* and the new economic understanding it developed is one example, the political theory of Montesquieu, Voltaire, Locke, and others, and the resulting French and American revolutions, another. Science, particularly the Newtonian example of the existence of natural laws susceptible to discovery by rational thought, fueled the movement toward political freedom for millions of human beings and the creation of the United States of America. Now, science is central to much of modern life, for good and for ill, directly and through its influence on technology. The reality of the twentieth century is unthinkable without the Scientific Revolution.

12. GALILEO

Galileo's telescopic discoveries in 1610 and 1613 and his 1632 *Dialogue*, in which a terrestrial physics compatible with motions of the earth was set forth in the language of laymen, gained for Copernicus's heliocentric astronomy its first widespread understanding and acceptance. During virtually the same years, modifications introduced by Johannes Kepler gave Copernican astronomy the sound technical basis that it had been lacking. Kepler's modifications assured acceptance of Copernicus's heliocentric astronomy among scientists, though more gradually so than was its public acceptance. Modern cosmology began half a century later with the Newtonian synthesis of Galileo's new physics and Kepler's basic laws of planetary motions.

What had to be overcome early in the seventeenth century to make way for modern cosmology was basically a metaphysical postulate: that the earth was absolutely fixed at the center of the universe. In complete agreement on this structural postulate were both Aristotelians, who dominated university science, and Platonists, who had gained a very large following among other intellectuals. The central, fixed earth was intuitively plausible. Moreover, the idea of a central, fixed earth seemed to receive support from the Bible. Thus from the outset, considerable antagonism existed among both Catholic and Protestant theologians against the writings of Copernicus, Galileo, and Kepler. Because the trial and condemnation of Galileo in 1633 by the Roman Inquisition seriously impeded the rise of scientific cosmology, it is religion rather than philosophy that has come to be regarded as the chief early obstacle to modern science. But this is a misapprehension. [See chapter 30, *Galileo and the Inquisition*, in section IX, *Cosmology and Religion*.]

Another widespread misapprehension is, in the words of the historian Alexandre Koyré, that "Galileo was a Copernican *ab initio*"—rather as if he had been born believing in the diurnal rotation and the annual revolution of the earth. But Galileo's earliest surviving composition, *De universo*, probably written toward the end of the 1580s, explicitly rejected the Copernican motions of the earth, for reasons that Galileo ascribed to philosophers and astronomers. Not until 1590, when he was professor of mathematics at the University of Pisa, did Galileo write commentaries (now lost) on Ptolemy's *Almagest*, commentaries in which Galileo apparently lent his support to a geo-heliocentric system resembling that proposed by Tycho Brahe. Within two years Galileo was revising the commentaries to allow the diurnal rotation of the earth on an axis tilted to the ecliptic. Only in 1595, after his move to Padua, did Galileo become convinced of the two principal Copernican motions of the earth. By this time Galileo was over thirty years old—hardly a Copernican *ab initio*.

To understand Galileo's cosmology, it is important to realize that he was primarily a mathematician and a physicist, not an astronomer, nor a philosopher. This is evident from all his earliest writings after his student days, writings which were on the specific gravities of substances, on motion, and on centers of gravity. An analysis of rotations of a material sphere, not astronomical considerations as such, led Galileo to a semi-Copernican position in 1591. His view that motions should be explained in terms not of forces but of other motions led Galileo in 1595 to his explanation of tides in terms of the combined rotation and revolution of the earth. Thus it was a physical rather than an astronomical phenomenon that first made him fully Copernican.

By a curious coincidence, it was also in 1595 that Kepler became convinced of the Copernican astronomy—but on purely mathematical and astronomical grounds. Kepler's approach, though original, was in the astronomical tradition, whereas Galileo's approach was not, and was far from being so. Physical reasoning had been barred to astronomers by the compromise of Geminus in antiquity, under which physics was left solely to philosophers as persons trained in causal analysis. Even before Geminus, Aristotle had drawn a sharp line between terrestrial phenomena and celestial science, on the basis of straight and circular motions.

Studies of motion and of mechanics occupied Galileo almost exclusively until 1609, when news of the invention of the telescope in Holland reached him. Galileo promptly proceeded to construct a telescope of his own. The moon attracted his attention first, and then the satellites of Jupiter, the Milky Way, nebulae, and the host of stars invisible to the naked eye. Cosmology promptly replaced "natural" (i.e., spontaneous) motion as Galileo's principal interest over nearly two decades, from 1610 to 1630. Yet he wrote little on cosmology as such, instead concentrating mainly on new branches of astronomy: satellites, sunspots, and comets. Concerning conventional planetary astronomy, Galileo wrote nothing, even in his private papers, beyond expressing his growing conviction that the Copernican system would eventually triumph. His only published statement to that effect was in an appendix to his 1613 *Letters on Sunspots*; he had found that motion of the earth must be assumed in satellite astronomy in order to calculate and predict eclipses. After 1616, Catholics were prevented by a church edict from ascribing motion to the earth. [See chapter 30, *Galileo and the Inquisition*, in section IX, *Cosmology and Religion*.]

Galileo's 1632 *Dialogue* (only after 1744 would it be titled *Dialogue Concerning the Two Chief World Systems, Ptolemaic and Copernican*) opens, as does his 1610 *Starry Messenger*, with praise of cosmology (study of the constitution or arrangement of the universe) as the highest and most noble inquiry open to mankind. Essential to the study of cosmology was a primary determination of which celestial bodies were in motion and which were at rest. Thus Galileo's original predominant interest—motion for its own sake—came again to the fore in his later predominant interest—cosmology. The organization of the *Dialogue* makes this evident, with Aristotle's analysis of motion first subjected to examination and criticism. In the second "day" of discussion, Galileo's new physics of the relativity of motion, conservation of motion, and independent composition of motions is set forth, and is supported in detail by references to terrestrial phenomena. Following this discussion of Galileo's new physics comes an examination of newly discovered celestial phenomena, with attention focused on their potential value in finding out which bodies move and which are at rest—the primary problem of cosmology. Finally, the hypothesis of Copernican motions of the earth is applied to the explanation of the tides, without the need of any appeal to force, but only to the obstructed flow of water.

Galileo's mature cosmology is more nearly an idealized form of the Copernican system (i.e., literally heliocentric) than a causal account. Like Aristotle, Galileo proposed an integrated physics and a descriptive astronomy, each internally consistent. Unlike Aristotle, Galileo did not separate the two sciences by an impassable barrier, into an "elemental" and a celestial sphere. That his proposals were mutually compatible in his view is shown by the cosmogony suggested in the first "day" of the 1632 *Dialogue*, in which the places and periods of the planets were explained by speeds acquired in uniform acceleration toward the sun from the place of the planets's creation.

In contrast, the astronomies of Copernicus, Tycho Brahe, Ursus, and Kepler were supposed to fit with the physics of Aristotle—an impossible condition. Galileo's contribution to cosmology was not a new astronomy but a new physics. The astronomy of Copernicus sufficed for Galileo, and he ignored Kepler's greatest single contribution: elliptical planetary orbits with the sun at one focus of the ellipse. Not until Newton's *Principia* in 1687 were the contributions of Galileo and Kepler united by the creation of dynamics to produce a truly modern scientific cosmology.

—*Stillman Drake*

BIBLIOGRAPHIC NOTE

Galileo's 1632 *Dialogue* is conveniently available in S. Drake, trans., *Dialogue Concerning the Two Chief World Systems, Ptolemaic and Copernican* (Berkeley: University of California Press, 1953). Other books dealing with Galileo and cosmology include: S. Drake, *Galileo at Work* (Chicago: University of Chicago Press, 1978); A. Koyré, J. Mepham, trans., *Galileo Studies* (Atlantic Highlands, New Jersey: Humanities Press, 1978); and W.R. Shea, *Galileo's Intellectual Revolution* (New York: Science History Publications, 1972). See also S. Drake, "Galileo's 'Platonic' Cosmogony and Kepler's *Prodromus*," *Journal for the History of Astronomy, 4* (1973), 174–191.

13. KEPLER

Breaking the circular orbits of Greek and Copernican cosmology would represent a major change in cosmological world views and was a key step in the Scientific Revolution of the sixteenth and seventeenth centuries. Galileo may have begun a revolution in physics, but with regard to the Greek astronomy of circular orbits, he was as conservative as was Copernicus. It remained for Johannes Kepler, using Tycho Brahe's observations of unprecedented accuracy, to break the 2,000-year domination of the circle. His setting of planetary astronomy on a sound technical foundation of elliptical orbits in turn set a new problem for cosmology in the seventeenth century—that of explaining what holds the planets in their orbits around the sun. Isaac Newton's answer completed the revolution.

Kepler's Life

Kepler was born on 27 December 1571 in Weil der Stadt in southern Germany to an undistinguished and penurious family. After studies in local schools, he matriculated at the University of Tübingen in 1587 with the support of a scholarship from the Duke of Württemberg. Here Kepler had the good fortune to study mathematics and astronomy with Michael Maestlin, who was thoroughly familiar with Copernicus's writings and seems to have accepted the heliocentric system. Kepler did well and, after receiving his master's degree, proceeded to the theological school, intending to become a Lutheran minister. In 1594, however, before completing the course, he was asked to take a position as a mathematics teacher at a school in Graz, a post he accepted despite a disinclination to do so.

While in Graz, Kepler devoted much energy to cosmological questions and wrote the *Mysterium cosmographicum*, published in Tübingen in 1596. This book brought Kepler to the attention of Tycho Brahe in Denmark, the foremost astronomer of the time. Brahe invited Kepler to join his research staff, an invitation Kepler accepted in 1600 after the school in Graz was closed and Brahe had moved to Prague.

Kepler's relations with Brahe were strained but fruitful. Kepler was set to work on the theory of Mars, an assignment that he later described as providential, since it was only for this planet that the orbit's departure from circularity had an appreciable effect upon observations. Kepler's Mars theory, perhaps his greatest accomplishment, was completed in 1605, although publication was held up until 1609 by Brahe's heirs (Brahe having died in 1601). Kepler's book, aptly titled *New Astronomy*, contains versions of what we now call his First and Second Laws.

Kepler had meanwhile succeeded Brahe as Imperial Mathematician at the Prague court, and although he was not well paid, he had the opportunity to devote nearly his full attention to various scientific works. Chief among these were *De stella nova* (published in Prague in 1606) on the 1604 supernova, and two highly original optical treatises. After a series of disasters in 1611, including the abdication of his patron Rudolph II and the deaths of his favorite son and his wife, Kepler moved in early 1612 to Linz in Upper Austria, where the governors had created a special post for him. Here he completed his *Harmonice mundi* (published in Linz in 1619), a work on cosmic harmony that had been conceived over twenty years before. This book contains the third of his planetary laws. He also wrote the *Epitome of Copernican Astronomy* (Linz and Frankfurt, 1618–1621), which was to become the most widely used Copernican textbook during the seventeenth century. Kepler's final major work, *Tabulae Rudolphinae* (Ulm, 1627), was printed in the extraordinarily difficult circumstances occasioned by the Thirty Years' War. This book, a set of planetary tables of unprecedented accuracy, did much to establish the superiority of Kepler's apparently odd planetary theories.

War and rebellion in Linz forced Kepler to find a new home, and in 1628 he moved to Sagan, in Silesia, where he was under the patronage of Albrecht von Wallenstein. In 1630, while in Regensburg attempting to collect debts owed him from the Imperial Treasury, Kepler became acutely ill, and he died on 15 November.

Sixteenth-Century Context

To understand Kepler's cosmology, one must realize that ideas of the cosmos current in the sixteenth century were radically different from our own. Furthermore, there was little consensus about many fundamental questions even among traditionalists, and nontraditional systems also abounded. With perhaps one exception, none of these cosmologies—whether traditional or not—had anything like what we would call a scientific basis. Indeed, it was partly through Kepler's work that cosmology became established as a science, that is, as an account of the perceptible universe that could give a precise enough account of phenomena to permit empirical testing. This emphasis upon precision and testability that so sets Kepler's cosmological system apart from those of his contemporaries tends to be obscured for modern readers by the more fanciful features of his theories.

It is usually said that the dominant cosmological system in the sixteenth century was Aristotle's. This statement is to some extent true, but some of the sixteenth-century versions of Aristotle's cosmological system would have surprised and dismayed Aristotle himself. Islamic commentators such as Averroes and medieval Western scholastics of various stripes had interpreted the master's writings to suit their own ends. Furthermore, since Aristotelian theories were usually prescribed by university statutes, there was a strong incentive, even when presenting innovative ideas, to do so in an Aristotelian context.

There were nevertheless some rather well-established competitors. The religious strife of the time tended to favor those who based their universe primarily on Scripture, including such features as the "waters above the heavens" of Genesis 1.1, which Aristotle would have found quite absurd. The rise of Renaissance humanism also had its effect, favoring the theories of the Stoics, which had originally been promoted by Cicero and adopted by many of his contemporaries. Many humanists preferred Plato to Aristotle, a trend furthered by the publication of the previously unavailable Greek texts of Plato's works. Plato also had his commentators and interpreters, such as Proclus in the fifth century, most of whom we now call Neoplatonists. Finally, there were writers who generally followed Aristotle but incorporated features of other authors in their systems.

Nearly all the cosmological systems proposed by these various schools of thought, Aristotelian and non-Aristotelian alike, differ dramatically from present-day theories in much the same way, namely in their treatment of the distinction between material and immaterial entities. In the sixteenth century, bodies, souls, spirits such as angels, and so on, all had their place (often literally so) in the cosmos. Thus we find Aristotle, in *On the Generation of Animals*, II, 3, comparing the heavens to the animal soul, a position that Thomas Aquinas, commenting upon Aristotle, extended by likening the celestial substance to the human intellect. It was almost universally believed that the motions of the planets were guided by angels, which were identified with Aristotle's moving intelligences of book 12 of his *Metaphysics*. The planetary bodies were embedded in invisible and intangible spheres, by which they were carried about. Although these spheres were usually thought to be real, it would be a mistake to ascribe to them material or corporeal qualities such as hardness or crystallinity. The celestial substance was quite literally of a more ethereal nature than "sublunar" matter, and did not—indeed could not, according to Aristotle (*On the Heavens*, I, 3)—possess the properties displayed by the elements. The whole cosmos was arranged according to a spatial scale of being, with the lowest, grossest parts in the center and the highest and most refined at the circumference. Not infrequently, the outermost region was the so-called "Empyrean heaven," where the angels and saints enjoy a state of beatitude and nearness to God.

Despite the prevalence of this cosmological format, one can also find the beginnings of alternative views. The two most important of these for our understanding of Kepler are what might best be called "speculative physics" and neoclassical humanism.

The speculative physical writers took their inspiration from a variety of sources, although the chief models were Plato's *Timaeus* and, to a lesser extent, the largely Stoic views of classical Roman writers such as Cicero and Galen. Plato's imaginative cosmogony was in better accord with Christian beliefs than was Aristotle's eternal and uncreated world, and a number of writers (notably Francesco Patrizi and Bernardino Telesio) followed Plato's example. They portrayed the cosmos as a natural result of the interaction of various entities and principles, both material and animate. And thus they might seem to have anticipated the scientific cosmologists of later times. But their theories were altogether lacking in the requisite precision, or the association (even in principle) with any kind of predictive power, that characterizes scientific cosmology. Indeed, Patrizi made no secret of his disdain for astronomers, who, he believed, created "monstrous" systems.

The neoclassicists, on the other hand, while espousing many different views, were united in their rejection of medieval traditions in favor of the Greek and Roman originals. Scholasticism, with its crude Latin and its emphasis on logic, seemed barbaric in comparison with the eloquence of classical antiquity. The aim of education shifted from the exploration of arguments to the emulation of the style and substance of classical models.

We can see an example of this trend in Vesalius's *De humani corporis fabrica* (1543), which was a reworking (*renovatio*) of Galen's physiology. In the realm of cosmology, a remarkable example is Girolamo Fracastoro's *Homocentrica* (1538), an attempt to revive Aristotle's belief that the planets maintain a constant distance from the center of the earth. But by far the most significant example of neoclassical *renovatio* is the reworking of Ptolemaic astronomy in Copernicus's *De revolutionibus* (1543).

It may seem surprising, to those unfamiliar with Copernicus's work, that it is largely a representation of Ptolemy. Copernicus followed his exemplar so closely that Kepler remarked that he "ever took it upon himself to express Ptolemy, not the nature of things." This emulation serves to emphasize dramatically Copernicus's departures from Ptolemy. The one for which he is chiefly known is, of course, his choice of center for the cosmos. Another—less notorious but equally radical (in the strict sense)—was his insistence that the uniform angular motions be carried out about the centers of the circles representing those motions. Both of these departures are themselves based on classical precedent. For his heliocentrism, Copernicus found ancient authority, and also used an allusion to Horace to criticize Ptolemy's "monstrous" universe, which had well-formed parts that had not been correctly put together. His insistence on regular motion about proper centers of circles was based upon Ptolemy himself, whom Copernicus criticized for violating his own principles. Thus even his innovations can be seen as expressions of neoclassicism.

There is another, perhaps more significant respect in which the Renaissance spirit infused Copernicus's work. Just as the humanists had shifted attention away from the logical minutiae of arguments and toward the study and emulation of classical models, Copernicus introduced a new criterion into the evaluation of theories of the universe: *symmetria*. This word does not simply mean "symmetry." Instead, Copernicus used the term primarily in its ancient Greek sense: "the quality of having a common measure." "Well proportioned" would be a good English equivalent. A well-proportioned model need not predict planetary positions more accurately than a badly proportioned one. It does, however, match its exemplar better in style and substance, and it also exhibits a beauty lacking in the other. Thus the criterion of *symmetria* exemplifies exactly those qualities that had been most highly valued by the humanists.

It is in this sense that Copernicus's system can be called the first truly scientific cosmology (it is the one possible exception mentioned in the first paragraph of this section). Copernicus combined two features that hitherto had been separate: he demanded that his theory provide mathematically precise results, and further, that it give an accurate depiction of the whole. The mathematicians had accomplished the first but not the second, whereas the speculative physicists mentioned above had tried to give a picture of the whole while ignoring, or even spurning, mathematical accuracy.

Tycho Brahe

A discussion of the background for Kepler's work must also include a discussion of Tycho Brahe, though this individual—who had, arguably, the greatest influence on Kepler's career—somewhat paradoxically believed in a very different cosmological system. Tycho's astronomical activities and his cosmological world view present an unusual problem for our understanding of the historical development of cosmology. His own planetary system served but briefly as a compromise between the Copernican planetary system, whose earthly motion was unacceptable to some, and the Ptolemaic planetary system, whose planetary positions, particularly for Mars, were found not accurate enough. Tycho's system did not constitute any lasting innovation in astronomical theory. Yet his work was both an essential and major part of the great revolution in cosmology beginning with Copernicus, running through Kepler, and culminating in Newton.

Tycho, who lived from 1546 to 1601, was born into the small fraction of the Danish nobility that had historically played significant roles in the administration, governance, and defense of the realm. The implications and ramifications of privilege help explain his aspirations and actions.

At a very early age, probably between his first and second birthdays, Tycho was abducted by his uncle Jørgen Brahe, who subsequently obtained permission to keep Tycho and raise him. Jørgen justified his action by pointing out that Tycho's father, Otto Brahe, had a second son, and that he should share his wealth of sons with his less fortunate brother. The consequences for Tycho's education were significant. While his younger brothers would serve prominent foreign lords to broaden their experience before taking up governance of major Danish fiefs, Tycho, beginning when he was fifteen, in 1562, traveled with a tutor among the universities of Europe, in the tradition of his foster-mother's father, Peder Oxe. Rather than the exercise of arms in a courtly atmosphere, the feudal pattern of castle and court education, Tycho spent his formative years in cities and universities.

Already interested in astronomy, probably by a solar eclipse, Tycho now began buying astronomy books and reading them in secret. The serious study of astronomy was not part of the educational program envisioned by his foster parents and supervised by his tutor. Rudimentary observations of the positions of stars and planets quickly showed him that the astronomical tables of the day were inaccurate, and Tycho soon cherished ambitions to rectify this sorry state of affairs. For that, he would need accurate observations, and appropriate instruments with which to make the observations.

Tycho returned to Denmark in 1566, just before his uncle's unexpected death. Both King Frederick II and Jørgen fell into the water under Amager Bridge near the royal castle at Copenhagen. Contemporary accounts suggest that both men had been drinking, that the king fell in first, and that Jørgen fell in while trying to fish out the king. Whether from the carousing or the ducking, Jørgen never recovered. He had planned to make Tycho his legal heir, but had only initiated the process, so instead his castle and income from hundreds of tenants reverted first to his wife, and after her death to a general distribution among the Brahe family.

His father encouraged Tycho to embark on a career of civil service to the King. Tycho, though, resisted, and resumed his university studies. It is at this time that a portion of his nose was hacked away in a duel. Dueling deaths were then common among the Danish nobility, as were disfiguring wounds. In later life Tycho took to wearing a nosepiece, probably one of gold and silver for dress occasions and a lighter one for everyday wear, kept in place by the adhesive salve that he always carried around with him in a little box. If not already interested in medicine, Tycho became so during his recuperation and rehabilitation, and he was receptive to the proposition that medicine, alchemy, astrology, and astronomy were linked in a great cosmic harmony.

Tycho returned to Denmark again in 1567, and then embarked on a third trip abroad, shortly after his twenty-first birthday. By now Peder Oxe had become lord high steward and head of the government in all but name, and he arranged that Tycho would receive the next vacant canonry at Roskilde Cathedral. Noblemen were barred by custom from virtually all scholarly positions in schools, universities, and the church, and crown pensions to men of learning were few and only for the duration of the royal pleasure. The canonries of the Lutheran cathedral chapters, though, could be awarded to noblemen as well as to commoners, and to men of learning as well as to government servants. Incomes of canonries were substantial, and there was no obligation to enter holy orders or live otherwise than as a secular lord. (Copernicus had spent his life as an unordained administrator of a cathedral chapter.) Tycho anticipated living as befit the dignity of a nobleman while at the same time pursuing a lifetime career as an astronomer and a scholar.

Meanwhile, he traveled, arriving in Augsburg in the spring of 1569. There he spent fourteen months. With the help of local craftsmen and money from the alderman of Augsburg, Tycho had constructed a huge astronomical instrument for measuring positions in the sky, so heavy that it required forty men to put it in place. But before many observations could be made, Tycho was called home in 1570 by his father, whose health was failing. He died the next year, leaving Tycho, after the estate was settled in 1574, sufficient wealth to live in financial independence and to do as he pleased. Two of Tycho's brothers made arrangements for marriages; Tycho made plans for another trip abroad.

He had already wooed and won his life's mate, in 1571, though she was not a noblewoman and thus they could not be formally married. An aristocratic marriage would have seriously compromised his interests in scholarly pursuits and his abhorrence of courtly life. During the early 1570s Tycho lived and worked at Herrevad Abbey, which was economically self-sufficient, including facilities for working wood and iron. Furthermore, a party of Venetian glassmakers was induced by Tycho to set up shop on the Abbey grounds, and Tycho's uncle was persuaded to establish a paper mill. Here Tycho began to conceive of and develop the prototype of the modern research institute.

On the night of 11 November 1572, returning from his alchemical laboratory for supper, Tycho noticed an unfamiliar starlike object in the sky. (It was a supernova, a star that had exploded and increased hundreds of millions of times in brightness.) For 1,900 years, Aristotelian cosmology had dictated that there could be no change in the heavenly spheres beyond the moon. Tycho's observation of a change in the heavens may have had little immediate impact on Aristotelian beliefs, but the decision to publish his discovery set his life's profession.

Concerned that political and social responsibilities would hamper his research efforts, Tycho decided to move to Basel. King Frederick offered Tycho the island of Hven, isolated and unassociated with any administrative obligations, and funds from the royal exchequer for the founding and maintaining of a proper establishment, whose work would redound to the credit of the country and the king (figure 1). Annual grants proved most generous, amounting to nearly one percent of the crown's total revenues. Tycho also received the rents from some forty farms on the island, and two workdays per week from each farm.

While building Uraniborg—his manor and observatory, the great instruments for observing stellar and planetary positions, a paper mill and a printing press, and the many other necessities for self-sufficiency—Tycho experienced the second of the two great astronomical events of his lifetime. The comet of 1577 was, Tycho indisputably showed with measurements of its position at different times, above the moon. The comet thus corroborated the anti-Aristotelian impli-

Figure 1. An engraving, dated 1587, of Tycho Brahe's famous mural quadrant, with a painting of Tycho pointing to the heavens. One assistant looks through the back sights, another reads off the time, and a third records the data. The dog at Tycho's feet is a symbol of sagacity and fidelity. In the background is a representation of the main building, with an alchemical laboratory in the basement, the library with its great globe and space for making calculations higher up, and above this instruments for observing. In a niche in the wall behind Tycho is a small globe, automated to show the daily motion of the sun and the moon and the phases of the moon. On either side of the small globe are portraits of King Frederick II and Queen Sophia, whose financial generosity made the whole thing possible. At the very top of the picture is a landscape and setting sun.

cations of the new star of 1572. (Furthermore, the comet was shown to move through the regions of the solar system previously thought fully filled by Aristotelian crystalline spheres carrying around the planets.) In presenting all his data, in acknowledging error, and in analyzing sources of error, Tycho was unprecedented.

There followed more than a decade of productive work on his long-envisioned renovation of the whole science of astronomy. Spectacular discoveries in the lunar theory were Tycho's most remarkable results.

By 1590, when he had finished writing and printing most of the ten projected chapters of his 850-page tome on the new star of 1572, Tycho was facing distractions that would plague him to the end of his life. As is so frequently the case in human affairs, many of the distractions were of Tycho's own making. A steady stream of curious tourists whose comings and goings had turned his summers into periods of negligible productivity probably was beyond his control. More within his control were relations with the government and the royal family, which Tycho shamefully neglected in a number of strikingly egregious ways. Changes were no doubt inevitable after Frederick's death in 1588, followed by a regency until the coronation of King Christian in 1596, but Tycho added to his own vulnerability. For some twenty years he had been funded to an almost unprecedented extent, and on a completely unprecedented and largely temporary basis. Even without having compiled a record of administration that virtually demanded some sign of official disapproval and having maintained a style of life that was offensive to at least some of the nobility and some of the clergy, Tycho's funding would surely have been the first place the new government of King Christian looked once a decision was made to recover control of the royal budget.

Deprived in 1596 of much of his previous share of government largesse, and perhaps also discouraged by a series of quarrels largely brought on by himself and the dubious legal status of his children, Tycho packed up his instruments and departed Hven in 1597. After a short stay in Copenhagen, probably to afford the Danish government an opportunity to restore his funding and keep him for the glory of the country, Tycho set sail for Germany with his household of two dozen persons and his instruments. Two and a half years were to pass before Tycho was again set squarely on his feet and his instruments set up for observations, in July 1599 in a castle near Prague under the very generous patronage of Emperor Rudolph II of the Holy Roman Empire. But his stalled professional working life never really started up in exile, and he died suddenly in 1601, seemingly of uremia due to hypertrophy of the prostate.

In the brief time in Prague before his death, Tycho established Kepler as his successor as imperial mathematician. In 1627 Kepler finally completed the planned book on planetary theories, under the name of the *Rudolphine Tables* and long after Rudolph's death in 1612. Tycho was opposed throughout his life to the idea that the earth revolved about the sun. He kept an unmoving earth in the center of his universe. Also, his world view and sense of physics was largely Aristotelian. Tycho's observations, though, including the appearance of a new star in 1572 and a comet in 1577 in the supposedly unchanging Aristotelian heavens above the moon did much to undermine faith in the Aristotelian cosmology. And it was Tycho's observations of planetary positions to a new degree of accuracy that enabled Kepler to discover the fatal flaw of the old planetary theories and to press on to inaugurate a new astronomy and a new cosmology, culminating in the work of Newton and the overthrow of the Aristotelian physics and world view after some 2,000 years of dominance over Western thought.

Tycho Brahe's Cosmology

Though undoubtedly an enthusiastic admirer of Copernicus, Tycho's admiration was for the geometry and perhaps also the astronomical data processing of Copernicus's planetary theory, not for Copernicus's cosmology or his observational prowess. The Tychonic planetary system is a rather trivial-looking inversion of the Copernican system. The sun and moon circle the earth in their orbits, and the distant sphere of the fixed stars centered on the earth revolves around the earth every twenty-four hours. The other planets then known, Mercury, Venus, Mars, Jupiter, and Saturn, were in orbits centered on the sun (figure 2). Tycho was unable to believe that the earth could be moving, and having once consciously adopted the geocentric world view, he subsequently did all of his technical thinking within that framework. Like all of his contemporaries, Tycho viewed the physical world as an articulated unit created by God. He subscribed for the most part to the account of its workings synthesized by Aristotle and adopted by the medieval church. Tycho had a strong alchemical bent, and also incorporated some Paracelsian features into his world system. Tycho's view of the astronomical world differed somewhat from the Aristotelian picture, but above the moon Tycho's view was traditional, except, of course, for his intrusion of the new star and comets and his rejection of spheres.

Tycho found that his new system, discovered or invented sometime around 1583, required that the orbit of Mars intersect the orbit of the sun. He apparently failed to notice that the orbits of Mercury and Venus would have to intersect the sun's orbit in exactly the same way. Whatever the illogicalities involved, though, even one required intersection was enough to ensure that it would take more than the flash of insight that had inspired the system to make Tycho a believer. But if Tycho was not prepared to go so far as Copernicans did to

NOVA MVNDANI SYSTEMATIS HYPOTYPOSIS AB
AUTHORE NUPER ADINUENTA, QUA TUM VETUS ILLA
PTOLEMAICA REDUNDANTIA & INCONCINNITAS,
TUM ETIAM RECENS COPERNIANA IN MOTU
TERRÆ PHYSICA ABSURDITAS, EXCLU-
DUNTUR, OMNIAQUE APPAREN-
TIIS CŒLESTIBUS APTISSIME
CORRESPONDENT.

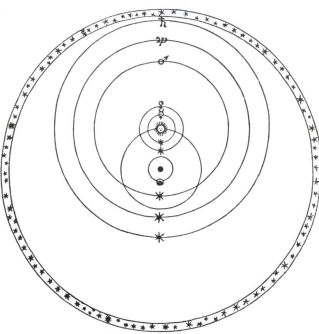

Figure 2. *The Tychonic system of the world, from an engraving in Tycho Brahe's* De mundi *of 1588. The earth is the large black dot at the center, about which are the orbits of the moon and the sun, and at the border of the diagram the sphere of the stars. Centered on the sun are the orbits of Mercury, Venus, Mars, Jupiter, and Saturn. Note that the orbits of the sun and Mars intersect, and that the sphere of the stars is placed immediately beyond Saturn's orbit.*

retain the aesthetic sense of unity and simplicity that in the sixteenth century constituted the sole grounds for taking seriously the motion of the earth, neither was he willing to abandon what he had labored so long to achieve. He must have spent many hours staring at the relationships between the geometry and the astronomy and looking for a means of escape. Unable to cope with the intersections initially, he may simply have refused to acknowledge them, to draw the system with the orbit of Mars arbitrarily enlarged so as to encompass the orbit of the sun completely and then hope for some kind of inspiration that would solve the technical problem of accounting for the observed positions of Mars.

In 1588 Nicolai Reymers Ursus, a mathematician, did publish a form of Tycho's system, with the orbit of Mars enlarged sufficiently so as not to intersect the orbit of the sun. Almost certainly Ursus had plagiarized the early version of Tycho's system, which could not account for the stations and retrogradations of Mars. Ursus had worked for Tycho briefly in 1584, before arousing Tycho's suspicion by snooping in the library, after which he was searched and then expelled by Tycho. To have made a clear case for intellectual theft in 1588, though, Tycho would have had to admit that he himself had been using the nonintersecting system in 1584, and thus to have confessed some incompetence of his own. To the end of his life, however, Tycho pursued Ursus with the charge of plagiarism and with attempts to have him punished for it. It may be that Tycho drew Kepler to his side more for potential help that Kepler could give regarding proof of Ursus's plagiarism than for any technical mathematical help Kepler could offer in determining planetary orbits.

Although it is difficult to resurrect the ontology encapsulated in medieval and Renaissance references to the celestial spheres, there can be no doubt that in the second half of the sixteenth century at least, intellectuals in general and Tycho Brahe in particular believed that something real existed in the heavens to carry the planets through their appointed rounds. Tycho confessed that he initially "could not bring myself to allow this ridiculous penetration of the orbs, so that for some time, this, my own discovery, was suspect to me."

Exactly how long Tycho's system languished in this status is a matter of dispute. It has generally been assumed that Tycho had resolved his difficulties by the fall of 1584. But as late as 1586 he had not made any plans to publish his system, suggesting that it was still in a form that he knew was unpublishable. In that year he received more observations of the comet of 1577, observations which showed that the comet had passed right through what Tycho and everyone else regarded as the Ptolemaic spheres of Mercury and Venus, and thus that those spheres could not be the solid objects everyone seems to have thought them to be. Furthermore, a manuscript Tycho was then reading stated that "the very motion of the comets is the strongest argument that the planetary spheres cannot be solid bodies."

If the comet really had moved in this fashion, there could be no solid spheres, and therefore no reason that the orbit of Mars could not intersect the orbit of the sun. When similar computations from his own data yielded similar distances, Tycho registered, in letters written in mid-January 1587, his first doubts concerning the existence of solid spheres. Tycho now wrote up an exposition of his system, had his engraver cut the illustration for it, and added it to his manuscript *De mundi aetherei recentioribus phaenomenis* . . . (Concerning the More Recent Phenomena of the Ethereal World . . .) as chapter 8, following seven chapters on the comet of 1577. The resulting book was published in 1588.

Literally to his dying day, Tycho regarded his system of the world as the most significant achievement of his career. Although his pride in his system may seem inordinate from a modern perspective, it is important to realize that in its day it represented the best of both worlds. Until the advent of the telescope, at the very earliest, the available evidence did not render belief in the mobility of the earth even plausible, let alone convincing. It is not necessary to deny either that the specter of a moving earth may have been the crucial incentive to a new physics of motion or that the traumatic seventeenth-century battle with the Roman Catholic church may have had the life-and-death implications for early modern science that its practitioners saw in the struggle. It is reasonable to insist, however, that neither of these issues had much to do with the progress of astronomy, at least during the generation or two after Tycho. The geometry of the Copernican system, on the other hand, represented a significant astronomical advance. By extricating it from the controversial, and perhaps even unrespectable, company of Copernicus's moving earth, Tycho provided an important technical service to his discipline and almost surely also hastened acceptance of the motion of the earth.

Tycho wrote in *De mundi* that it was the comet's orbit in space that had led him to explain his ideas about the construction of the universe. The ethereal world was of wonderfully large extent, with the distance to the farthest planet, Saturn, 235 times as great as the distance from the earth to the moon, which in turn was 52 times the radius of the earth. The distance from the earth to the sun was about 20 times that from the earth to the moon. It was in this vast space that the comet of 1577 moved. The Ptolemaic system for this space was too complicated, in using equant points to preserve the fiction of uniform circular motion. (In its detailed working out, though, Tycho's own system with epicycles, little rotating circles attached to the planetary orbits [deferents] and carrying the planets around with their motions, thus varying the planetary positions in complex ways, was also far from simple. Still, there was no egregious violation of uniform circular motion, as there had been in Ptolemy's system.) The Copernican system, while containing nothing contrary to mathematical principles, was in opposition to both physics and the authority of Scripture, namely in placing the earth in motion in orbit around the sun. Tycho's hypothesis, however, was in accordance with mathematical and physical principles, with observation, and with Scripture. As the orbits were not real, solid objects, but were merely geometrical representations, there was no absurdity in letting the orbits of Mars and the sun intersect each other.

One major difference between the Copernican and Tychonic worlds was their size. To explain away the absence of any measurable stellar parallax resulting from the motion of the earth, Copernicus had to postulate an immense distance from the earth to the sphere of the stars. Some of Tycho's unwillingness to consider Copernicanism stemmed from his inability to imagine that God would have created a universe containing as much wasted space as the Copernican scheme and the absence of annual parallax required. Because Tycho could not find any diurnal parallax for the stars, he had no direct way of getting information on their distances. He had no doubt, however, that they were situated just beyond Saturn.

Tycho was unable to believe that the earth could be moving, and having once consciously adopted the geocentric world view, he subsequently did all of his technical thinking within that framework. Like all of his contemporaries, Tycho viewed the physical world as an articulated unit created by God. He subscribed for the most part to the account of its workings synthesized by Aristotle and adopted by the medieval church. Tycho's view of the astronomical world differed somewhat from the Aristotelian picture, but above the moon Tycho's view was traditional, except, of course, for his intrusion of the new star and comets and his rejection of spheres.

Tycho did not expect his book *De mundi* to overthrow the Aristotelian world view, nor did it. Nor did a later volume by Tycho on the new star of 1572 result in the immediate overthrow of Aristotelian cosmology. Tycho did, though, succeed in *De mundi* in establishing the celestial nature of comets, thus helping set in motion a process that would end only with the complete overthrow of Aristotelian cosmology. The eventual result was so radical that it very likely would have left Tycho, had he lived to see it, with serious doubts about the wisdom of having set the process in motion in the first place.

Kepler's Background and Inclinations

The purpose of the above rather lengthy excursion into sixteenth-century cosmological theories, Tycho Brahe's and others, was not to give a complete picture of such theories. Rather, the discussion has the more modest aim of describing a few of the main trends that are of particular importance for understanding Kepler's cosmological ideas. Although Kepler realized that much of his work appeared revolutionary, he saw it as a continuation of a number of traditions and went to some trouble to show this. For example, in the introduction to Book IV of his *Epitome of Copernican Astronomy* (1620), Kepler invites comparison with Aristotle's *On the Heavens* and says that Aristotle "would never have scorned Tycho Brahe or even myself, if that fatal necessity of the generations had made us contemporaries." So we would do Kepler an injustice to ignore the traditions to which he refers.

Kepler received a sound classical education at the University of Tübingen, one of the foremost Lutheran universities of the time. Nevertheless, there was one unusual feature of his training: his mathematics professor, who also taught astronomy (a branch of mathematics), was Michael Maestlin, probably the only Copernican professor of the time. Kepler later claimed not to have had any particular inclination toward astronomy as a student. Nonetheless, he became friendly with Maestlin, and on at least one occasion argued a Copernican thesis in a public debate. He thus appears to have had a sound mathematical education, and he had the inestimable benefit of being the world's first "second generation" Copernican, having explored the issues of heliocentrism while still in school.

Although he had thoroughly studied the writings of Aristotle, Kepler was by inclination a Neoplatonist, with a particularly high regard for the mathematical Platonism of Proclus. From the very beginning, we find him arguing that the Creator made use of eternal mathematical archetypes in making the visible universe, just as the Demiourgos in Plato's *Timaeus* had done. Kepler nonetheless differed from these earlier writers in being a Christian. He was accordingly inclined to weave Christian motifs, particularly the Trinity, into his Platonic cosmos. Something of the flavor of his first cosmological thoughts is shown in his early letters to Maestlin (quoted extensively below). Here we find the curved surface of a sphere compared to God, while creatures are represented by straight lines. "The Trinity," Kepler continues, "is consequently in the globe: the spherical surface, the center, and the comprehended space. Thus in the world of rest are the fixed stars, the sun, and the aura or intermediate aether; and in the Trinity, the Son, the Father, and the Spirit." Although Kepler later criticized his early speculations, he never abandoned this Neoplatonist-Trinitarian icon, but used it as the basis for the further elaboration of his cosmology. It therefore makes sense to begin with an exposition of this early model.

Early Thoughts: The Cosmographical Mystery

The *Preface to the Reader* of the *Mysterium cosmographicum* (1595),

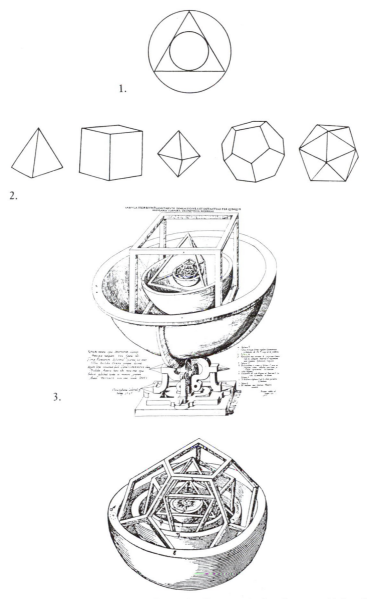

Figure 3. 1: *simplified two-dimensional example of an inner circular planetary orbit inscribed within a triangle and an outer orbit circumscribed around the triangle. 2: the five perfect, Platonic, Pythagorean solids, in each of which all faces are identical: tetrahedron (or pyramid) of 4 equilateral triangles; cube of 6 squares; octahedron of 8 equilateral triangles; dodecahedron of 12 pentagons; and icosahedron of 20 equilateral triangles. 3: Kepler's model of the universe with Saturn's sphere outermost. 4: detail of Kepler's model of the universe, from Mars' sphere inward to earth, Venus, Mercury, and the sun (in the center).*

which was Kepler's first major publication, gives an engaging account of how he hit upon the basic schema of his universe. He had been racking his brains for some time in search of an answer to the question of how the cosmos as we see it expresses the nature of its Creator as revealed in Christianity. "Almost a whole summer was wasted on this ordeal," he writes. "Eventually, a minor incident brought me closer to the truth of the matter. I have thought that it was by divine will that I hit by chance upon that which I was unable to attain by any amount of effort, and have been all the more convinced of this because I had always prayed to God that if Copernicus has spoken the truth, all would come out as it did."

The chance occurrence he mentioned was an idea that struck him in the middle of a class he was teaching, which was on quite a different subject. A diagram he was drawing suggested a new way of determining planetary distances, using inscribed polygons. Although this method was unsuccessful, it suggested an analogous line of approach (figure 3). Kepler continues:

> And the end of this fruitless attempt coincided with the beginning of the last, successful one. For I realized that, following this route, I would never [. . .] find a reason for there being six movable orbs, rather than twenty or a hundred. And nevertheless, the diagrams were satisfactory, in that they were quantitative, and were prior to the heavens. [. . .] Again I set to work. Why should there be plane figures between solid orbs? Solid figures should have been brought in instead. Behold, reader, this discovery, and the subject matter of this entire little book. For if anyone with the least knowledge of geometry were told this in as many words, he would immediately recall the five regular solids with the ratio of the circumscribed orbs to the inscribed ones, and there would immediately appear before his eyes that Euclidean scholium to Proposition 18 of Book 13, where it is proved that there cannot be, or be imagined, more regular solids than five. [. . .] As a memorial of the event, I present you with this idea, just as it occurred, in the words conceived at that very moment: "The Earth is the circle which is the measure of all. About it circumscribe a dodecahedron. The circle enclosing this will be Mars. About Mars circumscribe a tetrahedron. The circle enclosing this will be Jupiter. About Jupiter circumscribe a cube. The circle enclosing this will be Saturn. Now in the Earth inscribe an icosahedron. The circle inscribed in this will be Venus. In Venus inscribe an octahedron. The circle inscribed in this will be Mercury." Here you have the reason for the number of the planets.

It should be remarked that Kepler did not believe the Platonic solids to be physically contained in the heavens. Rather, they were conceived as archetypes that govern the disposition of the planetary orbits. An analogy from modern science is the set of principles governing the arrangement of orbitals around the nucleus of an atom.

The rest of the *Mysterium cosmographicum* is, unfortunately, not as concise as this account in the preface. However, Kepler covers much of the same material more briefly and effectively in his letters to Maestlin announcing his discovery. In the third of these (3 October

1595, Old Style), after giving the reason for the number of the planets, he unfolds the image further:

> *Accordingly, the sun, in the middle of the movables, being itself at rest and nonetheless the source of motion, bears the image of God the Father, the Creator. For what creation is to God, motion is to the sun. It moreover moves in the fixed stars, just as the Father creates in the Son. For unless the fixed stars had generated Place by their being at rest, nothing could have moved. I held this as an axiom even while I was at Tübingen. Now the sun disperses a moving power through the medium in which the movables are, and in just this way the Father creates through the Spirit or the power of His Spirit. And now, from the necessity of the presuppositions, it follows that the motions are proportional to the distances. I have therefore concluded that the opinion that preserves this same order in motions and distances is true; that which does not, is false.*

After going into the features of the nested polyhedra in some detail and asking for his teacher's comments, Kepler returns to the sun and to the dynamic role it plays in his cosmos:

> *I shall now add a summary of what follows from the written theses. As I have said, there is a moving soul in the sun. If an equality of motion, the same power, were to come from the sun into all the orbs, one would nevertheless go around more slowly than another because of the inequality of the circuits. The periodic times would be as the circles, for the length measures the motion. But the circles are as their semidiameters, that is, as the distances, and so we might easily set forth the mean distances given precisely known mean motions. But there is another cause that enters in, that makes the more distant ones slower. Let us take an example [experimentum] from light, since light and motion are conjoined in action just as they are in origin, and perhaps light itself is the vehicle of motion. So, in a small orb, and likewise in a small circle close to the sun, there is just as much light as in a greater and more remote one. So the light is more tenuous in a great circle, and more dense and powerful in a small one. And this strength is again in [inverse] proportion to the circles, or to the distances. Now if the proportion of the motion is the same (and nothing more probable than this could be posited), it would follow that the distances have a double effect in retarding the motion. Therefore, half the difference of two motions added to the lesser motion will be the distance of the more remote, while the lesser motion itself will be the distance of the nearer.*

Here we have Kepler's first step toward what we now call Kepler's Third Law, which describes the relation between a planet's mean distance and its periodic time of revolution around the sun. Although this first step is quantitatively incorrect, the qualitative argument is perfectly sound. And indeed, the numerical agreement between his "half difference" method and the Copernican distances is within a few percentage points. This apparent success led Kepler to wonder whether the fault lay in his theory or in errors in the Copernican numbers:

See how close I have come to the truth. And hitherto you doubt that I weep exceedingly whenever anything like this falls out. For, by Hercules, this is the a priori road to correction of the motions; and there is some hope if others, to whom observations are available, try it.

Evidently, in his youthful exuberance, Kepler believed that it would prove possible to deduce all the planetary motions from his archetypal principles.

In the *Mysterium* itself, in chapter 22, he carries the dynamic theme a step further, applying it to the variation of each planet's speed as its distance from the sun changes.

The cause and manner [of this nonuniformity of motion] *are shown more clearly by these arguments of ours, if, that is, the cause of slowness and swiftness in the individual orbs is the same as it was* [. . .] *for the world as a whole: the planet's eccentric path is slow when higher, fast when lower. It was in fact for displaying this motion that Copernicus posited epicycles, Ptolemy equants.*

Epicycles and equants were purely geometrical devices that had the effect described by Kepler. It was his particular genius to see that a dynamic principle could be substituted for the physically less plausible geometric one. Yet his cosmos still contained a strong animistic element. For after showing this equivalence, Kepler adds:

You now have the cause for the center of the equant's [position]: *be it that the whole world is filled with a soul, that carries along whichever of the stars or of comets it reaches, and with that nimbleness required by the distance of the place from the sun and the strength of the power there; and next, be it that each planet has its individual soul by whose aid the star would ascend in its circuit. And the same things will follow even with the orbs removed.*

Within a few years of writing this, however, Kepler had abandoned the moving souls. In a later edition of the *Mysterium*, he remarks:

If for the word "soul" you substitute the word "force," you have exactly the principle from which celestial physics is developed in the Commentaries on Mars [i.e., the *New Astronomy* of 1609] *and elaborated in Book IV of the* Epitome of Copernican Astronomy. *For I once believed that the cause moving the planets was entirely a Soul.* [. . .] *But when I considered that this moving cause weakened with distance, and that the sun's light too is attenuated with distance from the sun, I came to the conclusion that this is some kind of force, if not literally corporeal, then at least figuratively so. For we likewise say that light is some kind of corporeal thing, that is, an emanation* [species] *emitted by a body, although an immaterial one.*

Here may be seen the main features of Kepler's universe: the six planetary orbits whose spacing is determined by the five Platonic solids, with the sun in the center radiating light and moving power out into the movable world, in imitation of God the Father.

The Region of the Fixed Stars: Arguments Against Infinity

In his earliest writings, Kepler (like most astronomers of that time) gave little attention to the fixed stars and their region. However, the supernova of 1604, which Kepler studied carefully, raised many questions about the nature of the stars and provided a convenient occasion for writing about them.

By the time he wrote his book *De stella nova* (1606), he was aware of the infinite universe of Giordano Bruno, and undertook to refute the idea. Kepler recoiled from such vastness, saying, "The mere thought brings with it I know not what of secret horror, when one suddenly finds himself wandering in this immensity, which is deprived of boundaries, of middle, and therefore of all definite positions." On conceptual grounds, he objected to the idea that any actually existing thing could be infinite. The notion of an actual star whose distance from us is infinite seemed absurd. Here, his argument closely resembles Aristotle's treatment of the infinite in the *Physics*.

It is Kepler's physical arguments against infinity, however, that are the most remarkable. He made use of the accepted values for the apparent diameters of the fixed stars in an attempt to prove that the universe could not be visually homogeneous. He took the three stars in the belt of Orion as an example. An observer located on one of those would see the others as very large, not at all like the stars as we see them. Nor would it help to move them out to different distances, since their diameters would have to increase proportionally. The observer would then see many large stars on one side, and a great gulf containing the sun on the other. In conclusion, he wrote, "however large one assumes the world to be, the manner of distribution of the fixed stars is such, by the measure of our vision, that this place of ours among those fixed stars possesses something special, distinguished by clear evidence (i.e., a large space empty of fixed stars) from the place through which those stars are distributed." He also noted that the Milky Way surrounds us in an unbroken circle, which lends support to the idea that we occupy a special position in the universe.

Since this was written over three years before Galileo's first telescopic observations of the fixed stars, Kepler could not have known that the accepted values for apparent stellar diameters were wrong: the telescope showed that these apparent diameters were imperceptibly small. This considerably weakened the force of Kepler's argument. Nevertheless, in Book I of the *Epitome of Copernican Astronomy* (1618) he presented a modified version of the argument, without the stellar diameters. His approach here was to imagine how the universe would appear if the stars were about the same size and were distributed nearly uniformly. There would then be about twelve nearby stars, a larger number of stars twice as distant, a yet larger number of stars thrice as far, and so on. This progression, he argued, would quickly reach a distance at which the stars would no longer be visible, without including very many stars in the total number. Kepler believed that far too many stars are visible to fit this model, and so concluded that even if the stars were at various distances from the sun, there would nonetheless be a large empty space surrounding the sun.

It is especially remarkable that these arguments from *De stella nova* and the *Epitome* appear to be the only attempt in the entire first half of the seventeenth century to use observations to decide among different models of stellar distribution. Nor was there any response to Kepler's arguments: they appear to have been entirely ignored. This fifty-year period was crucial for the establishment of the idea of an infinite, or indefinitely large, universe in which the sun is in principle like the stars. The lack of any attempt to respond to arguments, based on phenomena, claiming to prove the finite extent and spherical structure of the universe, put forward by the most prominent astronomer of the time, shows how little the phenomena had to do with the establishment of certain features of the universe that have come to characterize all later cosmologies.

Fine Tuning: The Harmonics of the World

In his later work, Kepler retained the general schema he had worked out in the *Mysterium cosmographicum*, while he adjusted the predicted dimensions by adding new criteria. The most important of these was cosmic harmony. Shortly after writing the *Mysterium*, Kepler expressed his intention of writing a work "on the harmony of the world." Although he was unable to realize this intention for many years, he continued to work on it, making a detailed study of Ptolemy's *Harmonics* as well as more recent music theory. Meanwhile, as he worked with Tycho's observations, he came to realize that the planetary distances could not be made to fit the polyhedral model exactly, and so conjectured that some kind of harmonic principle, analogous to musical harmonies, might have been introduced into the planetary orbits by the Creator. This might be able to account for the discrepancies in the distances.

What Kepler discovered was that the angular motions of the planets, measured around the sun at their maximum and minimum distances, corresponded to the simple whole-number ratios of musical harmonies. The agreement was very close. The final piece of the puzzle was solved when he found what we now call Kepler's Third

Law, namely, that the cubes of the mean planetary distances are proportional to the squares of the periodic times. He was able to use this relation, together with the periodic times (which were known with much greater accuracy than the distances) to compute mean distances. Next, he used the harmonic proportions to determine the planetary eccentricities. The resulting aphelial and perihelial distances agreed very closely—usually within a fraction of one percent—with those obtained from the observations. It is curious to note that for some planets, Kepler's harmonically derived distances are actually closer to the modern values than are the observationally determined ones.

Reception of Kepler's Cosmological System

Kepler's cosmology has been aptly likened to the art and architecture of the Baroque period. The complexity of the system—with its Platonic, or better, Neoplatonic, hierarchy of interacting archetypes—accords well with the Baroque fondness for simple elements that interact and interweave to produce extremely elaborate, though symmetrical, forms. The system does not, however, appear to have found an appreciative audience. Although Kepler's planetary theories forced assent by their unprecedented accuracy, and there were a few who adopted the physical theories that accompanied them, Keplerian cosmology appears to have been a dead end. Even the Danzig brewer and astronomer Johannes Hevelius (1611–1687), probably the last true Keplerian, abandoned his master's harmonic and cosmological principles while retaining his physics. Whatever lasting influence Kepler had upon cosmology arose from two aspects of his work. The best known of these is his working out of the three principles of planetary motion that we now know as Kepler's laws. It was on the basis of these that Newton established universal gravitation, which is a crucial component of all subsequent cosmological thought. Less well known, though perhaps as influential, was Kepler's insistence that mathematical theories make physical sense, and that, in turn, even the most speculative and far-reaching physical and cosmological theories must stand up to observational testing.

—W.H. Donahue
With material on Tycho Brahe by Victor E. Thoren

BIBLIOGRAPHIC NOTE

The best account of Kepler's cosmology (nontechnical, well informed, and well written) is J. Field, *Kepler's Geometrical Cosmology* (London: The Athlone Press, 1988). The standard biography is M. Caspar, *Johannes Kepler*, translated by C.D. Hellman (New York: Collier, 1962). A new edition, with much-needed footnotes and index by O. Gingerich and A. Segonds, is in preparation. An excellent introduction to technical aspects of Kepler's astronomy is Curtis Wilson, "How Did Kepler Discover His First Two Laws?," *Scientific American*, *226:3* (1972), 93–106; reprinted in C. Wilson, *Astronomy from Kepler to Newton* (London: Variorum Reprints, 1989), section II, pp. 1–14. Technical aspects are treated very thoroughly in Bruce Stephenson, *Kepler's Physical Astronomy* (New York: Springer-Verlag, 1987). F. Hallyn draws the big picture, showing how Kepler's thought relates to literature and art of the Renaissance, in *The Poetic Structure of the World: Copernicus and Kepler*, translated by D. Leslie (New York: Zone Books, 1990). On Tycho Brahe, see Victor E. Thoren, *The Lord of Uraniborg: A Biography of Tycho Brahe* (Cambridge: Cambridge University Press, 1990).

14. NEWTONIAN COSMOLOGY

Isaac Newton's *Mathematical Principles of Natural Philosophy*, or *Principia* as it is universally known from the central word of its Latin title, expounds most of the features of Newtonian cosmology. Published toward the end of the seventeenth century, in 1687, the *Principia* represented the culmination of what is called the Scientific Revolution; Newtonian cosmology did likewise. The basic problem it addressed was the problem raised by Copernican astronomy, which placed the sun rather than the earth at the center of the universe. Early in the seventeenth century Johannes Kepler gave Copernican astronomy its enduring form with his three laws of planetary motion, and in doing so he set the critical problem at the heart of seventeenth-century science, the problem to which Newtonian cosmology offered the enduring answer. Geocentric cosmology had posited the existence of solid, crystalline spheres, which carried the planets in their motions around the earth and which furnished the physical structure of the universe. Kepler had convinced himself that the observations of Tycho Brahe, especially his observations of comets, proved that there were no crystalline spheres; as Kepler put it, Tycho had shattered the crystalline spheres. Kepler's elliptical orbits embodied that conclusion. The central problem of cosmology in the seventeenth century was to explain how the planets continue, millennium after millennium, to retrace the same paths around the sun through the immensity of space.

Kepler himself proposed an answer to the problem, but his answer was couched in terms drawn from Aristotelian mechanics and employed a set of concepts that no one following him found acceptable. René Descartes offered a different answer in terms that the century did find acceptable. His universe consisted of an infinite number of huge whirlpools, or vortices, of cosmic matter. Our solar system formed the center of one of the vortices; the planets, swimming in the matter of the whirlpool like twigs in a stream, were borne in closed paths around its center. The theory of vortices explained, in the mechanistic concepts that the seventeenth century espoused, why all of the planets move (the outer ones more slowly than the inner) in the same direction in about the same plane around a luminous central body. It explained comets, which it treated as erratic, nonperiodic bodies thrown from vortex to vortex, in a similar way.

Newton met both Keplerian astronomy and Cartesian natural philosophy while he was an undergraduate in Cambridge during the early 1660s. He had been born on Christmas day, 1642, in the village of Colsterworth in Lincolnshire, to an illiterate yeoman farmer, who was in fact already dead by the time his only son was born. Largely because his mother came from a family accustomed to learning, Newton was sent to grammar school in nearby Grantham. His mother intended that he would manage her estate, which had been considerably increased by a second marriage, but when it became apparent that he would be impossible in this role, he was sent to Cambridge in 1661, and here his genius unfolded. Casting aside the established, traditional curriculum, Newton discovered the new science, including expositions of Keplerian astronomy and the works themselves of Descartes. His cosmology, constructed over a period of twenty-five years as he gradually defined the issues with greater clarity, supplied a dynamic foundation to the first by destroying the system of the second.

All of this transpired in Cambridge, where Newton became a Fellow of Trinity College soon after completing his undergraduate career and not long after that the second Lucasian Professor of Mathematics. He was a lonely figure in Cambridge, pursuing his studies of mathematics, optics, and mechanics mostly in isolation from the academic community around him. The scientific subjects we associate with Newton today were not the only ones that occupied him. He immersed himself in alchemy, collected the vast literature it had produced, read deeply in it, and experimented in a garden laboratory he constructed outside his chamber. He plunged into theology as well, adopted the ancient heresy of Arianism (secretly, for knowledge of his views would have led to his immediate expulsion from the university), and composed interpretations of the Biblical prophecies, which he related to his Arianism. All of this bore upon his developing cosmology, and when a visit from Edmond Halley turned him back to topics he had encountered first during his student years, he wove all of his intellectual pursuits into the *Mathematical Principles of Natural Philosophy*.

Newton's great book made him the most admired scientist in England. Had he never published it, the final chapters of his life would have been entirely different. As it was, the government of

England appointed him Warden, and later Master, of the Mint, and he moved to London where he spent his final thirty years. The Royal Society elected him President in 1703, and by the time he died in 1727, Newton's authority in matters of science was recognized, not only in England, but throughout the Western world.

His initial exercise in the direction of his eventual cosmology began with a problem he found in Galileo's *Dialogue*. A standard objection against the Copernican universe held that the earth cannot turn daily on its axis because the centrifugal effect would fling loose bodies into space, and this does not happen. Newton saw that the objection involved a simple issue in quantitative science. He could calculate the centrifugal acceleration from the accepted size of the earth and its period of rotation, and he could compare that with the measured acceleration of gravity. To carry out the calculation Newton had to analyze the dynamics of circular motion, arriving at an expression for the radial force, which he thought of at that time as centrifugal, identical to the one still used in mechanics. In a paper composed about 1666 Newton substituted the relations of Kepler's third law into this expression and found that the centrifugal forces of the planets, with the simplifying assumption that the planets move in circles, vary inversely as the squares of their distances from the sun. Implicitly treating circular motion as a state of equilibrium between opposed forces, he compared the centrifugal acceleration of the moon (which would be equal to the opposite acceleration generated by the force that holds the moon in its orbit) with the acceleration of gravity at the surface of the earth. With the moon 60 earth radii away, an inverse square relation would have made the acceleration of gravity 3,600 times greater. Newton was using too small a measurement for the size of the earth, so that he arrived at a number considerably larger than 4,000. Significantly he chose to state the result as "4000 and more," diminishing the discrepancy, and years later, in speaking of this calculation, he said he found that the two quantities being compared answered "pretty nearly." At the time he thought some effect of the vortex, a notion to which he still subscribed, accounted for the difference. The calculation remained in his memory as a significant step toward universal gravitation.

More than ten years later, in a letter to Robert Hooke toward the end of 1679, Newton suggested an experiment that might reveal the diurnal rotation of the earth. If the earth turns on its axis, the top of a tower is moving faster than the base, and an object dropped from it ought to fall, not far to the west as old Aristotelian objections had it, but slightly to the east. Newton drew a diagram in illustration, treating the path of fall as part of a spiral that ended at the center of the earth. This was a serious error, as Hooke pointed out, and in the ensuing correspondence, in which both men assumed that the object was dropped at the equator and that the earth was split apart so that the object moved below the surface without resistance, the problem of fall on a turning earth became the problem of orbital motion. Indeed, Hooke stated that his initial objection derived from his theory of orbital motion, which assumed a rectilinear tangential motion that is constantly deflected by a force toward the center; and it seems clear that Newton's term *centripetal force*, which he coined soon after this correspondence and which replaced his earlier notion of a force to recede from the center exerted by a body moving in a circle, expressed his debt to Hooke—an inverted and thus corrected conceptualization of the dynamics of circular motion. When Newton sketched the path of a body dropped in a constant force field, Hooke also indicated that he was assuming a force that varies inversely with the square of the distance, and he challenged Newton to employ his known skill in mathematics to demonstrate the shape of an orbit in such a force field. As we have seen, Newton had earlier proved to himself that Kepler's third law demands an inverse square force. He later asserted that early in 1680, in response to Hooke's challenge, he demonstrated that when the force is directed toward one focus, an elliptical orbit entails an inverse square force. As Proposition XI of Book I, this demonstration would later be one of the crucial foundation stones of the *Principia*.

Nevertheless, Newton had not yet arrived at the concept of universal gravitation. In the winter of 1680–81, two spectacular comets appeared in the skies of Europe, the first one seen before dawn moving toward the sun, and then later one in the evening sky moving away from the sun. John Flamsteed, the astronomer royal, was convinced that the two were in fact a single comet that reversed its direction in the vicinity of the sun. In correspondence with Flamsteed, Newton, who had recently analyzed planetary motion and understood the impossibility of Flamsteed's fantastic magnetic explanation, refused to apply his orbital dynamics here and refused to accept Flamsteed's suggestion that the two appearances were a single comet. Comets had always been considered foreign bodies, which were not governed by the same principles as planets. In 1681 Newton still shared this point of view.

Three years later, in August 1684, he received a visit from Edmond Halley, who had been in a discussion with Hooke and Christopher Wren in London. From Kepler's third law and Christiaan Huygens's analysis of circular motion, the three men had concluded, much as Newton had twenty years earlier, that there must be a force toward the sun that varies inversely as the square of the distance, and they had discussed what the shape of an orbit in an inverse square force field would be. Newton informed Halley that he had demonstrated that it was an ellipse, and he promised to send him a copy of the proof. What Halley received three months later was a short treatise of nine pages with the title *On the Motion of Bodies in an Orbit*, known from the first two words of the Latin as *De motu*. *De motu* presented a dynamics of orbital motion in nine propositions (all of which made their way into the *Principia*), which demonstrated Kepler's three laws from basic principles. The tract contained two additional propositions on motion through resisting media, the beginning of what became Book II of the published work.

As the content of *De motu* suggests, something about Halley's problem seized Newton in a way that had not happened in his earlier encounters with orbital motion, and he refused this time to lay it aside. In the months following the original composition he drafted two successive versions of *De motu* and then some separate papers devoted to the laws of motion and to the definition of concepts necessary to a mathematical science of dynamics. Already his mind was racing beyond orbital dynamics. He now considered comets as planet-like bodies governed by the same laws. In the third version of *De motu* he mentioned a new correlation of the moon with the measured acceleration of gravity. Using the recent French measurement of the earth, Newton found that the correlation between the acceleration of gravity (determined with great precision from the length of a seconds pendulum) and the centripetal acceleration of the moon did not merely answer pretty nearly as the mistaken calculation in 1666 had; it now yielded an exact inverse square ratio. In correspondence with Flamsteed at this time, Newton asked if Saturn and Jupiter revealed any anomalies when they were in conjunction—a question clearly inspired by the notion that the two planets attract each other.

More suggestive yet was the definition of quantity of matter, or mass, a concept that entered physics at this point. Newton stated that mass is usually proportional to weight, and he suggested a device by which to compare the quantities of matter in two different substances. Hang equal weights of the two on pendulums of equal length, and the quantities of matter will be inversely proportional to the number of oscillations in equal periods of time. Newton crossed this passage out, and opposite it he reported that when tests were carefully made with gold, silver, lead, glass, sand, common salt, water, wood, and wheat, they resulted always in the same number of oscillations. That is, the earth must attract all the particles that compose these different substances in exactly the same way. Then Newton began to realize that

God was performing the same experiment for him in the heavens. If the planets obey Kepler's third law, the sun must attract each planet with a force exactly proportional to its quantity of matter. If the satellites of Jupiter also obey Kepler's third law, Jupiter must attract them with forces exactly proportional to their quantities of matter. If the satellites of Jupiter remain in orbits concentric with Jupiter, the sun must attract both them and Jupiter with forces exactly proportional to their quantities of matter. Sometime in the early months of 1685 the principle of universal gravitation appeared in full clarity to Newton. It was probably at that moment that he undertook to expand the tract *De motu* into the book he would call the *Mathematical Principles of Natural Philosophy*. During the following two years he devoted himself entirely to that task with a tenacity nothing short of ferocious.

The central feature of the cosmology that Newton spelled out in the *Principia* was mechanical—its treatment of celestial motions as problems in mechanics governed by the same laws that determine terrestrial motions. Although Newton's solution to the problem differed from earlier ones, his convictions that celestial motions are problems in mechanics and that the mechanics of the heavens differ in no way from terrestrial mechanics repeated fundamental themes of the entire Scientific Revolution. His mechanics, moreover, were quantitative, in a precise manner, so that the first requirement on his celestial dynamics was that it generate another earlier achievement of the Scientific Revolution, Kepler's mathematical kinematics of the heavens. Throughout the investigation that led to the *Principia*, Kepler's laws functioned as the quasi-empirical controls.

Newton's first law of motion stated that bodies (including planets) remain at rest or move uniformly in straight lines unless external forces alter their state. If in fact the planets follow closed orbits around the sun, there must be some force of attraction toward the sun that continually draws them away from rectilinear paths and holds them

---

Kepler's Laws of Planetary Motion

I. Each planet moves about the sun in an orbit that is an ellipse, with the sun at one focus of the ellipse.

II. The straight line joining a planet and the sun sweeps out equal areas in space in equal intervals of time.

III. The squares of the sidereal periods of the planets are in direct proportion to the cubes of the semimajor axes of their orbits.

in those orbits. He was able to demonstrate that any centripetal force, whatever its law in proportion to distance, leads to Kepler's second law; that is, the radius vector drawn to the center of force sweeps out areas that are proportional to time. As we have seen, he had demonstrated earlier that an elliptical orbit around a center of force located at one focus entails an inverse square force, and he extended the demonstration that Kepler's third law requires an inverse square force so that it applied to elliptical orbits and not just to circles.

Newton's dynamics bound terrestrial motions equally to the same system. Early in the seventeenth century, soon after Kepler defined his laws, Galileo, by defining the concept of uniformly accelerated motion and identifying free fall as such, laid the foundation of the science of motion that Newton extended. The genius of Newton's theory lay in the fact that Galileo's kinematics of free fall also emerged as a necessary consequence of Newton's science of dynamics. Galileo's observation that all bodies fall with the same acceleration was a consequence of universal gravitation fundamentally identical to the pendulums that Newton found to beat in synchrony.

Newton's cosmology represented more to him, however, than a system that bound together the empirical data summarized in Kepler's and Galileo's kinematics of celestial and terrestrial motions. In Book I of the *Principia* Newton consistently explored the entire range of force laws, that is, forces that vary in different proportions to distance. Among them, two proved to hold special interest from a purely theoretical point of view—forces that vary inversely as the square of distance, and forces that vary directly as distance. With such forces, and with such forces alone, bodies will orbit in ellipses—with the center of force located at a focus in one case and at the center in the other. With such forces, and with such forces alone (except for increasingly implausible proportions such as $-11/4$, $-26/9$, $-47/16$, and so on), the lines of apsides of orbits (i.e., the major axes of elliptical orbits) will remain stable in space. With such forces, and with such forces alone, composite bodies will attract according to the same law as the particles that compose them. In an infinite universe, as Newton believed our universe to be, forces that vary in direct proportion to distance would lead to physically impossible results, such as infinite accelerations and infinite velocities. A universe organized around inverse square attractions was then the only one that is rationally possible—the ultimate demonstration that our world was created by an omniscient God.

In Book I, Newton first demonstrated that Kepler's three laws follow from inverse square forces directed to abstract centers of force, what has been called (in respect to the first two laws) the single body problem. In fact, he was convinced that forces stem from physical bodies, and the principle of universal gravitation asserts that all the bodies in the universe exert forces of attraction upon each other. Will the mutual attractions not upset the observed regularities of Kepler's three laws? Newton's strategy was to attack this problem step by step. First he addressed the two-body problem, in effect the sun and one planet, and he demonstrated that for a given intensity of force the mutual attraction, by shifting the center of motion to the common center of gravity, alters the size of the ellipse and its period but leaves Kepler's laws intact. But of course the universe does not consist solely of the sun and one planet. In addressing the multi-body problem, Newton let the solar system represent the universe, tacitly assuming that the combined attractions of an infinite number of stars spread through infinite space exert no net effect on the motions of our planets. Likewise he used the three-body problem—a central attracting body, an orbiting body, and a perturbing body that is further removed—to reason about the multiple bodies of the solar system. He did not arrive at an analytic solution to the three-body problem; later mathematicians have demonstrated that such a solution is impossible. He did develop an analysis that broke the perturbing attraction of the external body down into two components, a radial one, and one parallel to the line between the central body and the perturbing body. Because Kepler's area law holds for all centripetal forces, the radial component does not upset it, but its addition to the attraction of the central body renders the centripetal force on the orbiting body no longer varying inversely as the square of the distance and thus does upset Kepler's first law. The other component upsets all three. Newton's analysis was a quantitative one, however, and it led him to a rational argument that in a universe such as ours, where the central sun is immensely larger than the planets, and where distances are large, most perturbations will be so small as to fall below the threshold of observation as it then was.

Book I offered a theoretical discussion of motions under the influence of various forces. In Book III Newton began with the observed phenomena of the solar system. Because the planets move in closed orbits around the sun as their radii sweep out areas proportional to time, there must be some force of attraction toward the sun; and because the orbits are elliptical with the sun at one focus and with lines of apsides at rest in space, and because the system of planets observes Kepler's third law, the force must vary inversely as the square of their distances from the sun. Because the satellites of Jupiter and Saturn also move in closed orbits, sweeping out areas proportional to time, and obeying Kepler's third law, Jupiter and Saturn must attract them with forces that vary inversely as the square of the distance. Because the moon moves in a closed orbit around the earth sweeping out areas proportional to time, the earth must attract it. The argument that established that the earth's attraction varies inversely as the distance squared and permitted Newton to attach the ancient word *gravitas* (weight or heaviness) to the cosmic inverse square force was the correlation of the acceleration of gravity with the centripetal acceleration of the moon. Calling then upon the experiments with pendulums that had been important in the development of his own thought and upon the argument drawn from the satellites of Jupiter and Saturn, Newton proceeded to derive the principle of universal gravitation.

Having utilized this limited range of phenomena to derive the principle, Newton then employed it to explain, with similar quantitative precision, a further set of phenomena. For this purpose he made extensive use of the analysis of the three-body problem he developed originally to justify the argument that most perturbations fall below the threshold of observation. He had realized that the same analysis could disclose the dynamic source of those perturbations that were above the threshold and had been observed. With the earth as central body, the moon as orbiting body, and the sun as external perturber, he was able to explain the dynamic source of a number of peculiarities of the lunar motion that had been known to astronomers, some since antiquity, others since Tycho. Imagining a ring of water around the earth as the orbiting body, with the moon as the primary perturber and the sun as secondary, he offered the first successful explanation of the tides. A similar argument applied to the bulge of matter at the equator (caused by the centrifugal effect of the earth's rotation) explained the conical motion of the earth's axis that produces the phenomenon called precession of the equinoxes. And at the end of Book III Newton succeeded in treating the great comet of 1680–81 as a body that orbits like a planet and in deriving the conical path that corresponds to the comet's observed locations.

The concept of attraction, a force operating at a distance and defined in precise quantitative terms in proportion to distance, which Newton made central to the *Principia*, raised problems for natural philosophers of the seventeenth century. I have said that the central feature of Newton's cosmology was mechanical. In seventeenth-century usage, the word *mechanical* was reserved for systems that admitted the existence only of particles of matter in motion. Mechanical philosophies, of which Descartes' was the most influential example, insisted that only the direct impact of one particle upon another can cause a change of motion, and that therefore all of the phenomena of nature must be explained in terms of particles in impact; from this point of view the concept of attraction at a distance was a reversion to modes of natural philosophy that mechanical philosophies had displaced. It does seem possible that Newton's study of alchemy helped lead him to this concept.

Book II of the *Principia*, meanwhile, devoted itself primarily to demonstrating that a cosmic material medium, which mechanical philosophers called upon through the theory of vortices, cannot exist. The argument had two prongs. One of them developed a theory of motion through resisting media. It assumed that resistance derives primarily from the inertia of the particles that compose the media through which bodies move, striking the particles and putting them into motion. A material medium without resistance is therefore impossible; no matter how finely divided, it would always have the same mass per unit volume. A medium such as Descartes proposed, an absolute plenum, would bring planets to rest almost immediately. The continuance of planetary motions unchanged through the millennia of human observation implied that the heavens are in fact wholly devoid of matter. In the final Queries that Newton added to the second edition of his *Opticks* in 1717, he admitted the possibility of an elastic aether that could vibrate to explain certain optical phenomena. So as not to have retarded planetary orbits enough to be observed over the centuries, the aether had to be extremely rare. Newton calculated that the ratio of its elasticity to its density had to be 49×10^{10} greater than that of air. Only an aether whose particles repel each other at a distance could have had this property; in no way did it compromise Newton's rejection of standard mechanical philosophies.

The second prong of his attack was an analysis of the dynamics of vortices based on the proposition that only the friction between layers of a vortex can drive it. Newton demonstrated that a vortex cannot be self-sustaining, as vortex theory assumed, and that the relation between tangential velocity and distance from the center in a vortex is different from the relation in Kepler's third law. As Newton concluded, the hypothesis of vortices "rather serves to perplex than to explain the heavenly motions."

Nevertheless, Newton stated elsewhere that the notion of one body acting on another at a distance, without the mediation of anything else to convey the action, is so great an absurdity that no one competent in philosophy could hold it. He made it clear, however, that in his opinion the medium that conveys the action could be immaterial. The omnipresence of God pervaded the Newtonian cosmos, operating as an immaterial aether that offered no resistance to bodies while it moved them at will. In one telling phrase he likened infinite space to the sensorium of God, and used the human capacity to move our bodies as an analog to God's capacity to act directly on bodies in the physical universe. All of this occupied the level of metaphysics. On the level of science, Newton accepted the existence of forces that act at a distance through empty space and hold the planets in their orbits around the sun.

In Newton's cosmology the solar system was set in infinite, absolute space. The traditional arguments used over the centuries to demonstrate that the universe is finite had all hinged on the geostatic conception, which Copernican astronomy had destroyed. Newton's conviction that the universe is infinite was another dimension of his heritage from earlier stages of the Scientific Revolution. His conviction that infinite space is an absolute order of places came from his own personal reaction against the relativism of Cartesian philosophy. Newton believed that Cartesianism was a recipe for atheism. Absolute space and absolute time were aspects of his passionate rejection of any suggestion that the material realm is autonomous and his assertion of the dominance of spirit. God is eternal and infinite, and "by existing always and everywhere, he constitutes duration and space." In another telling phrase Newton referred to God as pantocrator, the absolute ruler of the universe whose presence and activity pervade infinite space imposing rational order on the cosmos. The concept of God as pantocrator, and much of the theological cast of Newton's cosmology, came directly from his Arian theology.

Infinite suns are spread throughout infinite space, held in a pattern of equilibrium by their mutual attractions. Newton's concept of absolute space did not lead him to picture the universe as a static order, however. Quite the contrary, it was characterized rather by dynamic process. Light consists of material corpuscles, and suns will therefore exhaust themselves if they do not receive new fuel. At least once Newton speculated that comets may finally fall into suns, supplying the fuel. At another time, when he still accepted the existence of an aether, he thought of a constant process whereby it is digested in suns, with the resultant movement of aether seen as the force that holds planets in their orbits. He accepted that similar processes, by which fluids are digested into solids and solids transformed into fluids, take place constantly in the earth, so that his cosmology has been called, with some justice, alchemical.

In a famous passage Newton once speculated that the cosmic order is not self-regulating, but that irregularities from the mutual attractions of comets and of planets may increase over time until the system has to be reformed by the Creator. The notion is not typical of Newton's cosmological thought. He saw the presence and activity of God, not in isolated interventions, but rather in the constant maintenance of the order and regularity of nature. As he put it in the General Scholium to the *Principia*, this most beautiful system of sun, planets, and satellites, all moving in the same direction in the same plane, could only be the creation of an omniscient and omnipotent God. No idea was more central to his cosmology.

—*Richard S. Westfall*

BIBLIOGRAPHIC NOTE

There is an enormous volume of literature about Newton. None is more stimulating and more provocative than the essays in Alexandre Koyré, *Newtonian Studies* (Cambridge, Massachusetts: Harvard University Press, 1965). Bernard Cohen, *The Newtonian Revolution* (Cambridge: Cambridge University Press, 1980), is another excellent discussion. A very recent book by Betty Jo Teeter Dobbs, *The Janus Faces of Genius: The Role of Alchemy in Newton's Thought* (Cambridge: Cambridge University Press, 1992), contrary to what the subtitle might imply, is directly relevant to Newton's cosmology and gives the best available exposition of the centrality of religious considerations in it. My own book, *Never at Rest. A Biography of Isaac Newton* (Cambridge: Cambridge University Press, 1980), is the most recent full-scale biography; in its bibliography (as also in those of the other books mentioned in this paragraph) you can find a guide to the extensive further literature that is available.

V. Galaxies: from Speculation to Science

Introduction

A RATIONAL ORDER FOR THE COSMOS

elief in the orderliness of the universe followed from Newton's demonstration of the most beautiful system of sun, planets, and satellites all moving in the same direction in the same plane as the creation of an omniscient and omnipotent God. Central to Newton's cosmology was the idea of an absolute ruler of the universe whose presence and activity pervaded infinite space and imposed a rational order on the cosmos. This theological cast to Newton's cosmology dominated much of astronomical speculation and science through the eighteenth century and into the nineteenth. Astro-theology, the determination of the rational order of the cosmos, became an important theological, philosophical, and scientific endeavor. Much of that intellectual activity, whether now better understood as speculation or as science, focused on the nature of spiral-shaped nebulae in the heavens.

William Whiston, Newton's successor at Cambridge University in 1703, argued that the system of the stars, the work of the Creator, had a beautiful proportion, even if frail man were ignorant of the order. Whiston was unable to propose an order for the Milky Way, a dense band of stars, but the self-taught English astronomer Thomas Wright did, in 1750.

Immanuel Kant, inspired by an incorrect summary of Wright's book, later explained the Milky Way as a disk-shaped system viewed from the earth located in the plane of the disk. Thoroughly imbued with a belief in the order and beauty of God's work, Kant went on to suggest that nebulous patches of light in the Heavens are composed of stars and are other Milky Ways or island universes. The paradigm of the Newtonian solar system provided for Kant a model of the larger stellar system and a plausible physical explanation of its structure. The arrangement of the stars might well be similar to the disk structure of the planets around the sun, and the same cause that gave the planets their centrifugal force and directed their orbits into a plane also could have given the power of revolving to the stars and have brought their orbits into a plane.

Although Kant did not explicitly state that the nebulae are thus rotating, as is the analogous solar system, the line of reasoning leading to an assumption of rotation is obvious. Rotation of the nebulous patches of light was assumed even by some who doubted Kant's island universe interpretation. The French astronomer Pierre de Maupertuis believed the nebulae to consist each of a single rotating ellipsoidal star.

Kant's ideas now are well known, but they exerted little influence on the thoughts of others during the last half of the eighteenth and the first half of the nineteenth century. His manuscript of 1755 perished in the printer's bankruptcy. Only a condensed version appeared, hidden in the appendix of another of Kant's books, published in 1763. Johann Lambert, a self-taught German scientist, published a similar theory in 1761, and seems not to have learned of Kant's ideas until 1765. William Herschel published his own theories on the construction of the Heavens, beginning in 1784, apparently with knowledge of neither Wright's, Kant's, nor Lambert's work. Herschel's investigations constitute a monumental accomplishment based primarily on observation, in contrast to the philosophical and theological speculations of Wright and Kant.

If Herschel's astronomy contained within itself an implicit claim of independence from astro-theology, soon there was open rebellion against the central tenet of astro-theology. Reflecting the new atheistic approach to nature of some scientists of the French Enlightenment, the French astronomer and mathematician Pierre-Simon Laplace attempted to replace the hypothesis of God's rule with a purely physical theory that also could explain the observed order of the universe, particularly the remarkable arrangement of the solar system. Laplace was successful, at least in his own mind. According to legend, when Napoleon asked him whether he had left any place for the Creator, Laplace replied that he had no need of such a hypothesis. Instead, Laplace had his nebular hypothesis, which appeared in six versions between 1796 and 1835, the last after his death in 1827. The theory postulated that the sun had once been a hot fluid extending beyond the present orbits of the planets. As the fluid cooled and condensed, gradually shrinking to the present size of the sun, zones of material were left rotating around the sun. These rings of gas shed by the contracting sun condensed to form the planets. Thus was Laplace able to explain the structure and dynamics of the solar system as a direct consequence of physical laws.

Marking further movement from speculative astro-theology to scientific astronomy, the nineteenth century saw the development of new evidence concerning the nature of the spiral nebulae. The evidence came both from detailed telescopic examinations of nebulae and from the new science of spectroscopy.

Giant reflecting telescopes were constructed between the 1780s and the 1860s, all in Great Britain and Ireland, and coinciding with a period of world supremacy for British technology. Telescope makers tapped energies characteristic of contemporary British and Irish society. Herschel transformed the reflecting telescope from a toy into a serious working instrument, but most astronomers still preferred to use refracting telescopes (employing glass lenses to focus the light), because they were instruments of greater precision. However, big reflectors were dominant in cosmological investigations; for all their failings, they gave a raw power not attainable with refracting telescopes, whose lens size and consequent light-gathering power was limited by the ability to pour glass discs of large size *and* near perfect optical quality. Reflecting telescopes, in contrast, required only that the surface be accurately figured and polished (though this itself was not a simple task, because the metal reflecting mirror distorted under its own weight differently in different observing positions and also continually tarnished, requiring frequent reworking).

William Parsons, the third Earl of Rosse, predicted that his new giant reflecting telescope, through whose tube Dean Peacock of the Church of Ireland had walked at the official opening ceremonies wearing a top hat and carrying a raised umbrella, probably would disclose forms of stellar arrangements indicating modes of action never before contemplated. In the spring of 1845, observations of M51, object number 51 in the French astronomer Charles Messier's 1771 catalog of nebulae and star clusters, revealed for the first time the spiral arrangement of matter in that nebula. Observations the next spring revealed the spiral nature of M99, and indications of spiral structure in other nebulae raised to fourteen the number of nebulae that Rosse classified as spiral systems, with several more marked doubtful on his working list.

Observations required to form a surer groundwork for nebular research were beginning to be made, with the development of the new research tool of astronomical spectroscopy. Spectroscopy received attention from scientists in several fields of study after Robert Kirchhoff, professor of physics at the University of Heidelberg, showed in 1859 that each element has a characteristic absorption pattern, resulting in identifiable dark lines in the spectrum of light passed through a prism. Chemical constituents revealed themselves in spectra.

Amateur astronomers pioneered the use of the new technique. In 1854 William Huggins disposed of his silk and linen business and moved to Tulse Hill, a suburb of London. There, two years later, he set up his own observatory. Huggins began routine astronomical work, making drawings of planets and taking transits (timing the instant of passage of a star across the meridian). Vaguely dissatisfied with these lines of research, Huggins responded to news of Kirchhoff's work on the chemical constitution of the sun "like the coming upon a spring of water in a dry and thirsty land." By 1870 he had identified several elements in the spectra of stars and nebulae. Eventually spectroscopic studies would provide an empirical foundation for one of the basic premises of cosmology: the approximate chemical homogeneity of the universe and of the uniformity of the physical processes in different stars and galaxies.

With Huggins occurred both the beginning of astronomical spectroscopy and, near the end of his pioneering research, a loss of British leadership in cosmological studies. Astronomical entrepreneurship in America's gilded age at the end of the nineteenth century saw the construction of new and larger instruments and a shift of the center of astronomical spectroscopic research from England to the United States.

New spectroscopic techniques soon were employed to measure motions, at first the approach or recession of objects and later the rotation of objects. Developments in the spectroscopy of galaxies took place about half a century after corresponding work on stars, largely because distant galaxies are far fainter and thus far more difficult to study than are nearby stars. Much of this work is discussed in section VI, *The Expanding Universe.*

15. WILLIAM HERSCHEL

William Herschel, the famous British astronomer whose observatory was located first in Bath and later in Slough, England, at one point described his two chief goals as being "to carry improvements in telescopes to their utmost extent" and "to leave no spot of the heavens unexamined." Nonetheless, his researches seem more accurately characterized by his statement, made in an 1811 paper, that "A knowledge of the construction of the heavens has always been the ultimate object of my observations." In fact, Herschel's writings are so rich in both observational and theoretical results that a strong case can be made for ranking his contributions to astronomy above those of any other astronomer of the eighteenth or nineteenth century.

Stellar Astronomy

Herschel is widely recognized as the foremost pioneer of stellar astronomy. Before the 1780s, when he began to investigate the structure of the Milky Way and the nature of those faint patches labeled nebulae by the few eighteenth-century astronomers who took any interest in them, stellar astronomy scarcely existed. For a century after Herschel's death, astronomers struggled with various questions raised by his observations and analyses of the cosmos beyond our solar system.

In the period before Herschel, most astronomers were interested in the stars only as a backdrop against which planetary and cometary motions might be measured. It is true that in a book published in 1755, Immanuel Kant, on slim observational grounds, had speculated that the Milky Way is a disk-shaped structure and that the nebulae are other Milky Ways, a speculation that came to be known as the island universe theory. Yet interest—observational or theoretical—in nebulae was small. When, in 1780, Charles Messier published a list of the sixty-eight nebulae then known, he did so chiefly to help astronomers avoid confusing these objects with comets, in which there was intense interest.

Herschel made his first observation of a nebula, Orion, in 1774. He reexamined Orion on a number of occasions during the next seven years, concluding that it was changing, which implied that it must be relatively near. Nonetheless, he took essentially no interest in observing other nebulae until December 1781, when he received a copy of Messier's list of sixty-eight nebulae (expanded by Messier in 1781 to slightly more than one hundred). Although in a 1784 paper on these objects Herschel stated that immediately after receiving Messier's list he began to observe them all, his observation book shows that for at least seven months he did not observe any nebulae. By September 1782 he had made the first recorded observation of the type of object he named "planetary nebula," and in the summer of 1783 he began to observe nebulae systematically, using after October 1783 the 18.7-inch aperture reflecting telescope, which he had himself constructed.

Extraordinary results came from Herschel's examination of the nebulae. In a 1784 paper, he announced not only that he had discovered 466 new nebulae, but also that he had resolved "most of the nebulae" listed by Messier into individual stars, which indicated that they must be at vast distances. Moreover, using counts of the number of stars in various directions, he argued that the Milky Way must consist of an immense stratum of associated stars (figure 1). In short, Herschel, who almost certainly knew nothing as yet of Kant's 1755 book, was in effect rediscovering the island universe theory and providing it with a far more elaborate evidential base than Kant had supplied.

In his 1785 "On the construction of the heavens" Herschel not only discussed many particular forms of nebulae, including his planetary nebulae, but he also set out his cloven disk diagram of the Milky Way (figure 2). Moreover, he explicitly described the Milky Way as a "detached nebula" and commented that of the more than 900 nebulae he had observed, "there are many which in all probability are equally extensive with that which we inhabit. . . ." In 1786 Herschel published a catalogue listing 1,000 nebulae he had observed. A catalogue published by him in 1789 added another 1,000 nebulae, with a list of 500 more appearing in 1802. Herschel had thus increased the number of known nebulae by a factor of 25. The significance of what Herschel was reporting was vividly stated by the literary figure Fanny Burney, who, after visiting him in 1786, exclaimed that Herschel "has discovered fifteen hundred universes! How many more he may find who can conjecture?"

Included with his 1789 catalogue of nebulae was a discussion of the particular form of objects that Herschel designated "globular

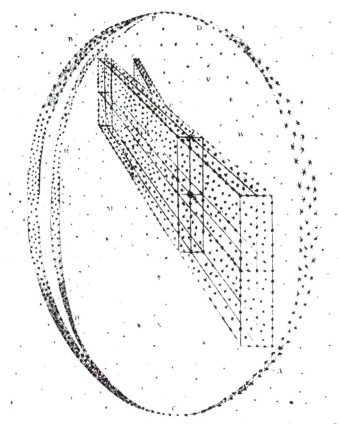

Figure 1. *An observer located at the center of a thin stratum of stars sees the surrounding stars projected as an encircling ring. From Herschel's 1784 paper on the construction of the heavens. He wrote: "A very remarkable circumstance attending the nebulae and clusters of stars is, that they are arranged into strata, which seem to run to a great length; and some of them I have already been able to pursue, so as to guess pretty well at their form and direction. It is probably enough, that they may surround the whole apparent sphere of the heavens, not unlike the Milky Way, which undoubtedly is nothing but a stratum of fixed stars."*

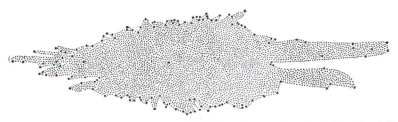

Figure 2. *A cross section through the Milky Way galaxy, as determined from Herschel's observations. From his 1785 paper on the construction of the heavens. He wrote that our Milky Way "is a most extensive stratum of stars" and, furthermore, that it is "a very extensive, branching, compound Congeries of many millions of stars."*

clusters," which he claimed show a spherical symmetry and are at extremely large distances. It is significant that in this study, he took not only a cosmological but also a cosmogonical approach, suggesting that the degree of compression in any globular cluster is an indication of its age.

Herschel's boldness in advancing new theories was matched by a willingness to back away from them when new evidence seemed to place them in jeopardy. This happened in a dramatic fashion after 13 November 1790, when he observed what is now known as NGC 1514 (object 1,514 in J.L.E. Dreyer's *New General Catalog*, first published in 1888). It appears as a roughly circular nebulous patch with a bright star at its center. Herschel took this to be conclusive evidence that truly nebulous material, incapable of resolution into stars, exists in the heavens. He reported this result in 1791 in his paper "On nebulous stars, properly so called," in which he stated that the material surrounding the central star of NGC 1514 consists of "a shining fluid of a nature totally unknown to us." The admission of this new conceptual entity into his cosmology placed in jeopardy his previous advocacy of the island universe theory. If truly nebulous matter exists, then possibly a large portion of the nebulae, rather than consists of vast numbers of stars located at immense distances, may be conglomerations of this "shining fluid" and relatively near us.

The degree to which Herschel had broken from the island universe theory was especially evident in an 1811 paper in which he suggested that many nebulae are nothing more than stars or even comets in the process of formation. Over the remainder of the nineteenth century, astronomers debated whether Herschel had totally or only partially surrendered the island universe interpretation of nebulae. Herschel's theory of the Milky Way proved hardier, yet even in this area, he realized that the fact that, with his giant 49.5-inch aperture reflector, he could see many stars invisible in lesser instruments implied that he could have no assurance that he had "fathomed" (i.e., observed) the farthest reaches of the Milky Way. This in turn forced him to the conclusion that his star counts provided no fully reliable basis for the figure he had assigned the Milky Way.

Because the distance of stellar objects is crucial to an understanding of the heavens, Herschel took a special interest in studying double and multiple stars, which he believed could provide assistance in distance determinations. This led him to compile catalogues of double stars, giving the angular separation, position angle, colors, and comparative brightnesses for each pair. His first catalogue, containing 269 such objects, was presented to the Royal Society in 1782. A second catalogue, listing 434 newly found doubles, followed in 1785;

and a final catalogue, bringing the total to 848, was published in 1822. A major result of this effort was that in 1803 and 1804 he published papers reporting that he had detected motions in some of the doubles, indicating that these doubles are "real binary combinations" (i.e., gravitationally linked).

One of Herschel's most important contributions to stellar astronomy did not in any way rest on the observational prowess of his telescopes. This was a 1783 paper in which he analyzed the proper motions of various stars as determined by earlier astronomers, drawing from his analysis the conclusion that the sun is itself in motion, moving approximately in the direction of the constellation Hercules. He returned to this difficult problem in papers published in 1805 and 1806, finding that new data on proper motions generally supported his earlier conclusion.

Variable stars were another long-standing interest pursued by Herschel, Mira Ceti being the subject of the first paper he published in the Royal Society's *Philosophical Transactions*. Between 1796 and 1799 he published, as an aid to the detection of variable stars, four catalogues consisting of sequences in which stars were arranged in order of brightness. Such sequences made it possible to detect whether any individual star in the sequence was varying in brightness, as in fact he found Alpha Herculis to be.

Possibly the best brief synopsis of Herschel's work in stellar astronomy is that he founded observational cosmology as a science. And it is a testimony to his remarkable achievements in this area that in treatises on stellar astronomy written even at the end of the nineteenth century (e.g., Agnes Clerke's 1890 *System of the Stars* and J.E. Gore's 1893 *Visible Universe*) his researches received as much attention as those of leading contemporary investigators.

The Solar System

Somewhat over half of the seventy major papers that Herschel published deal with the astronomy of the solar system. The best known of the numerous important contributions contained in these publications was his discovery of the planet Uranus, the first planet discovered in historic times. He first perceived the object on 13 March 1781, while making a systematic "sweep" of objects visible in his reflector. Noticing that the object showed a definable disk shape, he at first believed it to be a comet, as is indicated by the title of his discovery paper, "Account of a comet." Within a few months, however, and aided by other astronomers, Herschel recognized the object's planetary nature. In 1787, he announced his discovery of two satellites of Uranus (Titania and Oberon), and in 1798 he correctly claimed that the two moons revolve in a retrograde fashion. In the same paper he reported his detection of four additional Uranian moons, but this report has been disconfirmed.

At various times, Herschel carefully studied each of the other bodies in our solar system. For example, in 1789 he discovered the sixth and seventh Saturnian moons (Enceladus and Mimas). In a series of papers on Saturn extending to 1808, he argued for the solidity of the Saturnian ring system, noted Saturn's polar compression, and determined the rotation period of the planet. The system of Jupiter was treated in a 1797 paper, in which he argued that the Jupiterian satellites, like our moon, always keep the same side toward their primary. His papers on Mars, published in 1781 and 1784, included measurements of its period of rotation and a discussion of the seasonal changes in its polar caps, which he attributed to the sun's action. Venus and Mercury were also subjects of papers. His 1783 paper on Venus contained a critique of J.H. Schröter's claim that he had detected mountains on that planet. Herschel observed comets and speculated on their composition, but his sister Caroline, who discovered eight comets, occupies a larger place in the history of cometary observation.

The moon and sun were not neglected. In 1780, in one of his earliest publications, Herschel presented new measurements of the height of various lunar mountains. Herschel's solar observations led him to the claim (announced in 1795 and amplified in an 1801 paper) that the sun's heat and light derive from the sun having a glowing exterior, beneath which is a solid, cool, and possibly habitable core. This core, he believed, could at times be glimpsed through sunspots, which he interpreted as depressions in the exterior gaseous envelope.

On a more general level, Herschel viewed the objects in our solar system as analogous to the earth; in fact, he repeatedly referred to them as habitable. Herschel's attribution of inhabitants to the planets was not atypical of his times, but in claiming the habitability not only of the moon but also of the sun (and consequently of all stars), he went beyond current opinion and far beyond what his observations would justify. What was most innovative in his studies of the solar system was that, unlike most previous astronomers, Herschel did not concentrate on the positions of such objects but rather attempted, using his telescopes of unprecedented power, to learn what he could of their physical construction.

Conclusion

The researches of William Herschel transformed the cosmos of the astronomers. His discovery of Uranus nearly doubled the size of the solar system, whereas his sidereal researches delineated the structure of the Milky Way and populated the heavens with thousands of nebulae, for which he had supplied evidence, pro and con, concerning their possible status as island universes. He demonstrated that the heavens include globular clusters, planetary nebulae, true binaries, and possibly vast clouds of shining fluid. On the methodological level, he pioneered the use of giant reflecting telescopes, statistical studies, a cosmogonical approach, and stellar astronomy itself. The research program created by William Herschel gradually attained ever greater prominence in the nineteenth century, especially through the advancements in it that were made by his son, the distinguished astronomer Sir John Herschel (1792–1871). Such were William Herschel's achievements that they inspired the French astronomer François Arago, shortly after Herschel's death, to remark: "We may confidently assert, relative to the little house and garden at Slough, that it is the spot of all the world where the greatest number of discoveries have been made."

—Michael J. Crowe

BIBLIOGRAPHIC NOTE

William Herschel's papers can be conveniently consulted in J. L. E. Dreyer (ed.), *The Scientific Papers of Sir William Herschel*, 2 vols. (London: Royal Society and Royal Astronomical Society, 1912). Among the biographies of William Herschel, the most detailed is Constance Lubbock, *The Herschel Chronicle* (Cambridge: Cambridge University Press, 1933). See also Angus Armitage, *William Herschel* (Garden City, N. Y.: Doubleday & Co., 1963). A highly reliable shorter study is Michael Hoskin, "William Herschel," in C.C. Gillispie (ed.): *Dictionary of Scientific Biography, vol. 6*, (New York: Charles Scribner's Sons. 1972), pp. 328–36. See also Hoskin, *Stellar Astronomy: Historical Studies* (Chalfont St. Giles: Science History Publications, 1982).

16. RAW POWER: NINETEENTH-CENTURY SPECULUM METAL REFLECTING TELESCOPES

Instrumental developments often are overlooked in histories of cosmology, with the emphasis falling instead on conceptual breakthroughs. This focus has worked to downplay not only the importance of instrumental developments but also interactions between theory and scientific practice. The development of cosmology during much of the nineteenth century is largely a story not of conceptual breakthroughs but of big reflecting telescopes with speculum metal mirrors.

Technology

The age of large speculum metal reflectors began in the 1780s and lasted until the late 1860s. These decades encompass William Herschel's use of reflectors, Lord Rosse's construction of a 36-inch reflector in 1839 as well as his giant 72-inch "Leviathan of Parsonstown" in 1845, the fabrication of William Lassell's 24-inch reflecting telescope in 1845 and a 48-inch reflector in 1860, and the building by the Irish maker Thomas Grubb of the 48-inch Melbourne telescope in 1868.

All these telescopes were built in Great Britain and Ireland during a period when British technology was supreme. As the historian of astronomy J.A. Bennett has argued, the makers of these giant reflectors tapped energies characteristic of contemporary British and Irish society and applied the energies to the contemporary instrumental needs of cosmology. As Bennett points out, William Herschel made various trips to learn of new technological developments. He was an avid student of ironworks, furnaces, steam engines, and all sorts of machines. Rosse held for a time the Presidency of the Mechanical Section of the British Association for the Advancement of Science. Lassell was aided in his own telescope building by James Nasmyth, a superb engineer and inventor of the steam hammer, and the framework of the telescope tube for Lassell's 48-inch reflector owed much to developments employed in building the Britannia

Bridge. Mid-nineteenth-century cosmology as practiced with large reflecting telescopes owed a great deal to industrial developments.

What's the Driver?

With the evolution of giant reflecting telescopes was science led by technology or was technology led by science? Was the drive for bigger and more powerful reflectors the product of theoretical interests? And in particular, did the debate over the nature of nebulae propel the development of big reflectors?

The debate over nebulae focused on the following questions: First, were these dim patches of light remote star systems whose light merges to give a milky effect, or were the nebulae made of a luminous fluid? Second, if nebulae were groups of stars, were these star systems part of, or associated with, our galaxy, or were they themselves vast independent galaxies or "island universes"?

William Herschel at the end of the eighteenth century was the first to engage these questions with big reflectors. He crafted a research program that interwove theory and observation to investigate what he called the "natural history of the heavens." However, as Bennett has argued, it was Herschel's early success in building reflecting telescopes that in large part determined his eventual specialist interests in astronomy, and not the other way around.

For the two leading telescope makers in the mid-nineteenth century—Lord Rosse and William Lassell—the evidence is also strongly against their being driven primarily by theoretical interests. One of Rosse's later assistants recalled that Rosse's special interest in his 72-inch reflector seemed to have ceased as soon as the last nail was driven into it. And for Lassell, it was the fascination of devising such huge machines allied with the prospect of astronomical discoveries (whatever they might be) that fueled his telescope building. In important respects, therefore, we have to see big speculum metal reflectors as having lives of their own in which the logic of developing a technology and pushing it to its limits was key. The makers of big reflectors, in fact, often tended to push contemporary technology *beyond* its limits.

Problems and Opportunities

Difficulties with big reflectors are most apparent when we look at the telescopes as a complex and interlocking technological system of mirrors, tubes, mountings, sites, polishing machines, etc., devised to enable an observer to make astronomical observations. When we take into account how long it took to set up the big speculum metal reflectors for observations, how much time was needed to repolish and refigure the mirrors (speculum metal, an alloy of copper and tin, tarnished very easily and so needed regular repolishing), the energy and effort that had to be expended in maintaining the mounts, and the telescopes' usually poor siting (here we should note that Lassell was unusual in shifting both of his big reflectors to Malta in pursuit of better astronomical seeing), then we begin to understand better the challenges these telescopes presented.

A telling example of a telescope that failed in a systems sense is provided by William Herschel's giant reflector, which was completed in August 1789 with a 40-foot focal length and a 48-inch primary mirror. Even Herschel thought he had been too ambitious. Sir James South later recalled that toward the end of his life Herschel, though speaking with confidence that the reflecting telescope was only in its infancy, had nonetheless at the same time emphasized that his son John must never consider rebuilding the 40-foot telescope. Although the giant telescope was an object of widespread interest and astonishment, it was also an awkward instrument. Among other problems, Herschel eventually decided that to set up the telescope for observing simply required too much time, which, he noted, "in fine astronomical nights is too precious to be wasted in mechanical arrangements."

The primary mirrors of the giant telescope were also bothersome. They tarnished rapidly and needed frequent repolishing. Another major problem, and one that Herschel seems to have underestimated, was flexure: the primary mirror was pulled out of shape by gravity as it was directed to different parts of the sky. The speculum mirror was supported in an iron ring, resting there at its lowest point and confined by an iron cross over its back. In the opinion of later telescope makers, for a speculum weighing over a ton this support system was hardly adequate. Flexure would also prove to be a recurring issue for both Lord Rosse and Lassell, even with mirrors smaller than Herschel's 48-inch-diameter colossus.

When, in 1833, John Herschel transported his father's 20-foot reflector with an 18.7-inch primary mirror—probably William's most successful instrument—to the Cape of Good Hope, John employed woolen cloth to help support the primary. A mechanical system of levers was devised in 1835 by Thomas Grubb for Thomas Romney Robinson and a 15-inch reflector at Armagh Observatory. The Armagh design also provided the starting point for the design of the lever system for the primary mirrors of Lord Rosse's 36-inch and 72-inch reflectors.

Even with his 24-inch reflector, William Lassell faced flexure problems. His observing books and journals reveal that he regularly experimented with new support systems for his primary mirrors.

Usually, he was delighted at first with the new system, but soon he was as disappointed as ever. It may even be that flexure problems gave rise to what ultimately proved to be a spurious set of observations. The 24-inch was complete by late 1845, and, with its aid, Lassell in October 1846 glimpsed a moon (Triton) of the newly discovered planet Neptune (a discovery he confirmed the next year) and also purportedly identified a ring around Neptune. Several other observers with the 24-inch telescope also thought they saw a ring. Nevertheless, the supposed identification of a ring around Neptune troubled Lassell for several years until he finally decided in 1852 that the ring was such "that whatever may be the cause it is more intimately related to the telescope than [to Neptune]." None of the makers of large speculum metal telescopes ever fully mastered the problem of equably supporting the telescopes' mirrors, despite the use of various systems of hoops and levers.

The last of the great speculum metal reflectors to be built was a 48-inch completed in 1868 by Grubb for the Melbourne Observatory in Australia. It, however, was a failure. Thomas Romney Robinson had been intimately involved in its planning, and he blamed the incompetence of the Melbourne astronomers for its problems. This, however, was a harsh judgment, as the telescope's optical system was inadequate.

Despite the enormous challenges of building large speculum metal reflectors and successfully operating them, we must not lose sight of the point that for roughly a hundred years such telescopes provided capabilities that other telescopes could not match. In the hands of skilled makers and observers, they could exhibit a spectacular, if, to use the historian Albert Van Helden's term, "raw," and at times wayward, light-gathering power.

Observations

What kinds of observations, then, did astronomers make with these instruments? As already noted, William Herschel pursued a research program centered on the natural history of the heavens. Through his painstaking efforts to catalog nebulae (observed as fuzzy or misty spots on the sky), he became convinced that the nebulae progressed through a sequence from a highly diffuse, genuinely nebulous state to a more compressed condition. He also proposed that some star clusters were large enough to be counted as galaxies of stars, lying far beyond our own Milky Way.

In his speculations on nebulae and their life histories, Herschel presented an audacious and novel astronomical vision, but one that other astronomers found overly ambitious because, for them, it was not grounded in sufficient evidence. The general view was well put in an address to the British Astronomical Society (later the Royal Astronomical Society) in 1820:

> *Beyond the limits however of our own system, all at present is obscurity. Some vast and general views of the construction of the heavens, and the laws which may regulate the formation and motions of sidereal systems, have, it is true, been struck out; but, like the theories of geologists, they remain to be supported or refuted by the slow accumulation of a mass of facts.*

John Herschel, at his father William's urging, pursued the same sorts of astronomical observations as his father had. Hence, in the 1820s, John began to sweep the skies systematically for nebulae and star clusters. To do so, he employed his father's newly reconstructed 20-foot reflector. By 1833, he had compiled a catalog of 2,306 star clusters and nebulae that he had observed. But these included only those objects visible from England's northerly position and so did not constitute the fruits of sweeps of the entire heavens. To sweep the southern skies, Herschel and the 20-foot reflecting telescope embarked for South Africa in 1833. The results of these researches, which lasted from 1834 to 1838, were published in 1847. In his *Results of Observations Made During the Years 1834,5,6,7,8, at the Cape of Good Hope . . .* , Herschel again cataloged numerous nebulae and star clusters. He also revised his public opinions on the nebulae.

Before his research in South Africa, John Herschel had taken his father's position, arguing that the great majority of clusters and nebulae were star systems, although he was prepared to concede that some nebulae—for example the Orion Nebula and the Andromeda Nebula—were truly nebulous. His observations in South Africa, however, transformed John's thinking. Especially important for him were his observations of the Magellanic Clouds, which, until Herschel had trained the 20-foot reflector on them, had never been scrutinized by an observer with a large telescope. In 1847 he wrote:

> *The Nubecula Major, like the Minor, consists partly of tracts and ill-defined patches of irresolvable nebulae, and of nebulosity in every stage of resolution, up to perfectly resolved stars like the Milky Way, as also of regular and irregular nebulae properly so called, of globular clusters in every stage of resolvability, and of clustering groups sufficiently insulated and condensed to come under the designation of "clusters of stars."*

In the larger of the Magellanic Clouds, Herschel had viewed stars that differed greatly in brightness, and he also had viewed an extremely wide variety of different types of nebulosity. A key point was that the variations in the distances from the earth of these objects appeared to be far smaller than the distances of these objects from the earth; i.e., they could all be treated as at approximately the same distance from the earth. Herschel therefore regarded all of the large Cloud's stars and nebulae as essentially the same distance. Hence, the enormous variations he saw in the nebulae and stars in the Large Magellanic Cloud were, surely, due to genuine ranges in size and brightness. His father had argued that celestial objects belonged to individual "species" and that the variations within a species were small; John was now persuaded, however, that there was no limit to be set to the variety to be found in celestial objects. Whereas his father had ruthlessly exploited simplifying assumptions in his cosmological investigations (for example, that all stars have the same brightness), John shied away from doing the same.

John also decided that our galaxy is itself one member of a system of nebulae, and that the center of the system lies in the direction of a great concentration of many sorts of nebulae in the constellation Virgo. He did, nevertheless, accept that this theory had an anomaly. If our galaxy was indeed just one member of a system, why did other nebulae avoid the area of sky marked out by the plane of our galaxy? (This region on the sky later became known as the *zone of dispersion* or *zone of avoidance*.) What was it that tied our galaxy so intimately to the distribution of the nebulae?

By 1847, however, notwithstanding the publication of Herschel's Cape of Good Hope observations, the sixty-year dominance of William and John Herschel in studies of nebulae had already been shattered. In 1839 William Parsons (he was to become the third Earl of Rosse upon the death of his father in 1841) completed a 36-inch reflector at Parsonstown in central Ireland. Six years later, after many trials and tribulations and brilliant improvisations, the 36-inch—itself modelled on Herschel's telescope designs—was followed by the "Leviathan of Parsonstown," one of the scientific wonders of the age, with a primary mirror 72-inches in diameter.

Visitors to Parsonstown in its heyday were deeply impressed. One such was George Peacock, who was there in 1843 and witnessed some of the activities surrounding the construction of the Leviathan. He wrote:

> *Whatever met the eye was upon a gigantic scale: telescopic tubes, through which the tallest man could walk upright; telescopic mirrors whose weights are estimated not by pounds but by tons, polished by steam power with almost inconceivable ease and rapidity, and with a certainty, accuracy and delicacy exceeding the most perfect production of the most perfect manipulation; structures of solid masonry for the support of the telescope and its machinery more lofty and massive than those of a Norman keep; whilst the same arrangements which secure the stability of masses which no ordinary crane could move, provide likewise for their obeying the most delicate impulse of the most delicate finger, or for following the stars in their course, through the agency of clockwork, with*

a movement so steady and free from tremors, as to become scarcely perceptible when increased one thousand-fold by the magnifying powers of the eye-glass.

Observations began with the great telescope in March 1845 and were to provide extremely important and novel resources for the debates on the nature of the nebulae. To begin with, forty nebulae from the catalog compiled by John Herschel were examined by Thomas Romney Robinson and Sir James South. Shortly thereafter, Robinson proclaimed to the Royal Irish Academy that he:

> could not leave this part of his subject without calling attention to the fact that no real nebula seemed to exist among so many of these objects chosen without any bias: all appeared to be clusters of stars. [. . .] If it prove to be the case that all the brighter nebulae yield to this telescope, it appears unphilosophical not to make universal Sir J. Herschel's proposition, that "a nebula, at least in the generality of cases, is nothing more than a cluster of discrete stars."

A Nebulous Debate

The debate over the resolution of the nebulae, because it related directly to the larger debate over the possible existence of nebulous fluid in the universe, was also—at least for some people—highly charged politically. Laplace's nebular hypothesis and its postulation of nebulous fluid had been seized on by, among others, the political economist and astronomer John Pringle Nichol. For Nichol, the nebular hypothesis—with its developmental view of the solar system—provided a general model of universal progress, a model that could be pressed into service to help justify political reform, too. Reformers such as Nichol interpreted and exploited the evidence of the heavens to justify progressive changes in human society.

Some astronomers, particularly Robinson, strongly objected to Laplace's nebular hypothesis. Even before Rosse's Leviathan had been completed, Robinson had contended that the great telescope would undermine the nebular hypothesis by resolving nebulae into stars. It is hardly surprising, therefore, that Robinson quickly decided in 1845 that *all* nebulae were resolvable, given a telescope of sufficient light-gathering power. Particularly telling evidence came the next year with the news that the Leviathan had seemingly resolved the Orion Nebula, which earlier had resisted John Herschel's efforts to resolve it.

Confirmation of the purported resolution of the Orion Nebula was forthcoming in 1847. George Bond, the director of the Harvard College Observatory, was eager to publicize the quality of the Observatory's new 15-inch telescope which, along with the Pulkovo Observatory's 15-inch telescope, was one of the two largest *refracting* telescopes in the world. (Refracting telescopes pass incoming light through a lens to focus it, rather than focusing incoming light by reflection from a mirror.) Directing the Harvard College Observatory's telescope at the Orion Nebula, Bond claimed to have confirmed Rosse's resolution of the nebula into individual stars. Bond proudly reported to Harvard's President:

> *You will rejoice with me that the great Nebula in Orion has yielded to the powers of our incomparable telescope! [. . .] It should be borne in mind, that this Nebula and that of Andromeda have been the last strong-holds of the nebular theory; that is, the idea first suggested by the elder Herschel of nebulous matter in process of condensation into systems.*

Lord Rosse himself, however, was more cautious than some of his colleagues; he was not prepared to claim that all nebulae could be resolved into stars. Indeed, the observations made with the Leviathan carried ambiguous messages.

In examining the ebb and flow of the debate over the nebulae, we should also keep in mind that even by the 1850s no one had yet determined the distance to a single nebula or star cluster, and distances to only a very small number of stars were known with any degree of accuracy.

Evidence secured with the Leviathan and other big speculum metal reflectors did not settle the debate over the possible existence of nebulous fluid in the heavens. Even the claimed resolution of the Orion Nebula did not much disconcert such a staunch advocate of nebular fluid and cosmic evolution as Nichol. He had observed at Parsonstown and was deeply impressed with what had been done there; he even suggested a comparison between Tycho Brahe's legendary sixteenth-century observatory, Uraniborg, and Rosse's Parsonstown. But Nichol did not think the Leviathan had disproved the existence of nebular fluid; instead, he pointed to the finds Rosse and his observing colleagues had made of spiral nebulae as compelling pieces of evidence *for* nebular fluid.

The first spiral nebula so designated was Messier 51, a large nebula in the constellation of Ursa Major, and its spiral form was noticed by Rosse and his observing colleagues soon after the Leviathan went into operation in 1845. This find, publicized by Rosse's remarkable sketches and drawings of the nebula, also underlined the fact that the 72-inch mirror was capable of revealing more about this object than had the Herschels' telescopes; thus the observation of the spiral structure of Messier 51 was a significant indicator of the relative quality of the Leviathan. The spectacular Parsonstown observations of Messier 51 caused Nichol to write: "There is nothing to which we can liken it save a scroll unwinding, or the evolutions of a giant shell!"

Life on Other Worlds

The debates over the nature of the nebulae also had an important bearing on the widely discussed issue of the plurality of life. In the 1850s the British polymath William Whewell was a central figure in contests swirling about this issue. [See chapter 27, *Plurality of Worlds*, in section VIII, *Cosmology and Philosophy*.]

Many people had decided that the observations made at Parsonstown and with the Harvard refractor favored the existence of island universes. If island universes populated space, here were more domains for extraterrestrial life beyond the Milky Way. Whewell, however, was an ardent opponent of the existence of extraterrestrial life, and he endeavored to undermine the rationales of those who favored it. His most significant contribution to this end was his 1854 book *Of the Plurality of Worlds: An Essay*. Whewell wanted to restrict the visible universe to the Milky Way, and he therefore attacked the notion of nebulae as simply star systems in disguise. For Whewell, nebulous fluid truly existed in the heavens, and in support of this point he drew on John Herschel's observations of the Magellanic Clouds. To counter the argument that telescopes *had* resolved some nebulae into stars, Whewell would only concede that the nebulae had been resolved into "dots." In his opinion, it was a very bold assumption that these dots were in fact stars.

Additional Observing Techniques

In 1864 the debate over the nature of the nebulae took an unexpected turn from a novel direction. In that year, the British astrophysicist William Huggins directed his refracting telescope *and* its attached spectroscope at a planetary nebula in the constellation of Draco. What he found took him by surprise: not the spectrum he anticipated, but a single bright line. Such lines are characteristic of a luminous gas, not clusters of stars. Here was very strong evidence that at least some nebulae are indeed genuinely nebulous.

Results Huggins secured from other nebulae, however, were less definitive. Not all nebulae, nor even a majority of nebulae, possessed bright emission lines in their spectra. Although Huggins had shown to everyone's satisfaction that *some* nebulae are definitely not composed of collections of stars, the debate on the nature of the nebulae was far from settled. Huggins's results did, though, tell against the theory of island universes. His findings, when taken in conjunction with other pieces of evidence, led to a general decline of confidence in the island universe theory.

Huggins's spectroscopic investigations also saw to it that the debate over the nature of the nebulae would now engage more than raw telescopic power. The debate came to encompass a variety of scientific observing techniques. [See the following chapter, *Spectroscopy*.]

Conclusion

Large speculum metal reflectors were difficult to construct well and required huge reserves of energy, patience, and skill—along with no small amount of money—to build and operate. Before reflectors could be more than an exercise in somewhat unpredictable power and before they could take their place in the observatories of professional astronomers, a host of problems had to be overcome. A significant start would be made with the mastery of the silvering of glass mirrors, thereby avoiding the need to produce the much heavier speculum metal mirrors. But the mirrors were only one part of the technological system. The coming of age of the big reflecting telescope also waited upon the development of a new industrial infrastructure and the division of labor in telescope building, allied with the needs of the new and rising discipline of astrophysics. The beginning of the twentieth century marks the coming of age of the big reflector.

The new generation of large reflecting telescopes were built largely in the United States. As J.B. Morrell has noted, in the years between 1820 and 1880, American science advanced from colonial dependency to an independent professional maturity, shaping European models to the American republican context. American science, including astrophysics, flourished. Through foundation and other philanthropic support, large reflectors were built in the United States early in the twentieth century. In fact, during this period the rise of the U.S. as a leading economic power was mirrored by its growing importance in the manufacture and use of giant reflectors. The high point of this shift is manifest in the great 100-inch Hooker telescope on Mount Wilson, completed in 1917. Except for the optics, the Hooker telescope was chiefly the product of large-scale engineering by U.S. companies. Indeed, because of the success of American astronomers in securing research funds, the placing of expensive and powerful telescopes at good sites, and the increasingly sophisticated training of American astronomers, the U.S. emerged as the dominant power in observational cosmology. [See chapter 18, *Cosmology 1900–1931*, in section VI, *The Expanding Universe*.]

Despite the demise in the late nineteenth century of large speculum metal reflectors, their light gathering capacities had made them the principal instruments in the detection, as well as in the highly charged debates on the resolution, of nebulae during much of the nineteenth century. As such, they were enormously important for cosmological theory and practice.

—*Robert W. Smith*

BIBLIOGRAPHIC NOTE

On links between technology and reflecting telescopes, see J.A. Bennett, "The Giant Reflector, 1770–1870," in E. Forbes, ed., *Human Implications of Scientific Advance* (Edinburgh: Edinburgh University Press, 1978), pp. 553–558; and *Church, State, and Astronomy in Ireland. 200 Years of Armagh Observatory.* (Armagh: Armagh Observatory & Institute of Irish Studies, 1990). On John Herschel, see M.A. Hoskin, "John Herschel's Cosmology," *Journal for the History of Astronomy*, *18* (1987), 1–34. S. Schaffer discusses political and social aspects of the nebular hypothesis in his path-breaking article "The Nebular Hypothesis and the Science of Progress," in J.R. Moore, ed., *History, Humanity, and Evolution* (Cambridge: Cambridge University Press, 1989), pp. 131–164. Schaffer also presents an important account of William Herschel in "Herschel in Bedlam: Natural History and Stellar Astronomy," *British Journal for the History of Science*, *13* (1980), 211–239. A dated but still useful general work on the history of telescopes is H.C. King, *The History of the Telescope* (London: Charles Griffin & Co., 1955). For more background on the nebular hypothesis, see S.G. Brush, "The Nebular Hypothesis and the Evolutionary Worldview," *History of Science*, *25* (1987), 245–278.

17. SPECTROSCOPY

Early spectroscopic foundations on which much of modern cosmology is built are explored in this article. Three main developments in spectroscopy with significant impact on cosmology were:

- The discovery of the Doppler effect, its verification for light waves, and its application to stars and gaseous nebulae.

- The interpretation of spectral lines and the application of spectroscopy to celestial spectra to determine the chemical composition of the emitting source, and the demonstration that a common chemistry exists throughout the universe.

- The early spectroscopic observations of spiral "nebulae," which gave a vital clue concerning their extragalactic nature and provided the first evidence for the expansion of the universe.

The first two parts of this discussion relate to work whose origins were in the nineteenth century. The first astronomical Doppler shift determinations and the first qualitative analyses of astronomical spectra were made for nearby stars at a time when astronomers were blissfully ignorant of the true nature and extent of the universe. Interaction between spectroscopy and cosmology (in the present-day meaning of the term) was not therefore possible in the nineteenth century. Nevertheless, the solid foundation laid by spectroscopists, mainly in the second half of the nineteenth century, was absolutely essential for the development of cosmology in the twentieth century. Indeed, the measurement of redshifts for galaxies and the assumption that these are Doppler shifts is so fundamental to cosmology that it is sobering to consider how much more limited our knowledge of the universe would be if the Doppler effect had somehow escaped inclusion in the astronomers' toolkit. For this reason, the discussion of the origins of astronomical spectroscopy is particularly germane to the later development of cosmology.

The Doppler Effect

Doppler's Paper on Colored Stars

In May 1842 Christian Doppler, professor of mathematics in Prague, delivered a remarkable lecture to the Royal Bohemian Scientific Society. The gist of his paper was that the frequency of light waves should be changed by the motion of either source or observer toward or away from the other. Doppler made analogies with sound waves and with sea waves, and deduced elementary formulae that, in effect, predicted a frequency change in proportion to the line-of-sight velocity of either source or observer through the ether.

So far so good. But Doppler next made two major errors in the application of his effect to starlight. First, he supposed that the radiation from stars was largely confined to the visible spectrum. And secondly, he supposed that the motions of stars, including orbital motions in binary systems, could be a significant fraction of the velocity of light. As a result, he strived to explain stellar colors as a consequence of their radial motions (the component of motion in the line of sight). A star that partially shifted its radiation out of the visible spectrum would appear colored instead of white, while a yet higher velocity would render it completely invisible to the eye.

The different colors of visual binary stars and the changes in color of long-period variables and of novae could all be explained by invoking the supposed effect based on these faulty premises. Even some unreliable evidence that visual binary stars might undergo color changes was cited in support of his theory.

It is perhaps surprising that a paper on star colors with one correct premise followed by two faulty ones and some erroneous conclusions should have become so influential in astronomy and cosmology. Indeed, Doppler's paper provoked considerable controversy among physicists, who debated whether any frequency change occurred at all as a result of motion along the line of sight. But among astronomers, the Doppler effect was largely ignored for the next two decades.

In the meantime, two other scientists independently predicted the change in frequency (or wavelength) of light waves when either source or observer was in radial motion. (This prediction was little more than an introductory section to Doppler's paper, and was the only part of his work in which his conclusions were substantially correct.) The two scientists were the French experimental physicist A.H. Fizeau, in a lecture delivered in 1848 but not published until 1870, and the German physicist and philosopher Ernst Mach, in a paper published in 1860. Both of them had the foresight to see that a radial velocity would give rise to shifts of the lines in astronomical spectra. Given that up to 1860 the sun remained the only celestial body to have been examined in any detail by the spectroscope, the predictions of Fizeau and Mach concerning the possibility of measuring stellar radial velocities are all the more remarkable. It is unlikely either scientist had ever observed a stellar spectrum at the time of their pronouncements.

Confirmation of the Doppler Effect

The validity of the Doppler effect was a matter of considerable controversy for several decades. Initially, arguments for the effect mainly concerned sound waves, and the Dutch physicist C.H. Buys-Ballot in 1845 was among the first to demonstrate a change in the pitch of wind instruments on passing express trains. But he doubted the color changes predicted by Doppler for moving light sources, because light in the ultraviolet or infrared regions should be shifted into the visible and no color change would result. In spite of these acoustical experiments, J. Petzval in 1852 argued vehemently against them and Doppler's conclusions, citing a series of spurious objections.

Later, Mach and the German astrophysicist H.C. Vogel confirmed the Doppler effect for sound. Mach used a rotating whistle, while Vogel appealed to the change in pitch of the steam whistle heard from an approaching or receding locomotive. But no empirical confirmation for the optical Doppler effect was at first forthcoming. The Swedish astronomer and physicist A.J. Ångström conducted two negative experiments, which lent no support to Doppler. These were (1) the invariability of the appearance of emission lines in spark spectra with the direction of observation (even though the spark was assumed to travel at a high velocity across the electrode gap), and (2) the lack of any detectable change in the position of the solar spectrum lines with the season due to the eccentricity of the earth's orbit. These observations are nevertheless of historic importance, because they searched for spectral line shifts rather than visual color changes. Furthermore, the solar observations were one of the earliest attempts to apply the Doppler effect to astronomy.

The empirical confirmation of the Doppler effect for light came in the 1870s, when Vogel, then at the private Bothkamp Observatory near Kiel, measured a shift in the positions of the solar lines between the east and west limbs on the solar equator using a Zöllner prism spectroscope. The velocity difference deduced was in approximate accord with the known solar rotation rate of 2 kilometers per second (km/s) determined from sunspots. The measurement confirmed Father Secchi's qualitative observation at the observatory of the Collegio

Romano in the previous year of a detectable shift of the solar lines between the east and west limbs of the sun. Many observers repeated and refined this observation, including C.S. Hastings, C.A. Young, S.P. Langley, and H. Crew in the United States and L. Thollon, A. Cornu, and N.C. Dunér in Europe. Cornu's work on D-line shifts (from the element sodium) was notably precise, the measured shift being within 3 percent of the predicted value.

Thus, from the mid-1870s, the optical Doppler effect was quite well established. The planets provided a further testing ground for its confirmation, since planetary orbits were well determined. Tests on planets not only confirmed the effect, but also showed that even a moving body shining by reflected light gave a line shift, as Mach had correctly predicted. Various observers confirmed the Doppler effect using planetary spectra, including Vogel and the French astronomer-physicist Henri Deslandres.

The literature on the validity of the Doppler effect from the 1850s to 1870s is particularly rich and interesting, and often contentious. Two points that arise from the discussion are (1) whether the relationships between perceived color, refrangibility (refractive index), and wavelength remain the same for the light from stationary as well as from moving bodies, and (2) whether the flow of the medium (the ether or the air) past the observer should also induce a Doppler shift, as was claimed by Petzval for sound waves. W. Klinkerfuess in 1870 attempted to demonstrate that such a shift was caused by the earth's rotation. He observed absorption lines from bromine vapor at intervals of twelve hours in easterly and westerly directions, an experiment somewhat reminiscent of the famous experiment carried out later by Michelson and Morley. Needless to say, no shift caused by the flow of the ether past the earth was found!

Early Applications of the Doppler Effect in Astronomy

William Huggins in London and Secchi in Rome were the first observers to attempt to measure stellar radial velocities in the 1860s. Both used visual spectroscopes attached to equatorial refracting telescopes. Secchi reported an unsuccessful attempt to measure a Doppler shift of the H_β (hydrogen) line of the star Sirius in March 1868, while Huggins undertook the same observation in April of that year and claimed a positive result of 29.4 miles-per-second recession, a result since shown to be wrong in both amount and sign. The feasibility of measuring stellar Doppler shifts continued to be disputed by both of these observers.

The matter was finally settled by Vogel and Julius Scheiner at the Potsdam Observatory, where they commenced in 1888 a program of recording stellar spectra by photography. The range in probable error was reduced by a factor of at least ten relative to visual work, to about ± 2.6 km/s for late-type stars. By 1892 radial velocities for fifty-one stars had been measured.

One of the early results of this work was the discovery of periodic line shifts in the eclipsing binary star Algol that correlated with the light variations. Edward Pickering, director of the Harvard College Observatory, at about the same time found similar periodic shifts in the spectrum of ζ (zeta) Ursae Majoris. The discovery of such spectroscopic binary stars opened up a new and profitable avenue of research. It was also a convincing demonstration of the Doppler effect for extra-solar system bodies moving at velocities well under a thousandth of the speed of light.

Techniques developed by Vogel around 1890 for measuring stellar Doppler shifts photographically became the basis for most radial velocity work on stars and galaxies for much of the twentieth century, although many refinements in detail were introduced over time. Vogel's original method involved observations with the 30-cm Potsdam refracting telescope and two-prism spectrograph to record spectra in the neighborhood of the blue H_γ absorption line of hydrogen. A thin hydrogen Geissler discharge tube was mounted in the telescope to give an emission line for the zero velocity reference or comparison spectrum. The spectra were widened by trailing the telescope. The photographic plates were then measured with a traveling microscope to obtain the line positions. Sometimes the numerous stellar metallic lines near H_γ were compared with these lines in the solar spectrum, which acted as a secondary standard.

Although photography was an essential tool for the accurate measurement of the minute Doppler shifts in stellar spectra, one pioneering observer did obtain quite reliable visual measurements in the 1890s. This was James Keeler at the Lick Observatory, using the newly installed giant 36-inch refracting telescope. Keeler's notable achievements were his demonstration that the rings of Saturn show differential rotation and therefore could not be a solid body, and his measurement of the radial velocities of fourteen gaseous (mainly planetary) nebulae from their H_β emission lines, as well as the radial velocities of several bright stars. He concluded that the gaseous nebulae have space motions generally of the same order of magnitude as stars, a property later found to be in marked contrast to the spiral nebulae. Huggins, the pioneer of nebular spectroscopy, had reached a similar conclusion for gaseous nebulae in 1874, albeit from observations of lower accuracy.

Stellar Radial Velocities Early in the Twentieth Century

Once the validity of the Doppler effect had become fully accepted and photography had been established as an indispensable technique in stellar astronomy, many observatories undertook major programs in photographic Doppler shift measurements for stars. The observatories at Pulkova, Paris, Cambridge (England), Yerkes, Bonn, and the Cape of Good Hope were all active early in the twentieth century. However, W.W. Campbell at the Lick Observatory became the doyen of all the Doppler shift observers. From 1897, in collaboration with W.H. Wright and J.H. Moore, he established a vast program to obtain the velocities of all the stars in both hemispheres to visual magnitude 5.51. The Lick 36-inch refractor in California and a similarly sized reflector established near Santiago, Chile, were used for this program. The resulting catalog embraced 2,771 stars and was published in 1928. The reductions entailed measuring wavelengths of selected lines. Henry Rowland's solar spectrum wavelengths were used as a reference for later-type stars. Precision as good as ± 0.5 km/s probable error was obtained from the best spectra, while for early-type stars ± 3 km/s was typical.

It is largely as a result of the Lick data that Campbell was able to determine a reliable value for the sun's motion relative to the stars in the solar neighborhood. Another by-product was the discovery at Lick of many new spectroscopic binaries. The 1928 catalog listed 351 stars with variable velocities, most of them binaries.

When the Mount Wilson Observatory (in Pasadena, California) and the Dominion Astrophysical Observatory (in Victoria, British Columbia) entered radial velocity research with their large reflectors, Doppler shifts on many fainter stars could be obtained. Large programs were undertaken by Walter Adams and Alfred Joy at Mt. Wilson and by J.S. Plaskett at Victoria. These observations led to the discovery of the rare high-velocity stars (by Gustaf Strömberg and others) and then to rotation of the Milky Way galaxy by the Dutch astronomer Jan Oort.

Measuring stellar Doppler shifts was one of the most productive and important activities in stellar astrophysics during the first half of the twentieth century. By 1931 over 6,700 stars had had their radial velocities determined. When R.E. Wilson's *General Catalogue of Stellar Radial Velocities* was published in 1953, data for over 15,100 stars were available, mainly from observations at Mt. Wilson, Lick, and Victoria.

Although the principle of the Doppler effect was simple in concept, its application was difficult in practice. Radial velocity work was painstaking indeed, and involved many possible sources of systematic error that could vitiate the results. Temperature and flexure effects in spectrographs degraded the data, the nonlinear dispersion of prisms made the reductions more complicated, and the effect of blended lines could introduce systematic errors depending on spectral type. Many of these problems were reduced or overcome as a result of improvements in instrumentation and in reduction methods. Especially noteworthy were the use of new instruments, including diffraction gratings, coudé spectrographs, and spectrographic Schmidt cameras (Mt. Wilson Observatory was the principal pioneer); and the use of standard wavelengths for stars of different spectral type and standard radial-velocity stars in the reductions (here R.M. Petrie and J.A. Pearce at Victoria were the leaders).

Stellar radial velocities have given invaluable data on stellar masses, stellar pulsation, and galactic structure. From the cosmological viewpoint, the importance lay in the verification of the Doppler effect and the development of the technique of measuring small Doppler shifts from photographic spectra. These techniques were adapted with only minor changes to the task of obtaining velocities of the spiral "nebulae," or galaxies, which in turn has become the basis for much of modern observational cosmology.

Chemical Analysis by Spectroscopy

One of the tenets of cosmology is that the universe looks essentially the same no matter where the observer might be located and in which direction he might look. Therefore, the same chemical elements that comprise the material of the earth and the solar system should be found throughout the universe, and in roughly similar relative abundances as are observed for the solar system and for nearby stars. This conclusion, which follows from the so-called cosmological principle, has not been tested extensively on a quantitative basis for objects outside our Galaxy. However, locally within the Galaxy, the spectroscopy of stars and of gaseous nebulae reveals a remarkable uniformity of composition, and at least qualitatively, this uniformity appears to extend to other galaxies as well.

The cosmological principle is an important philosophical concept that underpins our understanding of the universe. Spectroscopy is one type of observation that could in theory invalidate this concept, although in practice it never has. Today, astronomers often take the cosmological principle for granted, yet the fact that astronomical spectra arise from the same elements as those occurring naturally on the earth was not always accepted automatically. In the following section of this paper we trace the origins of the use of spectroscopy for chemical analysis in astronomy and the realization that a common chemistry pervades the universe.

Early Pioneers in the Study of Spectral Lines

Although William Wollaston was the first to find dark features in a prismatic solar spectrum (in 1802 in England), more important was the careful work of the German optician Joseph Fraunhofer, who in 1814 observed and then cataloged a large number of absorption lines in the spectrum of sunlight. This observation was the precursor of several laboratory studies of flame and later arc and spark spectra from the 1820s to ascertain whether the emission features observed in the laboratory might coincide with those seen in absorption in the sun. The English scientists John Herschel and William Fox Talbot were two of those who suspected that the appearance of the emission spectra might provide a clue to the composition of the emitting gas, but this belief proved troublesome to establish conclusively, largely because of the chemical impurity of laboratory samples and the ubiquity of the D-lines due to small traces of sodium mixed with other elements.

Absorption lines generally were associated with the passage of white light through a cool gas, and indeed, the Scottish optician David Brewster's observation of numerous absorption lines produced by nitrous acid vapor in his laboratory led him to assert mistakenly that this same substance was to be found in the sun.

A key observation by the French experimental physicist Léon Foucault in 1849 was an important step toward understanding the origin of spectral lines. He showed that the bright orange lines seen in emission from a carbon arc appeared in absorption when sunlight was passed through the arc. Indeed, the solar D-lines now were stronger than when the arc was absent. He also found that the arc absorbed light from a continuous spectrum at the D-line wavelengths, which coincided with the bright emission lines of the arc by itself.

Foucault's observations were repeated a decade later by the German physicist Robert Kirchhoff, who came to the unequivocal conclusion that sodium, which produced the bright lines in the laboratory spectra, was also the cause of the solar D-lines in absorption, due to a cooler layer containing sodium gas around the sun. In fact, many researchers were involved in establishing that each chemical element produced its own unique pattern of spectral lines, including Ångström at Uppsala and David Alter in Pennsylvania. This work resulted in the detailed recording of the emission-line spectra of some of the commoner elements (lithium, sodium, potassium, calcium, strontium, and barium) by Robert Bunsen and Kirchhoff in Heidelberg, together with the clear statement that these same lines should be seen in absorption in the solar spectrum if these elements were present in the sun.

This work was extended to thirty elements by Kirchhoff using a four-prism spectroscope to compare the elements' spark spectra with a carefully drawn map of the spectrum of sunlight. From the line coincidences he concluded that iron, calcium, magnesium, sodium, nickel, and chromium are present in the sun, and thus he completed the first qualitative chemical analysis of a celestial body using spectroscopy.

Qualitative Chemical Analysis of Stars and Nebulae

In the 1860s several observers resumed work on the visual spectroscopy of stars, a field that had lain practically dormant since the pioneering but neglected stellar spectroscopy of Fraunhofer between 1814 and 1823. First the Italian astronomer Giovanni Donati in 1860, and then in 1862–63 the amateur American astronomer Lewis Rutherfurd, Huggins and his neighbor William Allen Miller, a chemist and spectroscopist, Secchi, and England's Astronomer Royal George Airy all ventured into the new field of stellar spectroscopy.

The work of Huggins at his private observatory in London (at first with the assistance of Miller) was especially significant, in that he undertook the first analyses of the compositions of celestial bodies outside the solar system. The observations entailed the careful visual recording of the bright lines of the spark spectra of some twenty-seven elements and the comparison of the dark lines in the spectra of bright stars with those observed in the laboratory. For the star Aldeberan the elements hydrogen, iron, sodium, magnesium, and calcium were found from seventy line positions, while nine other stars showed some of these elements. Huggins later wrote of this work: "One important object of this original spectroscopic investigation of the light of the stars and other celestial bodies, namely to discover whether the same chemical elements as those of our earth are present throughout the universe, was most satisfactorily settled in the affirmative; a common chemistry, it was shown, exists throughout the universe."

Huggins went even further in his major paper with Miller in 1864, writing: "It is remarkable that the elements most widely diffused throughout the host of stars are some of those most closely connected with the constitution of the living organisms of our globe, including hydrogen, sodium, magnesium and iron."

As early as 1864 Huggins turned his spectroscope toward the nebulae, and he quickly found eight nebulae, each with three or four bright emission lines at wavelengths of 5007Å and 4959Å (at first thought to be nitrogen, but incorrectly identified), as well as H_β and sometimes H_γ (hydrogen lines). This work was extended to some seventy apparently nebulous objects by 1868, including both planetary nebulae (each a single star surrounded by a shell of gas) and diffuse nebulae (clouds of gas) as well as unresolved globular clusters (groups of stars). About one third of these objects showed the emission lines, and Huggins concluded that they "must be regarded as enormous masses of luminous vapor or gas." He discounted the possibility that the bright-lined objects could be composed of stars. On the other hand, the remaining nebulous objects showed either continuous or stellar spectra, including the Andromeda nebula with a continuous spectrum. Even at this early stage, it was shown to be of a different nature from that of gaseous nebulae.

Secchi also made outstanding contributions to visual stellar spectroscopy. He devised a system of stellar spectral classification, and in the context of stellar composition, his discovery of the rare carbon stars is particularly noteworthy. He correctly suggested that the broad bands found in the spectra of these red stars resembled those of carbon seen in laboratory flame spectra.

Photographic spectroscopy greatly facilitated the difficult task of cataloging the positions of spectral lines observed in stellar and nebular spectra, and also extended the wavelength range into the ultraviolet region of the spectrum. Huggins himself was a pioneer in this field. Scheiner at Potsdam was also preeminent in the 1890s. He catalogued many lines in forty-seven stars covering a wide range in spectral type and gave element identifications wherever possible.

An even more influential contribution came from the lower-resolution objective prism spectrography undertaken as part of the Henry Draper Memorial at Harvard. Pickering, assisted by Williamina Fleming, Antonia Maury, and especially Annie Cannon, studied the photographic spectra of several hundred thousand stars. The behavior of spectral lines in stars of different spectral type was carefully cataloged, and lines of elements not hitherto found in stars were recorded, including, for example, strong lines of silicon or strontium found in certain A-type stars with peculiar spectra.

The two bright-green lines found in gaseous nebulae by Huggins were named the "nebulium" lines once it became clear that they corresponded with none of the lines found in laboratory spectra. The problem of their identification confounded many investigators, including Keeler and Wright at Lick, and became all the more acute once the spectra of all the elements in the periodic table had been intensively studied and it was realized that no new abundant elements remained to be discovered. For a while it seemed that Huggins's ideas of a common chemistry throughout the universe could be challenged. Not until 1927 did the American Ira Bowen resolve this problem and identify the mysterious nebulium lines as the forbidden lines of ionized oxygen, which only appear in the spectra of huge volumes of hot gas at very low densities, under conditions that cannot be reproduced in the laboratory.

Discovery of Helium

The discovery of helium was one of the early triumphs of astronomical spectroscopy. In 1868 the English scientist Norman Lockyer had found that an orange line from solar-prominence spectra occurred at a wavelength of 5876 Å, somewhat less than the well-known D_1 or D_2 lines of sodium. The new line had first been seen by G. Rayet and others at the solar eclipse of 1868. Lockyer's successful wavelength measurement relied on his mastering of the technique of observing the prominences outside of the time of eclipse. Since the wavelength did not correspond to sodium, Lockyer named the line D_3 and ascribed it to a new element he called helium, hitherto unknown on the earth.

Helium was not found terrestrially until 1895 when the physical chemist William Ramsay, then at University College, Bristol, isolated some of the gas from the mineral cleveite and showed that it produced the same D_3 line in a discharge tube. (In 1904 Ramsay, now at University College, London, would receive the Nobel Prize in chemistry for his discovery of the family of inert gases.) Meanwhile, Lockyer found that D_3 and other helium lines occur in absorption in the spectra of certain stars of spectral type B, including Bellatrix (γ Orionis). The lines of the cleveite gas were observed in a number of series whose relative strengths appeared not to be correlated. This finding led Lockyer and others to the belief that two gases were present in both the laboratory samples and in the stars; the two gases were named helium (which gave rise to the D_3 line) and parhelium (or sometimes, asterium). Later it was recognized that both helium and parhelium were due to a single element, but in either the singlet (opposed electron spins) or triplet (parallel electron spins) states whose spectra were formed independently.

The discovery of helium helped solve another significant astronomical enigma, the so-called Pickering series of lines found in the spectrum of ζ Puppis and other O-type stars. These lines were found by Pickering, who showed that they occurred at the wavelengths of the Balmer lines of hydrogen, but with additional lines lying between those of the normal Balmer series. Pickering therefore believed that hydrogen was responsible for these lines, even though hydrogen spectra in the laboratory never showed the additional lines. Only after the English astrophysicist Alfred Fowler observed the ζ Puppis lines in his laboratory in 1912 from a mixture of hydrogen and helium, did the Danish atomic and nuclear physicist Niels Bohr correctly identify the origin of the lines as helium at high temperatures. We now recognize this as the singly ionized element whose energy levels resemble those of the hydrogen atom.

Quantitative Analysis of Stellar Spectra

The approximate quantitative analysis, using the new Saha ionization theory, of a large number of stellar spectra of different spectral type by Cecilia Payne in 1925 for her doctoral thesis at Harvard was the first clear evidence for the near uniformity of composition for the great majority of stars. Moreover, the lighter elements were found to be the more abundant, with hydrogen and helium several orders of magnitude more preponderant than the next most abundant elements.

This result, which implied that the stellar universe was composed almost entirely of hydrogen and helium, was at first doubted by Payne herself. But Henry Norris Russell's exhaustive analysis at the Princeton Observatory in 1929 of the solar spectrum confirmed the very high solar abundance of hydrogen, while the high helium abundance was confirmed when Albrecht Unsöld in 1941 analyzed the spectrum of τ Sco, a bright southern B-type star, using the curve-of-growth technique to interpret the strengths of the stellar absorption lines. His result was H:He:other elements = 85:15:0.24 (by number) for this star. Further and more reliable evidence for the high abundance of hydrogen in the sun was obtained by B. Strömgren, who interpreted the solar line strengths in 1940 using the recently discovered continuous opacity of the solar photosphere based on the H^- ion.

In the last half century the techniques of stellar-abundance analysis have been considerably refined, from the groundwork done by Russell, Strömgren, Unsöld, Jesse Greenstein at Mt. Wilson, and others. An understanding of continuous opacity sources, the modeling of the structure of stellar atmospheres, the improved laboratory determination of line oscillator strengths, and the advent of linear solid-state detectors in astronomy to replace photographic plates have all resulted in much more reliable element abundance determinations in stars. However, the most significant general result has been the striking uniformity of composition from star to star, and the concept of a cosmic or universal abundance distribution for the elements, a point stressed by Otto Struve in the United States as early as 1948. Since that time, the analysis of stars in the galactic halo (including the so-called subdwarfs) has shown abundances for the heavier elements relative to hydrogen to be from ten to a thousand times deficient. These rare objects therefore form an exception to the rule of uniformity, but the relative abundances of the heavier elements among themselves are still similar to the solar values.

The data on stellar element abundances have provided clues of immense cosmological significance. The results are the essential evidence for the theory of stellar nucleosynthesis of the heavier elements (from carbon onwards in the periodic table) and for the cosmological origin in the Big Bang of the hydrogen and nearly all of the helium at present in the universe. This widely accepted theory was expounded by Geoffrey and Margaret Burbidge, W.A. Fowler, and Fred Hoyle in a 1957 treatise that discussed all the evidence then available in an attempt to account for the observed cosmic abundance distribution of the elements in stars, nebulae, and the solar system. This theory, although modified in detail since 1957, continues as the basis of current ideas on the formation of elements in the universe.

The vast majority of the approximately one thousand stars for which high-dispersion spectroscopic-abundance analyses have been undertaken are relatively nearby objects within the Galaxy, many in the solar neighborhood. The 1968 analysis of the extragalactic supergiant star HD 33579 in the satellite galaxy, the Large Magellanic Cloud, by Antoni Przybylski was therefore an event of some importance, since it was the first such analysis of any extragalactic body. A small deficiency of heavy elements was found but regarded as insignificant. Subsequent work has shown that this galaxy probably has a small overall deficit of heavy elements relative to the sun by about a factor of two. Therefore, the idea that the same nuclear processes have been occurring in stellar interiors in different galaxies has been supported by direct observational evidence. Spectroscopy has provided a foundation for one of the basic premises of cosmology, that of the approximate chemical homogeneity of the universe and of the uniformity of the physical processes in different stars and galaxies.

Early Spectroscopic Observations of Spiral Nebulae

Early developments in the spectroscopy of the spiral "nebulae" (galaxies) took place about half a century later than the corresponding work on stars, a consequence of the faintness of these objects and, especially, of their low surface brightness. This section reviews the first quarter century of this work from 1899, a time when the true nature of the spirals was still a matter for debate. Two major results came from the early spectroscopic work: (1) the discovery of solar-like spectra for spiral nebulae and (2) the discovery of large positive Doppler shifts for the great majority of these objects.

Discovery of Solar Spectra for Spiral Nebulae

The visual spectroscopy of Huggins in 1864 revealed only a continuous spectrum for the great Andromeda nebula, M31. This at least distinguished it from gaseous nebulae with their emission-line spectra. Huggins again observed the spectrum of the Andromeda nebula in 1888, this time using dry gelatine emulsion photography. The exposure of about four hours weakly showed several absorption lines, including those at H, K, and G, although he was uncertain whether these may not have been due to contamination by moonlight. The spectrum was not published until 1899, and Huggins only commented on it briefly in 1909.

Meanwhile, Scheiner in January 1899 obtained a 7½-hour exposure of the Andromeda nebula spectrum and found both the H-line and G-band to be clearly visible. He concluded that "the suspicion that spiral nebulae are star clusters is now raised to a certainty," and he made a simple comparison between the Andromeda nebula and the Milky Way system. This classic observation marked the beginning of a major new era in astronomy and provided a vital clue for the Great Debate of 1920 over whether spiral nebulae are galaxies independent of our own Galaxy. The evidence nevertheless was clouded by Vesto M. Slipher's observation at the Lowell Observatory in 1912 of a stellar spectrum from the nebulosity surrounding several of the bright stars in the nearby Pleiades galactic star cluster, which showed that a nebula with a starlike spectrum was not necessarily an extragalactic star system. (The Pleiades nebulae now are recognized as arising from scattering by circumstellar dust grains.)

Other early observers of spiral nebulae spectra were Max Wolf in Heidelberg and Edward Fath in California. Both observers confirmed the solar-type spectra for spirals. Fath used the Crossley reflector at Lick (with a specially designed spectrograph for nebular work) as part of his doctoral dissertation in 1909. He reported the measurement of fourteen absorption lines in the blue to ultraviolet regions from two exposures of M31, one of them lasting over eighteen hours. He also obtained spectra of seven other spirals, thus greatly extending the scope of Scheiner's work. One of these (NGC 1068) gave a mixture of emission and absorption lines, while the remainder showed faint traces of solar-like absorption features. Fath considered that these data lent strong support to the idea that spirals were unresolved star clusters, presumably at great distances. However, a faulty trigonometric parallax determination for M31 (which would have implied a small nebulous object near the sun!) apparently still left some doubt in his mind.

In hindsight, it is clear that the discovery of the solar spectra of the spirals was the best evidence available at that time for a much larger universe than defined by the bounds of the Milky Way star system. It also pointed to a general similarity in stellar compositions between the stars from one galaxy to another, thus extending to galaxies the same principles that Huggins had applied to stars half a century earlier.

First Radial Velocities for Spiral Nebulae

Slipher commenced a spectroscopic program on spiral nebulae in 1912, using the 24-inch Lowell Observatory refractor and prism spectrograph. His initial result from four long exposures of the Andromeda nebula gave the first Doppler shift for a spiral nebula, and his value of −300 km/s (velocity of approach) was the largest Doppler shift then known for any object.

Slipher soon extended this work to other spirals. The exposure times were generally from twenty to forty hours over several nights, occasionally even weeks, and the success of the program can be ascribed in part to his tenacity in observing such faint objects over such long exposures, and also to his use of a fast spectrograph camera (f/2.5) designed for spectrography of low-surface-brightness objects. The dispersion was a modest 140 Å/mm at Hγ, and an iron-vanadium spark spectrum was used as a comparison.

In 1917 Slipher reviewed progress in this difficult program. He had by then obtained Doppler shifts for twenty-five spiral nebulae with velocities between −300 and +1100 km/s. Of these, twenty-one had positive values indicating recession. By 1925, when the program was terminated, the radial velocities of forty-one spirals had been secured, and now the highest value was 1800 km/s (reported by Strömberg in 1925).

By far the most striking result from Slipher's measurements was the preponderance of very large positive Doppler shifts, indicating recessional velocities of spiral nebulae. The Andromeda nebula happened to be one of the few approaching the sun, as well as being the first to be observed and measured. But as Slipher extended the observations to fainter nebulae over the years, the trend of ever larger positive Doppler shifts was found for these fainter objects. The highest values were some ten times larger than the velocities of high-velocity stars within the Milky Way.

Another result of importance was that some spirals showed slanting lines, indicating rotation. This was found first by Slipher in 1914 for the spiral NGC 4594 in Virgo seen edge-on, and later for several others. The rotation sense was always that of spiral arms trailing, and for the Virgo spiral the rotation velocity was some 300 km/s at 2 arc minutes from the center. As early as 1755 the German philosopher Immanuel Kant had proposed the idea that spiral nebulae were "island universes" and had invoked rotation to account for their apparently flattened morphology. Slipher's spectroscopy gave striking support to Kant's bold prediction.

Slipher was practically the only observer to undertake this difficult work up to 1925. Pease had also obtained a spectrum of NGC 4594 at Mt. Wilson in eighty hours on the 60-inch telescope, and it confirmed Slipher's result for the large Doppler shift. Moore and Wright at Lick and Wolf in Heidelberg each made a handful of Doppler measurements on spirals, but Slipher's work still comprised some 90% of the available results in 1925.

The interpretation of large positive-wavelength shifts in spiral nebulae spectra as recessional velocities was explicitly made in all of Slipher's published results, and this in turn implied an expansion of the system of nebulae that comprise the universe. Moreover, such large velocities meant that spirals could not be local objects gravitationally bound to the Milky Way.

Carl Wirtz in Germany decided in 1918 to apply a K-term (or correction factor) to these large shifts, by analogy with the much smaller positive mean radial velocity of galactic B stars, which also appeared to be expanding away from the sun. This anomalous result for B stars was later found to be the result of a systematic error, but the very large and systematically positive trend for faint spirals was much greater than any conceivable error probable in the observations. Gravitational redshifts were also suggested, but it seemed implausible that spiral nebulae could be such small and compact objects required to give the shifts observed. The interpretation of the shifts as the result of recessional velocities was the only one that received any credence.

The prediction in 1917 by the Dutch astronomer Willem de Sitter of a correlation between velocity and distance in an expanding universe as a possible solution to Einstein's field equations motivated Wirtz, the Swedish astronomer Knut Lundmark, and Strömberg to look for such a correlation using Slipher's velocity data in 1924–25. Unfortunately, the distances of the objects, based either on angular diameters or apparent magnitudes, were so unreliable that the correlation was either weak or absent. This barrier to further progress had to await the American astronomer Edwin Hubble's much more reliable relative distances, based on his ability to resolve stars in the nearer spirals using photographs taken with the 100-inch Hooker telescope on Mt. Wilson, and to identify novae and Cepheid variables among these faint stars. Hubble's work not only completely confirmed the idea of extragalactic star systems or "island universes," but also showed a clear velocity-distance correlation in agreement with de Sitter's cosmology.

—John B. Hearnshaw

BIBLIOGRAPHIC NOTE

On the history of the Doppler effect and on the analysis of astronomical spectra, see J.B. Hearnshaw, *The Analysis of Starlight: One Hundred and Fifty Years of Astronomical Spectroscopy* (Cambridge: Cambridge University Press, 1986; paperback edition 1990). On the early spectroscopy of the spiral nebulae, see Edwin Hubble, *The Realm of the Nebulae* (New Haven: Yale University Press, 1936; paperback reprint, New York: Dover Publications, Inc., 1958), particularly chapter 5; and Vesto M. Slipher, "Nebulae," *Proceedings of the American Philosophical Society*, 56 (1917), 403–409.

VI. The Expanding Universe

Introduction

THE GREAT DEBATE

ne of the more dramatic episodes in the history of astronomy was the so-called "Great Debate" before the National Academy of Sciences in Washington, D.C., on 26 April 1920. Actually, we could discuss two debates, the oral presentations—heard by very few scientists—and the substantially different published versions—read by many more scientists. Both presentations pitted Harlow Shapley from the Mount Wilson Observatory against Heber D. Curtis of the Lick Observatory.

Neither set of discussions was exactly a debate, because the two scientists had come to argue somewhat different issues. Shapley discussed his new estimate of the size of our galaxy, whose boundaries now encompassed the spiral nebulae *if* purported measurements (now known to have been spurious) by Adriaan van Maanen of rotations of spiral nebulae were correct. Shapley's major concern was to defend his estimate of the size of our galaxy, not necessarily to encompass the spiral nebulae within our galaxy. Shortly before the meeting, he had written to a colleague that he was coming to Washington to discuss the scale of the universe but he did not intend to say much about spiral nebulae because he did not have a strong argument. Curtis, on the other hand, arrived in Washington intent on discussing the spiral nebulae and on showing that they are independent galaxies or "island universes" beyond the boundaries of our own galaxy.

If we call the meeting in 1920 the great debate, what should we call the centuries-long debate over whether spiral nebulae are extragalactic systems, in whose long history the Washington meeting appears as but a brief instant? This greater debate over whether spiral nebulae are island universes rose out of philosophical speculations in the eighteenth century. The construction of new and much larger telescopes and the development of the new observational techniques of photography and spectroscopy produced new and conclusive data, making possible resolution of the debate in the 1920s. Resolution however, scarcely came as an anticlimactic conclusion to two centuries of inquiry. In the mid-1920s the astronomical community suddenly was confronted with two sets of observations, each possessed of considerable claim to validity but mutually contradictory.

Juxtaposed on one side—the losing side, as became apparent only four years after the inconclusive debate between Shapley and Curtis—were van Maanen's reported measurements of the rapid rotation of spiral nebulae. These measurements, obtained over a period of years beginning in 1916 and extending into the early 1920s, initially were not so easily dismissed as they are now, with the aid of hindsight and historical perspective. Van Maanen had made the observations at the Mount Wilson Observatory, the world's premier observatory, and he was a respected astronomer, carefully following scientific procedures. His work commanded respect, even acceptance—were it not for contradictory evidence of comparable heritage. A matrix of supporting observational evidence, linkages between observation and theory, and personal considerations all argued for van Maanen's measurements, thus heightening the drama and the dilemma.

Equally sanctified were the observations of Edwin Hubble, another member of the Mount Wilson Observatory staff. In 1924 Hubble detected in several spiral nebulae a type of star called a Cepheid variable. These variable stars are invaluable distance indicators, and showed the spiral nebulae to be at great distances. Hubble's data placed the spiral nebulae beyond the boundary of our galaxy as decisively as van Maanen's data confined the spiral nebulae within our galaxy.

The result was a dramatic conflict between Hubble and van Maanen, embarrassing for the astronomical profession and particularly difficult for the two astronomers and for the Mount Wilson Observatory, the common institutional home of the major opponents of one of the most important scientific problems of the day.

Out of this conflict, though, came remarkable progress.

> Never in all the history of science has there been a period when new theories and hypotheses arose, flourished, and were abandoned in so quick succession as in the last fifteen or twenty years.

So reminisced the Dutch astronomer Willem de Sitter in 1931, looking back on two decades of discovery, turmoil, and upheaval of beliefs in science. In these glorious years astronomers finally managed to resolve the centuries-old debate over whether luminous nebulae scattered throughout space are galaxies composed of stars and comparable to our own stellar system. Furthermore, astronomers with their new discoveries undermined the generally held assumption that the universe is static, and instead demonstrated that the universe is expanding. Speculations regarding spiral nebulae have a long history, which we may briefly review here (at the risk of repeating some material from *A Rational Order for the Cosmos*, the introduction to section V, *Galaxies: from Speculation to Science*). Eighteenth-century belief in the orderliness of the universe made determination of that order an important theological, philosophical, and scientific endeavor. William Whiston, Thomas Wright, and Immanuel Kant are readily placed in the tradition of believing in the order and beauty of God's work. Kant suggested that nebulous patches of light in the Heavens are composed of stars and are other Milky Ways or island universes. Johann Lambert published a similar theory in 1761 and William Herschel published his own theories on the construction of the Heavens beginning in 1784, apparently with knowledge of neither Wright's, Kant's, nor Lambert's work. [See chapter 15, *William Herschel*, in section V, *Galaxies: from Speculation to Science*.] Reflecting the new atheistic approach to nature of some scientists of the French Enlightenment at the end of the eighteenth century, Pierre-Simon Laplace attempted to replace the hypothesis of God's rule with a purely physical theory that also could explain the observed order of the universe. Groups of stars—nebulae—might be explained by the hypothesis of a contracting nebulous mass.

Meanwhile, new evidence was accumulating from increasingly detailed telescopic investigations of nebulae. [See chapter 16, *Nineteenth-Century Reflecting Telescopes*, in section V, *Galaxies: from Speculation to Science*.] Two questions were foremost:

- Were nebulae aggregations of stars, or were they made of truly nebulous matter?
- Did the physical appearance of nebulae suggest rotation?

In 1845, observations of M51 by the Earl of Rosse revealed for the first time the spiral arrangement of matter in that nebula. Rosse thought:

> That such a system should exist without internal movement, seems to be in the highest degree improbable: we may possibly aid our conception by coupling with the idea of motion that of a resisting medium; but we cannot regard such a system in any way as a case of mere statical equilibrium.

Observations the next spring revealed the spiral nature of M99. Indications of spiral structure in other nebulae raised to fourteen the number of nebulae that Rosse classified as spiral systems, with several more marked doubtful on his working list.

Previous, implicit suggestions of the rotation of nebulae had met with indifference. Rosse's actual observations of the spiral form of some nebulae, however, were sufficiently dramatic to establish a historically continuous stream of thought regarding the possible, indeed probable, rotation of nebulae. Robert Grant, who would become the director of the observatory at the University of Glasgow, felt that both reason and analogy amply justified his belief "that the constituent bodies of every system are revolving in curvilinear orbits under the influence of their mutual gravitation." Similar enthusiasm for rotational motion rose up in the United States. Stephen Alexander, who taught at Princeton (originally the College of New Jersey), concluded, under the joint influence of Laplace's nebular hypothesis and Rosse's descriptions of nebulae, that the spiral nebulae might be "the partially scattered fragments of enormous masses, once rotating in a state of dynamical equilibrium; and that the separation of these fragments may still be in progress."

Astronomers were enthusiastic about Rosse's observations, but realized that more work was needed to convert speculation into established fact. Grant noted that the study of the displacements arising from the internal movements in spiral nebulae would constitute an assiduous course of observation and research over a long succession of ages. Until observations were amassed to form a sure groundwork of research, speculations about the mechanical structure of the more complex nebulae were premature. Equally cautious was J.L.E. Dreyer, in 1877 an assistant at Rosse's observatory. No previously alleged instance of variability of a nebula survived his critical examination. One could not "expect immediate and startling results, which would look well in popular books."

Popular books, though, were not long in appearing, even without new and startling observations. Greatly increased public interest in celestial astronomy warranted the publication of Agnes M. Clerke's *Popular History of Astronomy* in 1885 and in three subsequent editions and four printings. Clerke discussed Kant's and Laplace's speculations, Rosse's observations, and their interpretation by Grant and Alexander. The further observations needed to form a sure groundwork for nebular research, however, were only beginning to be made, with the development of two new research tools of astronomy: photography and spectroscopy.

The basic fundamentals of photography were known when man recognized the action of the sun's rays on chemicals, in the tanning of skin by light, and when the camera obscura, which focused light through a small round aperture onto a piece of paper, was described by Leonardo da Vinci. Fixing the image of the camera obscura by chemical means did not begin, however, until the nineteenth century. Louis Daguerre's invention caused a great sensation, and was widely used from 1839 into the 1850s. In 1851 a new process was introduced, using plates exposed in a wet condition. A few photographs of the brightest stars were taken, but not until the introduction of much more sensitive dry plates after 1878 did photography become common in astronomical studies.

Amateur astronomers took up the new invention of photography. Professional astronomers were, by definition, already fully employed in well-defined research projects. Also, there was a strong mathematical bias among professional astronomers, and they may have viewed photography as less than a rigorous discipline. The stated aim of the American Photographical Society was to show that its members were not "mere mechanics."

In the United States John Draper took the first known photograph of a celestial object, the moon, in the 1840s, and his son Henry took the first photograph of a nebula, the Orion Nebula, in 1880. Henry Draper's photography came to equal Rosse's drawings in delineating features of nebulae, and Draper anticipated that further progress would secure details entirely invisible to the naked eye.

Draper's optimism regarding the future of nebular photography was shared by Andrew Ainslie Common, an English sanitary engineer whose avocation was astronomy. Common's enthusiasm was based upon his own successful photograph of the Orion Nebula in 1882. Furthermore, Common realized that evidence of any change in a nebula (he was interested in changes of form or brightness) would have to come from a comparison of photographs taken over some interval of time.

Isaac Roberts, who had moved up through the building trades in England to become a partner in a large firm of contractors, took his first photograph with a telescope in 1883. Soon he developed a program to photograph star clusters and nebulae. After more than a decade of study, Roberts reported strong if not conclusive evidence of the evolution of stellar systems.

Meanwhile, results from another branch of astronomical research, also newly developed during the latter part of the nineteenth century, were providing further information about the spiral nebulae. [See chapter 17, *Spectroscopy*, in section V, *Galaxies: from Speculation to Science.*]

Spectroscopy received attention from scientists in several fields of study after Robert Kirchhoff, professor of physics at Heidelberg, showed in 1859 that each element has a characteristic absorption pattern, resulting in identifiable dark lines in the spectrum of light passed through a prism. Chemical constituents revealed themselves in spectra. William Allen Miller, a professor of chemistry at Kings College, London, had made observations in 1845 of the flame spectra of various chemicals when burned. Now, in the early 1860s, Miller became a proponent of spectrum analysis, and, through William Huggins, exerted a direct influence on astronomical investigations.

With spectroscopy as with photography, amateur astronomers pioneered the use of a new technique. In 1854 Huggins disposed of his silk and linen business and moved to Tulse Hill, a suburb of London. There, two years later, he set up his own observatory. Huggins began with the rather routine astronomical work of making drawings of planets and taking transits, timing the instant of passage of a star across the meridian. He was vaguely dissatisfied with these lines of research, and when he heard of Kirchhoff's work on the chemical constitution of the sun, the news was "like the coming upon a spring of water in a dry and thirsty land."

At about the same time, Huggins attended a meeting of the Pharmaceutical Society at which spectroscopes were shown. Seized by a sudden impulse, he suggested to Miller, also at the meeting and Huggins's friend and neighbor, that they return home together to commence observations of stellar spectra. At first Miller was skeptical about applying Kirchhoff's methods to the stars, whose light is so much fainter than that of the sun, but he did agree to go to Huggins's observatory on the first clear night for some preliminary experiments. By 1870 the two men, working together, had identified several elements in the spectra of stars and nebulae.

The new spectroscopic techniques soon were also employed for the measurement of motions. In 1841 Christian Doppler argued that the spectrum would be shifted by the motion of the source of light. A shift of the spectral lines from moving objects was not immediately accepted by all scientists, but the idea was discussed in the two decades before Huggins's investigations in the late 1860s and early 1870s, investigations that were to prove the existence of the predicted spectral shift. In 1868 Huggins found what appeared to be a slight shift for a hydrogen line in the spectrum of the bright star Sirius; by 1872 he had more conclusive evidence of the motion of Sirius and several other stars; and by 1874 he had extended his results to a few gaseous nebulae.

Instrumental limitations prevented Huggins from extending effectively his spectroscopic investigations to the spiral nebulae. He had a 15-inch refracting telescope and an 18-inch reflector borrowed from the Royal Society; astronomers in the United States had the 36-inch refractor at the new Lick Observatory, constructed in California in the 1880s. Because the light gathered from most objects is proportional to the area of the lens, which in turn is proportional to the square of the diameter of the lens, the Lick telescope was over five times as powerful as Huggins's refractor. In addition, the Lick telescope was equipped in 1890 with a grating (a flat surface with 14,438 lines per inch ruled on it). The grating gave a dispersion (separation of spectral lines) approximately twenty-four times as great as that obtained with a single 60-degree prism. Huggins, who had begun the measurement of velocities of nebulae, was left behind by the mainstream of spectroscopic work on nebulae. Astronomical entrepreneurship in America's gilded age saw the construction of new and larger instruments and a shift of the center of astronomical spectroscopic research from England to the United States.

Here in the mid-1920s new evidence favoring the argument for spiral nebulae as island universes would finally settle the centuries-long debate. Furthermore, unexpected new evidence from observations beginning in 1912 but not interpreted conclusively until 1929 would force upon astronomers in particular and upon our culture in general belief in an expanding universe.

18. COSMOLOGY 1900–1931

During the first three decades of the twentieth century, the modern science of cosmology was constructed from general relativity theory and new methods and instruments of observation, particularly large optical telescopes perched thousands of feet above sea level in California and Arizona. These radical changes in the theoretical and observational tools used by astronomers, physicists, and mathematicians accompanied important changes in cosmology itself. The new cosmology of the early 1930s encompassed a new set of questions about the physical universe as well as the novel framing of old questions, including questions about the age of the universe. It also included two key cognitive features generally absent from the thinking of astronomers at the turn of the century: first, that galaxies visible in earth-based telescopes exist outside our own stellar system, and second, that these galaxies evince the expansion of the universe.

By the early 1930s the practice of observational cosmology had also become in many ways a distinctly American activity, a major shift from the European dominance of the eighteenth and nineteenth centuries that was due to a mix of scientific, technical, and institutional developments. A division of labor had also developed, so that there were two main groups pursuing cosmological questions, what could be termed observational cosmologists and theoretical cosmologists.

Island Universes?

In 1900 very few astronomers accepted the existence of visible external galaxies. Our own galaxy, the Milky Way, was widely agreed to constitute the entire visible universe. In fact, cosmology as it would be understood in mid-century hardly existed in the early1900s. The term *cosmology* itself was very rarely used, and when it was, it was taken to mean how the universe is currently ordered. Even *universe* often meant something very different from what it would mean by the early 1930s. To astronomers in the first two decades of the century, the term often implied what we would now call our galaxy or

Milky Way system. This use of *universe* was therefore loaded with the generally held view that our galaxy is the only such visible aggregation of stars. (Although it is not our concern in this chapter, when the Dutch astronomer J.C. Kapteyn pursued his program to determine the size and structure of the galaxy, he was engaged in "statistical cosmology" through the use of stellar statistics.)

Astronomers had not always held such a restrictive view of the physical universe. For periods in the eighteenth and nineteenth centuries, there had been lively debates on the nature of the nebulae. The debates had focused on the following questions:

- First, were these dim patches of light remote star systems whose light merges to give a milky effect, or were they made of a luminous fluid?

- Second, if these dim patches of light were groups of stars, were these star systems part of or associated with our galaxy, or were they themselves "island universes"—vast independent galaxies?

Only two years before its close, the nineteenth century's verdict on the island universe theory was passed by a writer on the history of astronomy: "The island universe theory of nebulae, partially abandoned by Herschel after 1791 ... but brought into credit again by Lord Rosse's discoveries ... scarcely survived the spectroscopic proof of the gaseous character of certain nebulae." When, for example, some photographs of the Andromeda Nebula secured by the English amateur astronomer Isaac Roberts had been shown at the British Royal Astronomical Society in 1888, there were cries of "'Saturn,' 'the Nebular Hypothesis made visible,' and so on." The Society's members interpreted the nebula as a new solar system in the process of condensation from nebulous material in the manner of Laplace's nebular hypothesis.

At this time, too, the Andromeda Nebula was classed as a "spiral nebula." By 1900 and through the researches begun by the American astronomer James E. Keeler at the Lick Observatory in 1898, astronomers accepted the idea that there were vastly more visible nebulae than had previously been anticipated. Keeler's figure was 120,000, and he held that most possessed a spiral structure. This was an astonishing claim because, although spiral nebulae had first been observed by Lord Rosse and his colleagues half a century earlier, only a few dozen had been identified before Keeler's studies. The spiral nature of prodigious numbers of nebulae was now employed against the island universe theory. One prominent writer, for example, judged that:

> *The relationship between the various orders of nebulae is manifest. The tendency of all to assume spiral forms demonstrates, in itself, their close affinity; so that to admit some to membership of the sidereal system while excluding others would be a palpable absurdity. And since those of a gaseous constitution must be so admitted, the rest follow inevitably.*

In addition to the photographs of spiral nebulae interpreted in terms of the nebular hypothesis, astronomers reckoned three pieces of data to be particularly telling against the island universe theory:

- The distribution of nebulae.

  By the mid-1860s, it had become widely accepted that the nebulae were not equally distributed over the sky. Rather, the nebulae seemed to avoid the plane of the Milky Way. But if the nebulae were truly independent galaxies, it would be curious indeed for them to avoid just that region of the sky that the Milky Way marked out.

- The gaseous nature of some nebulae.

  Starting in 1864, astrophysicists (the first of whom was William Huggins) produced evidence quickly taken as conclusive that at least some of the nebulae were not aggregations of stars but were instead composed of a luminous gas.

- The "nova" of 1885 that had flared in the center of the Andromeda Nebula.

  The 1885 nova, or "new star," that blazed in the heart of the Andromeda Nebula provoked great interest. The nebula was one of the largest, and astronomers presumed nearest, of the nebulae. Whatever the cause of the nova, it now seemed inconceivable to astronomers that the nebula could be composed of millions of stars; at its brightest, the nova had attained a luminosity of roughly one-tenth of the entire nebula, and how, astronomers asked, could a single star rival the gigantic light output of an immense stellar assembly? (The new star was in fact a supernova, but at the time there was no reason to believe in the occurrence of such violent stellar outbursts.)

If in 1900 astronomers generally rejected the existence of *visible* galaxies, this does not mean that these same astronomers necessarily rejected the existence of galaxies. This was because by the end of the nineteenth century, as the historian of astronomy S.L. Jaki has emphasized, "it had become customary to picture the universe as consisting of two parts: one visible and confined to the Milky Way, and another, truly infinite, which was believed to be forever beyond the reach of visible observations." So, while astronomers had insisted that there was no observational evidence for external galaxies, some had been strongly inclined to think that such objects did in fact exist. In

1897, David P. Todd, director of the Amherst Observatory in the United States, for example, had speculated on "other universes than ours":

> *Are, then, the inconceivable vastnesses of space tenanted with other universes than the one our telescopes unfold? We are driven to conclude that in all probability they are. Just as our planetary system is everywhere surrounded by a roomy, starless void, so doubtless our huge sidereal cluster rests deep in an outer space everywhere enveloping illimitability. So remote must be these external galaxies that unextinguished light from them [. . .] cannot reach us in millions of years.*

Despite widespread rejection of the island universe theory, it still had some advocates. One was the German astronomer Julius Scheiner. In 1898 Scheiner secured a spectrogram of the Andromeda Nebula. Astronomers had known for years that spiral nebulae possessed faint, continuous spectra, but Scheiner detected dark lines that crossed the continuous band. Furthermore, Scheiner decided that a comparison of this spectrum with that of the sun, taken with the same apparatus, indicated that the sun and the Andromeda Nebula were surprisingly similar. For Scheiner, the inference to be drawn was clear: the spiral nebulae are star systems.

First at the Lick Observatory and then at the Mount Wilson Observatory in California in the years around 1910, E.A. Fath studied the spectra of more spiral nebulae. He too decided that the spectra indicated that the spirals are star systems.

Astronomers also speculated on the nature of our own galaxy, and a few, like Max Wolf at Heidelberg and Sir David Gill at the Cape of Good Hope, proposed that our galaxy is arranged on a spiral pattern. If our galaxy is a spiral, then it seemed more plausible to these astronomers and to some of their colleagues as well that the spiral nebulae are themselves galaxies.

In the opinion of most astronomers, however, there was no crushing evidence at hand to compel a shift of allegiance to external galaxies. Nevertheless, by about 1910, galaxies had become a topical issue in astronomy, and by now there had arisen two main versions of the island universe theory.

- One stated merely that there are independent stellar systems beyond the confines of our own galaxy.
- The second version might be better termed the "comparable-galaxy" theory, since it carried the extra implication that the island universes are comparable in size with our galactic system.

Until the end of the nineteenth century the distinction mattered little, because the majority of astronomers wanted to force all of the unresolved nebulae into one class: *either* comparable galaxies *or* clouds of truly nebulous matter. In the last decades of the nineteenth century, however, this desire for simplicity was on the wane. While the shift initially was hesitant, certainly by 1910 astronomers had thrown off the shackles that had bound their nineteenth-century predecessors to a restrictive, nonpluralistic belief in *either* comparable galaxies *or* clouds of nebulous matter, *but not both*.

Furthermore, by about 1910 it was assumed beyond doubt that the planetary and diffuse nebulae are within our stellar system. For astronomers, the evidence of stars associated with planetary and diffuse nebulae, the motions of planetary and diffuse nebulae, and their distribution in space (both the planetaries and diffuse nebulae were clearly concentrated toward the Milky Way) all pointed unambiguously toward these objects being part of our galaxy.

But in the early years of the century, astronomers judged that there were relatively few observational constraints on any theory of the spirals. In this situation, the acceptance of different theories, each in accord with much of the available evidence, turned out to be in considerable part a mix of personal scientific "taste" and institutional preferences. With the benefit of hindsight, we can see that what was required to settle the issue was a means of determining distances to the spirals that was widely accepted as accurate by the practitioners of nebular astronomy. Until such a method could be fashioned, the question of the existence of island universes would remain perplexing for many.

Radial Velocities of Spiral Nebulae

In 1912 a new trail that would eventually lead to a novel and immensely important method of calculating distances to spiral nebulae was opened by Vesto M. Slipher when he became the first to measure the radial velocity of a spiral. Slipher was a member of the staff of the Lowell Observatory in Arizona. In 1909 he was instructed by the observatory director, Percival Lowell, to investigate the spectrum of the Andromeda Nebula with the hope of discerning some clues to the origin of the solar system. Indeed, at this date Slipher was sure that the nebula was a single solar system in formation.

By December 1910 Slipher had detected features in the spectrum of the Andromeda Nebula that had not been seen previously. He continued to try out various observational techniques, including making various adaptations to his spectrograph. By 1913 he had four photographic plates on which he could measure the shifts of the Andromeda Nebula's spectral lines, and hence (interpreting the shifts as Doppler shifts) its radial velocity.

In 1911 W.W. Campbell, the director of the Lick Observatory in California, derived an empirical formula linking the age of a star and its radial velocity: the older a star, the higher its speed. Since many astronomers placed the spirals at the start of the evolutionary path of the stars, the spirals were expected to possess low radial velocities (in fact, on the order of 10 kilometers per second.) Even if an astronomer suspected that some of the larger spirals were not proto-stars but were star clusters associated with our galaxy, no doubt he or she would have anticipated the spirals to have velocities comparable to the velocities of other members of the stellar system—certainly not velocities far greater than the velocities of stars and planetary nebulae. But both of these expectations were shattered by Slipher's startling find that the Andromeda Nebula is rushing towards the sun at 300 km/sec—the highest speed then recorded for any astronomical body!

So surprising was this result that, to begin with, some of Slipher's colleagues were skeptical of its trustworthiness. But Slipher soon measured high velocities for more spirals, and other astronomers confirmed the result for the Andromeda Nebula. By 1914 Slipher had results for fifteen spiral nebulae, most of them velocities of recession, or redshifts. When in that year he presented these results to a meeting of the American Astronomical Society, he won a standing ovation.

At first Slipher interpreted the spiral nebulae as proto-solar systems. But a number of other astronomers explained Slipher's results in a very different manner. To these astronomers, the speeds of the spirals seemed altogether too great for them to be gravitationally bound to our stellar system, and the use of the radial velocities of the spirals to reinforce the claims of the island universe theory soon became common practice.

Evidence Against Island Universes

To be counted against the evidence of the spectra and radial velocities of the spirals were still the distribution of spiral nebulae in the sky and the 1885 nova. In addition, photographs of some of the larger and brighter spirals, such as those secured by the Mount Wilson astronomer G.W. Ritchey, disclosed what he termed "nebulous stars" in the outer regions of the spirals. The numbers of these nebulous stars in the spirals were such that if they truly were stars, the spirals were sparse clusters of stars and not entire galaxies.

Furthermore, starting in 1916, another Mount Wilson astronomer, the Dutchman Adriaan van Maanen, secured a different set of influential data on the spirals. He did this by measuring photographic plates of a particular spiral taken at two different times and then comparing the photographs to see if there had been any motion of the material in the spiral in the interim. He indeed found such motion for the giant spiral nebula M101. (See figure 2 in the following chapter, *Hubble's Cosmology*, page 350.) Later, such motions would be taken as evidence against the spirals as island universes, but to begin with, this was not so. The influential English mathematician and astronomer James Jeans also had expected the sorts of motions in spirals that van Maanen had found, and his speculations added credibility to van Maanen's observations.

For Jeans, the spirals were star systems in the process of formation. At this period astronomers were unsure about the size and structure of our galaxy, but the sun was universally placed close to the center of our galactic system. A diameter for the system of about 30,000 light-years also seemed a reasonable estimate for many astronomers. Therefore the distances obtained from van Maanen's measurements, on the order of 15,000 light-years, buttressed the theory of the spirals as galaxies by placing the spirals at the boundaries of or outside our galaxy. Van Maanen's measures did, nevertheless, help to undermine the then prevalent belief that the larger spiral nebulae are merely proto-solar systems or sparse clusters in the making. The measures indicated masses for the spirals of at least a thousand solar masses, and the combination of their apparent sizes and Jeans's distances (even if Van Maanen's measures were exaggerations indicating only minimum distances) made the spirals appear far too large to be incipient solar systems or sparsely populated clusters.

Though initially regarded by astronomers as support for the spiral nebulae as galaxies, van Maanen's measures conflicted with the idea of the spirals as galaxies comparable to our own Milky Way galaxy. This issue was thrown into even starker relief when in 1917 Harlow Shapley, then at the Mount Wilson Observatory, presented his radical new theory of the galactic system, the "Big Galaxy." If our galaxy were anything like as big as Shapley claimed (300,000 light-years) and if van Maanen's measures were accurate, the larger spiral nebulae were very much smaller than, and inside, the Milky Way. By 1920 Shapley was speculating that the peculiar distribution of the spirals and their systematic recession from our galaxy could be explained by supposing them to be masses of nebulous material repelled in some manner from the galactic system, which he presumed to be moving through a field of such nebulae of indefinite extent.

Shapley acquired a reputation as a leading critic of the island universe theory, and this reputation helps to explain his invitation to participate in the so-called Great Debate on the scale of the universe at the National Academy of Sciences in Washington, D.C., in 1920. His opponent in the debate was a Lick astronomer, Heber Doust Curtis. In many ways Curtis was the spokesman for the Lick Observatory points of view on spirals and the structure and size of our galaxy. At Lick a school had developed whose members shared, in particular, a commitment to the island universe theory and, in general, a conservative attitude toward new developments in astronomy. In fact, W.W. Campbell, the Lick director, had initially been proposed as the advocate of island universes, but after George Ellery Hale, one of the debate's key organizers, had seen a Lick volume that contained three papers by Curtis, it was Curtis who became the chosen speaker.

Three years earlier Curtis had also helped to take the discussion on island universes in a new direction, when he and also Ritchey at Mount Wilson announced the finds of faint novae in spiral nebulae. That bright novae had appeared in two spiral nebulae was well known. The 1885 nova, S Andromedae, that had flared in the Andromeda Nebula, had been such a phenomenon. In 1895 a celestial firework on the same scale, Z Centauri, had been observed in the spiral NGC 5253, but because it was detected when rapidly fading in brightness, it had failed to excite the same interest as had S Andromedae. In March 1917 Curtis had found three novae in spirals, and in July of the same year Ritchey found another. After the finds of Curtis and Ritchey, some astronomers were soon busily employed searching old plates for more novae. Many novae were tracked down, and the principal results they afforded were estimates of the magnitudes of the novae near maximum light.

Curtis fastened on this material to manufacture another means of determining the distance to the Andromeda Nebula. Although he did not invent this method, the fact that a sufficient number of much fainter novae in spirals had been observed persuaded Curtis to reject S Andromedae and Z Centauri on the grounds that they were anomalous objects from which conclusions should be drawn only with great care. Calculating distances to the spirals using the fainter novae now gave Curtis distances of millions of light-years and clearly placed the spirals far outside our galactic system. His calculational scheme rested on the assumption that the novae in our galaxy and the spirals reach similar absolute brightnesses. But astronomers were unsure how bright the galactic novae were, and there was still the 1885 nova in the Andromeda Nebula to explain. So, while the novae were widely seen by astronomers as furnishing strong evidence in favor of the spirals as island universes, the evidence was not regarded as conclusive by any means.

At the Great Debate, Curtis argued for the existence of external galaxies but against Shapley's new theory of the galaxy. The original intention of the organizers in proposing the debate was that the participants focus on island universes. However, Shapley sought to move the focus to his galactic theory. As he told a correspondent, "I am going to touch lightly on the spiral nebulae (I have neither time nor data nor very good arguments)." In any case, Shapley believed that the existence of external galaxies would stand or fall with the acceptance or rejection of his Big Galaxy. By proclaiming the correctness of his galactic model, he was by implication assailing the island universe theory, and so at Washington and in the papers published after the meeting Shapley focused his analysis on the size of our galaxy. The meeting in Washington also saw Shapley—at this stage in his career a poor public speaker—pitch his talk at a non-technical level and, in considerable part, direct his remarks to those members of the audience concerned with the future activities of the Harvard College Observatory, for whose vacant directorship Shapley was a candidate. Curtis, who had expected a technical presentation, was left throwing his verbal blows at a nonexistent protagonist.

It was indeed in the published papers of Shapley and Curtis that they really came to grips with each other's arguments. Much of Curtis's published paper was an attack on Shapley's galactic model. If the spirals were as large as our galactic system and if one assigned to them dimensions as great as 300,000 light-years, then, Curtis admitted, "the island universes must be placed at such enormous distances that it would be necessary to assign what seem impossibly great absolute magnitudes to the novae which have appeared in these objects."

Curtis and many other supporters of the island universe theory had actually gone further and embraced a comparable galaxy theory. Although Curtis commented that it was "entirely possible to hold both to the island universe theory and to the belief in the greater dimensions for our galaxy by making the not improbable assumption that our own island universe, by chance, happens to be severalfold larger than the average," he strove to avoid retreating to this position. In addition, Curtis pointed to the novae as favoring the island universe theory. Curtis also, as other Lick astronomers had done, argued that the apparently peculiar distribution of spiral nebulae on the sky was due to a dark lane or ring of obscuring material circling the heavens and blocking from view those spiral nebulae close to the galactic plane. According to Curtis, this lane explained the apparent clustering of the spirals around the galactic poles.

Despite the arguments of Shapley and the observations secured by van Maanen, it is likely that most astronomers at the time of the

Great Debate supposed the spiral nebulae to be independent galaxies outside our galaxy. But the combination of Shapley's galactic theory and van Maanen's measures was a fairly potent mix, and it persuaded some people to rethink their beliefs. Thus we find in 1922 a one-time advocate of island universes, Harold Spencer Jones, a future English Astronomer Royal, writing: "Much controversy has raged as to the nature of spiral nebulae. The view held recently, when our galactic system was thought to be of much smaller dimensions than is indicated by more modern evidence, was that they were separate galactic systems."

Instruments

Here it should also be noted that in the early years of the twentieth century the observational study of the spiral nebulae had become centered on the United States. In the middle of the nineteenth century observational cosmology was largely the province of the owners of giant reflecting telescopes. Such instruments were essentially British and Irish in conception, construction, and use. By the early twentieth century, however, the rise of the United States as a leading economic power was mirrored by its growing importance in observational astrophysics as well as in the manufacture and use of giant reflecting telescopes. The high point of this shift was represented by the great 100-inch Hooker telescope on Mount Wilson, completed in 1917. Except for the optics, the Hooker telescope was chiefly the product of large-scale engineering by U.S. companies. Indeed, because of the success of American astronomers in securing research funds (available largely through newly wealthy entrepreneurs and philanthropic foundations), the placing of expensive and powerful telescopes at good sites, the use of high quality auxiliary equipment such as Brashear spectrographs, and the increasingly sophisticated training of American astronomers, the U.S. emerged as the dominant power in observational cosmology.

This is certainly not to suggest that there was any attempt by American astrophysicists and astronomers to gear up early in the century for an attack on cosmology. Instead, other interests and concerns (chiefly, the astrophysics of stars and, most particularly, stellar evolution) both allowed and drove the leaders of American astrophysics to build up in support of their discipline an infrastructure that far outstripped that of European competitors. The result was that when cosmological questions (in particular, the study of spiral nebulae) did begin to loom large, Americans were excellently placed and equipped to tackle the questions. When the great observational cosmologist Edwin Hubble wrote in 1936 that the "conquest of the Realm of the

Nebulae is an achievement of great telescopes," he was surely ascribing too much importance to the instruments and too little to the people using them. He did, nevertheless, catch a crucial point, for by about 1920 the observational study of the spiral nebulae had become centered on the small group of astronomers tackling nebular problems at the observatories of Lick, Lowell, and Mount Wilson. It was a group endowed with powerful telescopes at good sites in the West of the United States, and whose members could concentrate on such research. By this time, no matter how gifted an observational astronomer might be, he or she needed easy access to large and costly telescopes to be at the forefront in studies of spiral nebulae. And such access was available only to a privileged minority, to which Edwin Hubble belonged.

Hubble and Distances

From his writings before, during, and after 1923, it is clear that ever since he was a graduate student, Hubble had been well aware of the ramifications of the theory of the spirals as external galaxies. Although the evidence is ambiguous, he probably supported the theory. For example, as a staff member of the Mount Wilson Observatory, Hubble spent a considerable amount of time examining what would now be termed a giant elliptical galaxy, M87, but which in the early 1920s was usually classified as a spiral nebula. Hubble was deeply influenced in the 1920s by the evolutionary scheme of James Jeans. Jeans interpreted the forms of nebulae in terms of the development of a single basic type—initially formed from a huge cloud of nebulous material—that was gradually converted into stars from the outer regions inwards. Hence, when Hubble detected objects he thought indistinguishable from star images in the outer regions of M87, he was inclined to go a step further and indeed view these condensations as stars.

Given the brightnesses of these apparent star images and comparing them with bright stars in the galaxy, one could estimate a distance to M87 that put it far outside our stellar system, although based on such estimates M87 was not a comparable galaxy. But if Hubble could demonstrate that the apparently stellar objects really were stars (they are now recognized as globular clusters), he could measure the distances to the larger spirals relatively easily. He investigated the condensations by taking photographs centered on the outer regions of the spirals, in this way avoiding the off-axis aberrations of his telescopes. As Hubble later recalled in his classic 1936 book *The Realm of the Nebulae*, he could indeed detect what he judged to be star-like condensations on the resulting plates. But this did not mean that Hubble was yet fully convinced that he could see individual stars; instead, he reckoned the images to be indistinguishable from stellar images. Until the images could *definitely* be shown to be stars through the display of some stellar characteristics, the case for the spirals as external galaxies would not be much strengthened.

In 1923 Hubble began, in his invariably thorough fashion, to concentrate upon a study of novae in the spiral nebulae. Eager to establish its distance accurately, he took numerous plates of the Andromeda Nebula. As already noted, Curtis and others had already detected faint novae in the Andromeda Nebula and in some other spirals, but the use of novae as distance indicators was problematic because of uncertainties in the measurements of the absolute brightnesses of novae in our galaxy and in the very wide range of apparent brightnesses of novae in the Andromeda Nebula. Comparing the apparent brightnesses of novae within our galaxy and employing the principle of uniformity of nature—according to which novae in the Andromeda Nebula should be, on average, as bright as novae in our galaxy—Hubble hoped to estimate the distance to the Andromeda Nebula with more certainty.

Several months after embarking on this program, Hubble made a momentous find. He wrote to Shapley: "You will be interested to hear that I have found a Cepheid variable in the Andromeda nebula. . . . I have followed the nebula . . . as closely as the weather permitted and in the last five months have netted nine novae and two variables. . . . The two variables were found last week." The first Cepheid variable had appeared on a plate taken in October 1923. Hubble's observing program was designed to discover novae, and he initially assumed that the variable was a nova. But after checking earlier plates and drawing a light curve (a plot of brightness versus time), he realized that the object displayed the characteristics of a Cepheid variable. He quickly calculated an approximate period for the star's variations in light. (See figures 3 and 4 in the following chapter, *Hubble's Cosmology*, pages 352 and 354.)

Armed with the period, he calculated the star's actual brightness by use of the period-luminosity relationship (discovered by Henrietta Leavitt and then calibrated by Hertzsprung and Shapley). Further assuming that throughout the universe Cepheids with the same periods have the same absolute luminosities, Hubble compared the observed Cepheid's apparent and actual brightness and secured an estimate of the distance to the Andromeda Nebula—about a million light-years. He rapidly discovered other Cepheids in the wake of the first. Within a year or so, Hubble had accumulated enough evidence from the Cepheids and other distance indicators to convince almost all astronomers that the outer regions of the Andromeda Nebula consisted of clouds of stars and that the nebula was indeed an external galaxy. Following these finds and similar ones for other large, and hence presumably nearby, spirals, the debate over the existence of visible galaxies was all but at an end.

An Expanding Universe

During the 1920s there was a slowly developing interest among astronomers, physicists, and mathematicians in another problem associated with spiral nebulae: Is there a relationship between the size of a spiral's spectral shift and its distance? Following various mathematical researches into the cosmological consequences of general relativity (particularly the widely known studies of the Dutch astronomer Willem de Sitter), there seemed to be good reason for suspecting there should be such a relationship. A few people even tried to determine observationally the form (if any) of the relationship. These included the German astronomer C.A. Wirtz and the Swedish astronomer Knut Lundmark. But the graphs of data that they published were far from convincing and, in the opinion of their colleagues, left open the question of the existence of such a redshift-distance relationship.

Hubble transformed this situation. At first, using a few radial velocities secured at Mount Wilson by Milton Humason and many other velocities obtained earlier by Slipher, together with his own estimates of distances to the galaxies, Hubble persuaded his colleagues that, at least in the first approximation, there was a linear relationship between spectral shift and distance.

This claim had been made earlier by, among others, Georges Lemaître in 1927. Why, then, did Hubble succeed when others had failed to convince their colleagues of the existence of a linear relation? Hubble's claims about a redshift-distance relation were generally regarded as much superior to those of earlier investigators. Not only did Hubble have estimates of distances to the galaxies regarded as more accurate than those used by others who had earlier sought to plot redshift against distance, but also Hubble's standing as a leading—if not *the* leading—student of galaxies (*extragalactic nebulae* in Hubble's terminology) and his use of the most powerful telescope in the world (the Mount Wilson 100-inch) were guarantees in the eyes of his colleagues of the credibility of his claims. Hubble was also very careful to distance himself from arguments that had been made earlier in the 1920s by Ludvik Silberstein about redshift-distance relations and which had tended to discredit such studies. When this position was combined with Hubble's own energetic and skillful advocacy of his own researches, the result was that, in the 1920s and afterward, the contributions of others to establishing the redshift-distance relationship have tended to be ignored.

Hubble's initial paper on the redshift-distance relation, published in 1929, was followed two years later by another, much more extensive one, coauthored with Humason, which presented many more redshifts measured by Humason. Some of the galaxies employed in the 1931 plot were also at much larger distances than those used in 1929, and the linear relationship was judged to be even clearer by astronomers. (See figures 5 and 6 in the following chapter, *Hubble's Cosmology*, pages 360 and 362.)

By the early 1930s, Hubble very much viewed himself as an "observer" who sought to test the various models proposed by "theorists." Such a distinction could not have been made earlier. In general, those who pursued theoretical researches in astrophysics were also schooled in, or actively pursued, observational matters. But with the sophisticated mathematics required to investigate the cosmological consequences of general relativity, a division of labor developed, exemplified by Hubble, the observer, and his sometime collaborator at the California Institute of Technology, R.C. Tolman, the theorist.

Whatever his private suspicions, Hubble was always careful in print to avoid definitely interpreting the redshifts as Doppler shifts. But the writings of A.S. Eddington and others soon meshed the calculations of Lemaître and various theorists with Hubble's observational research on the redshift-distance relation. The notion of the expanding universe was swiftly accepted by many, and the linear relationship between redshift and distance was later widely accepted as Hubble's law and the constant of proportionality as Hubble's constant.

A Beginning for the Universe

If one accepted the concept of the expanding universe, there immediately arose the puzzling and fundamental question of what started the expansion. One of those to stress this problem was Eddington, then the most influential astrophysicist in the world.

For Eddington, the "most attractive" case, to use his term, was that in which the mass of the universe is equal to the mass of the Einstein universe (that is, the mass of the universe in the static solution Einstein had discovered over a decade earlier for the field equations of general relativity). The idea was that the world had evolved from an Einsteinian universe and so had developed "infinitely slowly from a primitive uniform distribution in unstable equilibrium." So, there was a beginning of the expansion, but Eddington shied away from a creation of the universe, as seemed to be implied, when he considered the possibility of a universe of mass larger than the Einstein universe. This Eddington explicitly rejected because "it seems to require a sudden and peculiar beginning of things." He would contend that: "As a scientist I simply do not believe that the present order of things started off with a bang; unscientifically I feel equally unwilling to accept the implied discontinuity in the divine nature."

Nor in the early 1930s was Eddington alone in worrying about the problem of the start of the expansion and how the universe might have evolved from an Einstein universe. One reason for this was the so-called time scale problem. According to then fashionable theories of stellar evolution and dynamical theories of the evolution of double stars and star clusters, the ages of the stars seemed to be much greater than the age of the universe as estimated from the value of what had very quickly become known as the Hubble constant. The value of the constant in Hubble's 1929 paper was 500 kilometers per second per megaparsec, which—if one simply extrapolated the present expansion back in time and took no account of a possible acceleration in the expansion—implied that there was a time about 2×10^9 years ago when all the material in the universe had been packed much closer together than now.

The choice of a suitable time scale became "the nightmare of the cosmologists." As de Sitter pointed out, the "temptation is strong to identify the epoch of the beginning of the expansion with the 'beginning of the world,' whatever that might mean." If so, the "stars and the stellar systems must be some thousands of times older than the universe!" He concluded reluctantly that the "expansion has only been going on during an interval of time which is as nothing compared with the duration of the evolution."

Of course, one explanation for the expansion was that it really did start with the beginning of the entire universe. Lemaître introduced this concept into the cosmological practice of the 1930s. In his seminal 1927 paper on the expanding universe, he had envisaged the expansion starting very slowly from an equilibrium state, but in 1931 he proposed the first detailed example of what later became known as big bang cosmology. Unlike modern big bang theories, however, Lemaître's universe did not evolve from a true singularity but from a material pre-universe, what Lemaître referred to as the primeval atom. For Lemaître, a cosmical singularity is a nonphysical notion in which neither time nor space exists. Lemaître emphasized that cosmology could and should be understood in physical terms. So, for Lemaître writing in 1931, "The last two thousand million years are slow evolu-

tion: they are ashes and smoke of bright but very rapid fireworks."
[See chapter 20, *Big Bang Cosmology*.]

Here is an example of the changing vocabulary of cosmology in this period, the introduction of the beginning of the universe as a legitimate subject for scientific discourse, even if some, including Eddington, found it distasteful.

Nevertheless, as James Jeans had emphasized in 1929, contemporary astronomy "makes it clear that the present state of matter cannot have existed forever; indeed we can probably assign an upper limit to its age of, say, some such round number as 200 million million years. And, wherever we fix it, our next step back in time leads us on to contemplate a definite event, or series of events, or continuous process, of creation of matter at some time not infinitely remote."

An End for the Universe

As well as debate on the start of the expansion of the universe, there was also renewed interest in its fate. The idea of the "heat death" of the universe had been introduced in 1854 by Hermann Helmholtz, following his reading of a paper by William Thomson (later Lord Kelvin) dealing with Thomson's principle of the dissipation of energy. As Helmholtz put it, "We must admire the sagacity of Thomson, who . . . was able to discern consequences which threatened the universe, though certainly after an infinite period of time, with eternal death." In the late 1920s the heat death of the universe reached a very broad audience through popular writings and newspaper editorials and articles.

The possible heat death of the universe was particularly widely discussed when it became enmeshed in a debate on one of the hottest scientific topics of the late 1920s—the nature of cosmic rays. This debate was conducted chiefly between two scientific heavyweights, James Jeans and the Nobel-prize-winning U.S. physicist Robert Millikan. For Millikan, cosmic rays were the tell-tale signs of a creative process of the conversion of radiation into matter. Millikan adduced not only scientific evidence for this claim, but also what he identified as a philosophical reason: "In the radioactive process the heavier atoms are disintegrating into lighter ones. It is, therefore, to be expected that somewhere in the universe the building-up process is going on to replace the tearing-down process represented by radioactivity." This view led Millikan to develop a grand cosmological theory characterized by a continual cycle of the birth and death of the elements in a kind of steady state universe. Jeans challenged both Millikan's particular theory, as well as the attack on the unidirectionality of the evolution of the universe, which in turn was underpinned by the second law of thermodynamics. A major task some cosmologists later took upon themselves was the incorporation of the second law into cosmological models based on general relativity.

Conclusion

By the early 1930s cosmology was an altogether more extensive enterprise than it had been at the beginning of the century, and there had been established, through the redshift-distance relationship, an exchange between theory and data undreamt of only a few years earlier. Although pitifully meager by present-day standards, the achievements in hand by 1930 created a confidence among astronomers and mathematicians that they could discuss and ultimately explain the entire history of the universe. In large part, cosmology was now concerned with the nature and evolution of galaxies and their place in the expanding universe. This new state of affairs constituted a radical departure from the statistical cosmology that had held sway at the beginning of the century. In addition, through a variety of scientific, technical, and institutional developments, American astronomers now dominated observational cosmology. Indeed, in the first three decades of the twentieth century, the content of cosmology had been reshaped and extended, and the practice and the very nature of the enterprise of cosmology had been redefined.

—Robert W. Smith

BIBLIOGRAPHIC NOTE

Most of the topics treated in this chapter are examined in more detail in R.W. Smith, *The Expanding Universe. Astronomy's 'Great Debate' 1900–1931* (Cambridge: Cambridge University Press, 1982). On the Lick Observatory, see D.E. Osterbrock, J.R. Gustafson, and W.J. Shiloh Unruh, *Eye on the Sky: Lick Observatory's First Century* (Berkeley: University of California Press, 1988). On theoretical cosmology, see J.D. North, *The Measure of the Universe* (Oxford: Clarendon Press, 1965). For more on Slipher, see W.G. Hoyt, *Lowell and Mars* (Tucson: University of Arizona Press, 1976). There is also relevant material in R. Bertotti, R. Balbinot, S. Bergia, and A. Messina, eds., *Modern Cosmology in Retrospect* (Cambridge: Cambridge University Press, 1990). For references to books and articles on Hubble, see the following chapter.

19. HUBBLE'S COSMOLOGY

Edwin Hubble's cosmology is a monumental achievement by the foremost astronomer of the twentieth century. He changed our view of the world; he proved that other galaxies exist and he demonstrated that the universe is expanding. Instruments, observations, and theories available to Hubble were accessible to others as well, but it was Hubble who forged a brilliant synthesis, then explored it, developed it, and finally demonstrated it persuasively. He built his remarkable creation with intelligence, hard work, persistence, cooperation with scientists whose skills complemented and extended his own, and an excellent sense of strategy both in pursuing problems and in presenting solutions. Hubble's cosmology is one of the great works of the human intellect.

Classification of Nebulae
Edwin Powell Hubble was born in 1889, studied mathematics and astronomy at the University of Chicago, and went to Oxford in 1912 as a Rhodes scholar. There he studied Roman and English law, and he may have practiced law briefly upon his return to the United States, though recent research finds that he mainly taught in a high school and coached the basketball team. Soon back in school as a student, Hubble completed a Ph.D. at the Yerkes Observatory in 1917.

Much of Hubble's subsequent work on cosmological problems can be seen, in embryo, in his doctoral dissertation. It was based on photographs of seven clusters of nebulae taken with the Yerkes 24-inch reflecting telescope between 1914 and 1916. To the 76 known nebulae in clusters Hubble added 512 more.

The rapidly increasing number of nebulae observed and the estimated 150,000 within reach of existing instruments emphasized the need for a classification scheme. Hubble wrote that "extremely little is known of the nature of nebulae, and no significant classification has yet been suggested; not even a precise definition has been formulated." His classification scheme would be announced in the 1920s, his tuning fork diagram of the sequence of nebular types would be famous by the 1930s, and the Hubble atlas of galaxies would appear posthumously (figure 1).

Island Universes

Centuries of speculation over the possible existence of island universes similar to our galaxy were coming to a head early in the twentieth century, with attention focusing on spiral nebulae.

Favoring an extragalactic nature was the spirals' resemblance to the hypothetical spiral structure of our Milky Way system. In addition, V.M. Slipher's discovery at the Lowell Observatory beginning in 1912 of extraordinarily high radial velocities (velocity components in the line of sight) of spiral nebulae gave new life to the theory that these objects were distant stellar systems, although the later discovery of a few galactic stars also with high radial velocities would somewhat weaken the argument.

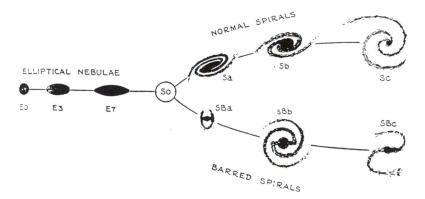

Figure 1. Hubble's "tuning fork" classification sequence of galaxies. Hubble explained that elliptical nebulae range from globular bodies, E0, through flattening, ellipsoidal figures to the limiting, lenticular form, E7. About one nebula in forty is irregular, lacking rotational symmetry around a dominating central nucleus. The sequence of spirals is double, one containing normal spirals and the other barred spirals. Each branch is divided into a, b, and c subsections, determined by the amount of material in the arms relative to that in the nuclear region, by the openness of the arms, and by the degree of resolution. Hubble explained that the diagram is a schematic representation of the sequence of classification. A few nebulae of mixed types are found between the two sequences of spirals. The transition stage, S0, is more or less hypothetical. The transition between E7 and SBa is smooth and continuous. Between E7 and Sa no nebulae are definitely recognized. Hubble called Sa and SBa spirals "early" and Sc and SBc "late," thus revealing that he attributed an evolutionary significance to the classification scheme.

Arguing against an extragalactic nature was the absence of observable spiral nebulae in low galactic latitudes, this *zone of avoidance* seemingly linking the spirals physically to our stellar system, although the observed distribution also could be secured by liberal use of a hypothetical encircling opaque material. Observations of bright novae in spiral nebulae further argued for short distances to the spirals, until consistently much fainter novae not in conflict with the island universe theory were detected, beginning in 1917.

The most compelling argument against the spiral nebulae as external galaxies came from Adriaan van Maanen's purported measurement at the Mount Wilson Observatory of internal motions in the spiral nebula M101 (figure 2). Harlow Shapley forcibly explained its import. A rough calculation easily demonstrates his reasoning. Given a diameter for our galaxy of 100,000 light-years (a light-year is the distance an object moving with the velocity of light travels in one year), the circumference of a comparable galaxy would be about 300,000 light-years (circumference is π [3.14 . . .] times the diameter). But van Maanen claimed to have measured a period of rotation for M101 of only 85,000 years. This meant that the outer edge of the spiral nebula must have been moving over a distance greater than 300,000 light-years in less than 85,000 years. To do so, it must have been moving with a speed several times greater than the speed of light. In conclusion, if spiral nebulae were extragalactic systems comparable in size to our galaxy, estimates of which Shapley had recently increased significantly, and rotating as rapidly as van Maanen claimed, the outer edge of the nebula would be traveling faster than the speed of light. Shapley's conclusion stood as a serious *reductio ad absurdum* argument against the view of spiral nebulae as extragalactic systems. Proponents of the island universe system could only ask for a suspension of judgment until further measurements either confirmed or refuted van Maanen's claim. Their doubts would be justified by later events, but meanwhile van Maanen's measurements were a compelling argument against the island universe theory for many astronomers.

Hubble had no conclusive evidence in 1917 regarding the nature of spiral nebulae. He thought, however, that "the great spirals, with their enormous radial velocities and insensible proper motions apparently lie outside our system."

Cepheid Variable Stars

Following service in World War I, Hubble went to the Mount Wilson Observatory. There photographic plates taken with the new 100-inch telescope revealed the presence of variable stars (stars that changed

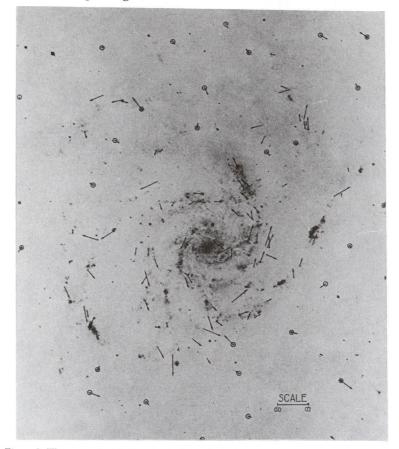

Figure 2. The arrows in this photograph of the spiral nebula M101 indicate the direction and magnitude of the mean annual motions for identifiable points of matter that Adriaan van Maanen purportedly measured from a comparison of photographic plates taken several years apart. Comparison stars are circled. We now know that the observations were spurious.

With a little calculation we can choose between the alternatives of a small, nearby nebula or a large, distant island universe comparable in size to our own galaxy. Note that the scale indicated on the photograph is 0.1 seconds of arc. Trigonometrically (as shown in many astronomy textbooks), with a very long distance and a relatively short base line, the arc moved through is approximately to the circumference of an entire circle as the angle of proper motion is to 360 degrees: (length of arc, that is, the time moving multiplied by the speed of the point in kilometers per second)/(2πr) = (annual mean motion in seconds per year)/(360 degrees). Remember that π is 3.14. . . and the formula for the circumference of a circle is $c = 2\pi r$. In this case, r (the radius of the circle) is also the distance of the object from us. Measure the annual mean motion from the photograph. Play around in the formula with speeds less than the speed of light (3 times 10^{10} cm/sec) and distances to the nebula less than and greater than the diameter of our galaxy (approximately 100,000 light-years.) Hint: multiply kilometers/second by 60 seconds/minute times 60 minutes/hour times 24 hours/day times 365 days/year to convert km/sec to km/year; divide centimeters by 100 cm/meter and meters by 1,000 meters/kilometer to convert centimeters to kilometers.

in brightness over time) in the irregular nebula NGC 6822 and in several spiral nebulae.

Cepheid variable stars would be the key to measuring distances and settling the debate over whether the nebulae are outside our galaxy. Henrietta Leavitt at Harvard, while determining the periods of variable stars in the Small Magellanic Cloud, had found it "worthy of notice that . . . the brighter variables have the longer periods." Since stars in the distant nebula all were at approximately the same distance from the earth, there existed a common scale factor that would convert the measured apparent magnitudes, a function of intrinsic brightness and distance, into absolute magnitudes (the intrinsic brightness as measured from a distance of 10 parsecs, or approximately 3.26 light-years).

The period-luminosity relation soon was calibrated, primarily by Shapley on a few nearby Cepheids whose distances were determined trigonometrically (figure 3). In a paper published in 1918, Shapley noted that:

> For some years we have known that the longer-period Cepheids in the galactic system are also giant stars, and a casual examination of their motions [yielding distances, and thus absolute magnitudes; see the caption to figure 3 for converting motion to distance] *indicates a fairly small dispersion in actual luminosity. The possibility is at once suggested, therefore, of utilizing the motions of galactic Cepheids to obtain a mean value of the absolute magnitude of such stars, which, when compared with observed magnitudes of analogous objects in globular clusters, permits an estimate of the distances of the clusters themselves. One obvious advantage of operating with a group of giant stars is that in many of the most distant clusters only the stars of highest luminosity are within reach of our greatest telescopic power.*

The average absolute magnitude for 11 Cepheids was -2.26 ± 0.22. As shown in figure 3, this corresponds to a period (time for the star's brightness to go from maximum through minimum and back to maximum) of slightly less than ten days (i.e., a logarithm less than 1, with $10^1 = 10$ days and $10^0 = 1$ day).

Then, in a scientific achievement as brilliant as it was bold, Shapley used the new and still unproven period-luminosity relation to measure distances to globular clusters and thus outline the framework or skeleton of our galaxy.

Distances to Spiral Nebulae

Writing to Shapley in 1923, Hubble informed him of the discovery of variable stars in NGC 6822 and his intention to hunt for more and to investigate their periods and light curves. Shapley wrote back: "What a powerful instrument the 100-inch is in bringing out those desperately faint nebulae."

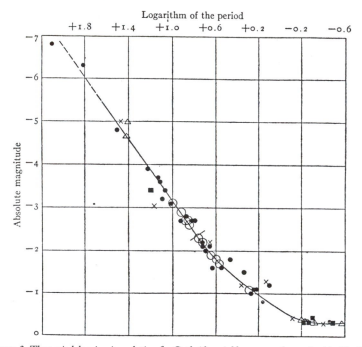

Figure 3. The period-luminosity relation for Cepheid variable stars, as determined by Harlow Shapley in 1918. Various symbols designate variable stars in different groups. Period, graphed as its logarithm, is in days. A logarithm is the exponent to which a base number, commonly 10, is raised to produce the desired number. The log of 100 is 2, since $10^2 = 100$. The log of 10 is 1, of 1 is 0 (any number to the zero power equals one), and of 0.1 is –1 (a negative log means the inverse, or $1/10^1$). Graphing the logarithm gives equal space for periods 0.1 to 1, 1 to 10, and 10 to 100 days. These would otherwise occupy 0.9%, 9%, and 90% of the graph and necessitate crowding together data points at shorter periods. The uniformly sloping line, until it bottoms out, reveals that period varies with magnitude by a common factor. From magnitude –6 to –5, period drops from roughly 63 to 25 days, while from –3 to –2 it drops from about 10 to 4 days. The changes are different, 38 days and 6 days, but share a common dividing factor, of approximately 2.5, from 63 to 25 and from 10 to 4.

Hubble's success did not follow automatically from access to the 100-inch telescope. Shapley, who moved from the Mount Wilson Observatory to the Harvard Observatory in 1921, might have extended his work to spiral nebulae. Indeed, there is a remarkable story— yet to be documented but not impossible—that around 1920 Milton Humason at the Mount Wilson Observatory had examined some plates of the Andromeda Nebula and had reported to Shapley the presence of Cepheid variables on the plates, only to have Shapley rub off the marks placed by Humason identifying the Cepheids and explain why the spots could not be Cepheids. Shapley accepted the Mount Wilson astronomer George Ritchey's opinion that the images beginning to be resolved in photographs of spirals were not stellar.

Astronomers at the Lick Observatory also could have searched for Cepheids in spiral nebulae, albeit only with the intractable Crossley Reflector. Unlike Shapley, they believed in the island universe theory. But they may have failed to appreciate the importance of Cepheids as distance indicators, or they may have lacked confidence in the period-luminosity relation.

It was Hubble, inspired by the island universe theory, who:

- Presciently focused his attention on a consequential problem.

- Examined the issue and its potential observational consequences.

- Found a variable star in a spiral nebula.

- Searched for more variable stars in a spiral nebulae.

- And used the variable stars in spiral nebulae to advance our understanding of the universe.

Using the period-luminosity relation for Cepheid variable stars to calculate the absolute magnitudes of such stars from their observed periods, and then comparing these estimated intrinsic magnitudes to the observed magnitudes (diminished due to the distances of the nebulae in which the Cepheid stars were embedded), Hubble derived distances that placed the spiral nebulae far beyond the boundary of our galaxy. Early in 1924 he wrote to Shapley:

> *You will be interested to hear that I have found a Cepheid variable in the Andromeda Nebula (M31). I have followed the nebula this season as closely as the weather permitted and in the last five months have netted nine novae and two variables [. . .] the distance comes out something over 300,000 parsecs [. . .] I have a feeling that more variables will be found by careful examination of long exposures. Altogether the next season should be a merry one and will be met with due form and ceremony.*

Figure 4 shows Hubble's period-luminosity relation for Cepheid variables in the spiral nebula M31. It can be compared with Shapley's calibrated curve (figure 3). Somewhat inconveniently, the two graphs are not drawn to the same scale, nor do they share the same choices of horizontal and vertical axes. Hubble's data points, however, can be replotted on transparent paper with Shapley's scale and axes, and then the revised Hubble graph placed on top of Shapley's and shifted about to find the best fit. Alternatively, and far more roughly and quickly, one of Hubble's data points, say at log p 1.25 and mag. 19, can be set on Shapley's graph at log p 1.25 and approximately mag. −4. The difference between the absolute magnitude on Shapley's curve and the apparent magnitude on Hubble's curve, the *distance modulus*, then can be translated into a distance for M31.

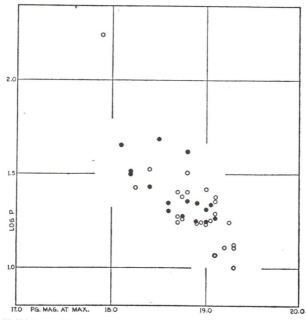

Figure 4. Hubble's period-luminosity relation among Cepheid variable stars in the spiral nebula M31. Photographic magnitude at maximum is plotted against the logarithm of the period (expressed in days). Cepheids in one region of the nebula are designated by circles, to show that the position in the nebula does not affect a Cepheid's measured magnitude or period.

The *absolute magnitude* of an object is the brightness it would appear to have (its *apparent magnitude*) at a standard distance. (By general agreement, the standard distance is taken to be 10 parsecs, with 1 parsec = 3.086 times 10^{18} centimeters, or 3.262 light-years, the distance light travels in a year at a speed of 2.99793 times 10^{10} centimeters per second). Also by definition, a difference in luminosity of a factor of 100 (or a difference in distance of ten times, since the luminosity decreases as the square of the distance) corresponds to a difference of five magnitudes, while a difference of one magnitude corresponds to a difference in luminosity of the fifth root of 100, or approximately 2.512 times. (Psychologically, observers perceive equal increases or decreases in the light *ratio* as approximately equal increases or decreases on the magnitude scale.)

When the absolute magnitude of an object is known, or can be estimated by comparison with presumed similar objects whose absolute magnitudes are known, then a comparison of the absolute and apparent magnitudes yields the distance of the object. With M the absolute magnitude and m the apparent magnitude, it follows from the definitions that the distance r in parsecs is given by the formula

$$m - M = 5 \log r / 10$$

How this formula works is easily demonstrated. When $m = M$, and thus the distance modulus ($m - M$) is zero, r is 10 parsecs (the standard distance). Remember that the logarithm is the exponent to which the base number 10 is raised to produce the desired number (see the caption to figure 3 for further explanation and examples). Any number to the zero power is one; thus the logarithm of $r/10$, or of 1 when $r = 10$, is 0; and thus both sides of the equation are equal to 0. When $m - M = 5$, $\log r/10$ must equal 1; the base number 10 to the one power is ten; hence r must be 100 parsecs.

It is often useful to run a few numbers through a formula, especially numbers permitting simplified calculations, as we have done here. The procedure enables us easily and quickly to check the plausibility of a formula and to get a feel for what the formula is doing.

In our very rough study, we have compared one of Hubble's data points, at log p 1.25 and apparent magnitude 19, with a point on Shapley's graph at log p 1.25 and absolute magnitude –4. From these very approximate values we obtain a distance modulus, $m - M$, of 23.

A distance modulus of 5 means a difference of 100 times in luminosity and also 10 times in distance. Here, also, a minimal amount of mathematics reveals that every time we increase the distance modulus by 5 we increase the luminosity by a factor of 100 and the distance by a factor of ten (for $r = 100$, $\log r/10 = 1$ and $m - M = 5$; for $r = 1,000$, $\log r/10 = 2$ and $m - M = 10$; for $r = 10,000$, $\log r/10 = 3$ and $m - M = 15$, etc.) Thus a distance modulus of ten corresponds to a distance of 1,000 parsecs, a distance modulus of fifteen signifies a distance of 10,000 parsecs, a distance modulus of twenty means a distance of 100,000 parsecs, a distance modulus of twenty-five corresponds to a distance of 1,000,000 parsecs, etc.

Our rough estimate or $m - M = 23$ indicates a distance for the Cepheid variable in the spiral nebula M31, and thus for M31 itself, of between a hundred thousand ($m - M = 20$) and a million parsecs ($m - M = 25$). Hubble more precisely finds a distance modulus of 22.2 and a corresponding distance of 275,000 parsecs. (Remember that each difference of one magnitude corresponds, by definition, to a luminosity factor of the fifth root of 100, or 2.512, and also a distance factor of the square root of 2.512, or 1.58. Similarly, a difference of two magnitudes corresponds to a luminosity factor of the fifth root of 100 squared, or 2.512 squared, and also a distance factor of the square root of 2.512 squared, or 2.512. Thus if we start with a distance modulus of 20, corresponding to a distance of 100,000 parsecs, and increase it by two magnitudes, we multiply the 100,000 by 2.512 and obtain a distance of 251,200 parsecs. For each 0.1 magnitude change the distance by about 5%.)

More than Luck

Hubble had first marked in his observation notebook the presence of a nova, and the discovery in M31 of the first definitely recognized extragalactic Cepheid was somewhat serendipitous. But Hubble, while searching for novae also had Cepheids in mind. In 1922 John Duncan at the Mount Wilson Observatory had recognized variable stars in the area covered by the spiral nebula M33, though he had not been able to determine their nature. And in 1923, before examining M31, Hubble had already found in NGC 6822 indications of Cepheids, to be fully established as such after more observations in 1924.

When Hubble's letter arrived, Shapley remarked: "Here is the letter that has destroyed my universe." To Hubble, Shapley replied that his letter "telling of the crop of novae and of the two variable stars in the direction of the Andromeda nebula is the most entertaining piece of literature I have seen for a long time." Shapley wrote that he was sure that the large number of plates Hubble had obtained furnished adequate assurance that the stars were genuine variables.

By August, Hubble had more variables and was beginning to consider the significance of his data. Shapley also recognized that the straws now were pointing toward the island universe hypothesis. He replied to Hubble:

> *Your very exciting letter received here* [. . .] *What tremendous luck you are having* [. . .] *I do not know whether I am sorry or glad to see this break in the nebula problem. Perhaps both. Sorry because of the significance for the measured angular rotations, and glad to have something definite and interesting come to hand.*

In retrospect, Hubble's work may appear inevitable. But no contemporary scientist realized a similar vision. Considerable courage— even lack of normal scientific prudence—was required. While hindsight justifies the overall endeavor, it also reveals that more than one of Hubble's assumptions required subsequent modification. What Hubble thought were stars around the nebula M87 were, in fact, globular clusters. And the comparison of Cepheid variables in spiral nebulae with Cepheids in our galaxy was not valid. Yet Hubble's working assumptions were necessary for his plan of scientific investigation to proceed. Justification lies not in any continuing agreement with advancing factual knowledge but in the *necessity of the assumptions* if extrapolations were to be made from the known to the possibly knowable.

Hubble's discovery of Cepheids in spiral nebulae and the distance determination confirming that spiral nebulae are independent galaxies were announced on 1 January 1925 at an American Astronomical Society meeting. This preliminary paper was followed by further work presented in convincingly voluminous and thorough detail over the next four years. Hubble strengthened his case further with evidence from novae, from the observed colors of the brightest stars, and from star counts, leaving van Maanen's purported rotations as the only shadow on an otherwise consistent picture of spiral nebulae as island universes. A good part of Hubble's genius and the extent to which his revolutionary conclusions commanded acceptance are to be attributed to hard work—and lots of it.

Advocacy in Science

Another aspect of Hubble's genius—and an important if sometimes overlooked factor in the acceptance of his cosmological vision—was his sense of strategy in presenting results. There is more to the advance of science than new observations and new theories. Ultimately, people must be persuaded. Advocacy is an integral part of science.

Slow to publish, Hubble explained to a colleague:

> *The real reason for my reluctance in hurrying to press was, as you may have guessed, the flat contradiction to van Maanen's rotations. The problem of reconciling the two sets of data has a certain fascination, but in spite of this I believe that the measured rotations must be abandoned.*

For a decade Hubble largely ignored the discrepancy, except in a talk prepared for colleagues at the Mount Wilson Observatory. In this talk Hubble brutally stripped away van Maanen's claimed corroboration. As Mrs. Hubble remembered the general situation, however:

> *It was, the contradiction, not very important in the long run. The work [Hubble's] had become so obvious and so extensive that van Maanen's measurements could be ignored.*

But things changed in the early 1930s. Hubble explained:

> *They asked me to give him time; well, I gave him time, I gave him ten years. When a speaker at the R.A.S. [Royal Astronomical Society] announced if it were not for van Maanen's measurements, Hubble's results might be accepted, I decided to make the measurements.*

At this time an astronomer's photographic plates were considered private property, and Hubble would have to take new photographs to repeat van Maanen's measurements. Hubble's own willingness to share observational material was as unusual as it is commendable.

Finding flaws was easy, and Hubble wrote several papers sharply criticizing van Maanen's work. Publishing the criticism was more difficult. Frederick Seares, editor of Mount Wilson publications, ruled:

> For two men in the same institution there is opportunity for personal contact and for direct examination of each other's results, and hence for the private adjustment of differences in opinion. The institution itself it seems to me, is under obligation to see that all adjustment possible be made in advance of publication. If agreement cannot be attained it may be necessary for the institution to specify how results shall be presented to the public. In that event, however, there must be opportunity for the expression of individual opinion, but any such expression should be concerned only with the scientific aspects of the question at issue.

Hubble apparently was not happy with the decision, because Seares added:

> The institution has, I think, the right to enforce this procedure; but in certain cases it may be wiser to waive its technical right and say to a dissatisfied individual, "print what you like, but print it elsewhere."

Faced with unsatisfactory alternatives of leaving van Maanen's work unchallenged or challenging it directly in an ungentlemanly, unscientific, and unacceptable manner, Hubble ingeniously resolved the dilemma. He included in the remeasurement Seth Nicholson, whose earlier measures van Maanen had claimed as corroboration. Nicholson did not explicitly recant his earlier testimony, but jurors could easily have formed that impression. Hubble published a brief note pointing to van Maanen's results as the outstanding discrepancy in the conception of nebulae as extragalactic systems and presenting a table of his own remeasures of four spiral nebulae, results establishing the existence of some systematic errors in van Maanen's rotations. On the source of the error, Hubble was silent. Privately, he had tried mightily but ultimately failed to attribute the erroneous measurements to random errors. Random errors would have produced random results, but van Maanen reported seven spirals as all winding up. With honest intention but mistakenly, van Maanen had found what he had expected to find, even when it didn't exist. In a note immediately following Hubble's, van Maanen presented new measurements in partial confirmation of his own while conceding that "it is desirable to view the motions with reserve." It was more a plea of *nolo contendere* than of innocent or guilty. Obviously the revised pleading had been negotiated with Hubble, who thus masterfully defended his distance determinations.

Velocity-Distance Relation

Establishing that spiral nebulae are island universes, a great accomplishment in itself, was but a starting point for Hubble. He seized upon distances to open a new phase of astronomical investigation, that culminated in an expanding universe. Hubble also employed a skillful strategy to overcome ingrained suspicion against the possible existence of a velocity-distance relation.

During much of human intellectual history, the universe was believed to be static. With the rise of relativity theory early in the twentieth century, astronomers initially sought only static solutions to Einstein's field equations. Even the idea of an expanding universe seemed senseless and irritating to Einstein.

The Dutch astronomer Willem de Sitter found a static model among whose observational consequences was an apparent—but not real—velocity of recession greater for objects at greater distances. The mutual incompatibility of mass and stability in de Sitter's model eventually would eliminate it from consideration as a real representation of the universe, but not before it stimulated searches for a velocity-distance relation. After the demise of de Sitter's theory, redshifts in the spectra of light from spiral nebulae would generally be interpreted as Doppler shifts indicative of real motions in an expanding universe.

In 1920 de Sitter knew of measurements of spectra of 25 spiral nebulae, most by Slipher, and 22 showed the predicted redshift. Distances, however, were yet to be determined. Both velocities and distances were known for globular clusters, and in 1924 Ludvik Silberstein, a mathematical physicist, attempted to demonstrate a velocity-distance relation. He reworked de Sitter's derivation and found the apparent velocity proportional to plus or minus the distance—conveniently so, inasmuch as redshifts (positive, receding velocities) and blueshifts (negative, approaching velocities) were commonly observed among globular clusters. For seven clusters Silberstein found a correlation between velocity and distance.

He had, however, excluded three clusters from his study because of "suspiciously small" radial velocities—velocities much smaller than predicted from his velocity-distance relation. Data for more globular clusters were available, and Gustaf Strömberg and Knut Lundmark at the Mount Wilson Observatory noted that the complete body of available evidence did not support a velocity-distance relation. They were careful to distinguish between Silberstein's now discredited work and the general hypothesis of a velocity-distance relation. A conclusive test of de Sitter's hypothesis would have to wait upon distances to spiral nebulae, because the globular clusters might not be distant enough to yield a measurable velocity-distance relation. Nonetheless, theoretical preconceptions of a velocity-distance relation such as had poisoned Silberstein's analysis were suspect.

Hubble was aware of de Sitter's cosmological model and its predicted velocity-distance relation, but his initial comprehension of the theory was uncertain. After discussing problems with colleagues at the 1928 International Astronomical Union meeting in Holland, however, Hubble returned to Mount Wilson determined to test de Sitter's solution. Hubble had Humason, a meticulous and gifted observer, systematically observe faint and more distant nebulae to determine if they had larger velocities than closer nebulae.

Using Humason's data, Hubble was able—step by systematic step—to estimate distances to increasingly more distant nebulae and establish the existence of a velocity-distance relationship (figure 5).

- Using the period-luminosity relation for Cepheids, Hubble first determined distances directly to five nebulae and indirectly to a sixth, a physical companion of one of the five.

- Then Hubble calibrated the absolute magnitude of the brightest stars in the six nebulae and, from observations of the apparent magnitudes of the brightest stars in fourteen more nebulae, estimated their distances.

- Next Hubble found an average absolute magnitude for all twenty nebulae and compared that value to the apparent magnitudes of

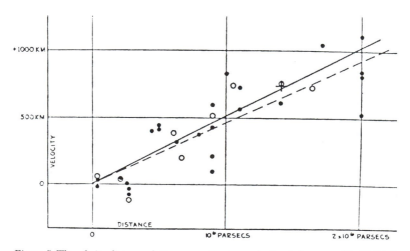

Figure 5. The velocity-distance relation among 46 extra-galactic nebulae reported by Hubble in 1929. Black discs and the full line represent the solution, taking individually the 24 nebulae for which individual distances were determined. The circles and broken line are for the 24 nebulae combined into 9 groups according to proximity in direction and in distance. The cross represents the mean velocity for 22 nebulae whose distances could not be estimated individually.

four nebulae in the Virgo Cluster of galaxies, thus obtaining distances to them. The distances combined with velocities showed a linear velocity-distance relation.

- For the remaining 22 nebulae with known radial velocities, but unknown distances and too far away for observations of Cepheids or bright stars, Hubble measured the apparent magnitude of each nebula, calculated a mean apparent magnitude for all 22 nebulae, compared that value to the mean absolute magnitude for nebulae whose distances were known, obtaining a mean distance for the 22 nebulae, and then showed that the mean distance and the mean velocity of the 22 nebulae agreed well with the velocity-distance relation determined from the first 24 nebulae.

The data base was skimpy and the interpretation shaky in detail, but it was a brilliant and bold extrapolation outward into space.

The technique owed much to Shapley's pioneering work on Cepheids, globular clusters, and the frame of our galaxy. Many of the radial velocities were due to Slipher. Humason carried the measurements farther out into space. And de Sitter's theory had directed attention to the possibility of a velocity-distance relation. Establishment of the velocity-distance relation, though, is justly credited to Hubble. He wrote to de Sitter:

> The possibility of a velocity-distance relation among nebulae has been in the air for years—you, I believe, were the first to mention it. But our preliminary note in 1929 was the first presentation of the data where the scatter due to uncertainties in distances was small enough as compared to the range in distances to establish the relation. In that note, moreover, we announced a program of observations for the purpose of testing the relation at greater distances—over the full range of the 100-inch [telescope], in fact. The work has been arduous but we feel repaid since the results have steadily confirmed the earlier relation. For these reasons I consider the velocity-distance relation, its formulation, testing and confirmation, as a Mount Wilson contribution.

Hubble explained to de Sitter that he had used "the term 'apparent' velocities in order to emphasize the empirical features of the correlation. The interpretation, we feel, should be left to you and the very few others who are competent to discuss the matter with authority." Soon Hubble, in collaboration with a colleague knowledgeable in the subtleties of relativity theory, would attempt an interpretation of the velocity-distance relation. But initially, Hubble focused his efforts on establishing the relation empirically. Not until the final paragraph of his 1929 paper did Hubble mention de Sitter or theory, and then he simply noted that the velocity-distance relation might represent the de Sitter effect and might be of interest for cosmological discussion. Such was Hubble's understated introduction of the key to the scientific exploration of the universe!

Hubble was pursuing a conscious strategy to convince his scientific peers of the reality of the velocity-distance relation, as is apparent in his otherwise incomprehensible emphasis on an otherwise insignificant matter. Hubble's new data indicated a linear correlation between velocity and distance regardless of whether the velocities were corrected for the solar motion. Yet Hubble emphasized the correction, even reporting that he'd had Strömberg check it. Hubble did not state that Strömberg accepted the velocity-distance relation, though readers could easily have leapt to that implication. Having returned to Sweden, Lundmark could not be brought into court as easily, but Hubble did cite Lundmark's solution for the solar motion. No matter that it differed from Hubble's solution, nor that Lundmark had used a velocity-distance relation different from Hubble's simple linear proportionality. Certainly Hubble required neither Strömberg's nor Lundmark's assistance to check the straightforward calculation of

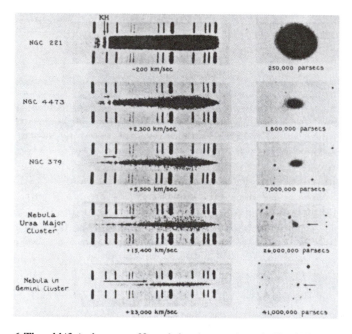

Figure 6. The redshifts in the spectra of five nebulae. Arrows point to the H and K lines of calcium and show the amount of displacement toward the red end of the spectrum. Photographs of the nebulae, all to the same scale, illustrate the decrease in observed size and brightness of the nebulae with increasing redshift (Doppler velocity). Presumably the decrease in observed size and brightness indicates proportionally increasing distance.

the solar motion. The appearance of Silberstein's critics, though, implied that Hubble's work did not suffer from the same faults.

Hubble orchestrated evidence for maximum effect. It was not a matter of fabrication or falsification or fudge factors; the data were honest. If Hubble tilted the playing field on which his interpretations battled others, it often was to restore a balance, to overcome prejudice against his ideas or in favor of others, and to attain a fair hearing.

Though begun in a climate of skepticism and suspicion created by an injudicious earlier attempt to establish a velocity-distance relation, Hubble's effort proceeded smoothly and quickly. By 1935 he and Humason had velocities for a hundred additional nebulae at distances as much as thirty to forty times farther than the Virgo Cluster (figure 6). The scientific case was outstanding, and it did not hurt to have it argued by a skilled barrister.

A Relativistic, Expanding, Homogeneous Universe

With the velocity-distance relation firmly established, Hubble moved on to a theoretical interpretation of the empirical relationship.

It would be a joint scientific effort, of which Hubble was an early exemplar. Cooperation was an important and distinctive feature of nebular research at Mount Wilson. Working in intimate association with colleagues at the California Institute of Technology, the scientists combined resources in particular investigations and interpreted the results in the light of constructive criticism from the group as a whole. Mrs. Hubble remembered:

> I think it was in the 1930s when about every two weeks some of the men from Mount Wilson and Cal Tech came to the house in the evening: H.P. Robertson, Richard Tolman, Baade, Minkowski, Milton Humason, Zwicky, and others, astronomers, physicists, mathematicians. They brought a blackboard from Cal Tech and put it up on the living-room wall. In the dining-room were sandwiches, beer, whiskey and sodawater; they strolled in and helped themselves. Sitting around the fire, smoking pipes, they talked over various approaches to problems, questioned, compared and contrasted their points of view—someone would write equations on the blackboard and talk for a bit and a discussion would follow.

The discussions no doubt were enlivened by the appearance of new cosmological models elicited by a crisis in astronomy in the early 1930s. At a meeting of the Royal Astronomical Society in January 1930, de Sitter commented on the inability of existing models to represent adequately the observed universe. Next, the English astronomer Arthur Eddington asked why there should be only two solutions. "I suppose," he answered himself, "the trouble is that people look for static solutions."

The Belgian astronomer Georges Lemaître, Eddington's former student, subsequently brought his nonstatic solution to Eddington's attention. Eddington immediately recognized in Lemaître's paper, on a homogeneous universe of constant mass and increasing radius accounting for the radial velocity of extragalactic nebulae, a solution to the dilemma following from the observational rejection of both static models. De Sitter also praised Lemaître's ingenious solution.

An alternative to Lemaître's expanding space was proposed by the astronomer E.A. Milne at Oxford. His budding interest in cosmology, nurtured by Einstein's Rhodes Lectures at Oxford in 1931, had burst into full flower upon reading letters to the editor of the *Times* of London. The requirement in relativity theory that space bend back upon itself ran counter to the common sense of some readers, one of whom suggested that an astronomer should bend the powers of his own mind to come up with a more comprehensible theory. Particularly bothering Milne, in addition to the curvature of space (at a time when philosophy professors at Oxford knew *a priori* that space was Euclidean, and were equally convinced that the theory of relativity was false), was the contention that we probably would never know why space is expanding rather than contracting, both being equally probable within relativity theory. Milne sought to understand why the universe is expanding and why the velocities are proportional to distances. In his common-sense explanation, Milne rejected the notions of curved space and the expansion of space. He proposed instead an initial group of nebulae moving in Euclidean space in random directions with different velocities. Such a model would soon acquire the appearance of an expanding group, nebulae of highest velocity naturally having receded farther from the starting point. Milne wrote to his brother that he had come up with "a thoroughly satisfactory picture of the universe."

A third major cosmological model was proposed by Fritz Zwicky, a physicist at the California Institute of Technology and, if Mrs. Hubble's memory is reliable, one of the men sitting around the fire at Hubble's home, eating sandwiches, drinking whiskey, and talking over various approaches to problems. He suggested that the universe might not be expanding, and that the observed redshifts could be the result of interactions between light quanta and matter in space rather than indications of real motion. A gravitational drag on light would explain the apparent velocity-distance relation, since it would increasingly affect the light from increasingly distant objects. Zwicky did not hesitate to advance a theory requiring a new principle of physics. Indeed, he perceived that requirement as an asset rather than a liability. According to his philosophy of morphological research, one should explore all implications of any problem without prejudice. "This means that at the start of a morphological investigation *all limitations must be discarded* and *no valuations must be introduced* until a total perspective is gained." Distinctive anomalies in the behavior of light might well lead to an improved understanding of physical laws. The morphological approach to thought and action would uncover the existence of unknown cosmic objects and phenomena, while other scientists were arriving at stagnation points, thinking they knew everything. For astronomers, Zwicky believed, 1930 was such a stagnation point with respect to the nature of the observed redshifts.

Hubble ostensibly took up the problem of discriminating on the basis of observations between Lemaître's, Milne's, and Zwicky's possible models of the universe. He was joined by Richard Tolman, a theoretical physicist at the California Institute of Technology who had developed the mathematical foundations of relativistic cosmology and who was one of the men engaged in the give-and-take of evening discussions at Hubble's home.

Hubble understood modern science to differ from medieval science as experiment does from dialectics. In ruminations on the nature of science, if not always in actual practice, Hubble viewed scientific theories as subject to empirical verification or refutation. A scientist "naturally and inevitably [. . .] mulls over the data and guesses at a solution." He proceeds to "testing of the guess by new data—predicting the consequences of the guess and then dispassionately inquiring whether or not the predictions are verified."

Hubble and Tolman began with the observed fact of redshifts in the light from distant nebulae, which they were inclined to interpret as Doppler shifts due to recessional motions. Not wanting to neglect prematurely the possibility of other causes, however, they continued to use the phrase "apparent velocity of recession." They discussed methods of distinguishing between recessional motion and some other cause of the redshifts. Before the link between theory and observation could be forged, various complicating factors in the treatment of the data also had to be dealt with. Theoretical relations were calculated in each case for real and for apparent velocities of recession. Hubble and Tolman formulated methods for interpreting the nature of the redshifts, but in their published paper they did not report a definite conclusion. Given the many observational problems, any conclusion could only be tentative.

As late as the 1920s, theorists had developed possible world models with little reference to actual observations, while observational astronomers had proceeded largely in ignorance of theory. In the 1930s Hubble brought together theory and observation. For him:

> *Mathematicians deal with possible worlds, with an infinite number of logically consistent systems. Observers explore the one particular world we inhabit. Between the two stands the theorist. He studies possible worlds but only those which are compatible with the information furnished by observers. In other words, theory attempts to segregate the minimum number of possible worlds which must include the actual world we inhabit. Then the observer, with new factual information, attempts to reduce the list further. And so it goes, observation and theory advancing together toward the common goal of science, knowledge of the structure and observation of the universe.*

The joining together of theory and observation by Hubble marks a turning point in science.

Observation vs. Philosophical Values

Cosmology, for centuries consisting of speculation based on a minimum of observational evidence and a maximum of philosophical predilection, had become with Hubble's work an observational science, its theories now subject to verification or refutation to a degree previously unimaginable. Yet human values are far from extinct in twentieth-century cosmology.

In writings on the philosophy of science, if not always in actual practice, Hubble deferred to observation. He stated that science dealt with facts and events, on which it was possible to obtain universal agreement, and that the necessity for such agreement completely barred science from the great world of values. Yet philosophical principles caused Hubble to choose Lemaître's relativistic, expanding, and homogeneous universe, even though his preliminary data better fit Zwicky's nonexpanding model, and the raw data uncorrected for temperature effects favored Milne's expanding but nonrelativistic and nonhomogeneous model.

Theory and observation might advance together in Hubble's philosophy, but theory also had an independent authority. Before recourse to observation, the number of logically consistent systems to be compared against observations could be reduced by application of fundamental principles. For Hubble, there were two fundamental principles:

- General Relativity Theory

  Which gives both an unstable universe, either expanding or contracting, and a universe whose space is curved in the vicinity of matter.

- The Cosmological Principle

  Acknowledged by Hubble to be "pure assumption," the principle states that on a grand scale the universe will appear the same from any position, homogeneous and isotropic, with neither center nor boundaries.

Milne's and Zwicky's models were logically consistent systems, but each contradicted beliefs held by Hubble to be fundamental. He might more frankly have rejected Milne's and Zwicky's models before recourse to observation.

Even if Milne's model with an *a priori* Euclidean space was to agree with observation, it would have been unacceptable to Hubble because relativity theory dictated a curved space. Milne's charge that the curvature of space was an *ad hoc* hypothesis held no weight with Hubble. Nor was Milne's charge of incompleteness in the matter of predicting either expansion or contraction of concern to Hubble: it was a simple matter for observation to settle. The issue of homogeneity, possible in Lemaître's model but not in Milne's, also constituted sufficient reason for Hubble to reject Milne's model. The part of the universe that Hubble observed must be representative of the whole universe if he were to be able to make sound inferences. Homogeneity and the uniformity of nature were necessary prerequisites for a methodologically sound exploration of the universe.

Zwicky's stable model also was unacceptable to Hubble on philosophical grounds. It was incompatible with relativity theory, which entailed instability. Furthermore, Zwicky's proposal of a new principle of physics was contrary to respectable scientific methodology, which abhorred the introduction of *ad hoc* principles. Most cosmologists, with the exception of Zwicky, agreed on an abhorrence of *ad hoc* explanations, though there could be considerable disagreement over what was or was not *ad hoc*.

Falsifiability in Science

Zwicky, unable to use the 100-inch telescope before 1948, later charged:

> Hubble, W. Baade and the sycophants among their young assistants were thus in a position to doctor their observational data, to hide their shortcomings and to make the majority of the astronomers accept and believe in some of their most prejudicial and erroneous presentations and interpretations of facts.

Hubble, though, in papers following the 1935 collaboration with Tolman, did reveal discrepancies between his data and the cosmological model that he preferred. Hubble conceded in 1936:

> The observations can probably be accounted for if red-shifts are not velocity-shifts.

And he acknowledged in 1937:

> Suspicions are at once aroused concerning either the interpretation of red-shifts as velocity-shifts or the cosmological theory in its present form.

Combining his data with the assumption of receding nebulae produced an expansion of the universe greater at greater distances (i.e., earlier times). If the expansion rate were slowing, estimates of the age of the universe based on the rate of expansion needed to be reduced, and the resulting age was then less than the age of the earth. Either the data were unreliable or the redshifts were not the result of velocities of recession in an expanding universe, Hubble conceded.

The ability to test a theory against physical evidence—first to make predictions from a theory, next to observe whether the predictions are fulfilled, and finally to reject or at least modify the theory should it not stand the test—has furnished a useful demarcation between science and pseudo-science. Were there a single indispensable characteristic unfailingly distinguishing scientific theories from the nonscientific, a philosopher of science might well posit falsifiability as the *sine qua non* of scientific-ness. The criterion of falsifiability is often taken to be both a necessary and a sufficient condition for being scientific. Yet, as historians of science recognize, there is also a long tradition of recalcitrant scientists refusing to accept falsification of their theories by contrary evidence. Direct observational and experimental tests of hypotheses have, in some instances, been accorded less weight than the conformity of the hypotheses with general theoretical superstructures.

Refusal to accept the falsifiability of theory by observation or experiment is not unusual in relativity theory. Einstein wrote:

> I do not by any means find the chief significance of the general theory of relativity in the fact that it has predicted a few minute observable facts, but rather in the simplicity of its foundation and in its logical consistency.

And Eddington wrote:

> For those who have caught the spirit of the new ideas the observational predictions form only a minor part of the subject. It is claimed for the theory that it leads to an understanding of the world of physics clearer and more penetrating than that previously attained.

It was this spirit, shared with Einstein and Eddington, that caused Hubble to refuse to accept the apparent falsification of a relativistic, expanding, homogeneous model of the universe by prediction and observation. Instead, he pursued unrelentingly his scientific vision. Ultimately, Hubble barred neither himself nor his science from the great world of values, and his cosmology was the better for it.

Epilogue

World War II interrupted Hubble's work on cosmology, and a heart attack ended his life in 1953, soon after the 200-inch telescope was completed on Palomar Mountain and too soon for conclusive answers from the research program planned by Hubble. Subsequent results have removed some of the questions surrounding his choice of a relativistic model. Various corrections resulted in a quadrupling of distances, sufficient to yield an age for the universe calculated from its rate of expansion greater than the age of the earth. Hubble's work, though, is not to be judged primarily on the basis of some number of answers currently believed to be correct. His cosmology should be appreciated more for the vistas it opened and for the opportunities it raised for future research.

Some facets of Hubble's science might be grouped under the labels "insight" or "intuition." These terms, however, have an unsatisfactory, mystical connotation; they suggest an acquisition of knowledge and understanding without any evident intervention of rational thought and inference. Analysis of Hubble's cosmology clears away some of the mystery of his success and leads to a fuller understanding of his achievement. Dissection need not destroy the beauty of his work; under close examination, Hubble's cosmology is all the more remarkable.

—Norriss S. Hetherington

BIBLIOGRAPHIC NOTE

For an overview of cosmology early in the twentieth century, see: R. Berendzen, R. Hart, and D. Seeley, *Man Discovers the Galaxies* (New York: Science History Publications, 1976); and R. Smith, *The Expanding Universe* (Cambridge: Cambridge University Press, 1982). The best introduction to Hubble's cosmology is in his own writings, particularly E. Hubble, *The Realm of the Nebulae* (New Haven: Yale University Press, 1936; reprinted in paperback by Dover Publications, 1958). See also Hubble's "Cepheids in Spiral Nebulae," *Popular Astronomy*, 33 (1925), 252–255; "A Relation between Distance and Radial Velocity among Extra-Galactic Nebulae," *Proceedings of the National Academy of Sciences*, 15 (1929), 168–173; *The Observational Approach to Cosmology* (Oxford: Clarendon Press, 1937); and "The Problem of the Expanding Universe," *American Scientist*, 30 (1942), 99–115. Several of Hubble's papers are presented, with annotations and historical introduction, in N. Hetherington, *Hubble's Cosmology: Guided Studies of Great Texts in Science* (forthcoming), while otherwise unpublished Hubble manuscripts can be read in Hetherington, *The Edwin Hubble Papers: Previously Unpublished Manuscripts on the Extragalactic Nature of Spiral Nebulae. Edited, Annotated, and with an Historical Introduction* (Tucson, Arizona: Pachart Publishing House, 1990).

20. BIG BANG COSMOLOGY

Evolving from a super-dense state or from a singularity in space-time, such a temporally finite universe is referred to as a big bang universe, and cosmological theories concerning such universes are referred to as big bang cosmologies. The distinguishing feature of big bang cosmology is thus a violent beginning or *creation* of the universe, a first event before which there existed neither time nor space. The name "big bang" has been in general use since the mid-1960s. Originally it was used as a nickname by opponents of the theory, and it was in this sense that it was first introduced by Fred Hoyle in 1950. But with the victorious development of big bang theories in the mid-1960s, the pejorative meaning of *big bang* has disappeared. Before then, the class of theories were referred to as *explosion*, *creation*, *evolution*, or *catastrophe* theories.

The development of big bang cosmology is closely linked to that of the expanding universe, but there is no one-to-one correspondence between the two theories. Models of the expanding universe predate the first versions of big bang cosmology, and rival theories such as steady state cosmology also accept a universe in expansion, a concept that does not necessarily imply a cosmic beginning. In other words, big bang cosmology is a subclass of expanding universe cosmologies.

Primeval Atom Hypothesis

Although the 1922 theory of the Russian mathematician and physicist Alexander Friedmann included a formal description of a universe expanding constantly from a singularity to its present state, it was not conceived of as a realistic world model with a beginning in space-time, either by Friedmann or by his contemporary cosmologists. For Friedmann and for other cosmologists in the 1920s, the age of the universe was more a mathematical curiosity than a possible physical reality; during this first phase of the theory of the expanding universe, there were no attempts to suggest a true beginning of the world. This

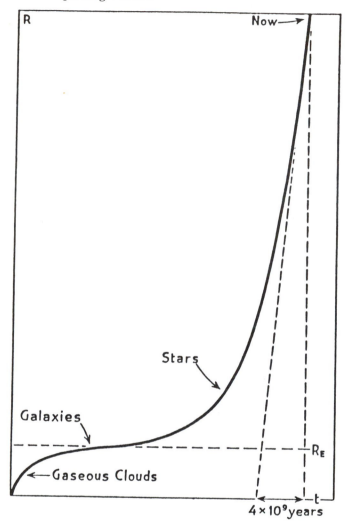

*Figure 1. Lemaître's world model in 1950. The intersection between **R** = 0 and the inclined dashed line to the right gives the Hubble time in terms of current values. The age of the world is several times the Hubble time.*

generalization—that there were no attempts to suggest a true beginning of the world—also holds true for the theory proposed by the Belgian abbé Georges Lemaître in 1927. His theory of an expanding universe was rediscovered by the English astronomer Arthur Stanley Eddington in 1930 and became a generally accepted concept in the early 1930s. The view then held by Eddington (sometimes called the Lemaître-Eddington model) was that the present universe had evolved gradually by expansion from an already existing pre-universe of mass and size equal to those of the equilibrium Einstein universe, the constant radius of which was given by $\mathbf{R_E} = 1/\lambda^2$, with λ the cosmological constant introduced by Einstein in 1917. The Lemaître-Eddington model had a logarithmic past (the radius approaching $\mathbf{R_E}$ for $t \to -\infty$) and thus was not a proper big bang theory (figure 1).

The revolutionary concept of a beginning of the world in a physically realistic sense was first introduced by Lemaître in 1931. There exist earlier suggestions of creation cosmologies, but they are vague philosophical or religious suggestions, not contributions to scientific cosmology. For example, the English astronomer James Jeans contemplated in a popular 1929 book "a definite event . . . of creation of matter at some time not infinitely remote," but he did not commit himself to the idea nor did he attempt to incorporate the idea of creation of matter into a scientific context.

In response to Eddington's statement in 1931 that the concept of a beginning of the universe was "repugnant," Lemaître suggested that the world had begun with the violent fragmentation of a simple quantum of energy. He speculated:

> *We could conceive the beginning of the universe, in the form of a unique atom, the atomic weight of which is the total mass of the universe. This highly unstable atom would divide in smaller and smaller atoms by a kind of super-radioactive process.*

This suggestion is the essence of what later would be called the primeval atom hypothesis, the first example of a big bang theory. Lemaître imagined the primeval atom as a spatially extended object and not as a true singularity (which is point-like, with no extension in space or time).

Later in 1931 Lemaître elaborated his audacious idea at the British Association meeting, where he argued for a "fireworks theory" of cosmic evolution, in which cosmic rays were understood to be remnants of the initial explosion. He continued to develop the theory after 1931, and in 1935 he supplied the theory with a geometrical framework based on his 1927 relativistic cosmology.

Lemaître then abandoned work on cosmology for a decade, only to resume work in 1945. The following year he published a semi-popular book on the primeval atom hypothesis and made the theory known to a wider audience.

Lemaître's cosmology did not change essentially after 1946, but remained close to the foundation he had laid down fifteen years earlier. A brief summary of the theory as it stood in 1946 follows.

The geometric world model (the temporal variation of the cosmic scale factor) consisted of three phases.

- Phase 1: An initial explosion or very rapid expansion of the primeval atom, estimated to have a nuclear density and radius of about one astronomical unit. Lemaître argued that this process was indeterministic and somehow governed by the laws of quantum mechanics, but he did not offer any explanation of how it happened. The initial fragmentation yielded high-energy cosmic radiation of photons and particles, the remnants of which were still observed.

- Phase 2: A long period of slowing-up or "stagnation," in which condensation processes took place and galaxies were eventually formed. In this phase, gravitational attraction was roughly balanced by the cosmic repulsion resulting from the cosmological term in Einstein's field equations. (Lemaître always argued for a nonvanishing cosmological constant.) Because of the stagnation phase (of a somewhat indefinite length), the model had the advantage of allowing a time-scale much longer than the Hubble time and thus avoiding the paradoxical situation arising from a universe younger than its parts (the time-scale difficulty).

- Phase 3: A period of accelerated expansion leading to the present era.

Many other aspects of the universal expansion were studied by cosmologists in the 1930s, largely independently of Lemaître's primeval atom model. Most of the studies focused on mathematical or conceptual problems, and did not consider the physical circumstances of the early universe. Nor did they attempt to interpret realistically what was only in a formal sense big bang cosmology. For example, in the British astrophysicist E.A. Milne's influential and controversial theory, developed in 1931–1935, the "origin of the universe" had only a conventionalist meaning. By an appropriate time transformation, the world could be made temporally infinite. Milne considered such a world picture to be as legitimate as the creation picture was.

Important contributions to what would later appear as mainstream big bang cosmology were made in the United States by Howard P. Robertson and Richard Tolman, among others. Robertson (and, independently, A.G. Walker in England) deduced in 1935 the most general form of the metric for a space-time satisfying the cosmological principle: the postulate that the universe is spatially homogeneous in its large-scale appearance. This metric became generally known as the Robertson-Walker metric. Together with Tolman, Robertson pioneered the study of thermodynamics in the theory of the expanding universe. Between 1931 and 1934 Tolman studied the adiabatic expansion (occurring without loss or gain of heat) of a radiation-filled universe and derived the important result that the temperature would vary inversely to the scale factor (the characteristic distance between two galaxies).

Yet, in spite of their fundamental contributions to cosmological theories of the big bang type, neither Robertson nor Tolman followed Lemaître in equating their theoretical models with physical reality. Characteristically, Tolman concluded his seminal 1934 textbook:

> *It is difficult to escape the feeling that the time span for the phenomena of the universe might be most appropriately taken as extending from minus infinity in the past to plus infinity in the future. [. . .] The discovery of models, which start expansion from a singular state of zero volume, must not be confused with a proof that the actual universe was created at a finite time in the past.*

This conclusion was shared by most cosmologists in the 1930s but not, of course, by Lemaître.

Phoenix Universe

It can be argued that creation cosmology does not necessarily imply a temporally finite universe. Already in 1922, Friedmann had shown that one class of solutions to Einstein's field equations corresponds to what he called a "periodic world" (i.e., a world model in which the radius of a closed universe oscillates between zero and a maximum value). Such a model implied an infinity of expansions and contractions, an everlasting number of big bangs and "big squeezes."

The oscillatory solution, picturesquely referred to as the "phoenix universe," was advocated by Einstein in 1931 and investigated in detail by Tolman in the early 1930s. Tolman argued that in a realistic case, the singularities that divided the cycles would be replaced by very small and dense universes, somewhat similar to Lemaître's primeval atom. The phoenix model was usually considered to be a fanciful—though conceivable—hypothesis, and was either ignored or rejected on philosophical grounds. Lemaître, among others, refused to accept the strange idea of an infinite succession of primeval atoms, and Eddington found it "rather stupid to keep doing the same thing over and over again," as he expressed himself in 1928. A few scientists, including the French physicists Jean Perrin (in 1941) and Alexandre Dauvillier (in 1963) and the Estonian astronomer Ernst Öpik (in 1960) defended the phoenix universe. In the United States, George Gamow played with the idea in 1950 and coined the term *big squeeze*. The idea of a phoenix universe also fascinated Robert Dicke, who discussed the oscillatory model in the early 1960s. In the 1970s the idea was taken up by John A. Wheeler, but most modern cosmologists refuse to take a phoenix universe seriously.

Philosophically and religiously motivated forms of the phoenix universe go far back in time and are part of the oldest known cosmological thoughts. Speculations on an eternal succession of worlds were common in Greek and Roman antiquity, where "the Great Year" symbolized the vast period after which the history of the world would repeat itself. The idea of an eternal succession of worlds was revived in the nineteenth century, when it was expounded by philosophers and social reformers, including Louis Blanqui, Friedrich Engels, and Friedrich Nietzsche. Their speculations, interesting as they are in their own right, had no impact, however, on twentieth-century scientific cosmology.

Physics Enters Cosmology

Lemaître's theory was not much discussed during the 1930s. It failed to attract the interest of physicists because of the speculative nature of the primordial atom and the very early universe. Although Lemaître vaguely talked of the initial explosion as some kind of quantum phenomenon and imagined that future progress in quantum mechanics would make the initial explosion physically comprehensible, he did not attempt to combine his cosmology with the new sciences of nuclear and particle physics. When he proposed his model in 1931, nuclear physics had barely existed and neither the neutron, the positron, nor the neutrino (three elementary particles) had yet entered physics. This situation changed drastically over the next two decades, but the fundamental change in physics is scarcely evident in Lemaître's subsequent presentations of his theory.

The fruitful development that eventually established cosmology as a branch of physical science took its start in the late 1930s, when a few nuclear physicists turned to problems relevant to astrophysics, such as stellar energy production, the abundance distribution of elements, and states of extremely dense matter. This development took place independently of Lemaître's theory and was at first not related to cosmology.

Gamow, the Russian-American pioneer of nuclear theory, soon emerged as a leading force in the development, and he was the first to apply arguments of nuclear physics to cosmology. He did so in a 1942 paper on the origin of the chemical elements, in which he proposed that the abundance distribution of elements might be understood as the result of an explosive fragmentation process taking place in a hypothetical proto-universe of extreme density. He postulated that this initial state contained a large amount of elements heavier than uranium, and he sought to explain qualitatively the formation of elements by fission processes (splitting of nuclei) and subsequent beta decay (expulsion of electrons from the nuclei caused by transformation of neutrons into protons).

This 1942 paper was important not for the suggested mechanism of element formation, which was soon abandoned, but for the very idea that element formation took place in the early universe as a cosmological process and not (as in the earlier theories of Robert d'Escourt Atkinson, Charles Critchfield, Hans Bethe, and others) in equilibrium processes in the interior of stars. The German physicist Carl Friedrich von Weizsäcker had postulated in 1938 that most elements were formed in explosions similar to that taking place in Lemaître's primeval atom, but he did not state clearly whether the birthplace was cosmological or stellar. The turning point came in 1946, when Gamow discussed quantitatively the role of nuclear processes in relativistic cosmologies. His brief 1946 paper on the expanding universe and the origin of elements marks the beginning of modern big bang theory.

Ylem and Nucleogenesis (1946–1953)

Gamow's new approach was to study nuclear building-up processes in an early universe of the Friedmann-Lemaître type that would result in an element distribution corresponding to the distribution found empirically. He imagined the early universe as consisting of protons and neutrons, with nucleogenesis occurring by successive neutron capture (in which neutrons combine with the atomic nucleus, followed by beta decay). Reliable data on the cosmic abundance distribution of elements had appeared in the late 1940s, including experimental data on the reaction rates of neutron capture processes (their cross-sections) obtained by Donald J. Hughes, a physicist at Brookhaven National Laboratory. This data indicated an inverse correlation of an element's neutron capture cross-section with its cosmic abundance, suggesting that Gamow was on the right track.

Gamow's 1946 paper set the direction that modern big bang cosmology would take, and a much improved theory was worked out in the years between 1948 and 1953. This work, which resulted in a coherent and quantitative theory of the early universe, was carried out almost exclusively in the United States and relied heavily on America's superiority in nuclear physics. It is interesting to note that the appearance of the mature big bang theory in 1948 coincided with the emergence of steady state theory in England. [See *Steady State Cosmology* immediately following this chapter on *Big Bang Cosmology*.]

The pioneers in the development of the theory of the early universe were Gamow, Ralph Alpher, and Robert Herman, who started a fruitful collaboration in 1947. Alpher was then a graduate student at

George Washington University and Herman worked at the Johns Hopkins Applied Physics Laboratory. Their professional backgrounds scarcely indicated that they would make pioneering contributions to cosmological theory. Alpher's earlier work had been in supersonic aerodynamics, and Herman had studied spectroscopy related to combustion reactions. We might also note here that Thomas Gold came to steady state cosmology from studies of the acoustics of the ear.

Another noteworthy feature of the approach to cosmology in the United States, distinguishing it from that of the British, was its quantitative nature. In the United States, scientists made detailed calculations of nuclear processes in models of the early universe, much in the same pragmatic and instrumentalist spirit that characterized postwar U.S. quantum and nuclear physics in general. To solve their equations numerically, they used some of the first electronic digital computers. The philosophical and qualitative aspects that characterized the British tradition, and are so visible in the works of Hermann Bondi and Gold, are absent in the works of big bang theoreticians in the United States.

The approach in the United States also differed markedly from that followed in Belgium by Lemaître, who—strangely—never referred to the papers of Gamow and his collaborators nor paid any attention to the Mayer-Teller theory of 1949 (discussed below), although this theory in many respects was consistent with Lemaître's own views. Although Lemaître was aware of developments across the Atlantic, he chose to ignore them. The American way was foreign to him, and he neither appreciated nor mastered nuclear and particle physics. Lemaître, a believer in simplicity and logic, found particle physics to be at an "entomological" level. Gamow, on his side, recognized Lemaître's fundamental work as a predecessor of his own theory.

In 1948 Alpher coined the word *Ylem* to designate the primordial hot and dense state of the world, a state corresponding to Lemaître's primeval atom. However, whereas Lemaître's original atom was a primordial nuclear fluid of comparatively low temperature (not breaking the nuclear bonds), Ylem was a primordial very hot gas.

Gamow, Alpher, and Herman saw it as their primary task to calculate the formation of elements that presumably took place right after the neutronic matter of Ylem had started expanding and decaying into protons and electrons. A big step forward was achieved with the so-called alpha-beta-gamma ($\alpha\beta\gamma$) theory of 1948, worked out by Alpher and Gamow. Bethe was added mainly to complete the alphabetical sequence. Gamow even tried to make Herman change his name to Delter, which would have transformed their theory into the $\alpha\beta\gamma\delta$ theory. Although Gamow failed in this effort, he nonetheless referred in a 1949 paper to "the neutron-capture theory of the origin of atomic species recently developed by Alpher, Bethe, Gamow and Delter."

The αβγ paper, occupying but a single page in the *Physical Review*, presented more of a program than a fully worked-out theory. But the paper did indicate the road to follow, and details were soon elaborated.

According to the theory, "nuclear cooking"—to use Gamow's phrase—had to take place within the first half hour of expansion (figure 2). The basic mechanism of such cooking was neutron capture, which required a very high neutron density. Right after the big bang, neutrons would decay into protons, and protons would then combine with electrons to form neutrons. With continued expansion

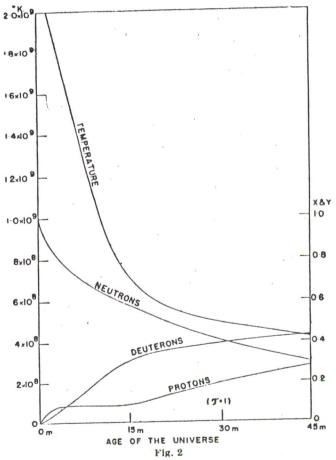

Figure 2. *The first 45 minutes of the universe, according to the Gamow-Alpher-Herman theory in 1948.*

and increases in temperature and density, the creation of neutrons would become rare, and soon stop altogether. Neutron decay rates (a radioactive property) would remain unchanged. The formed protons would then combine with neutrons to form deuterons, and higher nuclei would be built up by further neutron-capture processes. The building-up process was assumed to begin at about twenty seconds after t = 0. By working out this scenario quantitatively, Gamow and Alpher found a reasonably close fit with the abundance curve. Their calculations predicted that about a quarter of the mass of the universe would be helium, a figure that agreed remarkably well with observations both then and later.

The following years saw a rapid development of the 1948 big bang theory, primarily by Alpher and Herman, but also with contributions from Gamow, Enrico Fermi, Anthony Turkevich, and the Japanese scientist Chushiro Hayashi. Five years later, most of the details of the αβγ theory had changed but not the general approach nor the basic features. The theory received wide popular attention with Gamow's best-selling *The Creation of the Universe*, which gave a popular yet detailed account of big bang cosmology.

In 1949 another theory of the origin of elements was proposed, by German-born Maria Goeppert Mayer and by Hungarian-born Edward Teller, both then at the University of Chicago. Their theory relied on assumptions similar to those made by Lemaître (i.e., the fragmentation of a primordial, neutron-rich nuclear fluid). However, Mayer and Teller did not relate their work to Lemaître's model, and their "polyneutron" was not the primordial state of the world but a fragment of stellar size. Assuming that the polyneutron disintegrated through fission and beta decay, Mayer and Teller found a distribution of heavier nuclei consistent with that known empirically.

The Mayer-Teller theory was not further developed, and turned out not to be a viable alternative to the hot nuclear gas theory of Gamow, Alpher, and Herman. But her work on this theory did stimulate Mayer to study nuclear structure and to propose the shell theory of nuclei. For this work she became, in 1963, only the second woman to win a Nobel Prize in physics.

Already in 1948 it was realized that the original assumption of a matter-dominated early universe was untenable and that the first phase of expansion was in fact governed almost entirely by radiation. Photons (quanta of electromagnetic radiation), and not nucleons, became the stuff of the earliest universe, and other elementary particles were included in the primordial mix. The hypothetical neutrinos (only detected in 1956) became especially important, and weak interaction physics, which deals with beta decay and neutrinos, entered as an important ingredient of cosmology. In general, Alpher and Herman developed the theory by:

- Improving the accuracy of calculations.

- Applying updated empirical data.

- Taking into account more advanced theory, including Fermi's theory of weak interactions (such as beta decay) and relativistic quantum statistics.

- By including the most recent progress in particle physics.

In a 1948 paper on the evolution of the universe, Gamow had argued that there would exist a characteristic, universal "decoupling" time at which the universe would change from being radiation-dominated to being matter-dominated, and hence would become transparent to radiation. This result followed from the fact that in a Friedmann-Lemaître universe the radiation density varies as R^{-4} while the matter density varies as R^{-3}, with R the scale factor measuring the expansion of the universe.

The decoupling was studied the same year by Alpher and Herman, and they predicted the existence of a cosmic background radiation as a remnant of the decoupling. The spectrum of this radiation would be distributed in the same way as that of an ideal black body, which was first determined by Max Planck in 1900. Alpher and Herman calculated the present temperature to be 5 K. Their result, although easily available to scientists then, and now accorded the utmost significance in cosmological theories, surprisingly attracted virtually no interest when it was first announced.

In 1953 Alpher and Herman, now collaborating with James Follin (also at the Johns Hopkins Applied Physics Laboratory), provided a detailed and comprehensive analysis of the early universe. This work brought big bang theory to a position that would not change significantly over the following years. The big bang scenario was now pushed back to a beginning time of only 10^{-4} seconds after the initial explosion, and the scenario continued until the decoupling time some 100,000 years after the initial explosion (table I, *Timetable of Events in the Early Universe*, and figure 3, *The Evolution of the Universe in the Gamow-Alpher-Herman Theory*). During this period, R (the scale factor measuring the expansion of the universe) was found to expand by a factor of 10^8 and the temperature was found to drop from about 10^{13} to 10^3 degrees K.

The earliest universe was no longer the Ylem imagined a few years earlier, composed of neutrons or of neutrons plus radiation. The earliest universe now consisted of photons, neutrinos, and elec-

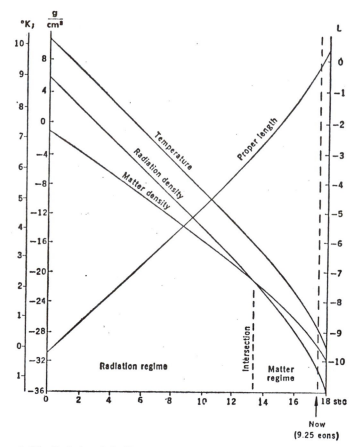

Figure 3. The Evolution of the Universe in the Gamow-Alpher-Herman Theory, based on empirical parameters as known in 1967. One "eon" equals one billion years. The horizontal time scale is logarithmic.

Table I. Timetable of Events in the Early Universe. The e^+e^- annihilation refers to the process in which a positive and a negative electron annihilate under the production of photons of energy corresponding to the mass of the two electrons. Adapted from a paper by Alpher, Follin, and Herman (1953).

Temperature (MeV)	(K)	Time	Events and particle content
>100			doubtful validity of field equations
≈100	10^{12}	6×10^{-6} sec	thermodynamic equilibrium; photons, neutrinos, electrons
≈10	10^{11}	8×10^{-3} sec	neutrinos frozen in; photons, neutrinos, electrons
≈2	2×10^{10}	0.2 sec	e^+e^- annihilation starts; photons, neutrinos, electrons
0.05	5×10^8	600 sec	e^+e^- annihilation; nucleogenesis; photons, neutrinos
0.03	3×10^8	30 min	nucleogenesis essentially complete
0.015	1.5×10^8	10^8 yr	galaxy formation

trons (positive as well as negative), and also muons, or μ-mesons (heavy electrons discovered in 1937). Only a very small part of the early universe was now believed to be made up of nucleons. After a few minutes of expansion, most electrons would have vanished by annihilation, producing an environment of mainly photons and neutrinos. Alpher, Herman, and Follin found that while neutrons and protons would initially exist in almost equal numbers, the neutron-to-proton ratio would decrease with the expansion. They calculated the ratio to lie between 1 to 4.5 and 1 to 6.0 when nucleosynthesis started. This furnished big bang theory with a useful basis for testing and further calculations.

By 1953, then, big bang cosmology had become a sophisticated quantitative theory and had scored significant successes in predicting abundances of elements. However, there were persistent problems in bridging the "mass gaps" at mass numbers 5 and 8 (there are no stable nuclei with these masses), and it became a problem to continue the building-up process to nuclei heavier than atomic weight 5. Several ingenious proposals were made, but the problem remained a stumbling block. This problem of the mass gaps was used to argue against the big bang theory by advocates of the steady state theory (figure 4).

Controversy and Interpretations

By 1953, when big bang theory had reached maturity, it attracted increasing interest both within and without the scientific community. Part of the interest was due to controversy between advocates of the big bang theory and advocates of the steady state theory.

The controversy included religious and philosophical aspects, which were much discussed during the 1950s, mostly in Europe. In England, Milne and the mathematician Edmund Whittaker drew theological consequences from modern cosmology. Pope Pius XII in 1951 utilized big bang theory to support the affirmation of a transcendent creator. Modern cosmology, the Pope stated, "has confirmed the contingency of the universe and also the well-founded deduction as to the epoch when the cosmos came forth from the Hands of the Creator." [See the last chapter in this book, *Cosmology and Religion.*] Gamow, always fond of breaking conventions, quoted the Pope's approval of big bang theory in a paper in the *Physical Review*. Lemaître, though a Catholic priest, maintained at the 1958 Solvay Congress that his theory remained "entirely outside any metaphysical or religious question." By and large, the occasional association of big bang cosmology with Christian theology played no significant role in the scientific debate.

NEW GENESIS

In the beginning God created radiation and ylem. And ylem was without shape or number, and the nucleons were rushing madly over the face of the deep.

And God said: "Let there be mass two." And there was mass two. And God saw deuterium, and it was good.

And God said: "Let there be mass three." And there was mass three. And God saw tritium and tralphium, and they were good. And God continued to call number after number until He came to transuranium elements. But when He looked back on his work He found that it was not good. In the excitement of counting, He missed calling for mass five and so, naturally, no heavier elements could have been formed.

God was very much disappointed, and wanted first to contract the Universe again, and to start all over from the beginning. But it would be much too simple. Thus, being almighty, God decided to correct His mistake in a most impossible way.

And God said: "Let there be Hoyle." And there was Hoyle. And God looked at Hoyle . . . and told him to make heavy elements in any way he pleased.

And Hoyle decided to make heavy elements in stars, and to spread them around by supernovae explosions. But in doing so he had to obtain the same abundance curve which would have resulted from nucleosynthesis in ylem, if God would not have forgotten to call for mass five.

And so, with the help of God, Hoyle made heavy elements in this way, but it was so complicated that nowadays neither Hoyle, nor God, nor anybody else can figure out exactly how it was done.

Amen.

Figure 4. George Gamow's commentary on the inability of big bang theory to explain the build up of nuclei heavier than atomic weight five and on Fred Hoyle's attempt at an explanation. (Tralphium was Gamow's name for the helium isotope He³.)

The religious aspect of the big bang theory may have contributed to its lack of acceptability in the Soviet Union. The steady state theory, though, did not enjoy a corresponding gain. During the Stalinist era, the party line held that the universe is infinite in space and time. Consequently, big bang cosmology was often branded as reactionary bourgeois idealism. A further consequence was that Soviet scientists were reluctant to engage in cosmology; they largely accepted the party line that the study of the universe as a whole was unscientific.

The cosmological debate during the 1950s was curiously one-sided. Big bang theoreticians in the United States devoted their efforts to developing their theory and to gathering new data. They were unwilling to engage in debate, an activity that they gladly left to the Old World. The cosmological debate took place mainly in Great Britain.

Here the debate was fueled largely by the emergence of the steady state theory, rather than by considerations from the big bang theory. Indeed, a large part of the debate in Britain was not concerned with big bang cosmology at all, but occurred between relativistic evolution cosmology and steady state theory. For example, Martin Ryle's arguments based on data from radio astronomy were directed *against* steady state theory but not put forward *for* big bang theory. None of the British steady-state antagonists advocated big bang theory in the form developed by Gamow, Alpher, and Herman. In Britain, their theory played only a minor role in the controversy, mostly as a target for Hoyle's criticism. When the British Broadcasting Corporation arranged a symposium on cosmological theories, published in 1960 as *Rival Theories of Cosmology*, none of the participants as much as mentioned the big bang theory of Gamow and his coworkers.

The same neglect of modern big bang theory characterized philosophically-oriented debates during the 1950s. Philosopher-scientists— including Herbert Dingle, Gerald Whitrow, and Rom Harré in Great Britain and Adolf Grünbaum, Milton Munitz, and Richard Schlegel in the United States—discussed the conceptual state of modern cosmology without referring to the pioneering works of Gamow, Alpher, and Herman.

One of the weak points in the Gamow-Alpher-Herman theory lay in the difficulty it encountered in accounting for the formation of heavier nuclei. In the 1950s several scientists argued that element formation did not require the special conditions of a big bang, but would take place in supernovae and stars. Furthermore, in 1952 Edwin Salpeter at Cornell University showed that the mass gaps at mass numbers 5 and 8 could be bridged in the interior of stars by a double helium burning, in which the highly unstable beryllium-8 nucleus fused with an alpha particle. The conditions necessary if this process were to occur were not satisfied in the early universe of big bang theory.

In 1957 Margaret and Geoffrey Burbidge, William Fowler, and Hoyle reviewed the entire field of nucleosynthesis and concluded that, contrary to the big bang view, "the stars are the seat of origin of the elements." (See figure 4.) The cosmic abundance of helium, however, could not be satisfactorily explained as the result of stellar processes. Hoyle admitted as much in 1964, when he concluded that helium was produced under conditions far more dramatic than those governing stellar interiors. Either a hot big bang or hypothetical compact objects (black holes) would be needed to produce helium. By the early 1960s it was generally admitted that element formation was best accounted for by a combination of cosmological and stellar processes, and that the failure of big bang theory to explain all the details of nucleosynthesis during the epoch of the early universe was not a telling argument against the theory.

Cosmic Background Radiation

By and large, the 5 K background radiation predicted by Alpher and Herman in 1948 was ignored. Even when Gamow, Alpher, and Herman repeated the calculation several times between 1948 and 1956, their work still failed to attract attention. There were neither attempts to detect the radiation nor suggestions of how to detect it. One reason for the lack of attention may have been that Gamow, Alpher, and Herman believed that the microwave radiation manifested itself mainly as an increased cosmic energy flux, and in this case the microwave radiation might not be separated from starlight and cosmic rays.

What has been called the most important observation in cosmology since Edwin Hubble's discovery of the redshift-distance relation in 1929 would not be made until 1965, and then serendipitously and independently of its 1948 prediction. In Princeton, Robert Dicke and P.J.E. Peebles were investigating models of an oscillating universe (with alternate phases of expansion and contraction). Dicke suggested that the last big bang might have left traces of cosmological radiation, and Peebles calculated the relic blackbody radiation. He estimated its temperature to be 10 K. The Princeton theorists started a collaboration with Peter Roll and D.T. Wilkinson, who had constructed a radiometer to measure thermal radiation at a wavelength of 3 centimeters.

Before they had obtained any results, however, they learned about experiments at the Bell Laboratories that indicated the existence of the predicted background radiation. E.A. Ohm, a Bell Labs physicist, had constructed a radiometer in 1961 to receive microwave signals reflected from the Echo balloon (a giant plastic balloon launched by NASA to measure the density of the atmosphere at high altitudes, but also a convenient reflector of radio waves). Ohm found an excess temperature of 3.3 K in the antenna, but the result attracted no attention. Two years later, Arno Penzias and Robert Wilson, also at the Bell Labs, started preparing the radiometer for use in radio astronomy. They found an antenna temperature of 7.5 K, when it should have been only 3.3 K. Since their measurements were very reliable (Ohm's were not), the result constituted a problem that threatened to make the instrument's use in radio astronomy dubious. Penzias and Wilson therefore spent most of a year trying to understand the reason for the discrepancy, by carefully checking all hypotheses and possible sources of error (which included a pair of pigeons roosting in the antenna). Only when they saw a preprint of Peeble's work did they realize that the noise might be of cosmological origin.

The discovery of the microwave background radiation might have taken place as early as 1941. In that year an Australian astrophysicist, Andrew McKellar, interpreted spectroscopic observations to indicate a rotational excitation of cyanogen radicals (CN molecules) in interstellar space. The energy of the rotating molecule would be released in the form of electromagnetic radiation. From the measured wavelength, he found an excitation temperature (corresponding to the rotational energy of the molecule) of about 2.3 K. However, no one thought of interpreting the result cosmologically, and it was only much later, after the measurements of Penzias and Wilson, that the significance of these early observations was recognized.

The Bell and Princeton physicists published their work in the spring of 1965 as companion papers. While Penzias and Wilson simply reported their finding of an excess temperature of 3.5 ± 1.0 K at a wavelength of 7.3 cm, the Princeton group discussed the cosmological implications. Neither of the papers mentioned the earlier work of Alpher and Herman. Penzias and Wilson received Nobel Prizes in 1979 for their discovery of the cosmic background radiation. There was also much new work extending and confirming the conclusion that the radiation was blackbody-distributed and isotropic (the same in all directions). After all, Penzias and Wilson had only measured it at a single wavelength. By the early 1970s, new measurements over a wide range of wavelengths strongly suggested that the 1965 conclusions were indeed correct; it was generally accepted that there exists a universal, isotropic blackbody radiation of temperature 2.7 K left over from the period in which matter and radiation decoupled. Since the phenomenon was predicted by standard big bang theory, but could only be accounted for in alternative theories by means of *ad hoc* hypotheses of little plausibility, the discovery of the cosmic background radiation was a big success for the big bang theory.

Standard Model of Cosmology
With the success of big bang theory and the elimination of its steady state rival, there emerged in the mid-1960s a consensus among cos-

mologists on the main problems to be solved and the criteria to be used. The hot big bang relativistic theory became the paradigm of cosmology, while alternative interpretations were marginalized. At the same time that cosmology became cognitively institutionalized, it achieved a social institutionalization that made the subject a full-time professional occupation rather than a part-time hobby for astronomers, mathematicians, and physicists. There was a growing integration of the subject into university departments, and courses and textbooks in cosmology became common. For the first time, students were taught standard cosmology and brought up in a research tradition with a shared heritage. The national differences that had characterized earlier cosmology also disappeared. Originally, big bang theory had been an American theory, while steady state theory had belonged to the British, and the Russians had hesitated doing cosmology at all. The field now became truly international; it was no longer possible to tell an author's nationality from the cosmological theory he or she advocated.

The victorious cosmological theory developed since the mid-1960s is sometimes called the "standard model" because of its general acceptance. It was developed by many researchers, continuing the existing big bang theory based on the works of Gamow, Alpher, and Herman. Prominent contributors to the standard model have included Peebles, Yakov Zel'dovich, Steven Weinberg, Joseph Silk, and Denis Sciama (the latter a long-time advocate of steady state theory). Basically, the standard model is an extension and refinement of the big bang theory prevailing around 1950.

Like the earlier theory, the standard model incorporates the cosmological principle and Einstein's field equations with a zero cosmological constant. And of course the standard model postulates a universe that originated in a hot big bang with a thermal history leading to the formation of elements, stars, and galaxies.

The main differences between the standard model and the Gamow-Alpher-Herman theory are quantitative. For example, in the standard model the history of the universe is traced much further back in time, toward the incredibly small "Planck time" (10^{-42} second after the big bang), in which the usual laws of physics are supposed to break down.

Cosmology has become increasingly integrated with elementary particle physics, and the early universe is seen as a laboratory for the physics of extremely high energies. Since no terrestrial accelerator will ever produce the ultra-high energies required for testing some modern theories of physics, the "ultimate laboratory" of the early universe has become a *sine qua non* for particle physicists.

The standard model has been further extended by new empirical findings, many of which were made possible by rocket flights and satellite observations. Improvements in optical and radio astronomy—leading to such new discoveries as X-ray stars (1962), quasars (1963), pulsars (1967), and cosmic gamma rays (1968)—also have nourished further development of the standard model.

In spite of the paradigmatic status of the standard big bang model, alternative views have remained popular among a minority of cosmologists. Versions of steady state theory (the Hoyle-Narlikar theory) are still defended, and models operating with a varying gravitational constant (such as Dirac's theory and the Brans-Dicke theory) have attracted a certain interest. Another alternative, worked out by Swedish physicists Oskar Klein and Hannes Alfvén in 1962, assumes the existence of entire galaxies made up of anti-matter. These and other alternative theories have not been serious competitors to mainstream big bang theory, and have either been abandoned or marginalized. The most fruitful and interesting developments in cosmology since the late 1960s have taken place within the standard model, inspired by problems generated from within the paradigm.

—Helge Kragh

BIBLIOGRAPHIC NOTE

The origin of Lemaître's first big bang model is analyzed in its historical context in H. Kragh, "The Beginning of the World: Georges Lemaître and the Expanding Universe," *Centaurus*, *31* (1987), 114–139. O. Godart and M. Heller in *Cosmology of Lemaître* (Tucson: Pachart Publishing House, 1985) offer a broader description of Lemaître's work and career, including a biographical sketch. George Gamow in *The Creation of the Universe* (New York: Viking Press, 1952) presents a fascinating, although not always historically accurate, account of the "new" big bang theory, written by its originator in his characteristic witty and provocative style. The early work on big bang cosmology is excellently reviewed by Ralph Alpher and Robert Herman in "Early Work on 'Big-Bang' Cosmology and the Cosmic Blackbody Radiation," in B. Bertotti, et al., eds., *Modern Cosmology in Retrospect* (Cambridge: Cambridge University Press, 1990), pp. 129–158. This book also contains chapters on other aspects of the development of modern cosmology, including Robert Wilson's "Discovery of the Cosmic Microwave Background" (pp. 291–308), and a section on the debate between big bang and steady state cosmologies. A popular, yet informative and historically sensitive account of the development of the standard big bang model is given in Steven Weinberg, *The First Three Minutes: A Modern View of the Origin of the Universe* (New York: Basic Books, 1977).

21. STEADY STATE THEORY

Steady state theory, as first suggested in 1948, mandates that the large-scale features of the universe do not change with time. In this theory, the observed expansion of the universe is exactly matched by the creation of new matter, at a rate of formation that preserves a stable situation, i.e., a constant density. For most but not quite all scientists, steady state theory died in 1965, with the discovery of the cosmic microwave background radiation, and the theory is now primarily a subject of historical and philosophical interest. Still, the controversy over the steady state theory during its brief ascendency tells us much about the underlying assumptions of our cosmology.

Background

Cosmology scarcely existed as a unified scientific discipline in 1948, when the steady state theory of the universe first appeared. Instead of a scientific field defined by a set of common goals, methods, and basic presuppositions, cosmology was then characterized by radically different and competing views. Issues such as what should constitute cosmology and how knowledge should be obtained about the universe were still being argued.

Most cosmologists—not professional cosmologists, as such, but rather physicists, mathematicians, and astronomers sometimes engaged in cosmology—supported one version or another of the relativistic expanding universe. According to this theory, Einstein's general theory of relativity described the large-scale features of a universe that had expanded for at least a billion years. The rate of expansion was generally estimated from Edwin Hubble's redshift-distance data. (Hubble had found that the velocities of galaxies are proportional to the distances of the galaxies from us: $v = Hr$, with H the Hubble constant.) In the relativistic interpretation, the Hubble constant provided a value for the "age" of the universe. This age (the characteristic time of cosmic evolution) was usually identified with the inverse value of the Hubble constant $(1/H)$, and in 1948 the

Hubble age of the universe was believed to be about 1.8 billion years. (The distance, r, that an object has moved equals the velocity v it is moving with multiplied by the time during which it has been moving, t, giving us the equation: $r = vt$. Thus $t = r/v$, or $t = 1/H$.)

The relativistic theory of the universe as worked out by Alexander Friedmann, Georges Lemaître, Arthur Stanley Eddington, Howard P. Robertson, and others did not, however, lead to a unique solution among possible universes. Because of observational uncertainty and free parameters appearing in the theory, it was possible to defend different models within the general paradigm of the relativistic expanding universe.

One possible relativistic theory was Lemaître's idea that the universe had evolved from an exploding "primeval atom," an idea which soon was developed to become the big bang theory. Although Lemaître's catastrophic theory built on general relativity, some rival cosmologies denied that Einstein's theory could be applied to the universe as a whole. Instead, these theorists argued, cosmology should be presented in a theoretical framework that did not rely on extrapolations of ordinary physics. The most powerful of the rival cosmologies was Edward Milne's kinematic relativity, developed during the 1930s and still defended by Milne as late as 1948. Other cosmological schemes in opposition to relativistic theory included "Newtonian cosmology" (developed by Milne, W.H. McCrea, A.G. Walker, and others) and the "large-number cosmologies" of Paul Dirac and Pascual Jordan.

Whatever their differences, all these cosmological theories agreed that the universe was expanding and that it had a definite age given by the inverse Hubble constant. Most of the theories also accepted the cosmological principle, according to which the large-scale properties of the universe at any time are independent of the location of the observer.

Relativistic cosmology, along with rival theories, had several serious problems, problems that both cast doubt on the theories themselves and also cast doubt on the enterprise in general. In contrast to its high standing now, cosmology in earlier years commanded little academic prestige or respectability. One problem was the inability of cosmological theories to account for the formation of galaxies. Another problem was the so-called time-scale problem. While the inverse of the Hubble constant indicated, in the late 1930s and through the 1940s, an age for the universe of less than two billion years, reliable radioactive dating measurements indicated that the earth was about three billion years old—more than a billion years older than the universe as a whole!

Origins of Steady State Theory

As far back as the 1920s, William D. MacMillan, an astronomer at the University of Chicago, had suggested that the universe maintains a stationary state as a result of a balance between matter-to-energy conversion in the interior of stars and creation of atoms from radiant energy in empty space. But his hypothesis, advanced before the expansion of the universe was established, soon fell into oblivion, and it had no influence upon later steady state theories. It is to the year 1948 that we must look for the beginning of modern steady state theory.

In 1948 the *Monthly Notices of the Royal Astronomical Society* contained two papers in which three young Cambridge physicists proposed a new cosmological theory, soon to become known as the steady state model. One of the papers was written by Thomas Gold and Hermann Bondi; the other paper was written by Fred Hoyle. The three had become acquainted with each other during World War II, while doing research on radar. (Gold and Bondi were Austrian citizens—both born in Vienna—and they also spent a year in internment.) After the war, when the three returned to Cambridge University, they continued to meet and discuss problems of physics. None of them had previously shown any interest in cosmology, but now they agreed that relativistic cosmology was unsatisfactorily developed and began to think about it.

In 1946 or 1947, Bondi, Gold, and Hoyle together went to a movie based on a ghost story. The movie ended in the same way that it started out, inspiring (according to Hoyle's reminiscences) Gold to suggest that perhaps the universe was unchanging yet dynamic. This germ of an idea and the discussions that followed may have been the beginning of the steady state theory.

It was soon realized that Gold's idea required creation of matter out of nothing, and early in 1948 Gold and Bondi worked out one version of the steady state universe while Hoyle independently worked out another version. Although the two versions were independent and rather different, the authors knew about each other's work before publication.

The fundamental idea of a universe that expands but does not change because of continuous creation of matter was due to the 28-year-old Gold, who was then completely unknown within the astronomical community. His only previous research had been on a very different subject—the acoustics of the ear.

The 1948 Papers

Although Hoyle's approach differed considerably from that of Bondi and Gold, the two resulting theories had so much in common that it is reasonable to speak about a single steady state theory.

- Both of the 1948 papers were characterized by philosophically based objections to standard (Friedmann-Lemaître) cosmology. Hoyle argued that "creation-in-the-past theories" went against the spirit of scientific enquiry because the creation could not be causally explained. Bondi and Gold objected to the large number of free parameters that made it impossible to deduce a definite model of the universe from general relativity.

- Both versions of the steady state theory adopted as a starting point what Bondi and Gold called *the perfect cosmological principle* and what Hoyle called *the wide cosmological principle*. According to the perfect cosmological principle, the universe is not only spatially *but also temporally* homogeneous; the universe looks the same at any location *and at any time*. The perfect cosmological principle naturally eliminated the time-scale difficulty, since the steady state universe had no finite age.

- The three young cosmologists objected that known physical laws could not be extrapolated to the super-dense state of the early big bang universe. As Bondi and Gold phrased the argument, "it is only in such a [steady state] universe that there is any basis for the assumption that the laws of physics are constant."

- In order to make a stationary universe agree with the observed (or inferred) recession of the galaxies, it was assumed that matter is continually created throughout the universe. This feature of steady state cosmology was the most controversial, and was often seen as the main characteristic of the theory—which consequently was sometimes referred to as "continuous creation cosmology." The idea of continuous creation of matter, however, was not the invention of steady state theorists. The idea figured prominently in Dirac-Jordan cosmology, and had also been suggested by Reginald Kapp in a little-known book from 1940. The creation of matter in steady state theory had to take place at an exceedingly slow rate (about three new atoms of hydrogen per cubic meter per million years), and the process was thus impossible to observe directly.

- From the assumptions mentioned previously, the geometrical properties (the "metric") of the steady state universe were deduced, leading uniquely to an accelerated expansion of the so-called de Sitter type.

The Bondi-Gold theory differed from the Hoyle theory primarily in considering the perfect cosmological theory as a fundamental axiom from which physical results should be deduced. If theoretical extrapolations from terrestrial experiments conflicted with the principle, Bondi and Gold were quite willing to reject such extrapolations. In particular, they found general relativity theory unacceptable because it insisted on the conservation of mass. They wanted to replace the relativity postulate with Mach's principle, according to which (in one of its many versions) inertia owes its origin to the rest of the matter in the universe.

Hoyle, on the other hand, did not appreciate the general and qualitative character of the Bondi-Gold theory. He wanted to formulate steady state theory within the framework of general relativity theory. For this purpose, Hoyle added to the Einstein field equations a term, the C-field, that represented the non-Einsteinian concept of matter creation. The C-field was introduced in a rather *ad hoc* manner, with its value adjusted to agree with Hubble's law. Through the means of this mathematical trick, Hoyle was able to deduce an expression for the average density of the universe, an important quantity that did not follow from the Bondi-Gold theory.

Bondi and Gold were not happy with Hoyle's theory, which they described as "unsatisfactory and unacceptable." They admitted that a field formulation of steady state theory was necessary, but they rejected Hoyle's extrapolating approach on philosophical grounds. They saw Hoyle's work as a betrayal of the foundation of the steady state theory. Hoyle, meanwhile, maintained that his less radical, extrapolating approach was more fruitful. He argued that the perfect cosmological principle should follow as a consequence of a modification of general relativity, and that the revolutionary approach of Bondi and Gold was too ambitious to be of any real use.

In the controversy that followed, though, steady state cosmology generally was perceived as a single theory with Bondi, Gold, and Hoyle fighting on the same team. Their disagreements were largely of a philosophical nature, and far less significant than the disagreements between steady state cosmology and mainstream relativistic cosmology.

Further Developments: 1949–1960
Shortly after the pioneering work of Bondi, Gold, and Hoyle, the steady state theory ran into strong opposition. It was response to this

opposition, rather than potentialities within the steady state theory itself, that led to much of the development of the steady state theory. Consequently, many developments took place in what may seem an incoherent fashion.

A proper school of steady state theory never formed, and at no time did the theory command a following of more than a dozen or so researchers. Remarkably, all the supporters of steady state theory were Britons. Apart from the original trio, steady state cosmology secured support from two important figures in British astronomy: Harold Spencer Jones and William McCrea. Spencer Jones, professor at the Royal Institution and Astronomer Royal, supported the theory in a talk in 1952 but did not himself engage in its further development. McCrea, professor at London University and a former collaborator with Milne, converted to steady state theory in 1950 and contributed to its development into the early 1960s. Few young scientists joined the new cosmology, one of the exceptions being Denis Sciama, who had studied under Dirac and Hoyle. Bondi, Gold, Hoyle, McCrea, and Sciama produced most of the developments in steady state theory during the 1950s.

The defense of steady state cosmology was carried out following two lines of argument, both approaches often found in scientific controversies.

• One strategy pursued in the defense of steady state theory was to modify the theory in order to counter various philosophical and observational objections.

Because of the rigid structure of the Bondi-Gold version, which allowed virtually no change in the original scheme, modifications were limited to Hoyle's version. The most important modification probably was Hoyle's formulation in 1960 of a covariant law for the C-field. This he claimed offered certain advantages over the original steady state theory. Hoyle's C-field technique and the relationship between steady state cosmology and the general theory of relativity were examined by several researchers, and in papers between 1951 and 1953 McCrea argued that continuous creation could be incorporated into standard general relativity and that this version of the theory promised a unification of quantum theory and cosmology.

• The second strategy pursued in the defense of steady state theory yielded indirect support for the theory. There were attempts to weaken its main rival, big bang relativistic theory, both by emphasizing inadequacies of big bang cosmology and by showing that its successes could be matched by steady state theory.

Of particular importance were the mechanism of galaxy formation and the origin of elements. Hoyle in 1953, Sciama in 1955, and Gold and Hoyle in 1959 examined the gravitational condensation of stars and galaxies and concluded that the process could not take place in an expanding big bang universe but only in a universe of the steady state type. Hoyle, collaborating with astrophysicists in the United States, produced in 1957 a successful nuclear theory of the origin of elements that did not presuppose the physical conditions of a hot, superdense universe in the past. Although this theory did not rely on steady state assumptions explicitly, it was generally regarded as a triumph of steady state cosmology because it contradicted big bang assumptions. Hoyle and his collaborators concluded that:

> In contrast with the other theories which demand matter in a particular primordial state for which we have no evidence, this latter theory [Hoyle's] is intimately related to the known fact that nuclear transformations are currently taking place inside stars. This is a strong argument, since the primordial theories depend on very special initial conditions for the universe.

Responses and Debate

Most members of the astronomical community simply ignored the steady state theory. Outside Great Britain, the theory received little attention. Mainstream cosmologists, including George Gamow, Robertson, Richard Tolman, and Lemaître, rejected steady state theory without even bothering to examine it seriously. They claimed that steady state theory was in conflict with observations. Furthermore, it was, as Gamow said, "artificial and unreal."

A solution to the time-scale difficulty made it easier for many astronomers to dismiss steady state theory. By the mid-1950s new observations and a reanalysis of Hubble's data had yielded a much longer time-scale—with the age of the universe now greater than the age of the earth.

Arguments philosophical and religious in nature played a significant role in the controversy between big bang and steady state cosmologies. Bondi and Gold had used methodological and epistemic arguments in favor of their new theory; so their opponents now marshaled similar arguments. Foremost among the critics was Herbert Dingle, an astrophysicist and a philosopher of science. Since the late 1930s he had fought a crusade against the rationalistic "cosmythology" of Eddington, Milne, and Dirac and for the empirical-inductive conception of science. He now used similarly strongly worded rhetoric to warn against steady state theory, which he accused of being dogmatic and plainly nonscientific. Other philosophically based criticism came from the philosopher Milton Munitz and the astronomer Gerald

Whitrow. Characteristic of the debate was its foundational nature, concerning the meaning of science as such. Is steady state theory more or less "scientific" than big bang theory? Is cosmology a "science" at all? Such broad questions were part and parcel of the debate.

In an obscure manner, steady state theory was sometimes intimated to be less "materialistic" than its rivals, and continuous creation was associated with values such as personal freedom and anti-communism. Such labels were part of the zeitgeist of the Cold War. However, Hoyle's forced attempt to present steady state theory as part of the anti-communist movement was as misleading as was Gamow's intimation that Soviet astronomers found the theory in agreement with the party line. In fact, official Soviet astronomy rejected categorically both big bang and steady state cosmologies. Both were seen as equally absurd and idealistic. The study of the universe as a whole was rejected as unscientific, and Russian scientists cautiously avoided supporting any cosmological theory.

For Hoyle, Catholicism and communism were associated world views, in the sense that both were dogmatic. Hoyle also found dogmatism to be characteristic of mainstream evolutionary cosmology. During the 1950s the relationship between religious thought and the rival big bang and steady state cosmologies was often discussed, but no consensus emerged from the discussions. In 1952 Pope Pius XII stated in a talk to the Pontifical Academy that modern big bang cosmology affirmed the notion of a transcendental creator and thus was in harmony with Christian dogma. Although the Pope did not mention steady state cosmology, and leading big bang cosmologists (including Lemaître, a Catholic priest) were unhappy with the Pope's argument, the widely publicized talk may have given some the impression that steady state theory was associated with atheism. Hoyle seems to have accepted the association; for him it was another argument against big bang cosmology—as an irrational process beyond the realm of science.

Philosophical, political, and religious components of the controversy between big bang and steady state cosmology were hotly debated. The course and outcome of the controversy, however, were not decisively shaped by these factors. Rather, it would be scientific arguments of a more conventional nature that pretty much settled the issue, at least among scientists.

Observational Tests

Despite fundamental disagreements, both scientific sides in the debate between steady state and big bang cosmology agreed that, ultimately, the question would have to be settled by observations. Bondi, inspired by Sir Karl Popper's falsificationist philosophy of science (that to be scientific, scientific theories must be in principle falsifiable), declared that steady state theory would have to be abandoned if observations contradicted even a single one of its predictions. But Bondi and other steady state theorists also emphasized that in any conflict between observation and theory, observation was as likely to be found at fault as was theory.

Further developments soon confirmed the generalization that observations were too readily and uncritically accepted as established fact while theories were correspondingly undervalued. In 1948 steady state theory faced its first major observational challenge, in the so-called Stebbins-Whitford effect. In the United States the astronomers Joel Stebbins and Alfred Whitford reported what appeared to be a reddening in the spectra of distant galaxies, in excess of that expected from the ordinary velocity-dependent redshift. In the context of evolutionary cosmology, their observation was interpreted as indicating older ages for more distant galaxies. But in steady state theory, the average age of galaxies does not vary with distance or with time. Big bang cosmologists used the Stebbins-Whitford effect to discredit steady state theory. In a critical review of the data, however, Bondi, Gold, and Sciama showed in 1954 that the effect was spurious. Two years later, Whitford admitted that the "observations" were wrong and thus of no value in deciding between the rival cosmological theories.

A relatively straightforward observational test that might provide an empirical basis for choosing between big bang and steady state theory seemingly lay in the variation of the speed of recession of galaxies with their distances. Whereas evolutionary cosmology predicted that the rate of recessional velocity would be disproportionately larger for distant (older) galaxies, the steady state theory predicted that the velocity would increase in direct proportion to distance. Data obtained by Milton Humason and Alan Sandage at the Mount Wilson Observatory indicated a slowed-down expansion, in agreement with evolutionary cosmology. But the data was too uncertain for scientists to accept its indications as decisive, except those scientists already convinced of the truth of big bang cosmology.

Much the same occurred with Gamow's argument against steady state theory. Galaxies consisting of old stars apparently showed an age variation, which was at odds with steady state theory. Proponents of steady state theory, however, argued convincingly that the determination of the galaxies' ages was highly uncertain, and that more sensitive measurements were needed for a decisive falsification of their theory.

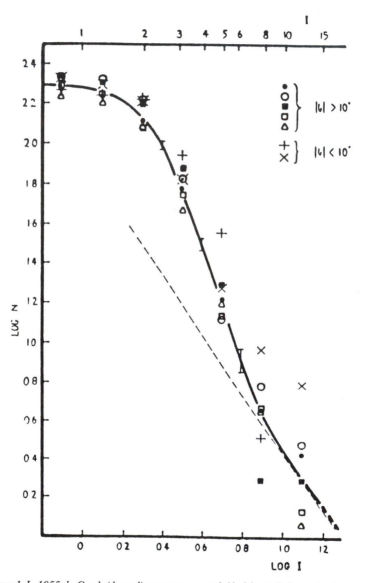

Figure 1. In 1955 the Cambridge radio astronomy group led by Martin Ryle completed a survey and catalog of almost 2,000 radio sources (the 2C survey) and plotted in a double-logarithmic diagram the number of sources N (vertical scale) with a minimum intensity I (horizontal scale). They found the main part of the sources to lie on a straight line with slope −3. The steady state theory, however, predicts a slope of −1.5 (the dashed line in the diagram).

The most serious observational challenge to steady state theory came from the new science of radio astronomy, which had evolved in Great Britain since the first discrete "radio star" had been discovered in 1946. The leading radio astronomer, Martin Ryle of Cambridge University, soon became the most active opponent of steady state cosmology. In 1955 the Cambridge group completed a survey and catalog of almost 2,000 radio sources (the 2C survey) and plotted in a double-logarithmic diagram the number of sources with a minimum intensity, $N(I)$ against the intensity, I. They found the main part of the sources to lie on a straight line with slope -3 (figure 1). This result was recognized as holding great cosmological significance, because steady state theory predicted a slope of -1.5 or more. Big bang cosmology did not lead to a similar unique prediction, so although the number count could falsify steady state cosmology, it could not be used to verify big bang theory. Ryle concluded that the Cambridge data were consistent with an evolutionary universe and that "there seems no way in which the observations can be explained in terms of a Steady-State theory." Ryle's conclusion, however, was premature, and it was not accepted by steady state proponents (figure 2). Once again, they questioned the certainty of seemingly valid observations. Indirect support for steady state proponents was soon forthcoming from Sydney, Australia, where radio astronomers in 1957 criticized the 2C results and argued that the slope of the **log N** vs. **log I** curve was about -1.8. After taking uncertainties into account, this value was too indecisive for a conclusive choice between rival cosmological theories. The discussion over the interpretation of radio source count measurements continued for some years, but in the early 1960s consensus was obtained among radio astronomers that the slope was -1.8, with an uncertainty of no more than 0.1. Ryle now reinstated his case against steady state theory and concluded again that the theory had been falsified. This time the results were accepted by most astronomers as finally falsifying steady state theory.

Capitulation

Bondi, Gold, and Hoyle now admitted that they were fighting an uphill battle, but they still believed that they could win the battle. Steady state supporters, including Sciama in 1963, produced alternative explanations of the radio source counts that made the disagreement between theory and observation disappear. However, these alternatives were *ad hoc* hypotheses and were not taken seriously by the majority of astronomers.

By 1964 attempts to rescue the simple steady state theory had ended. New theories, however, were put forward in attempts to avoid

"Your years of toil,"
Said Ryle to Hoyle,
    "Are wasted years, believe me.
The steady state
Is out of date.
        Unless my eyes deceive me,

My telescope
Has dashed your hope;
    Your tenets are refuted.
Let me be terse:
Our universe
    Grows daily more diluted!"

Said Hoyle, "You quote
Lemaître, I note,
    And Gamow. Well, forget them!
That errant gang
And their Big Bang—
    Why aid them and abet them?

You see, my friend,
It has no end
    And there was no beginning,
As Bondi, Gold,
And I will hold
    Until our hair is thinning!"

"Not so!" cried Ryle
With rising bile
    And straining at the tether;
"Far galaxies
Are, as one sees,
    More tightly packed together!"

"You make me boil!"
Exploded Hoyle,
    His statement rearranging;
"New matter's born
Each night and morn.
    The picture is unchanging!"

"Come off it, Hoyle!
I aim to foil
    You yet" (The fun commences)
"And in a while"
Continued Ryle,
    "I'll bring you to your senses!"

Figure 2. Commentary on Ryle versus Hoyle by
Barbara Gamow, wife of George Gamow.

capitulation to big bang orthodoxy. The most important of these new steady state theories was constructed by Hoyle in collaboration with Jayant Narlikar in a series of publications starting in 1961. They built up a complicated general relativistic field theory that could be brought to agree with the radio source counts and yielded a temporal variation of the gravitational constant. Although the Hoyle-Narlikar theory differed considerably from the original steady state theory, it nonetheless belongs to the class of steady state theories because it avoids a violent beginning of the universe and includes the continuous creation of matter. But Hoyle and Narlikar's theory, though worked out in many different versions between 1961 and 1975, was never generally accepted as a serious alternative to big bang cosmology. Ingenious and ambitious the new theory was, but it was too far removed from mainstream physics to make much of an impact on scientists. Few physicists were impressed by its emphasis on action-at-a-distance electrodynamics (at a time when field theories dominated microphysics). And astronomers, confident that standard big bang cosmology was true, felt no need to examine the cosmological implications of a rival theory.

For all practical purposes, the death knell for steady state theory rang in 1965, with the unexpected discovery of the 3K microwave background radiation. The finding was immediately and successfully interpreted in the big bang theory as the remnant of the decoupling between matter and radiation. And by implication, the discovery was consequently accepted by most astronomers as a decisive blow against steady state cosmology.

By the late 1960s Bondi, Gold, McCrea, and Sciama had stopped defending steady state theory. The last rear-guard battle was now fought over the version of the theory proposed by Hoyle and Narlikar. Whatever the version, by the early 1970s steady state theory ceased even to provoke discussion, and it was no longer considered worthy of attention in scientific journals or at conferences. After more than twenty years of tumultuous existence, steady state theory became exclusively a subject for historical and philosophical inquiry.

—*Helge Kragh*

Bibliographic Note

There is no satisfactory historical work dealing with the development of steady state theory. J. Narlikar, *The Structure of the Universe* (London: Oxford University Press, 1977), written by a leading steady-state cosmologist, gives a broad account of modern cosmology with some emphasis on steady state theories. J. Singh, *Great Ideas and Theories of Modern Cosmology* (New York: Dover Publications, 1970) is a popular, semi-historical summary of cosmological theories, including a good chapter on "Cosmology and Continuous Creation" (steady state theory).

The best source for the early history of the subject is Y. Terzian and E.M. Bilson, *Cosmology and Astrophysics: Essays in Honor of Thomas Gold* (Ithaca: Cornell University Press, 1982), which includes retrospective articles on the subject by Hoyle, Gold, and Bondi. Fred Hoyle's contribution is also published separately, as *Steady-State Cosmology Re-visited* (Cardiff: University College Cardiff Press, 1980). Among the few historical works on steady state theory is J.M. Sánchez-Ron, "Steady-State Cosmology, the Arrow of Time, and Hoyle and Narlikar's Theories," in B. Bertotti, et al., eds., *Modern Cosmology in Retrospect* (Cambridge: Cambridge University Press, 1990), pp. 233–245. The same book contains recollections of Bondi, William McCrea, and Hoyle. A vivid impression of the cosmological debate is obtained from H. Bondi, et al., *Rival Theories of Cosmology* (London: Oxford University Press, 1960), with contributions from W.B. Bonnor, Bondi, R.A. Lyttleton, and G.J. Whitrow.

VII. Particle Physics and Cosmology

Introduction

THE ULTIMATE ACCELERATOR

A great split has opened between conventional cosmology and the new cosmology of particle physics. The major action in cosmology has shifted from observatory domes to the innards of giant computers, where a new generation of cosmologists run simulations of hypothetical universes. Often it becomes the observers' job to reconcile observations with the numbers spewed out by theory, not the other way round.

Conversely, as theories outstrip the energies attainable in particle accelerators, particle physicists must turn to astronomical observations. The universe, a relic of historical events that happened at energies far beyond the wildest dreams of accelerator builders, is the poor man's particle accelerator.

Observational astronomy, itself, also has changed. The construction of new telescopes has broken the Mount Wilson Observatory's near monopoly on cosmology, while new technology has left individual observing skill less important. Now spectra are recorded on computer disks and analyzed automatically by software, while high density scanners automatically measure the blackness of images on photographic plates point by point, five million points to a plate, with the resulting information more fodder for a computer. No human eye nor hand leaves its subjective mark on the study; intuition is neither needed nor wanted here; it is astronomy as a physicist would do it.

Perhaps the most remarkable concept of the new cosmology is the realization that the visible luminous universe is just a scrim. Observational astronomers have spent centuries dancing in the dark, oblivious to dark matter comprising some 90% of the universe. In contrast to the certainty that much of the universe consists of matter not emitting light or other electromagnetic radiation, there is great uncertainty over the nature of this dark matter. It could be in the form of compact objects made of ordinary matter; it could be any of a wide range of exotic particles predicted by theoretical physics but yet to be seen in the laboratory; it could even consist of some form of topological defect remaining from the very early universe.

Philosophically minded scientists find several aspects of the standard big bang model of the universe unsatisfactory, even though each of the individual problems (colorfully labeled "the flatness problem," "the horizon problem," and "the smoothness problem") can be accounted for within the standard theory. When the standard model is extended back to the first fraction of a second of the history of the universe, it is found to require several extremely stringent initial conditions if it is to reproduce the universe currently observed. To the extent that it does not explain how the necessary initial conditions came to exist, the model is judged incomplete, and the way is open for a rival model explaining how some or all of the initial conditions came into being.

The flatness problem arises from the fact that a universe with a ratio of energy density to critical energy density (that density just sufficient eventually to halt the expansion of the universe) *not* equal to unity must either collapse to an infinite-density "big crunch" or expand to a zero-density "big chill" in a very short time. Of course, we can set this ratio almost exactly equal to zero in the standard big bang model, and undoubtedly many people would be happy with this solution and get on to more important matters. But for cosmologists, hardly anything could be more important than coming up with a theory in which the ratio of energy densities is naturally very close to 1 rather than an arbitrary value left to be selected by chance.

The horizon problem similarly follows. The standard big bang theory fails to explain (again assuming that such explanation is a requirement for a satisfactory theory, rather than merely a matter better left for observation to determine) the large-scale uniformity of the observed universe. In the standard model, the universe evolves so quickly that it is impossible for this uniformity to be created by any physical process, because of the impossibility that information or physical process can propagate faster than a light signal. At any given time there is a maximum distance, the horizon distance, that a light signal could have travelled since the beginning of the universe. The horizon distance in the standard model has been much smaller than the radius of the observed universe for most of its history, making it difficult to comprehend how the large-scale uniformity of the universe came to be. We can, of course, assume uniformity in the initial conditions, and the universe then will evolve uniformly.

The standard big bang model requires yet another assumption to explain the *nonuniformity* observed on smaller scales. We, ourselves, are small-scale inhomogeneities in an otherwise remarkably homogeneous universe. We also observe inhomogeneities on the scale of galaxies, clusters of galaxies, and superclusters of clusters. These are large-scale by our human standards, but small on the cosmological scale. To account for this observed clumping of matter, we must assume a spectrum of primordial inhomogeneities as part of the initial conditions. Lack of any explanation for this necessary spectrum of inhomogeneities is itself a problem. The problem becomes more pronounced when we realize that incipient clumps of matter will develop rapidly with time as a result of their gravitational self-attraction. Indeed, to reproduce the universe as we observe it now, the matter in the initial universe must have started in a peculiar state of extraordinary but not quite perfect uniformity. This peculiarity of the initial state of matter required by the standard model is called the smoothness problem.

Questions left unanswered by the standard big bang theory—initial conditions that must be specified to reproduce what is now observed rather than being determined from the theory—involve the mass density of the universe, large-scale homogeneity, and smaller-scale density perturbations. New ideas from elementary particle theory have led cosmologists to new ideas about the behavior of the very early universe, and plausible answers to the above questions now appear to be within our intellectual reach.

The answers seemingly are found in a new theory regarding the very early behavior of the universe, a theory called the inflationary model. Alan Guth first proposed in 1981 the idea of an inflationary universe, and the following chapter by him serves as an excellent introduction to modern cosmology and the growing role of particle physics within it. The following chapters on large-scale structure, cosmic strings, and quantum cosmology are also by leading practitioners in the field. Our new cosmology has been created and developed by superluminaries such as Guth and David Schramm. They are conscious of the historical, philosophical, and scientific background of their work and they are able to communicate the full sweep of this exciting intellectual adventure to general readers.

22. THE INFLATIONARY UNIVERSE

A small band of particle physicists began in about 1978 to dabble in studies of the early universe. They were motivated partly by the intrinsic fascination of cosmology, and also by developments in particle physics itself. The motivation arose primarily from the advent of a new class of particle theories known as "grand unified theories." These theories were invented in 1974, but it was not until about 1978 that the theories became a topic of widespread interest within the particle physics community. The theories are spectacularly bold, attempting to extrapolate our understanding of particle physics to energies of about 10^{14} GeV (one GeV is a billion electron volts, approximately the rest energy of a proton).

This amount of energy is not terribly impressive by the standards of your local power company—it's about what it takes to light a 100-watt bulb for a minute. The grand unified theories, however, attempt to describe what happens when this much energy is deposited on a *single elementary particle*. This constitutes an extraordinary concentration of energy, exceeding the capabilities of the largest existing particle accelerators by eleven orders of magnitude (by 100,000,000,000 times).

To get some feel for how tremendous this amount of energy is, imagine trying to build an accelerator that might reach such energy levels. It can be done, in principle, by building a very long linear accelerator. The largest existing linear accelerator, at Stanford University, is about two miles long and is capable of depositing on a single elementary particle a maximum energy of about 40 GeV. The output energy is proportional to the length of the linear accelerator, and a simple calculation shows that we would need an accelerator nearly a light-year in length to achieve an energy of 10^{14} GeV.

However astronomical the current national debt is, it is infinitesimally small compared with the number of dollars that would be required to build a light-year linear accelerator, even if we could find a spot big enough to put it and enough people to build it and enough materials with which to build it. If there is the slightest shred of sanity left in our government, the U.S. Department of Energy will not be receptive to proposals for funding a light-year linear accelerator.

Consequently, if we want to see the most dramatic new implications of the grand unified theories, we must turn to the only laboratory in the universe with the required energies, and that is the universe itself—in its very infancy. According to the standard hot big bang theory of cosmology, the universe had a temperature corresponding to a mean thermal energy of 10^{14} GeV at about 10^{-35} second after the big bang. This is why particle theorists have become interested in the very early universe.

Big Bang Theory

The development of cosmology in the twentieth century owes much to the work of such scientific giants as Einstein and Hubble. Beginning with belief in a static universe, our understanding of the universe has progressed to an expanding universe and an initial big bang. And we now have evidence to back up our new beliefs, in the form of Hubble's law, the cosmic background radiation, and big bang nucleosynthesis calculations.

Einstein and a Static Universe

Cosmology in the twentieth century began with the work of Albert Einstein. In March 1916 he completed a landmark paper, "The Foundation of the General Theory of Relativity." The theory is in fact nothing more nor less than a new theory of gravity. It is a complex but very elegant theory, in which gravity is described as a distortion of the geometry of space and time. Unlike Newton's theory of gravity, the theory of general relativity is consistent with the ideas of special relativity, which Einstein had introduced in 1905.

While the rest of the world waited to be persuaded, Einstein was immediately convinced that he had found the correct description of gravity, and he proceeded to apply it to the universe as a whole. Less than a year later, in February 1917, he published the results of his studies. In carrying out these studies, Einstein discovered something that surprised him a great deal. He found that it was impossible to build a static model of the universe that was consistent with general relativity theory. Einstein was perplexed by this fact. Like his predecessors, he had looked into the sky, noticed that the stars appeared motionless, and erroneously concluded that the universe was static.

The problem that Einstein discovered in the context of general relativity also exists in Newtonian mechanics, although it was not appreciated until the work of Einstein. The problem is fairly simple to understand: if masses were distributed uniformly and statically throughout space, then everything would attract everything else and the entire configuration would collapse.

Einstein nonetheless remained convinced that the universe was obviously static. He therefore modified his equations of general relativity by adding what he called a *cosmological term*, which amounts to a kind of universal repulsion that prevents the uniform distribution of matter from collapsing under the normal force of gravity. The cosmological term, Einstein found, fits neatly into the equations of general relativity; it is completely consistent with all the fundamental ideas on which the theory was constructed.

Einstein's ideas remained viable for about a decade, until astronomers began to measure the velocities of distant galaxies. They then discovered that the universe is not at all static. To the contrary, the distant galaxies are receding from us at high velocities.

Hubble's Law

The pattern of cosmic motion was codified at the end of the 1920s by Edwin Hubble, in what we now know as Hubble's law. [See chapter 19, *Hubble's Cosmology*, in section VI, *The Expanding Universe*.] Hubble's law states that each distant galaxy is receding from us at a velocity that is, to a high degree of accuracy, proportional to its distance. We can write the equation:

$$v = H \cdot l$$

with v the recession velocity, l the distance to the galaxy, and the quantity H the Hubble "constant." Though called a constant by astronomers because it presumably remains approximately constant over the lifetimes of astronomers, the value of H changes as the universe evolves; from the point of view of cosmologists, the Hubble constant is not a constant.

The value of the Hubble constant is not well known. The recession velocities of the distant galaxies are no problem—they can be determined very accurately from the Doppler shift of the spectral lines in the light coming from the galaxies. The distances to the galaxies, on the other hand, are very difficult to determine. These distances are estimated by a variety of indirect methods, and the resulting value of the Hubble constant is thought to be uncertain by a factor of about two. It is believed to be between 0.5 to 1 divided by 10^{10} years (i.e., between 0.00000000005 and 0.0000000001/year):

$$H = \frac{0.5 \text{ to } 1}{10^{10} \text{ } years}$$

Note that the Hubble constant has the units of inverse time; when the Hubble constant is multiplied by a distance, the result has the units of distance per time, or velocity. In particular, if the expression for H is multiplied by a distance in light-years, the result is a velocity measured as a fraction of the velocity of light. Alternatively, we can use the fact that H is approximately equal to (15 to 30) km/sec per million light-years

$$H \approx (15 \text{ to } 30) \text{km/sec per million light-years}$$

to obtain an answer in kilometers per second.

The development of cosmology in the twentieth century was somewhat confused by the fact that Hubble badly overestimated the value of the Hubble constant, reporting a value of 150 km/sec per million light-years. This mis-measurement had important consequences. In the context of the big bang model, an erroneously high value for the expansion rate implies an erroneously low value for the age of the universe. Indeed, Hubble's value for the Hubble constant implied an age of about 2 billion years for the universe—which was in conflict with geological evidence indicating that the earth is significantly older: approximately twice as old as the age of the universe estimated from the Hubble constant! Not until 1958 did the measured value of the Hubble constant come within the currently accepted range, primarily through the work of Walter Baade and Alan Sandage at the Mount Wilson Observatory.

Once it is noticed that the other galaxies are receding from us, there are two conceivable explanations. The first is that we might be in the center of the universe, with everything moving radially outward from us like the spokes on a wheel. In the early sixteenth century, such an explanation would have been considered perfectly acceptable. Since the time of Copernicus, however, astronomers and physicists have become instinctively skeptical of this kind of reasoning. Alternative explanations are sought, and in this case an attractive alternative has been found.

The alternative explanation can be called *homogeneous expansion* (figure 1). According to this explanation, there is nothing special about our galaxy, or about any galaxy. All galaxies are approximately equivalent, and are spread more or less uniformly throughout all of space. In this description, there is no center and no edge to the distribution of galaxies. As the system evolves, all intergalactic distances are enlarged. Thus, regardless of which galaxy we are living on, we would see all the other galaxies receding from us. Furthermore, this picture leads immediately to the conclusion that the reces-

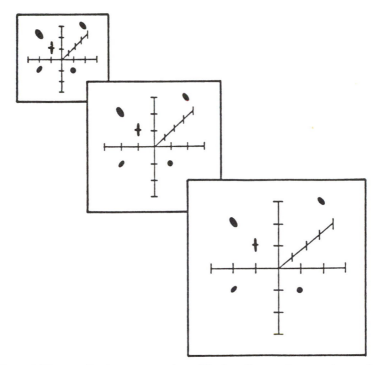

Figure 1. The expanding homogeneous universe. The three diagrams show successive snapshots of a region of the universe. Each is essentially a photographic blowup of the previous diagram, with all distances enlarged by the same percentage. There is nothing special about our galaxy or about any galaxy; all galaxies are approximately equivalent, so we might imagine that we are living on any galaxy in the diagram. The galaxies are spread more or less uniformly throughout space. Accordingly, there is no center and no edge to the distribution of galaxies. As the system evolves (from one diagram to the next), all intergalactic distances are enlarged. Thus, regardless of which galaxy we are living on, we will see all the other galaxies receding from us. Since all distances increase by the same percentage as the system evolves, the larger distances increase by larger amounts. The apparent velocity of a galaxy is proportional to the amount by which the distance from us increases, and hence it is proportional to the distance. Thus the recession velocities obey Hubble's law.

sion velocities obey Hubble's law. Since all distances increase by the same percentage as the system evolves, the larger distances increase by a larger amount. The apparent velocity of a galaxy is proportional to the amount by which the distance from us increases, and hence it is proportional to the distance.

The Big Bang

When we extrapolate this picture backward in time, we find that there is a certain instant in the past when the density of the universe would have been infinite. Such an event is called a singularity. In this case, the singularity is the instant of the big bang itself. [See chapter

20, *Big Bang Cosmology*, in section VI, *The Expanding Universe*.] The big bang occurred between 10 billion and 20 billion years ago. The time is very uncertain for two reasons: we do not know the Hubble constant very accurately, and we are uncertain about the mass density of the universe. The mass density is important in calculating the history of the universe, because it determines how fast the cosmic expansion is slowing down under the influence of gravity.

A warning is in order here. The calculation implying an infinite density at the instant of the big bang is not to be trusted. As we look backward in time, with the density of the universe going up and up, we are led further and further from the conditions under which the laws of physics as we know them have been developed. Thus, it is quite likely that at some point these laws will become totally invalid. Then it is a matter of guesswork to discuss what happened in the universe at earlier times. However, as the history of the universe is extrapolated backward using the laws of physics as we know them, we can still say that the density increases without limit. Most cosmologists today are reasonably confident in our understanding of the history of the universe back to one microsecond (10^{-6} second) after the big bang. The goal of the cosmological research involving grand unified theories is to solidify our understanding back to 10^{-35} second after the big bang.

Cosmic Background Radiation

Having discussed the key features of the big bang theory, we can now ask what evidence is found to support the theory. In addition to Hubble's law, there are two significant pieces of observational evidence in favor of the theory. The first is the observation of the cosmic background radiation.

To understand the origin of this radiation, we can begin by recalling that the temperature of a gas rises when the gas is compressed. For example, a bicycle tire is warmed when it is inflated by a hand pump. Similarly, a gas cools when it is allowed to expand. In the big bang model, the universe has been expanding throughout its history. Thus the early universe must have been much hotter. (Indeed, a mathematical treatment indicates that the temperature would have been infinite at the instant of the big bang. This infinity, however, like the infinite density, should not be considered convincing.)

Hot matter has the universal property that it emits a glow, just like hot coals in a fire. The hot matter of the early universe would therefore have emitted a glow of light that would have permeated the early universe. As the universe expanded, this light would have redshifted. Today, the universe should still be bathed by that early radiation, a remnant of the intense heat of the big bang, now redshifted into the microwave part of the spectrum. This prediction of a cosmic background microwave radiation was confirmed in 1964, when Arno Penzias and Robert Wilson of the Bell Telephone Laboratories discovered a background of microwave radiation with an effective temperature of about 3 degrees Kelvin (3 K).

At the time of their discovery, Penzias and Wilson were not looking for cosmic background radiation. Instead they were searching for astronomical sources capable of producing low-level radio interference. They discovered a hiss in their receiver which they carefully tracked down, verifying that it was an external source of radiation and not simply electrical noise from their circuitry. They also found, to their surprise, that the mysterious radiation was arriving at the earth uniformly from all directions in the sky. Later, the spectrum of the radiation was measured, and it was found to agree exactly with the kind of thermal radiation that would be expected from the glow of hot matter in the early universe (figure 2).

Big Bang Nucleosynthesis Calculations

The second important piece of evidence supporting the big bang theory is related to calculations of what is called "big bang nucleosynthesis." This evidence is somewhat more difficult to understand than the cosmic background radiation, and it is therefore much less discussed in the popular scientific literature.

To make any sense out of this argument, we must first understand that the big bang theory is not just a cartoon description of how the universe may have behaved. On the contrary, it is a very detailed model. Once we accept the basic assumptions of the big bang theory, knowledge of the laws of physics allows us to calculate how fast the universe would have expanded, how fast the expansion would have been slowed by gravity, how fast the universe would have cooled, and so on. Given this information, knowledge from nuclear physics allows us to calculate the rates of the different nuclear reactions that took place in the early history of the universe.

The early universe was very hot, so hot that even nuclei would not have been stable. At two minutes after the big bang, there were virtually no nuclei at all. The universe was filled with a hot gas of photons and neutrinos, with a much smaller density of protons, neutrons, and electrons. (The protons, neutrons, and electrons were very unimportant at this time. Later, though, they became raw materials for the formation of stars and planets.) As the universe cooled, the protons and neutrons began to coalesce to form nuclei. From the nuclear reaction rates, we can calculate the expected abundances of

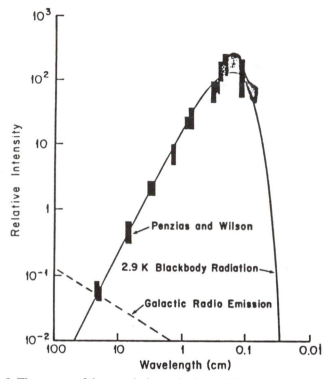

Figure 2. The spectrum of the cosmic background radiation. The vertical black bars and the shaded region at the top of the curve show the results of measurements of the intensity of the cosmic background radiation at several wavelengths. The solid line is a theoretical curve corresponding to thermal radiation at a temperature of 2.9 degrees Kelvin. Note that all the actual measurements agree with the theoretical curve.

the different types of nuclei that would have formed. We find that most of the matter in the universe would have remained in the form of hydrogen. About 25% (by mass) of the matter would have been converted to helium, and trace amounts of other nuclei would also have been produced.

Most of the types of nuclei that we observe in the universe today were produced much later in the history of the universe, in the interiors of stars and in supernova explosions. The lightest nuclei, however, were produced primarily in the big bang, and it is possible to compare the calculated abundances with direct observations. Such a comparison can be carried out for the abundances of helium-4, helium-3, hydrogen-2 (otherwise known as deuterium), and lithium-7.

The comparison is complicated by the fact that we do not know all the information necessary to carry out the calculations. In particular, the calculations depend on the density of protons and neutrons in the universe, a quantity that can be estimated only roughly by astronomical observations. Thus, the calculations have to be carried out for a range of values for the density of protons and neutrons. We then ask whether there exists a plausible value for which the answers turn out right.

The results of this comparison are shown in figure 3. The horizontal axis shows the present density of protons and neutrons, and the curves indicate the results of the calculations. The observations, with the estimated range of their uncertainties, are shown by the shaded horizontal bars. Note that there is a range of values for the density of protons and neutrons (indicated by the cross-hatched region) for which each of the four calculated curves agrees with the corresponding observations. Even though the abundances of these

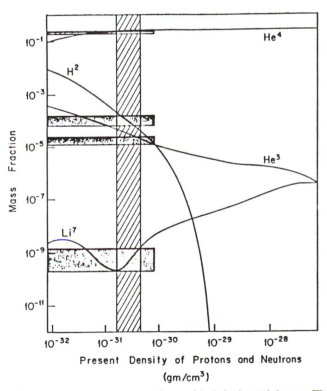

Figure 3. Big bang nucleosynthesis and abundances of the light chemical elements. The curves show how the predicted abundances of helium-4, helium-3, deuterium (hydrogen-2), and lithium-7 depend on the present density of protons and neutrons. The shaded horizontal bars show the abundances observed (with estimated uncertainties indicated by the vertical thickness of each bar). Note that for densities between roughly 10^{-31} and 10^{-30} grams per cubic centimeter on the horizontal axis of the graph (marked by the vertical cross-hatched bar) excellent agreement is obtained between predicted abundances (curves) and measured abundances (horizontal bars) for all four types of nuclei.

nuclei are not known with high precision, the success of the comparison is very impressive. Notice that the abundances span nine orders of magnitude. If there had never been a big bang, there would be no reason whatsoever to expect that helium-4 would be 10^8 times as abundant as lithium-7; it might just as well have been the other way round. But this ratio can be calculated in the context of the big bang theory, and it works out just right!

Unanswered Questions

Each piece of evidence we have discussed—Hubble's law, the cosmic background radiation, and the big bang nucleosynthesis calculations—probes the history of the universe at a different period of time.

- The observation of Hubble's law probes the behavior of the universe at times comparable to the present, billions of years after the big bang.

- The cosmic background radiation, on the other hand, samples the conditions in the universe at about 100,000 years after the big bang, when the universe became cool enough for the plasma of free nuclei and electrons to condense into neutral atoms. The plasma that filled the universe at earlier times was almost completely opaque to photons, which would have been constantly absorbed and re-emitted. With the formation of neutral atoms, however, the universe became highly transparent. Thus, most of the photons in the cosmic background radiation have been moving in a straight line since 100,000 years after the big bang, and they therefore provide an image of the universe at that time.

- The big bang nucleosynthesis calculations probe the history of the universe at much earlier times. The processes involved in determining the abundances of the light nuclei occurred at times ranging from about one second after the big bang to about four minutes after the big bang.

The big bang theory is a very successful description of the evolution of the universe for the whole range of times discussed, from about one second after the big bang to the present. Nonetheless, the standard big bang theory has serious shortcomings; a number of obvious questions are left unanswered. New ideas from particle physics have led to a radically new picture for the very early behavior of the universe, a picture that provides plausible answers to questions otherwise unanswered.

The Ratio of Protons and Neutrons to Photons

One question left unanswered by the standard big bang theory involves the number of protons and neutrons in the universe relative to the number of photons. The photons are mainly in the cosmic background radiation, whereas the protons and neutrons form the atomic nuclei of the matter that makes up the galaxies. The observed universe contains about 10^{10} photons for every proton or neutron. The standard big bang theory does not explain this ratio, but instead assumes that the ratio is given as a property of the initial conditions.

Large-Scale Homogeneity

A second question left unanswered by the standard big bang theory involves the large-scale homogeneity, or uniformity, of the observed universe.

Any discussion of homogeneity must be qualified, because the universe that we observe is in many ways very inhomogeneous. The stars, galaxies, and clusters of galaxies make a very lumpy distribution. [See the following chapter on *Large-Scale Structure and Galaxy Formation*.] Cosmologically speaking, however, all of this structure in the universe is very small-scale. If we average over very large scales, of 300 million light-years or more, the universe appears to be very homogeneous.

This large-scale homogeneity is most evident in the cosmic background radiation. Physicists have probed the temperature of the cosmic background radiation in different directions and have found it to be extremely uniform. It is just slightly hotter in one direction than in the opposite direction, by about one part in a thousand. And even this small discrepancy can be accounted for by assuming that the solar system is moving through the cosmic background radiation at a speed of about 600 km/sec. When the effect of this motion is subtracted out, the resulting temperature pattern is uniform in all directions to an accuracy of about one part in one hundred thousand.

Since the cosmic background gives an image of the universe at about 100,000 years after the big bang, we conclude that the universe was very homogeneous then. (The observed small-scale inhomogeneities are believed to have formed later, by the process of gravitational clumping.) The standard big bang theory cannot explain the large-scale uniformity; instead, the uniformity must be postulated as part of the initial conditions.

The difficulty in explaining the large-scale uniformity is a quantitative question, related to the rate of expansion of the universe. Under many circumstances, a uniform temperature would be easy to understand—after all, anything will come to a uniform temperature if it is left undisturbed for a long enough period of time. In the standard big bang theory, however, the universe evolves so quickly that it is impossible for the uniformity to be created by any physical process.

In fact, the impossibility of establishing a uniform temperature depends on none of the details of thermal transport physics. Instead, it is a direct consequence of the principle that no information can propagate faster than the speed of light. We can pretend, if we like, that the universe is populated with little purple creatures, each equipped with a furnace and a refrigerator, and each dedicated to the cause of trying to create a uniform temperature. We can then show by a straightforward calculation that the purple creatures would have to communicate at more than ninety times the speed of light to achieve their goal of creating a uniform temperature across the visible universe within the period of 100,000 years after the big bang.

The puzzle of explaining why the universe appears to be uniform over such large distances does not arise from any genuine inconsistency in the standard big bang theory. If the uniformity is assumed in the initial conditions, the universe will evolve uniformly. The question arises because one of the most salient features of the observed universe—its large-scale uniformity—cannot be explained by the standard big bang theory. Instead, the observed large-scale uniformity must be assumed as an initial condition. Since the goal of a physical theory is to explain as much as possible, the standard big bang theory seems to be subject to improvement.

Mass Density of the Universe

A third question left unanswered by the standard big bang theory involves the mass density of the universe. This mass density is usually measured relative to a benchmark, the *critical mass density*, which is defined in terms of the expansion of the universe. If the mass density exceeds the critical density, then the gravitational pull of everything on everything else will be strong enough to eventually halt the expansion. The universe will collapse, resulting in what is sometimes called a *big crunch*. If the mass density is less than the critical density, on the other hand, then the universe will go on expanding forever.

Cosmologists typically describe the mass density of the universe by a ratio designated by the Greek letter Ω (omega), defined as the mass density divided by the critical mass density:

$$\Omega \equiv (\text{mass density}) \div (\text{critical mass density})$$

Omega is very difficult to determine, but its present value is known to lie somewhere between 0.1 and 2.

This may appear to be a broad range, but consideration of the time development of the universe leads us to a very different point of view. The situation in which Ω exactly equals 1 represents an unstable equilibrium point of the evolution of the standard big bang theory; in other words, a universe with Ω exactly equal to 1 resembles the situation of a pencil balancing on end.

The phrase *equilibrium point* implies that if Ω is ever exactly equal to 1, it will remain exactly equal to 1 forever—just as a pencil balanced precisely on end will, in principle, according to the laws of classical physics, remain forever in the vertical position.

The word *unstable* means that any deviation from the equilibrium point, in either direction, will rapidly grow. If the value of Ω in the early universe was just a little bit above 1, it would rapidly rise toward infinity. If Ω in the early universe was just a tiny bit below 1, it would rapidly fall toward zero.

For Ω to be anywhere near 1 today, it must have been extraordinarily close to 1 in earlier times. For example, we can consider the universe at one second after the big bang, when the processes related to big bang nucleosynthesis were beginning to take place. For Ω today to be somewhere in the allowed range, of 0.1 to 2, at one second after the big bang Ω must have been equal to 1 to an accuracy of fifteen decimal places (i.e., between 0.999999999999999 and 1.000000000000001). If we go further, and consider the time of 10^{-35} second after the big bang, the time when thermal energies were typical of the energy scale of grand unified theories, at that time Ω must have been equal to 1 to an accuracy of forty-nine decimal places. (We won't attempt to write this number out!)

In the standard big bang theory there is no explanation whatever for the value of Ω in the early universe being so very nearly equal to 1. Robert Dicke and James Peebles of Princeton University have emphasized the lack of an explanation. At one second after the big bang, Ω could have had any value—except that most possibilities lead to a universe very different from the one in which we live. Like the large-scale homogeneity, the nearly perfect equality of the mass density and the critical density cannot be explained by the standard big bang theory. Instead, the fact that Ω began very nearly equal to 1 must be postulated as part of the initial conditions from which the universe evolved.

Density Perturbations

A fourth question left unanswered by the standard big bang theory involves the origin of the density perturbations that are responsible for the development of the small-scale inhomogeneities: galaxies, clusters of galaxies, and so on. Although the universe is remarkably homogeneous on very large scales, there is, nonetheless, a very complicated structure on smaller scales.

The existence of this structure undoubtedly is related to the gravitational instability of the universe. Any region that contains a higher-than-average mass density will produce a stronger-than-average gravitational field, thereby pulling in even more excess mass. Thus small perturbations are amplified to become large perturbations. Once such perturbations have begun, we understand—more or less—how they evolve, even if the details get rather complicated.

However, for galaxies to evolve, the early universe must have contained primordial density perturbations. The standard big bang model offers no explanation for either the origin or the form of these perturbations. Instead, an entire spectrum of primordial perturbations must be assumed as part of the initial conditions.

Grand Unified Theories

The four questions discussed above—the ratio of protons and neutrons to photons, large-scale homogeneity, the mass density of the universe, and density perturbations—involve some of the most basic and obvious features of the universe. New ideas from elementary particle theory have led cosmologists to new ideas about the behavior of the very early universe, and plausible answers to these four questions have now been proposed. Before discussing these new ideas in cosmology, however, it is necessary to summarize the recent advances that have taken place in elementary particle physics.

Elementary particle physicists use the word *interaction* to refer to any process that elementary particles can undergo, whether it involves scattering, decay, particle annihilation, or particle creation. All of the known interactions of nature are divided into four types. From the weakest to the strongest, the interactions are gravitation, the weak interactions, electromagnetism, and the strong interactions.

- **Gravitation**

  The force of gravity appears to be strong in our everyday lives because it is long-range and universally attractive. Thus we are accustomed to feeling the force that acts between all the particles in the earth and all the particles in our own bodies. The force of gravity acting between two elementary particles, however, is incredibly weak. It is much weaker than any of the other known forces—so weak, in fact, that it has never been detected.

- **Weak Interactions**

  The weak interactions are much stronger than gravity, but they are not noticed in our everyday lives because they have a range that is roughly a hundred times smaller than the size of an atomic nucleus. They are seen primarily in the radioactive decay of many kinds of nuclei, and they are also responsible for the scattering of particles called neutrinos, a type of experiment that is now routinely carried out at high-energy accelerator laboratories.

- **Electromagnetism**

  Electromagnetism includes both electric and magnetic forces, and is responsible for holding the electrons of an atom to the nucleus. Light waves, radio waves, microwaves, and X rays are also electromagnetic phenomena.

- **Strong Interactions**

  The strong interactions, which have a range of about the size of an atomic nucleus, account for the force that binds the protons and neutrons inside a nucleus. They also account for the tremendous energy release of a hydrogen bomb, as well as the interactions of many short-lived particles that are investigated in particle accelerator experiments.

The strong, the weak, and the electromagnetic interactions all appear to be accurately described by theories developed during the early 1970s. The strong interactions are described by quantum chromodynamics (QCD), a theory based on the hypothesis that all strongly interacting particles are composed of constituents called quarks. The theory provides a detailed description of the interactions that bind the quarks into the observed particles, and the residual effect of these quark interactions can account for the observed interactions of the particles. Unfortunately, our ability to extract quantitative predictions from the theory is very limited. QCD theory is very intricate, and at present only some of its consequences can be reliably calculated. Nonetheless, the evidence for the theory is strong enough that most particle physicists are convinced that the theory correctly describes the strong interactions over the full range of available energies.

The weak and electromagnetic interactions are successfully described by the unified electroweak theory, also known as the Glashow-Weinberg-Salam model (named for Sheldon Glashow of Harvard

University, Steven Weinberg of the University of Texas at Austin, and Abdus Salam of the International Center for Theoretical Physics in Trieste, who shared the 1979 Nobel Prize in physics for this work). Standard calculational techniques are very effective in extracting predictions from the unified electroweak theory, owing to the inherent weakness of the interactions being described.

Although a quantum theory of gravity remains to be developed, we nonetheless believe that general relativity provides the correct description of gravity at the level of classical physics—that is, in the approximation that the effects of quantum theory can be ignored. The effects of gravity, however, are noticeable only when the number of elementary particles is very large. The classical approximation is incredibly accurate in these situations, and the theory of general relativity is therefore sufficient to describe all the observed properties of gravity.

Quantum chromodynamics and the unified electroweak theory, taken together, form the standard model of elementary particle physics. Embedded in the standard model are three different types of fundamental interactions, labeled by the symbols $U(1)$, $SU(2)$, and $SU(3)$. (These symbols are actually the names of mathematical symmetry groups that determine the form of the interactions. But for our purposes, the symbols can be taken simply as labels for the three interactions.) The $U(1)$ and $SU(2)$ interactions are the fundamental ingredients of the Glashow-Weinberg-Salam theory, and they combine together in a somewhat complicated way to describe the weak and electromagnetic interactions. The $SU(3)$ label refers to the strong interactions described by quantum chromodynamics. Thus, since the early 1970s, elementary particle physics has been in a state of unprecedented success. The electromagnetic, weak, and strong interactions are successfully described by the standard model of particle physics in terms of three fundamental interactions. It appears that all known physics can be described by the standard model of particle physics, the theory of general relativity, or both.

Grand unified theories—often referred to by their acronym, GUTS—emerged from this atmosphere of enormous success. The first grand unified theory, the minimal $SU(5)$ model, was proposed in 1974 by Glashow and Howard Georgi.

The basic idea of grand unification is that the $U(1)$, $SU(2)$, and $SU(3)$ interactions of the standard model of particle physics are actually components of a single unified force. At first this idea seems impossible, since the strengths of the three types of interactions are very different. The interaction strengths cannot be determined theoretically, but instead must be fixed by experiment. The theory, however, implies that the interaction strengths depend on the energy of the particles that are interacting. Once the strengths of the three interactions are measured at one energy, the theory allows us to calculate the strengths at any other energy. The important feature is that all three interaction strengths appear to meet rather accurately at a single point, at an energy somewhere between 10^{14} and 10^{15} GeV (figure 4). It is this calculation—first carried out by Georgi, Weinberg, and Helen Quinn—that determines the enormous energy scale of the grand unified theories.

According to the grand unified theories, there is really only one interaction, not three. If we were able to do experiments in the energy range of 10^{14} or 10^{15} GeV, we would see clearly that there is only one interaction. At lower energies, however, a mechanism—spontaneous symmetry breaking—causes the one interaction to look as if it were three interactions. Spontaneous symmetry breaking is not new with grand unified theories; it has been used successfully in the Glashow-Weinberg-Salam theory of the electroweak interactions, and similar phenomena are known to occur in condensed matter physics.

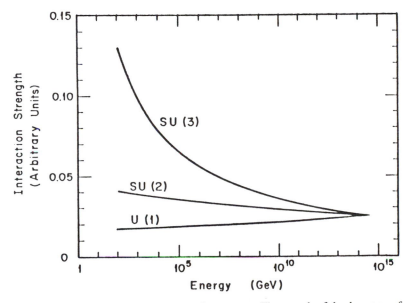

Figure 4. Dependence of interaction strengths on energy. The strengths of the three types of fundamental interactions—U(1), SU(2), and SU(3)—are measured at energies of about 100 GeV. The curves show the strengths of these interactions at higher energies, as calculated from the standard model of particle physics. The three curves meet in a point at an energy between 10^{14} and 10^{15} GeV, suggesting that the three interactions are unified at this energy level.

Although the standard model of particle physics is well established, the same cannot be said for grand unified theories. Even if the idea of grand unification is correct, we certainly do not know which of the many conceivable grand unified theories is likely to be the right one. Nonetheless, grand unified theories are considered highly attractive, for two major reasons:

- First, the grand unified theories are the only known theories that predict that the charges of the electron and the proton will be equal in magnitude. This is impressive. Prior to grand unified theories, nobody had even the fuzziest idea why the two charges should be equal. Each charge could be any number whatever; it was merely an experimental coincidence that the charges were equal to within one part in 10^{20}. Grand unified theories contain a fundamental symmetry that relates the behavior of electrons to the behavior of the quarks that make up the protons, and this symmetry guarantees that the charges of the electron and the proton are equal. Furthermore, if this symmetry was violated by even the smallest amount, then the theory would no longer be mathematically well defined. If the charge of the electron was found to differ in magnitude from the charge of the proton by even one part in 10^{24}, then the grand unified theories would have to be abandoned. If successively more accurate experiments continue to confirm that the two charges are equal, this result will have to be considered as further evidence in favor of grand unification.

- Second, the grand unified theories are considered attractive because the three interactions of the standard model of particle physics arise from a single fundamental interaction. The three curves describing the interaction strengths must all meet at a point. Thus, if any two of the interaction strengths are measured, then the third can be predicted by the criterion that its curve must pass through the point where the first two curves cross. This prediction of the grand unified theories is found to work very well—the experimental result agrees with the prediction to an accuracy of about three percent, well within the estimated experimental uncertainty of the test.

Grand unified theories suffer from one important drawback, the *hierarchy problem*. This problem is largely aesthetic, but is nonetheless taken seriously by the particle physics community. The problem arises from the enormous energy scale of the grand unified theories—10^{14} GeV. This value has to be *put in by hand*, by which we mean that there is no known *a priori* reason why this energy scale is so many orders of magnitude larger than the other energy scales of importance to particle physics.

There is, however, a clear *experimental* reason for believing that the energy scale of unification is very high. According to the dependence of interaction strengths on energy, the scale of unification must be very high to account for the large differences in the strengths of the three interactions observed at the energies of experimental particle physics (figure 4).

To understand the attitude of particle physicists, we must realize that the grand unified theories are not seen as the ultimate fundamental theory of nature.

- First, the ultimate fundamental theory obviously must include gravity, which grand unified theories do not.
- Second, the grand unified theories are considered too inelegant to be serious candidates for the ultimate fundamental theory of nature. In particular, even the simplest of the grand unified theories contains over twenty free parameters (numbers such as the charge of an electron, that must be measured experimentally before the theory can be used to make predictions).

Thus particle theorists expect that some day the correct grand unified theory will be derived as an approximation to the ultimate theory, which will contain few—if any—free parameters. In this context, the energy scale of grand unification will be calculable. When theorists say that the energy scale of grand unified theories is put in by hand, they are really saying that they do not at present see any reason why a future calculation will give such a large number. Advocates of grand unified theories hope that someday the reason will be found.

Baryon Number of the Universe

As we have seen, questions regarding:

- The ratio of photons to protons and neutrons in the universe.
- Large-scale homogeneity.
- The mass density of the universe.
- The origin of density perturbations.

are left totally unanswered by the standard big bang model. Plausible answers are, however, forthcoming from particle physics theories.

An answer to the first question, why the ratio of the number of photons to the number of protons and neutrons is about 10^{10}, comes from grand unified theories.

The idea that particle physics could provide an answer to this question was first suggested by Andrei Sakharov, the Russian scientist known as the father of the Soviet hydrogen bomb and for his courageous efforts in the cause of human rights. More detailed calculations in the context of grand unified theories were carried out by Motohiko Yoshimura of Tohoku University in Japan and by Weinberg in the United States. This was the first application of grand unified theories to cosmology, and the subject remains crucial to our understanding of cosmology in this context.

Particle physicists use the word *baryon* to refer to either a proton or a neutron. More precisely, particle physicists define the *baryon number* of a system by the equation:

baryon number ≡ (number of protons) + (number of neutrons) − (number of antiprotons) − (number of antineutrons) + . . .

The dots at the end of the equation denote the contributions from other particles that are very short-lived and therefore irrelevant to the questions we are now discussing.

It is useful to have a single word to refer to either a proton or a neutron, because in the early universe the two types of particles rapidly interconverted, by processes such as:

proton + electron ↔ neutron + neutrino

The baryon number is left unchanged by the reaction above, since the proton and the neutron each have a baryon number of 1. In fact, all physical processes observed up to now obey the principle of baryon number conservation: the total baryon number of an isolated system cannot be changed. This principle implies, for example, that the proton must be absolutely stable; because it is the lightest baryon, it cannot decay into another particle without changing the total baryon number. Experimentally, the lifetime of the proton is now known to exceed 10^{32} years.

Note, however, that the principle of baryon number conservation does not forbid the production of baryons, provided that equal numbers of antibaryons are also produced. For example, at high energies the reaction:

electron + positron → proton + antiproton

is frequently observed.

To estimate the baryon number of the observed universe, we must ask whether all the distant galaxies are composed of matter, or whether some of the distant galaxies might be formed from antimatter. This question has not been definitively answered, but there is a strong consensus that the universe probably is made entirely of matter. This belief is motivated mainly by the absence of any known mechanism that could have separated matter from antimatter over the large distances that separate galaxies. Assuming that this belief is true, then the total baryon number of the visible universe is about 10^{78}, corresponding to about one baryon per 10^{10} photons.

If the principle of baryon number conservation were absolutely valid, the baryon number of the universe would be unchangeable. Under this assumption, there would be no hope of explaining the baryon number of the universe—it would always have had a value of 10^{78}, a value necessarily fixed by the postulated initial conditions of the universe.

Grand unified theories, however, imply that baryon number is not exactly conserved. At low temperatures, the conservation law is an excellent approximation, and the observed limit on the proton lifetime is consistent with many versions of grand unified theories. At temperatures of 10^{27} K and higher, however, processes that change the baryon number of a system of particles are expected to be quite common.

Thus, when grand unified theories and the big bang picture are combined, the net baryon number of the universe can be altered by baryon nonconserving processes. However, to explain the observed baryon number, it is necessary that the underlying particle physics make a distinction between matter and antimatter. This distinction is essential, since any theory that leads to the production of matter and antimatter with equal probabilities would lead to a total baryon number far smaller than that observed. For many years, it was thought that matter and antimatter behave identically, but a small difference between the two was discovered experimentally in 1964 by Val Fitch of Princeton University and James Cronin of the University of Chicago (who shared the 1980 Nobel Prize in physics for this discovery). The inherent distinction between matter and antimatter has since been incorporated into many particle theories, including the grand unified theories. Thus, in the context of grand unified theories, the observed excess of matter over antimatter can be produced naturally by elementary-particle interactions at temperatures near 10^{27} K.

Finally, then, we come to the crucial question: do grand unified theories give an accurate prediction for the baryon number of the observed universe? Unfortunately, we cannot tell. The grand unified theories depend on too many unknown parameters to allow a quantitative prediction. We can, however, say that the observed baryon number can be obtained with what seems to be a reasonable choice of values for the unknown parameters. Thus grand unified theories provide at least a framework for answering the question of why the ratio of the number of photons to the number of protons and neutrons is about equal to 10^{10} rather than some other number. Sometime in the future, if the correct grand unified theory and the values of its free parameters become known, it will be possible to make a real comparison between theory and observation.

Inflationary Universe

Answers for the three other questions discussed above—large-scale homogeneity, the mass density of the universe, and the origin of density perturbations—all are to be found in a new theory regarding the very early behavior of the universe, a theory called the inflationary model

I first proposed the idea of an inflationary universe in 1981, but the model in its original form did not quite work. It had a crucially important technical flaw, which I pointed out but was unable to remedy in my 1981 paper. A variation on the inflationary model that avoids this flaw was invented independently by Andrei Linde of the Lebedev Physical Institute in Moscow and by Andreas Albrecht and Paul Steinhardt of the University of Pennsylvania. This chapter will discuss the Linde-Albrecht-Steinhardt version of the inflationary model, which is called the new inflationary universe.

Phase Transition

The key ingredient of the inflationary universe model is the assumed occurrence of a phase transition in the very early history of the universe. Grand unified theories imply that such a phase transition occurred when the temperature was about 10^{27} K. This phase transition is linked to the spontaneous symmetry breaking: at temperatures higher than 10^{27} K there is one unified type of interaction, while at temperatures below 10^{27} K the grand unified symmetry is broken and the $U(1)$, $SU(2)$, and $SU(3)$ interactions acquire their separate identities (figure 4).

When the universe cooled down to the temperature of this phase transition, either of two things may have happened: the phase transition may have occurred immediately, or it may have been delayed, occurring only after a large amount of supercooling. The word *supercooling* refers to a situation in which a substance is cooled below the normal temperature of a phase transition without the phase transition taking place. Water, for example, can be supercooled to more than 20 K below its freezing point, and glasses are formed by rapidly supercooling a liquid to a temperature well below its freezing point. If the correct grand unified theory and the values of its parameters were known, there would be no ambiguity about the nature of the phase transition; we would be able to calculate how quickly it would occur. In the absence of this knowledge, however, either of the two possibilities appears plausible. Calculations show, however, that only an extremely narrow range of parameters leads to an intermediate situation; in almost all cases the phase transition is either immediate or strongly delayed.

If the phase transition occurred immediately, its cosmological consequences would be very problematical. A large number of exotic particles called magnetic monopoles would be produced and the mass density of the universe would come to be strongly dominated by these particles. According to most grand unified theories, these monopoles would survive to the present day, leading to predictions that are grossly at odds with observation. Furthermore, in the case of an immediate phase transition, our questions concerning large-scale homogeneity, the mass density of the universe, and the origin of density perturbations would remain unanswered.

False Vacuum

The inflationary universe model is based on the other possibility, that the universe underwent extreme supercooling. The cosmological consequences of this assumption appear to be very attractive.

As the gas that filled the universe supercooled to temperatures far below the temperature of the phase transition, the gas would have approached a very peculiar state of matter known as a *false vacuum*. This state of matter has never been observed. Furthermore, the energy density required to produce it is so enormous—about sixty orders of magnitude larger than the density of the atomic nucleus—that it clearly will not be observed in the foreseeable future. Nonetheless, from a theoretical point of view, the false vacuum seems to be well understood. The essential properties of the false vacuum depend only on the general features of the underlying particle theory and not on any of the details. Even if grand unified theories turn out to be incorrect, it is still quite likely that our theoretical understanding of the false vacuum will remain valid.

The false vacuum has a peculiar property that makes it very different from any ordinary material. For ordinary material—whether gas, liquid, solid, or plasma—the energy density is dominated by the rest energy of the particles of which the material is composed. (The rest energies of the particles are related to their masses by the famous Einstein relation, $\mathbf{E} = \mathbf{mc}^2$.) If the volume of an ordinary material is increased, then the density of particles decreases, and therefore the energy density also decreases. The false vacuum, on the other hand, is the state of lowest possible energy density that can be attained while remaining in the phase for which the grand unified symmetry is unbroken. This energy density is attributed not to particles, but rather to fields called Higgs fields, which are included in the theory to produce the spontaneous symmetry breaking. These fields are some-what analogous to electric and magnetic fields, except that they do not specify a direction in space. Remember that we are assuming that the phase transition occurs very slowly, so for a long time (by the standards of the very early universe) the false vacuum is the state with the least possible energy density that can be attained. Thus, even as the universe expands, the energy density of the false vacuum remains constant.

Gravitational Repulsion

When this peculiar property of the false vacuum is combined with Einstein's equations of general relativity, we encounter a dra-matic result: the false vacuum leads to a gravitational repulsion. Throughout the rest of the history of the universe, gravity has acted to slow down the cosmic expansion. But when the universe was caught in the false vacuum state, gravity caused the expansion to accelerate. The form of this repulsion is identical to the effect of Einstein's cosmological constant, except that the repulsion caused by the false vacuum operates for only a limited period of time.

The gravitational repulsion would have produced a very rapid expansion, far in excess of the expansion of the standard big bang model. In the inflationary model essentially all the momentum of the big bang was produced by the gravitational repulsion. (In the stan-dard big bang theory, in contrast, all the momentum of the big bang is incorporated into the postulated initial conditions.) The universe would double in size in about 10^{-34} second, and would continue to double in size during each successive interval of 10^{-34} second, for as long as the universe remained in the false vacuum state. During this period the universe expanded, or inflated, by a stupendous factor. A factor of at least 10^{75} (in volume) is necessary to answer the cosmo-logical questions (large-scale homogeneity, the mass density of the universe, and the origin of density perturbations), but the actual infla-tion factor depends on highly uncertain details of the underlying particle theory, and could well have been many orders of magnitude larger.

Eventually the phase transition would have occurred. When it did, the energy density of the false vacuum would have been released.

(In the language of thermodynamics, this energy is the *latent heat* of the phase transition.) This input of energy would have produced a vast number of particles, and would have reheated the universe back to a temperature comparable to the temperature of the phase transition, about 10^{27} K. (The precise number is actually about one-half or one-third of the temperature of the phase transition—but such a relatively minute correction is inconsequential in order-of-magnitude estimates.) The baryon-number-producing processes would have taken place during and just after the reheating, as any baryon number density present before inflation would have been diluted to a negligible value by the enormous expansion. At the end of the phase transition, the universe would have been uniformly filled with a hot gas of particles, exactly as postulated in the initial conditions for the standard big bang theory. Here the inflationary model merges with the standard big bang theory; the two models agree in their description of the evolution of the universe from this time onward (figure 5).

Energy Conservation

In the inflationary model virtually all the matter and energy in the universe were produced during the inflationary process. This seems strange, because it sounds like an unmistakable violation of the principle of energy conservation. How could it be possible that all the energy in the universe was produced as the system evolved?

The inflationary universe model is consistent with all the known laws of physics, including the conservation of energy. The loophole in the conservation-of-energy argument is associated with the peculiar nature of gravitational energy. Using either general relativity or Newtonian gravity, we find that *negative* energy is stored in the gravitational field.

A simple way to make this fact seem at least plausible is to imagine two large masses, separated by a very large distance in an otherwise empty space. Now imagine bringing the two masses together. The masses will attract each other gravitationally, which means that energy can be extracted as the masses come together. We can imagine, for example, attaching the masses to fixed pulleys, and the wheels of the pulleys driving an electrical generator. Once the two masses are brought together, however, their gravitational fields will be superimposed, producing a much stronger gravitational field.

The net result of such a process is the extraction of energy and the production of a stronger gravitational field. If energy is conserved, then the energy in the gravitational field apparently goes down when its strength goes up. If the absence of a gravitational field corresponds to no energy, then a non-zero field strength must corre-

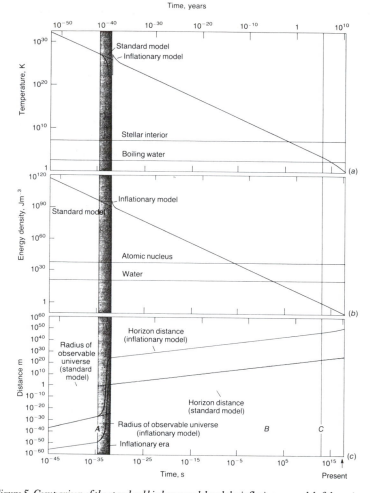

Figure 5. Comparison of the standard big bang model and the inflationary model of the universe. The time after the big bang is shown in seconds (bottom) and years (top) on the horizontal axis. Temperature and energy density (vertical axis, top two sections) are the same for the big bang and inflationary models except during a very brief period around 10^{-35} second (the vertical dark band). A key feature of the inflationary model is the prolongation of the phase transition (called the inflationary era, and indicated by the vertical dark band and the letter A near the bottom left of the graph). During this inflationary era, the universe expands by an extraordinary factor, shown here on the vertical axis (bottom section) as 10^{50}, but it could be much larger. While the universe is inflating, the temperature is plunging (note in the top section of the graph the dip within the vertical dark band), until it is stabilized at about 10^{22} K by quantum effects that arise in the context of general relativity. Near the bottom of the graph and further to the right along the horizontal time axis, B indicates the period when the lightest atomic nuclei were synthesized and C indicates the time when the universe became transparent to electromagnetic radiation. On the bottom section of the graph (with spatial dimensions on the vertical axis), the big bang and inflationary universe models are each represented by two curves, one showing the region of space that evolves to become the observed universe and the other showing the horizon distance: the total distance a light signal could have traveled since the beginning of the universe.

spond to a negative energy. Gravitational energy is usually negligible under laboratory conditions, but cosmologically it can be significant.

The energy stored in the false vacuum became larger and larger as the universe inflated. Then it was released, when the phase transition took place at the end of the inflationary period. At the same time, however, the energy stored in the cosmic gravitational field—the field by which everything in the universe is attracting everything else—became more and more negative. The total energy of the system was conserved; it stayed at a value at or near zero.

Thus inflation allows the entire observed universe to develop from almost nothing. The inflationary process could have started with an amount of energy equivalent to only about ten kilograms of matter. And even this small amount of energy could conceivably have been balanced by an equal contribution of negative energy in the gravitational field. Thus, if the inflationary model is correct, it is accurate to say that the universe is the ultimate free lunch.

Negative Pressure and Gravitational Repulsion

The mechanism that drives the accelerated expansion of the inflationary model of the universe is gravitational repulsion. It can be explained, though only crudely, without resorting to the formalism of general relativity.

The false vacuum is the state of lowest possible energy density that can be attained while remaining in the phase for which the grand unified symmetry is unbroken. Thus, as a region of false vacuum expands, the energy density remains constant. The constancy of the energy density is related to a very peculiar property: the false vacuum has a pressure that is large and negative.

To understand the connection between a constant energy density and a negative pressure, consider the fact that when a normal positive-pressure gas is allowed to expand, it will push on its surroundings and thereby lose energy to its surroundings. Both steam and gasoline engines operate on this principle. For the false vacuum, however, the situation is reversed. We can imagine a region of false vacuum that expands, but the expansion occurs at a constant energy density. The energy of the false vacuum region therefore increases as the volume increases, which means that the region is taking energy from its surroundings. Thus the region must create a negative pressure, or suction, so that energy is being supplied by whatever force is causing the expansion. By considering the energy balance involved in the expansion of a region of false vacuum, it is possible to determine the pressure uniquely. The pressure is equal to the negative of the energy density (when the pressure and the energy density are measured in the same units).

According to Newton's theory of gravity, a gravitational field is produced by a mass density. In a relativistic theory, the mass density can be related to a corresponding energy density by the relationship $E = mc^2$. According to Einstein's theory of general relativity, however, a pressure can also produce a gravitational field. When Einstein's equations are used to describe a homogeneously expanding universe, they show that the rate at which the expansion is slowed down is proportional to the energy density plus three times the pressure. Under ordinary circumstances, the pressure term is a small relativistic correction, but for the false vacuum the pressure term overwhelms the energy-density term and has the opposite sign. So the bizarre notion of negative pressure leads to the even-more-bizarre effect of a gravitational force that is effectively repulsive.

Remaining Questions

We still must deal with issues raised by large-scale homogeneity, the mass density of the universe, and the origin of density perturbations.

Large-Scale Homogeneity

In the standard big bang theory, the large-scale homogeneity cannot be explained because the universe did not have enough time to come to a uniform temperature. Consider, now, the evolution of the observed region of the universe, which has a radius today of about ten billion light-years. Imagine following this region backward in time, using the inflationary model. Follow the region back to the instant immediately before the inflationary period. Since the inflationary theory predicts a tremendous spurt of expansion during the inflationary period, we infer that our observed region of the universe was incredibly small before the expansion began. In fact, the region was more than a billion times smaller than the size of a proton. (Note that the universe *as a whole* was not necessarily small. The inflationary model makes no statement about the size of the universe *as a whole*. The universe might, in fact, be infinite.)

While our observed region of the universe was very small, there was plenty of time for it to have come to a uniform temperature. So in the inflationary model the uniform temperature was established before inflation took place, and in a *very, very small* region. The process of inflation then stretched this very small region, making it large enough to encompass the entire observed universe. Thus the sources of the microwave background radiation arriving at the earth today from all directions in the sky were once in close contact. While in close contact, they had time to reach a common temperature, before the inflationary era began.

Mass Density

The inflationary model also provides a simple resolution of the problem concerning the mass density. Recall that the ratio of the actual mass density to the critical density is called Ω (omega) and that the problem arose because the condition $\Omega = 1$ is unstable. Ω is always driven away from 1 as the universe evolves, making it difficult to understand how the value of Ω today can be anywhere near 1.

During the inflationary era, however, the peculiar nature of the false vacuum state results in some important sign changes in the equations that describe the evolution of the universe. During this period, as we have seen, the force of gravity acts to accelerate the expansion of the universe rather than to retard the expansion. It turns out that the equation governing the evolution of Ω also has a crucial change of sign, with the result that during the inflationary period the universe is driven very quickly and very powerfully *toward* a critical mass density.

So a very short period of inflation can drive the value of Ω very precisely to 1, no matter what the starting value of Ω was. Thus there is no longer any need to assume that the initial value of Ω was incredibly close to 1.

Furthermore, an interesting prediction comes out of this analysis. The mechanism that drives Ω to 1 almost always overshoots. This means that even today the mass density should be equal to the critical mass density to a high degree of accuracy. More precisely, the inflationary model predicts that the value of Ω today should equal 1 to an accuracy of about one part in ten thousand. (Deviations from 1 are caused by quantum effects, which we will discuss shortly.) Thus the determination of the mass density of the universe will be an important test of the inflationary model.

Unfortunately, it is very difficult to estimate reliably the mass density of the universe. Part of the problem in estimating the mass density lies in the fact that most of the mass in the universe is in the form of *dark matter*, matter that is totally unobserved except for its gravitational effects on other forms of matter. Since we do not even know what the dark matter is, it is difficult to estimate how much of it exists. Most of the current estimates, I must admit, give values that are distinctly below 1. Values ranging between 0.1 and 0.3 are the most common. These estimates, however, are highly uncertain, and there appears to be no compelling observational evidence to rule out the possibility that $\Omega = 1$. Furthermore, recent observations of large-scale motions of galaxies are difficult to explain unless Ω has a value near one.

Density Perturbations

Finally, we come the last of our four questions, concerning the origin of the primordial density perturbations in the universe.

The generation of density perturbations in the new inflationary universe was addressed in the summer of 1982 at the Nuffield Workshop on the Very Early Universe, held at Cambridge University. Several theorists were working on this problem, including James Bardeen of the University of Washington, Stephen Hawking of Cambridge University, So-Young Pi of Boston University, Paul Steinhardt of the University of Pennsylvania, Michael Turner of the University of Chicago, A.A. Starobinsky of the Landau Institute of Theoretical Physics in Moscow, and myself. We found that the new inflationary model, unlike any previous cosmological model, leads to a definite prediction for the spectrum of perturbations.

Basically, the process of inflation first smooths out any primordial inhomogeneities that may have been present in the initial conditions. For example, any particles that may have been present before inflation are diluted to a negligible density. In addition, the primordial universe may have contained inhomogeneities in the gravitational field, which is described in general relativity in terms of bends and folds in the structure of space-time. Inflation, however, stretches these bends and folds until they become imperceptible, just as the curvature of the surface of the earth is imperceptible in our everyday lives.

For a while, we were worried that inflation would give a totally smooth universe, which would obviously be incompatible with observation. It was pointed out, however, first by Hawking, that the situation might be saved by the application of quantum theory.

An important property of quantum physics is that nothing is determined exactly—everything is probabilistic. Physicists are, of course, accustomed to the idea that quantum theory, with its probabilistic predictions, is essential to describe phenomena on the scale of atoms and molecules. On the scale of galaxies or clusters of galaxies, on the other hand, there is usually no need to consider the effects of quantum theory. But inflationary cosmology implies that for a short period the scales of distance increased very rapidly with time. Thus the quantum effects that occurred on very small particle-physics-length scales were later stretched to the scales of galaxies and clusters of galaxies by the process of inflation.

Therefore, even though inflation would predict a completely uniform mass density by the rules of classical physics, the inherent probabilistic nature of quantum theory gives rise to small perturbations in the otherwise uniform mass density. The spectrum of these perturbations was first calculated during the exciting three-week period of the Nuffield workshop. After much disagreement and discussion, the various working groups came to an agreement on the answer.

First, we calculated the shape of the spectrum of the perturbations. The concept of a spectrum of density perturbations may seem a bit foreign, but the analogy of sound waves is a good one. People familiar with acoustics understand that no matter how complicated a sound wave is, it is always possible to break it up into components that each have a standard wave form and a well-defined wavelength. The spectrum of the sound wave is then specified by the strength of each of these components. In discussing density perturbations in the universe, it is similarly useful to define a spectrum by breaking up the perturbations into components of well-defined wavelength.

For the inflationary model, we found that the predicted shape for the spectrum of density perturbations is essentially scale-invariant; that is, the magnitude of the perturbations is approximately equal on all scale lengths of astrophysical significance. Although the precise shape of the spectrum depends on the details of the underlying grand unified theory, the approximate scale-invariance holds in almost all cases. It turns out that a scale-invariant spectrum was proposed in the early 1970s as a phenomenological model for galaxy formation by Edward Harrison of the University of Massachusetts at Amherst and by Yakov Zel'dovich of the Institute of Physical Problems in Moscow, working independently.

Unfortunately, there is still no way of inferring the precise form of the primordial spectrum from observations, since we cannot reliably calculate how the universe evolved from the early period to the present. Such a calculation is very difficult in any case, because it is further complicated by uncertainties about the nature of the dark matter. Nonetheless, the scale-invariant spectrum appears to be at least approximately what is needed to explain the evolution of galaxies. Thus this prediction from the inflationary model appears to be successful, at least so far. Galaxy formation is currently a very active subject of research, and a better determination of the spectrum of primordial density perturbations may be developed. Such a result will provide an additional test of the inflationary universe model. The density perturbations can also be probed by observing the very small nonuniformities in the cosmic background radiation. These nonuniformities were detected for the first time in 1992 by the Cosmic Background Explorer satellite, and the observed spectrum was in agreement with the predictions of inflation.

During the Nuffield workshop we also calculated the predicted magnitude of the density perturbations. But the implications of these results were much less clear. We found that the predicted magnitude, unlike the shape of the spectrum, is very sensitive to the details of the underlying particle theory. At that time, the minimal *SU(5)* theory, the first and simplest of the grand unified theories, was strongly favored by nearly all of us. We were, therefore, disappointed when we found that the minimal *SU(5)* theory led to density perturbations with a magnitude 100,000 times larger than that required for the evolution of galaxies. This result constituted a serious incompatibility with the inflationary universe model.

The credibility of the minimal *SU(5)* grand unified theory has since diminished. The theory makes a rather definite prediction for the lifetime of a proton, and a variety of experiments have been set up to test this prediction, by looking for proton decay. So far, no such decays have been observed. The experiments have pushed the limit on the proton lifetime to the point where the minimal *SU(5)* theory is now excluded.

With the exclusion of the minimal *SU(5)* theory, a wide range of grand unified theories becomes plausible. All of the allowed theories seem a bit complicated, so apparently we will need some kind of new understanding to decide which, if any, of the theories is correct.

A variety of grand unified theories that predict an acceptable magnitude for both the proton lifetime and the density perturbations have been constructed. Thus, although the inflationary model cannot be credited with correctly predicting the magnitude of the perturbations, it also cannot be criticized for making a wrong prediction. This situation is similar to the calculation of the net baryon number of the universe: the inflationary model provides at least a framework for calculating the magnitude of the density perturbations. If sometime in the future the correct grand unified theory and the values of its free parameters somehow become known, it will then be possible to make a real theoretical prediction for the magnitude of the perturbations.

A common feature of the known models that lead to acceptable density perturbations is the abandonment of the idea that inflation can be driven by the Higgs fields that break the grand unified symmetry. It appears that any Higgs field that interacts strongly enough to break the grand unified symmetry leads to density perturbations with a magnitude that is far too large. Thus we must assume that the underlying particle theory contains a new field—a field that strongly resembles the Higgs fields in its properties but that interacts much more weakly than the Higgs fields.

Unfortunately, all of the known theories that give acceptable predictions for the magnitude of the density perturbations look a little contrived. In fact, the theories were contrived—with the goal of getting the density perturbations to come out right. The need for this contrivance can certainly be used as an argument against the inflationary model. In my opinion, however, this argument is considerably weaker than the arguments in favor of inflation. Even if we ignore cosmology, any grand unified theory that is consistent with the known properties of particle physics appears to be rather contrived. Clearly there are some fundamental principles at work here that we do not yet understand.

Allusions to fundamental principles beyond the grand unified theories are not based on idle speculation—they are based on active and energetic speculation. Since the invention of grand unified theories in 1974, particle theorists have been vigorously working on attempts to construct the ultimate theory of nature—an elegant theory that will include a quantum description of gravity. The characteristic energy scale of such a theory is presumably the Planck scale, 10^{19} GeV, the scale at which the gravitational interactions of elementary particles become comparable in strength to the other types of interactions. We then hope that a grand unified theory will emerge as a low-energy approximation.

The latest and most successful of these attempts is a radically new kind of particle theory, known as a *superstring theory*. Superstring theories represent a dramatic departure from conventional theories in that particles are viewed as ultramicroscopic strings (length $\approx 10^{-33}$ centimeter). Furthermore, according to the theory, the universe has nine spatial dimensions. Early in the history of the universe, when the temperature cooled below 10^{32} K, all spatial dimensions but the three we know today stopped expanding and remained curled up with an unobservably small radius. As bizarre as this theory may sound, it has been shown to possess a number of unique properties crucial to a quantum theory of gravity, and it has totally captured the attention of a large fraction of the worldwide particle theory community.

So far we know very little about the behavior of superstring theories at energies well below the Planck scale. Nevertheless, encouraging progress is under way toward embedding the idea of grand unification into a larger framework. Superstring theories are highly constrained, which leads us to hope that someday we may be able to make rather definite predictions concerning physics at the energy scale of grand unified theories and beyond. If we are successful, the calculation of the predicted spectrum of density perturbations will provide a rigorous test of the inflationary cosmology model.

Cosmic Strings

Quantum effects during the inflationary era are not the only source of primordial density perturbations that particle physics can provide. There is also the possibility that the seeds for galaxy formation may have been objects called *cosmic strings* (not related to superstrings). [See chapter 24, *Cosmic Strings*.] Cosmic strings are predicted to exist by some, but not by all, grand unified theories. The strings would form in a random pattern during the grand-unified-theory phase transition. As their name suggests, cosmic strings are very thin, spaghetti-like objects. They can form infinite curves or closed loops of astrophysical size. A cosmic string has a thickness of about 10^{-29} centimeter and a mass of about 10^{22} grams for each centimeter of length. (In astronomical terms, the mass is about 10^7 solar masses per light-year.) In most theories, the density of the strings would be diluted to negligibility by the process of inflation. It is possible, however, to construct theories in which the cosmic strings survive by forming either after the inflation or at the very end of the inflation. Cosmic strings are an active topic of current research, and it is beginning to appear that a number of features of galactic structure can be explained naturally in terms of cosmic strings. Models of this type still make use of inflation to answer the questions concerning the large-scale homogeneity of the universe and the mass density, and also to smooth out any small-scale inhomogeneities that may have been present in the initial conditions.

Conclusion

The inflationary universe model has been highly successful in describing the broad, qualitative properties of the universe. In particular, the model provides attractive answers to four questions left unanswered by the standard big bang model—the ratio of protons and neutrons to photons, large-scale homogeneity, the mass density of the universe, and density perturbations. Although the inflationary model is speculative, some of us nonetheless feel that in its broad outline the inflationary universe model is essentially correct.

The inflationary model makes two observationally testable predictions: the mass density of the universe and the shape of the spectrum of primordial density perturbations. Although neither of these predictions is straightforward to check, it seems likely that significant progress will be made in the foreseeable future.

Even if the inflationary model is correct, however, it must still be emphasized that the model it not a completed project. The inflationary model is not a detailed theory, but is just an outline for a theory. Michael Turner calls it the *inflationary paradigm*. To fill in the details, we will need to know much more about the details of particle physics at the energy scales of grand unified theories, and perhaps beyond.

It appears to me that the fields of particle physics and cosmology will be closely linked for years to come, as physicists continue their efforts to understand the fabric of space, the structure of matter, and the origin of it all.

—Alan H. Guth

BIBLIOGRAPHIC NOTE

This article was adapted from A.H. Guth, "The Birth of the Cosmos," in D.E. Osterbrock and P.H. Raven, eds., *Origins and Extinctions* (New Haven: Yale University Press, 1988, pp.1–41). Some general discussions of cosmology and the inflationary universe include L.F. Abbott and S.Y. Pi, *Inflationary Cosmology* (Singapore: World Scientific, 1986); J. Bernstein and G. Feinberg, *Cosmological Constants: Papers in Modern Cosmology* (New York: Columbia University Press, 1987); S.K. Blau and A.H. Guth, "Inflationary Cosmology," in S.W. Hawking and W. Israel, eds., *300 Years of Gravitation* (Cambridge: Cambridge University Press, 1987); R.H. Dicke and P.J.E. Peebles, "The Big Bang Cosmology— Enigmas and Nostrums," in S.W. Hawking and W. Israel, eds., *General Relativity: An Einstein Centenary Survey* (Cambridge: Cambridge University Press, 1979); J. Gribbon, *In Search of the Big Bang: Quantum Physics and Cosmology* (London: Heinemann, 1986); A.H. Guth and P.J. Steinhardt, "The Inflationary Universe," in P. Davies, ed., *The New Physics* (Cambridge: Cambridge University Press, 1987); D.N. Schramm, "The Early Universe and High-Energy Physics," *Physics Today* (April 1983), 27–33; J. Silk, *The Big Bang: The Creation and Evolution of the Universe* (New York: W.H. Freeman, 1980); P.J. Steinhardt, "Inflationary Cosmology," in M.J. Bowick and F. Gürsey, eds., *High Energy Physics* (Singapore: World Scientific, 1986); M.S. Turner, "Cosmology and Particle Physics," in P. Ramond and R. Stora, eds., *Architecture of Fundamental Interactions at Short Distances* (Amsterdam: North Holland, 1987); and S. Weinberg, *The First Three Minutes: A Modern View of the Origin of the Universe* (New York: Basic Books, 1977).

23. LARGE-SCALE STRUCTURE AND GALAXY FORMATION

The most exciting and most active problem in cosmology today is that of structure formation. This is the central arena in which the juggernaut of early-universe theories inspired by particle physics collides with a growing mass of real observational data. Because of a wealth of new data, modern cosmology has changed from a branch of mathematics or philosophy into a true, experimentally testable science. The collision of ideas and data is reflected in numerous headlines in popular media, many of them misleading and even inaccurate.

Recent observations and experiments have created a remarkable amount of confidence in the general theory of the basic hot big bang universe; hereafter any reasonable model for structure formation in the universe must operate in the big bang framework. Contrary to some recent headlines, recent work is not challenging big bang theory; rather, specific models of structure formation are being tested and challenged. These models have difficulties; by necessity, though, they operate within the big bang framework, just as earthquake predictors have difficulties but the difficulties do not lead us to cease operating in the framework of a round earth.

Any structure formation model must have certain generic features:

- Matter.
- And seeds, to clump the matter.

The bulk of the matter is dark (nonshining). Some of the dark matter must be nonshining ordinary matter—in, say, brown dwarfs of some other low-luminosity form. The bulk of the dark matter, however, probably exists in some new exotic form, such as low-mass neutrinos, "axions," or super-symmetric "neutralinos." The seeds can be either small, random-density fluctuations, or they might be something more exotic—like cosmic "strings," "walls," or "textures." [See the following chapter on *Cosmic Strings*.]

Observations and experiments are beginning to test various hypothetical combinations of matter and seeds. Different combinations predict different patterns for the resultant structure of the universe and different levels and distributions for residual fluctuations in the cosmic microwave background radiation. We'll examine the current status of different theories, determining which combinations of matter and seeds have been eliminated as possible representations of the observed universe and which still look promising. We'll conclude with a discussion of future observations and experiments that might be able to resolve the theoretical problems and help us to converge on a model for how structure forms in the universe.

Establishing the Hot Big Bang

Edwin Hubble established the concept of an expanding universe in the 1920s and 1930s [see chapter 19, *Hubble's Cosmology*, in section VI, *The Expanding Universe*], work that was followed by the development of big bang cosmology [see chapter 20, *Big Bang Cosmology*]. This, in turn, led to modern physical cosmology and the hot big bang [see the preceding chapter on *The Inflationary Universe*]. Our modern theoretical understanding of the universe naturally focuses on two key quantitative tests:

- The cosmic microwave background radiation.
- Big bang nucleosynthesis and the light-element abundances.

The magnificent agreement of the 1990 Cosmic Background Explorer (COBE) satellite measurements, which were summerized recently by J.L. Mather and his colleagues with a perfect 2.73 K blackbody radiation spectrum has been well discussed in the press. We should remember that this spectral shape is exactly what the hot big bang predicts; no other theory naturally yields such a precise blackbody shape with only one free parameter, T, the temperature.

A second precise test of the standard hot big bang model is the consistency of light-element abundance measurements, which were summarized recently by T. Walker, G. Steigman, K. Olive, H. Kang, and me in 1991, and also the recent accelerator measurements of the number of neutrino species using predictions of nucleosynthesis calculations from the big bang model. Figure 1 shows the abundances produced in the standard calculation as a function of the fraction of the critical density (the density required to make the universe's gravitational binding energy equal to its expansion kinetic energy) in baryons, Ω_b, (*baryon* is the generic term for neutrons and protons, the normal matter out of which we are made). The vertical band in figure

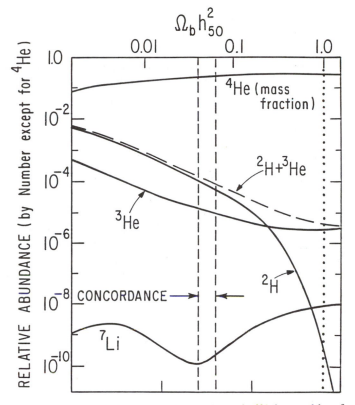

Figure 1. The abundances of light isotopes produced in the standard big bang model as a function of the fraction of the critical density in baryons, Ω_b. Note that the agreement with all abundance determinations is for $\Omega_b \sim 0.06$.

1 represents the allowed values that are simultaneously consistent with the observed light-element abundances of ^{4}He, ^{2}H, ^{3}He, and ^{7}Li (helium of atomic weights 3 and 4, hydrogen, and lithium) extrapolated to their primordial values unassociated with any heavier elements. Since ^{2}H cannot be produced significantly in any non-cosmological process—but only destroyed—the present abundance of ^{2}H puts an upper limit on the baryon density. Conversely, ^{3}He is made in stars, and since the bulk of the excess cosmological ^{2}H over the present value burns to ^{3}He in stars, the sum of ^{2}H and ^{3}He provides a lower bound on the baryon density. The allowed range of baryon density consistent with these bounds requires ^{7}Li to be at the minimum in its production curve (as shown in figure 1). Measurements by M. and F. Spite, reported in 1981 and subsequently verified by others, giving a ratio for ^{7}Li/H $\sim 10^{-10}$ in primitive, Population II stars further substantiates the standard hot big bang model. The light elements with

abundances ranging from approximately 24% to one part in 10^{10} all fit with the cosmological predictions, with the one adjustable parameter giving the baryon density $\Omega_b \sim 0.06$.

Recent attempts to find alternatives to this conclusion by introducing variations in the assumptions have ended up (once the models are treated in detail, as was done by H. Kurki-Suonio, R. Matzner, Olive and me) reaching essentially the same constraint on Ω_b as in the standard model. Thus the conclusions have proven remarkably robust.

Added to the impressive agreement of the abundances has been the measurement of the number of neutrino families, $N^v = 2.98 \pm 0.06$ using high-energy colliders. Nucleosynthesis arguments developed in the 1970s by Steigman, Jim Gunn, and me show that the cosmological ^{4}He abundance is quantitatively related to N_v. The current parameter values yield the cosmological prediction $N_v \leq 3.3$, specifically ruling out any light neutrinos beyond e, μ, and τ, and consistent with the collider measurements. This experimental particle physics test of the cosmological model is a "first" and effectively "consummates the marriage" of particle physics and cosmology. It also gives us even further confidence that we understand cosmological nucleosynthesis and thus know the cosmological baryon density, as well as giving us confidence in the basic hot big bang model of the universe.

Dark Matter Requirements

The narrow range in baryon density for which concordance occurs is interesting. Note that the constraint on Ω_b means that the universe *cannot be closed with normal matter*. If the universe is truly at its critical density, nonbaryonic matter is required.

The arguments requiring some sort of dark matter fall into separate and possibly distinct areas. The visible matter in the universe (stars) yields but a small fraction of the critical density, only about 0.007. This can be compared with the implied densities using Newtonian mechanics applied to various astronomical systems. These arguments (flat rotation curves, dynamics of binary galaxies, etc.) are summarized in figure 2. They reliably demonstrate that galactic halos seem to have a mass ~ 10 times the visible mass.

Note, however, that big bang nucleosynthesis requires that the bulk of the baryons in the universe be dark, since $\Omega_{vis} \ll \Omega_b$ and $\Omega_b \sim \Omega_{halo}$. Thus the dark halos could, in principle, be baryonic (and if they are not, there is an interesting coincidence between Ω_b and Ω_{halo}). However, when similar dynamical arguments are applied to larger systems, such as clusters of galaxies, the implied Ω rises to about 0.2. This same value of Ω can also be obtained from gravitational lensing

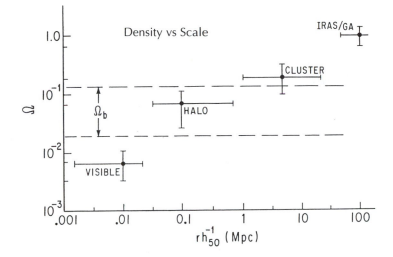

Figure 2. The inferred density in units of the critical density as a function of the scale on which it is "measured." Note the increase in Ω toward unity as larger scales are probed. Note also that Ω_b agrees with densities on the scale of galactic halos and is greater than the amount of visible matter.

of distant quasars and galaxies by intervening clusters of galaxies. Although the uncertainties might marginally allow an overlap between Ω_b and $\Omega_{cluster}$ at ~ 0.1, the central values are already hinting that, on the scales of clusters of galaxies (about 1 to 10 Mpc), there appears to be more than baryonic matter.

A new and dramatic development on even larger scales than clusters now suggests that on these very large scales (50 to 100 Mpc) the density approaches the critical value (Ω ~ 1). This new development uses the combined velocity and distance estimates for galaxies out to and slightly beyond the so-called "Great Attractor," discovered in the late 1980s by a group of astronomers who called themselves the Seven Samurai (D. Burstein, R. Davies, A. Dressler, S. Faber, D. Lynden-Bell, R. Terlevich, and G. Wegner). They determined the so-called peculiar velocities for galaxies out to about 100 Mpc by estimating the distance and then using it to determine the cosmological expansion velocity. The difference between a galaxy's actual velocity as determined by the redshift and the inferred expansion velocity is known as the "peculiar velocity." From an analysis of peculiar velocities, it became apparent that there is a large flow of galaxies (including our local group) toward something. The mysterious something is now called the Great Attractor. Recently, this flow has been mapped out in considerable detail using redshifts measured for the catalog of galaxies found by the Infrared Astronomy Satellite (IRAS). This data has been analyzed by teams from the Massachusetts Institute of Technology, Israel, Toronto, England, Stony Brook, Berkeley, and Fermilab. The universal conclusion, at least so far, is that the observed dynamics on this scale require that $\Omega = 1 \pm 0.3$. This result, if true, forces upon us a necessity for some sort of nonbaryonic dark matter.

Of course, theoretical cosmologists have long assumed that Ω is unity, so these recent (and still preliminary) results may prove to be a confirmation of this theoretical assumption. The theoretical argument is essentially that the only long-lived natural value for Ω is unity and that inflation (or something like it) provided the early universe with the mechanism to achieve this value and thereby solve the so-called flatness and smoothness problems. (The flatness problem is simply the fact that universes with Ω not equal to unity will rapidly either collapse to an infinite-density "big crunch" or expand to a zero-density "big chill" in a very short time. The $\Omega = 1$ solution corresponds to a flat Euclidean space-time. The smoothness or horizon problem is the uniformity of the microwave background radiation on scales that are farther apart than the distance light could have traveled in the age of the universe.)

Before turning to exotic nonbaryonic matter, we should note that some baryonic dark matter must exist, since the lower bound from big bang nucleosynthesis is greater than the upper limits on the amount of visible matter in the universe. We do not know what form this baryonic dark matter is in. It could be either in condensed objects in the halo, such as brown dwarfs and jupiters (objects ≤ 0.08 solar masses, so they are not bright shining stars), or in black holes (which at the time of nucleosynthesis would have been baryons). Or, if the baryonic dark matter is not in the halo, it could be in hot intergalactic gas, hot enough not to show absorption lines but not so hot as to be seen in X-rays. Evidence for some hot gas is found in clusters of galaxies. However, it is not clear how fairly these rich clusters sample the universe. Another possible hiding place for dark baryons would be failed galaxies, large clumps of baryons that condensed gravitationally but did not produce stars.

The more exotic nonbaryonic dark matter can be divided into two major categories for cosmological purposes: hot dark matter and cold dark matter. The former is moving near the speed of light until just before the epoch of galaxy formation; the best example consists of low-mass neutrinos with $m_\nu c^2 \sim 25\text{eV}$. The latter is moving slowly at the epoch of galaxy formation. Because cold dark matter is moving slowly, it can clump on very small scales; hot dark matter tends to have more difficulty being confined on small scales. Examples of cold dark matter could be massive neutrino-like particles with masses greater than several times the mass of a proton, or the lightest super-symmetric particle that is presumed to be stable and might also have a mass of several GeV. Following Michael Turner, all such weakly interacting massive particles are given the acronym WIMPS and, in the case of the super-symmetric candidates, they are also referred to as "neutralinos," or INOS, for short. Axions are very light but would also be moving very slowly and, thus, would clump on small scales. Or, for cold dark matter, there are nonelementary-particle candidates, such as planetary-mass black holes or "nuggets" of strange quark matter. Note that cold dark matter would clump in halos, thus requiring the dark baryonic matter to be out between galaxies, whereas

---

Table 1. Matter

Baryonic ($\Omega_b \sim 0.06$)

    Visible
        $\Omega_{vis} \lesssim 0.01$
    Dark
        Halo
            Jupiters
            Brown dwarfs
            Stellar black holes
        Intergalactic
            Hot gas at $T \sim 10^5$K
            Stillborn Galaxies

NonBaryonic ($\Omega_{nb} \sim 0.94$)

    Hot
        $m_{v_\tau} 25$eV
    Cold
        WIMPS/Inos ~ 100GeV
        Axions $\sim 10^{-5}$eV
        Planetary mass black holes

---

hot dark matter would allow baryonic halos. (It is also possible to have hybrid models with both hot and cold dark matter.) Table 1 summarizes the various dark matter candidates, both baryonic and nonbaryonic.

    A few years ago, the favorite dark matter candidate was probably a WIMP with a mass of a few GeV. The lack of discovery of any new particles in the high-energy collider experiments now means that the only massive particles that could serve as cold dark matter must have masses greater than about 20 GeV and interactions weaker than that of a neutrino. While discussing dark matter candidates, it is worth noting that recent hints from new solar neutrino observations suggest that neutrinos may indeed have small masses, as J. Bahcall has argued. Although the mass directly implied is too small to yield an Ω of unity, reasonable "see-saw" scaling of the results to the less constrained τ neutrino would put its mass in the range where it could yield an Ω of unity. This has created a renewed interest in hot dark matter models.

Seeds for Making Structure

In addition to matter, all models for making galaxies and larger structures require some sort of "seeds" to stimulate the matter to clump. The seeds can be divided into two generic categories:

- Random density fluctuations.
- Topological defects (cosmic strings, walls, textures, etc.).

Both random density fluctuations and topological defects are assumed to be generated by some sort of vacuum phase transition in the early universe. A familiar phase transition is water freezing to ice; the little white lines in an ice cube are equivalent to topological defects generated at the transition. Small quantum fluctuations in the position of water molecules in the ice crystals would be the random fluctuations. For the universe, the medium undergoing the phase transition is the vacuum itself. Proposed transitions are associated with the unification of forces. For example, the Grand Unified Transition (GUT) can, in principle, create both types of seeds when the universe was at a temperature of about 10^{28} K. Recently, it has also been proposed by C. Hill and me that a cosmological phase transition may occur as late as a temperature of ~ 100 K (after the decoupling of the cosmic background radiation) and also be able to generate either type of seed.

It is interesting to realize that *all* models for generating structure in the universe require some new fundamental physics, both in the form of exotic matter and some vacuum phase transition to produce seeds. Thus the study of the structure of the universe should teach us new physics as well as new astronomy.

It may be useful to put current ideas into the framework of the discussion of seeds and structure formation of twenty years ago. Prior to the introduction of grand unified or microphysics models for generating fluctuations, we merely noted that density fluctuations in matter could be divided into two general classes:

- Adiabatic.

- "Isocurvature" (or, almost equivalently, "isothermal," since in the early universe $\rho_b \ll \rho_{rad}$. (The symbol ρ represents density.)

In the adiabatic case, the ratio of baryon density, n_b, to radiation density, n_γ, is unchanging, so any variation in n_b varies; and in the isocurvature case, the total energy density (which yields cosmic curvature) is fixed so that variations in the energy density are accompanied by opposite compensating variations in the energy density of photons, ρ_γ. But since $\rho_\gamma \ll \rho_b$, the variations in ρ_b don't really affect ρ_γ so isocurvature behaves just like isothermal. Because of the development of grand unified models and particularly the realization that baryons probably were produced by some variant of the Sakharov process, Turner and I noted in 1979 that adiabatic fluctuations were preferred for baryon density fluctuations. If baryons are generated by temperature-dependent microphysics processes, then a constant isothermal temperature everywhere would result in the same baryon density everywhere and would yield no baryon density fluctuation. A way around this would be to have the "seed" not be a matter density fluctuation itself, but instead be some separate physical seed that is the function of a topological defect. Such a defect does not alter the thermal background, so in the old classification it is isothermal or isocurvature. However, topological seeds do not yield a Gaussian distribution but, instead, are fractal. Thus if we wish to use the old language, the random quantum seeds are the old Gaussian adiabatic fluctuations and topological seeds are the old isothermal/isocurvature seeds with the added constraint of being non-Gaussian. The key new point is that these models are motivated by fundamental physics ideas rather than just mathematical formalism.

Figure 3 shows how density fluctuations grow as the universe expands. If the seed is produced by a phase transition prior to the decoupling of the cosmic microwave background radiation, then the observed isotropy of that radiation constrains the initial fluctuation amplitude so that it is quite small, and small fluctuations grow slowly, as indicated. Such a slow growth means that the bulk of the objects form relatively late when the average fluctuation size is comparable to the average density itself. This slow growth is a serious constraint on such models and is one of the motivations behind recent models, which have a late phase transition occurring after the decoupling of the background radiation. In this latter case, the growth can be much faster without violating the isotropy limits.

The favorite structure formation model until recently has been a combination of cold dark matter and random density fluctuations with a spectrum of equal amplitude on all scales expected from quan-

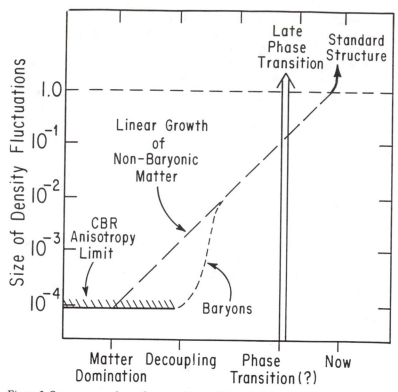

Figure 3. *Structure growth as a function of the redshift epoch. Note that any model that starts with primordial seeds that are constrained by the isotropy of the cosmic background radiation produces most of its structure relatively late.*

tum fluctuations at the end of inflation. Although the model is known simply as the "cold dark matter model," it is important to remember that a critical (and perhaps fatal) part of this model is its assumption about the nature of the seeds. (The model also requires something known as "biasing," so that only a small fraction of the baryons ends up in shining regions.) The alternative of random density fluctuations with hot dark matter fails because it doesn't produce "small" objects like galaxies fast enough. A similar problem may occur for the cold dark matter model, given recent observations of large numbers of high-redshift objects. However, hot dark matter (and cold dark matter, also) can avoid this problem if the seeds are topological (or if there is a late-time phase transition).

Large-Scale Structure Observations

Selecting among different structure formation models will be made possible, we hope, by large-scale structure observations.

Except for recent observations, the cold dark matter model with random fluctuation seeds has done a remarkably good job of explaining most extragalactic observations, including the basic observed properties of individual galaxies. Even bizarre "cosmologies" that fail to fit the 3 degree Kelvin background radiation or the light-element abundances and are designed in an *ad hoc* way to make galaxies (the so-called "plasma cosmology" comes to mind) do not do as good a job as the cold dark matter model in this regard.

Key recent large-scale structure observations include:

- The cosmic background isotropy.
- Quasars found at large redshifts.
- Large coherent velocity flows.
- Structures with scales of ≥ 100 Mpc.
- Large correlations of clusters of galaxies.

Cosmic Background Isotropy

The initial limits on the microwave anisotropy from the COBE (Cosmic Background Explorer) were already noted in figure 3. The limits marginally allow structures to form by the present epoch. COBE made the exciting announcement in April 1992 of $\Delta T/T \sim 10^{-5}$ on very large ($\theta \geq 7°$) angular scales. This has provided the large-scale normalization of the primordial spectrum and shows that the basic framework for structure formulation is reasonable. However, to compare directly the microwave fluctuations with the seeds that produced the observed structures requires measurements of angular scales of $\theta \sim 1°$. Independent cosmic anisotropy studies are being carried out at the South Pole by a Chicago-Princeton team and by a University of California team at these important angular scales. The April 1992 COBE large-scale result has been confirmed more recently by a balloon-born experiment.

Quasars at Large Redshifts

Pushing the opposite direction on the "zone of mystery" epoch between the background radiation and the existence of objects at high redshift is the discovery of objects at higher and higher redshifts. The higher the redshift of objects found, the harder it is to have the slow growth in figure 3 explain their existence. A few high-redshift objects could be dismissed as statistical fluctuations if the bulk of the objects still formed late, but we may already be running into problems. During 1990 alone, the number of known quasars with redshifts greater than 4 rose to thirty, with one having a redshift as large as 4.9. Furthermore, there appears to be no significant intergalactic gas near these quasars. Thus either the bulk of the gas has already been incorporated into objects (contrary to the slow growth picture), or the gas has somehow been heated and kept hot enough to be ionized (but not so hot as to emit X-rays), or both.

Large Coherent Velocity Flows

We have already discussed above the large velocity flows with regard to the implication of $\Omega = 1$ on scales of ~ 100 Mpc. To generate structures as large as the Great Attractor and the associated high-velocity flows on those scales can be a problem, since it tends to require large amplitude fluctuations if the seeds are random fluctuations. But the cosmic microwave background radiation limits go in the opposite direction.

Structures with Scales of ≥ 100 Mpc

Large-scale observations recently receiving considerable publicity are direct maps made in 1989 by M. Geller and J. Huchra of the large structures in the universe. In particular, their maps show objects such as the "Great Wall" that stretch for over 100 Mpc. Furthermore, the deep pencil beam surveys by R. Broadhurst, R. Ellis, D. Koo, and A. Szalay (figure 4) show that the great walls appear to be ubiquitous in the universe and may have a quasi-regular spacing of about 100 Mpc. Thus, again, we see indications of significant structure on scales of about 100 Mpc.

Although these maps certainly show large-scale structure in a graphic way, the question up until 1990 had been, "What's the statistical significance?" In other words, could these big things be relatively rare statistical flukes, or are they common?

Random-seed models with cold dark matter and a spectrum that has equal-size fluctuations on all scales can give occasional large structures, but is there more "power" on large scales than such a spectrum can yield? The answer to this question has come from some new large surveys of galaxy positions. In particular, the Automatic Plate Measuring (APM) survey headed by Efstathiou at Oxford University, the Queen Mary-Durham-Oxford-Toronto (QDOT) survey of Infrared Astronomy Satellite galaxies, and the second Palomar sky survey (POSS II) analysis by A. Picard at the California Institute of Technology all now have statistically significant samples that show that, indeed, there is more power on large scales than can be accommodated by the seed spectrum assumed in the so-called cold dark matter model (figure 5). Note that it is the seed part of the model that has encountered difficulties, not the matter itself.

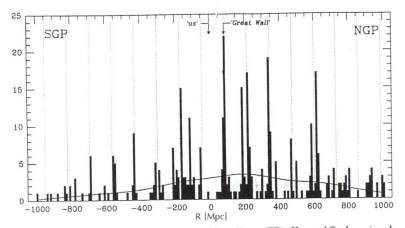

Figure 4. Data from the pencil beam surveys of Broadhurst, Ellis, Koo, and Szalay using the Anglo-Australian Telescope for the South Galactic Pole (SGP) pencil and the Kitt Peak Telescope for the North Galactic Pole (NGP) pencil. The plot shows numbers of galaxies versus distance. The solid line with a shaded region under it represents what we would find for a random distribution of galaxies. The spikes appear to show the pencil penetrating walls of galaxies. The average spacing of the spikes is about 130 Mpc.

Of course, a complete, statistically significant mapping out of the structures requires the three-dimensional positions of far more galaxies than any of the current surveys provide (~ 10,000 at most). The University of Chicago, Princeton University, Johns Hopkins, the Institute for Advanced Study, and Fermilab are now building a dedicated telescope that will obtain the three-dimensional positions of a million galaxies. This will, to some extent, fill in the pencil beams and enable us to see how regular the structures really are.

Large Correlations of Clusters of Galaxies

The last of the key recent large-scale structure observations is the apparent predilection that clusters of galaxies have to be near each other rather than randomly distributed. In fact, J. Bahcall and R. Soneira showed in 1984 that a cluster is more likely to be found near another cluster than a galaxy is likely to be found near another galaxy. Were gravity alone at play, clusters should not be so strongly correlated with each other. Scientists first tried to get around this point by arguing that projection effects might explain the apparent correlation. However, in 1991 M. West and S. Van Den Bergh showed that the centers of these clusters really are strongly correlated. Efstathiou and G. Bernstein, using the Automatic Plate Measuring survey, have not found as strong a correlation as that found by Bahcall and Soneira in 1984, although they still seem to find more power on large scales

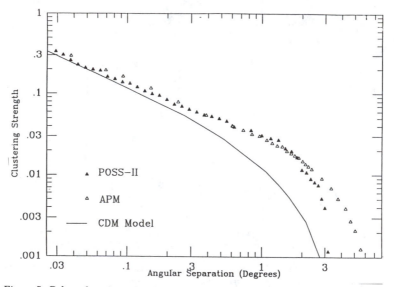

Figure 5. Galaxy clustering strength versus angular separation for the second Palomar Sky Survey (POSS II) and for the Automatic Plate Measuring (APM) Survey. The solid curve represents observations predicted from the cold dark matter theoretical model with a flat Gaussian initial fluctuation spectrum.

than a flat, random-seed spectrum would give. Complete resolution will require the new million-galaxy surveys or cluster correlations using clusters identified by their X-ray emission from the ROSAT and AXAF satellites. If correlations are stronger than random, we will have to conclude that galaxies and clusters do not form from just random seeds and gravity but, instead, that the seeds are laid out in some pattern. A pattern is exactly what topological defect models tend to predict.

Conclusion

Galaxy and structure formation is a very active scientific field. By necessity, the models work within the big bang framework. Details for the models all invoke new fundamental physics, both for the generation of seeds and for the nonbaryonic dark matter. Which new physics is right remains to be seen. The model with cold dark matter and random seeds was the front-runner, but it is running into problems with the new large-scale structure observations. However, variants on this model, putting larger amplitude fluctuations on large scales, may still survive. Other models with late phase transitions generating the seeds, or with topological defects as seeds, are also looking attractive. These latter models may work with either hot dark matter or cold dark matter (although hot dark matter is preferable).

Fortunately, a battery of experiments and observations soon will be carried out and should resolve the problem. In addition to the million-galaxy maps, improved cosmic background radiation measurements, and X-ray satellite observations, we also can look forward to observations by new large ground telescopes and by the Hubble Space Telescope of galaxies near the time of their formation. Furthermore, new dedicated telescopes are being developed to search for dark baryonic matter in the galactic halo, using gravitational microlensing techniques. Indeed, scientists are using more and more dedicated telescopes rather than general-purpose telescopes to attack cosmological problems. Nor are cosmological problems any longer tackled with telescopes alone. Experimental particle physicists also have gotten into the the game. Direct search experiments are being built to try to detect WIMPS and axions. In addition, new accelerator experiments, including work with the super collider just now in the planning and early construction stage and still facing political and funding perils, will put new, tighter constraints on WIMPS and may find the mass of the τ neutrino through its mixing with other neutrinos. Many of these questions should be resolved before the end of the 1990s.

—David N. Schramm

Bibliographic Note

Good introductions to modern cosmology in general and to the question of large-scale structure and galaxy formation in particular include S. Weinberg, *Gravitation and Cosmology: Principles and Applications of the General Theory of Relativity* (New York: Wiley, 1972); P.J.E. Peebles, *Large-Scale Structure* (Princeton: Princeton University Press, 1980); E. Kolb and M. Turner, *The Early Universe* (Redwood City, California: Addison-Wesley, 1990); and M. Riordan and D. Schramm, *The Shadows of Creation* (New York: W.H. Freeman, 1991). For references to many papers in scientific journals, see the entry on "Large-Scale Structure and Galaxy Formation" in *The Encyclopedia of Cosmology: Historical, Philosophical, and Scientific Foundations of Modern Cosmology* (New York: Garland, 1993).

24. COSMIC STRINGS

Thin lines of concentrated energy, called cosmic strings, are predicted by some theories of elementary particles. Such strings could have been formed shortly after the big bang, and might still be present today in distant regions of the universe. Cosmic strings, if indeed they exist, could provide an explanation for the existence of stars, galaxies, and larger cosmic structures. They can also produce a number of unusual observational effects, such as double images of distant galaxies and quasars, discontinuous changes in the intensity of the cosmic microwave radiation, and a broad spectrum of gravitational waves.

Symmetry Defects

Our modern story of Genesis tells us that the universe began roughly fifteen billion years ago in a great explosion, the big bang. The universe is still expanding from the force of that explosion, with distant galaxies moving away from the earth at very high speeds. By combining astronomical observations with verified laws of particle physics, physicists can trace the history of the universe back to the fraction of a second immediately following the big bang. In those moments, there were no stars or galaxies or atoms. Immediately following the big bang, the universe was a hot, dense fireball of elementary particles, such as electrons and photons.

Underlying the particles and determining their interactions was the vacuum. Far from connoting nothing, a vacuum is understood by physicists to be a state of minimum energy obtained in the absence of particles.

The relation between elementary particles and the vacuum is similar to the relation between sound waves and the material in which they propagate: the types of waves and the speed of propagation are different in different materials. Because the properties of the vacuum have not always been the same, the properties and interactions of elementary particles have also changed.

The early vacuum had an enormously high energy as well as a high degree of symmetry. There were no distinctions between the interactions of fundamental particles. The electromagnetic, weak, and strong nuclear forces were manifested as parts of a single, unified force. Today, however, the vacuum energy is zero and the elementary forces are distinct both in strength and in character. There is little left of the original unity. An important question for cosmology is how the very high early symmetry between the elementary particles was broken.

While the universe expanded and cooled after the big bang, the vacuum went through a rapid succession of phase transitions. The most familiar phase transitions are those that water undergoes when it is cooled from steam to liquid and finally to ice. Phase transitions can also be described in terms of symmetry breaking, often reducing symmetrical states to asymmetrical states. A crystal, for example, is less symmetrical than a liquid: the liquid looks the same in all directions, while different directions in a crystalline lattice are not all equivalent.

The exact number of cosmological phase transitions depends on details of elementary particle physics, but there must have been at least two. One would have occurred at an energy of about 10^{16} GeV [one GeV equals one billion electron volts], where the strong interaction became distinct from the electroweak interaction. Another would have taken place at about 100 GeV, where the electroweak symmetry was broken. However many phase transitions occurred in the young vacuum, all of them probably took place within the first second after the big bang.

Crystallization starts independently in different regions of space, with different directions of the crystalline axes. When these regions come together and try to match, it often happens that the matching is not possible in a continuous way. Defects are the result.

Quite similarly, phase transitions in the early universe can give rise to defects because the directions of symmetry breaking are not correlated between different regions of space. The logical reason is that correlations cannot be established at a speed faster than the speed of light, and in a universe that is only a fraction of a second old, light has had enough time to travel only a very short distance. Inside the defects symmetry is not broken and the earlier, more symmetric vacuum is trapped.

Different types of defects are predicted by different particle theories. Some theories predict that the defects will assume the form of surfaces; other theories predict lines; and some predict points. These defect types are called domain walls, strings, and monopoles, respectively. In addition, there are hybrid defects: monopoles connected by strings and domain walls bounded by strings.

Cosmic Strings

Of the three possible types of symmetry flaws in the continuity of the vacuum—domain walls, monopoles, and cosmic strings—strings seem to be the most likely to exist. Indeed, both domain walls and monopoles are incompatible with current cosmological models.

The high energy vacuum trapped inside defects is, by the Einstein mass-energy relation, extremely massive. Consequently, defects can have a profound influence on the evolution of the universe. A single domain wall stretching across the present-day universe would have far more mass than all the matter in the universe combined and would induce galaxies to cluster much more than they are observed to do.

Unlike a single domain wall with its highly observable consequences, a single monopole might escape detection. However, theories that predict monopoles also predict them in very large numbers. If monopoles existed, the universe should be swarming with them. They would be difficult to ignore. Yet monopoles have not been observed.

Nor have hybrid defects (monopoles connected by strings and domain walls bounded by strings) been observed. That is not surprising, because they decay rapidly and leave little if any trace. These defects are not ruled out by the absence of observations, but are rather uninteresting from the cosmological point of view.

Nor has anyone yet seen a cosmic string. But physicists do not expect cosmic strings to be as obvious as domain walls or monopoles.

Work on cosmic strings was pioneered by T.W.B. Kibble of the Imperial College of Science and Technology in London. Kibble studied how strings might be formed in the early universe, and in a 1976 paper he discussed some aspects of their evolution. Little attention was given initially to the notion of cosmic strings, but in the early 1980s Yakov Zel'dovich at the Institute of Physical Problems in Moscow and I at Tufts University independently realized that cosmic strings might be able to explain the clumping of matter in the universe. Our ideas inspired a small group of investigators to explore cosmic string theories in greater detail.

The physical properties of cosmic strings turned out to be fascinating and unique. Cosmic string theory quickly developed an attraction for physicists akin to the attraction the strings themselves are said to exert on stars and galaxies. In the late 1980s an avalanche of papers on cosmic strings descended on the scientific community. No direct experimental evidence for the existence of cosmic strings has yet been found. Even in the absence of empirical data, however, physicists have managed to put together an impressive profile on cosmic strings. Some properties of cosmic strings depend on the particle theory used to derive them, while others are common to all theories.

Cosmic strings are thin tubes of symmetrical, high-energy vacuum. They do not have ends; they either form closed loops or extend to infinity. A string's physical character is determined by the energy of the vacuum trapped inside it. The strings with the most symmetrical vacuum, in which strong, weak, and electromagnetic forces are united, are the thinnest and most massive. These are also the most conspicuous strings, and they are of the greatest cosmological interest because they could be responsible for the formation of galaxies.

Such strings have a thickness on the order of 10^{-30} centimeter and an astonishingly large mass: one inch of string would weigh 10 million billion tons! The tension in these strings is a match for their mass. The tension causes closed loops of string to oscillate fiercely at speeds approaching the speed of light. For example, a loop one light-year in length would complete an oscillation in a little more than a year. (A light-year, a unit of length, is the distance light travels in one year: about five trillion miles.)

When two strings intersect, they reconnect (figure 1). The sharp angle formed at the point of reconnection splits into two, and the two kinks propagate at the speed of light along the string. This process allows closed loops to fragment. An oscillating loop may self-intersect and break into two, and then the fragments may further break (figure 2). It has been shown that this process is finite, and that each loop fragments a finite number of times until, in the end, there are loops that never self-intersect.

Evolution of Cosmic Strings

The cosmic strings produced during a phase transition weave a tangled network that pervades the entire universe (figure 3). The evolution of this cosmic web is rich in physical processes. Although on the average the network is uniform, its individual strings are quite irregular and convoluted. Tension in curved strings makes them wiggle violently, and they often cross themselves and one another. Intersecting strings break at the crossing point and reconnect the other way. Long, twisted strings cross themselves many times and shed their coils in the form of closed loops. As a result, the strings are becoming straighter. The loops can self-intersect and further fragment, but eventually the fragmentation ceases and each loop leaves behind a family of nonintersecting daughter loops.

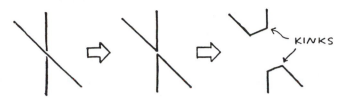

Figure 1. *When two cosmic strings intersect, they break and reconnect the other way. The sharp angle formed splits into two, and the two kinks propagate at the speed of light along the string.*

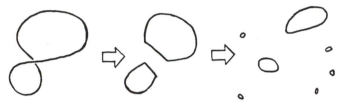

Figure 2. *A closed loop of cosmic string oscillates under the action of tension. It may self-intersect and break into two. The fragments may break further.*

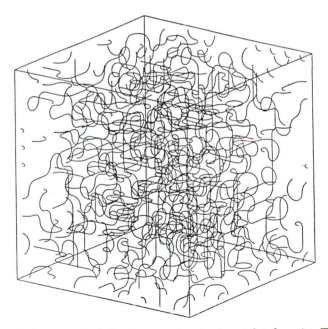

Figure 3. *A computer simulation of a system of cosmic strings at their formation. [From B. Allen and E.P.S. Shellard, 1990.]*

It takes longer to make large loops than to make small ones, because a cosmic string must fold back on itself to shed a loop, and it must fold back further for a large loop. The size of the loop that can be liberated at any moment is limited by the amount of time that has elapsed since the big bang. In particular, given that the cosmic strings move at about the speed of light, the loop can be no larger than the distance light has traveled since the birth of the universe: the *horizon length*. Hence smaller loops of cosmic string are characteristic of a younger universe, while loops of cosmic string created today are much larger. Correspondingly, the strings now are much straighter than they used to be at earlier times.

This does not mean that the current network of cosmic strings looks very different from the one that was initially established. Indeed, the evolution of cosmic strings includes an interesting feature called self-similarity, that preserves the statistical constancy of the network over time. If the cosmic string network were to be photographed at two different times, the main difference between the two pictures would be in overall scale, which is set by the horizon length.

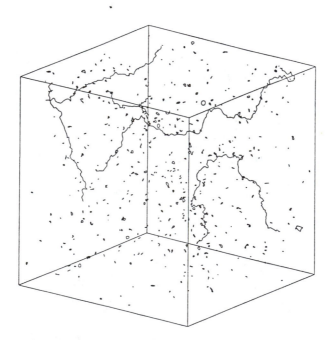

Figure 4. A computer simulation of a system of cosmic strings at an instant during their evolution. [From B. Allen and E.P.S. Shellard, 1990.]

Magnifying the first photograph by the ratio of the two horizon lengths would yield a picture quite similar to the second.

A computer simulation of string configuration in a cubic volume, of size about a third of the horizon length, is shown in figure 4. Note that long strings have zigzag shapes. This is due to sharp angles or "kinks" formed at the points of string reconnections in the course of previous evolution. Note also that the sizes of closed loops are very small. The typical size of the loop appears to be determined by the size of the smallest zigzags.

If closed loops that are being continuously produced by the network were allowed to accumulate, the universe would be swimming with them. What, then, happens to the loops? Theoretical analysis shows that, as they oscillate, loops generate rhythmic pulses of gravitational energy that propagate at the speed of light. These pulses are known as gravitational waves, and they sap the energy of a loop until it shrinks and eventually disappears. The lifetime of a cosmic string loop made of the highest-symmetry vacuum is about 10,000 oscillations, regardless of its size. Because the period of a single oscillation is greater for larger loops, they live longer than smaller ones. Likewise, loops of lighter, lower symmetry string last longer than the heavy, high-symmetry loops.

Cosmic Strings and Galaxy Formation

Observations of the universe find it to be rather lumpy; stars gather into galaxies, and galaxies in turn form clusters. With time the universe grows lumpier, as the gravitational pull of clusters of galaxies attracts other galaxies from neighboring regions. Modern theories of galaxy formation assume that in the past the universe was much smoother than it is today, and that all galaxies and clusters of galaxies have grown out of small incongruities in an otherwise nearly uniform distribution of matter. We still do not know, however, what were the initial incongruities, nor where they came from.

The initial incongruities around which matter in the universe first coalesced may have come from cosmic strings—if, indeed, cosmic strings exist and if they are sufficiently massive. There are basically two ways in which strings can produce cosmic structure. The first is through the gravitational attraction of matter to oscillating loops (figure 5). The second is through the formation of wakes behind rapidly moving long strings. Figure 6 shows a string perpendicular to the paper and moving to the left. In the wake of the string there is a higher density of matter. The very small loop sizes obtained in string simulations imply that the role of wakes is probably greater than that of the loops. It is possible that the wakes could explain the

Figure 5. Accretion of matter around an oscillating loop.

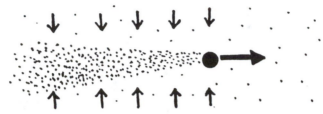

Figure 6. A perturbation is produced by the wake of a long cosmic string moving through a homogeneous medium. In the wake of the cosmic string there is a higher density of matter.

observed sheet-like distribution of galaxies with gigantic voids having almost no galaxies at all. A quantitative comparison of string theory with observations should await the results of numerical simulations, currently underway, that combine the string-evolution computer code with a simulation of the gravitational clustering of matter.

The Search for Cosmic Strings

Even if a cosmic string model can account for observed cosmological features, physicists will find it difficult to believe in cosmic strings until their existence is confirmed by direct observation. Thus attention shifts to possible observational effects of cosmic strings and the question becomes: If cosmic strings are really there, how can we detect them?

Of the heavy, highly symmetrical cosmic strings, the one closest to the earth should be about three hundred million light-years away. These elusive objects can possibly be detected at such vast distances by their bizarre gravitational properties. There might be lighter cosmic strings closer to the earth, but their presence would probably be even less conspicuous.

In considering the gravitational character of cosmic strings, it is convenient to begin with the case of an idealized stationary cosmic string lying along a straight line in space. According to Einstein's general theory of relativity, gravitation is synonymous with the curvature of space and time. Here we need consider only the curvature of space. Cosmic strings distort space in a very peculiar way. In Euclidean geometry, the ratio of a circle's circumference to its diameter is equal to the number π (≈ 3.14156). For a circle drawn around a cosmic string, this ratio is a tiny bit less, the difference appearing only in the fourth decimal place. The space around a cosmic string has a conical nature. To visualize it, imagine cutting a small angular wedge from the Euclidean space with its tip at the cosmic string. Then glue the exposed surfaces together, not by stretching but by bending the space around them. The result is that all planes perpendicular to the string become cones.

The angle of the wedge thus removed is called the deficit angle. In cosmic strings, the deficit angle corresponds to a few seconds of arc. All objects passing by the cosmic string—photons, atoms, and stars—will be deflected from their original direction of motion by an angle comparable to the deficit angle. Two objects moving along symmetrical, parallel paths on opposite sides of the cosmic string will collide after they pass the string (figure 7). To a person sitting on one of these objects, the other object would initially appear to be at rest. When the cosmic string passes in front of it, it would suddenly start

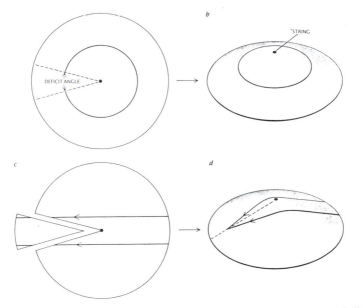

Figure 7. Cones of space around heavy cosmic strings illustrate their peculiar gravitational effects. This drawing shows the distortion of space caused by an idealized straight cosmic string. The distortion can be represented by cutting a wedge out of a plane of space perpendicular to the cosmic string (a) and pulling the two edges of the plane together to form a cone (b). Then, if two objects traveling along parallel paths pass a cosmic string on opposite sides (c), they will be deflected and will collide on the far side of the cosmic string (d).

moving toward the hapless observer at a speed equal to 0.00002 times the velocity of the cosmic string. Since cosmic strings move at nearly the speed of light, they induce velocities of about four miles per second.

What would happen if a cosmic string passed through a person? The effect is not difficult to picture. As the cosmic string cuts across the individual's waist, his head and feet start moving toward each other at a speed of four miles per second. This experience would, of course, be unhealthy, but there is no reason to panic: the probability of a cosmic string traveling through the solar system is extremely small.

The conical-space rendering of the gravitational properties of cosmic strings applies only to straight cosmic strings. The gravitational effects of curved cosmic strings and closed loops are much more complicated. Small segments of curved cosmic strings, though, can be thought of as being approximately straight. By combining the analyses of many small segments, it can be shown that at great distances from an oscillating closed loop the average effect of all segments is an ordinary gravitational attraction like that associated with the earth or the sun.

The distortions that cosmic strings induce in space might betray their presence. For example, because they bend space, cosmic strings can act as gravitational lenses. They will deflect the light from a distant galaxy so that it reaches the earth by two different paths. As a result, observers on the earth would see two images of the same galaxy, separated by an angle comparable to the deficit angle of the cosmic string (figure 8). Astronomers have in fact found several pairs of galaxies, and also pairs of extremely bright, distant objects called quasars, in which the two members of the pair show a compelling resemblance to one another and are therefore considered to be double images of the same object.

Ordinary galaxies and clusters of galaxies can also act as gravitational lenses, and thus an additional test must be applied to multiple images to ascertain their cause. Nick Kaiser of the University of Cambridge and Albert Stebbing of Fermilab have pointed out that cosmic strings should have a rather unusual effect on the cosmic microwave radiation. This radiation, a kind of afterglow from the big bang, fills the entire universe and comes from all directions with equal intensity. In the wake of a cosmic string, however, some radiation would gain extra momentum in the direction of the earth and would therefore approach the planet with a greater intensity. While other gravitational entities can cause smooth changes in microwave intensity, the change wrought by a cosmic string would be quite

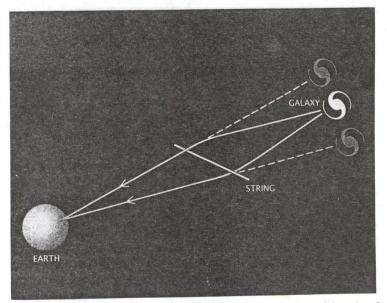

Figure 8. Double images of a single galaxy could result from the gravitational distortion of an intervening cosmic string. The cosmic string would bend light from a single source, giving to an observer on the earth an impression of two sources. If two galaxies look unusually similar, physicists can check for evidence of a cosmic string or of another gravitational lens between the pair. Cosmic strings can be distinguished from other such lenses because they cause an abrupt change in background microwave radiation. So far, none have been found.

abrupt. This sudden change of intensity should occur along a line drawn between the two images representing a single galaxy. The magnitude of the change may be only one part in 100,000, but detection, although difficult, should not be impossible.

In addition, it might someday be possible to detect cosmic strings by looking for evidence of their gravitational waves. Waves from loops of all shapes and sizes add up to a background of gravitational noise, the collective swan song of dead and dying cosmic strings. The intensity of this noise is high compared with that of the gravitational waves emanating from other sources. Gravity, however, happens to be the weakest of all forces in nature, and the predicted level of noise from the cosmic strings is still slightly below current observational limits.

Cosmic Superconductors

Until recently, physicists assumed that cosmic strings could interact with matter only through gravitational forces. But then Edward Witten of Princeton University proposed that cosmic strings might be endowed with a property highly regarded here on earth: superconductivity. Witten showed that some theories of elementary particles suggest that cosmic strings have unusual electromagnetic qualities of the kind that would make them behave like superconducting wires. More recently, it was shown that this property could produce dramatic cosmological effects.

Strings could be superconducting because the symmetrical vacuum trapped inside them changes the behavior of particles. In particular, some charged particles, like electrons, might have no mass inside a cosmic string. Therefore, it is possible with very little energy to create particle pairs in which the members have opposite charges and travel in opposite directions. The total charge and total momentum of a pair are equal to zero. The only energy input required is that necessary to impart motion. Because the particles are without mass, they move at the speed of light and cannot venture outside the string, where their mass would be greater than zero. Hence particles zoom through a trapped vacuum, carrying an electric current and encountering no resistance. This is the hallmark of superconductivity.

Jeremiah Ostriker of Princeton and his student Christopher Thompson collaborated with Witten to put forward an alternative string model of galaxy formation. The current in a superconducting cosmic string produces electric and magnetic fields that, in empty space, would propagate away from the cosmic string as electromagnetic waves. But the interstellar and intergalactic space is not exactly empty. It is filled with a dilute gas of electrons and charged atoms that prevents waves from leaving the vicinity of the cosmic string. As the energy of the radiation accumulates, it develops tremendous pressure and starts blowing a bubble, sweeping surrounding matter into a hot, expanding shell of gas. The expanding bubble is not much different from a huge explosion. In this scenario, galaxies are formed where bubbles collide.

Ostriker and his colleagues have had to posit that the universe was magnetized soon after the big bang, because a magnetic field had to be available to launch the particle pairs. No one knows for certain how this magnetization might originate. But if it was present soon after the big bang, residual magnetization should persist in intergalactic space today. The theory can therefore be tested by looking for evidence of weak, delocalized magnetization in the universe.

Another prediction from the theory is that the spectrum of the cosmic microwave radiation should get a characteristic distortion due to its interaction with the expanding shells of hot gas. No such distortion has yet been observed, and the explosion theory consequently is now under considerable observational pressure.

Aside from its role in theories of galaxy formation, the concept of superconducting cosmic strings suggests another way in which cosmic strings might be found. Even if magnetic fields were absent from the early universe, they are eventually generated by the rotational energy of the galaxies through a phenomenon known as the galactic dynamo effect. Today, the strength of the magnetic field in a typical galaxy is just one-millionth of the earth's magnetic field. The current such a field would induce in a superconducting loop of cosmic string is too weak to prompt an exploding bubble of radiation. However, calculations show that the interaction of this current with charged particles in interstellar space could generate radio waves. A loop of light superconducting string in our galaxy could therefore be detected as a line-like radio source.

Conclusion

Even as empirical trials of cosmic string theory begin, physicists are tempted to use the rich and unusual properties of the hypothetical cosmic strings to account for all kinds of mysterious phenomena. Strings have already been suggested as possible sources of cosmic rays, ubiquitous but unexplained streams of energetic particles in space. Strings might also be the origin of powerful bursts of gamma rays, which are regularly observed but poorly understood. Cosmic strings are even suspected of being the power engines behind quasars. The reasons given for these attributions are not particularly compelling, and most will probably turn out to be wrong. Nevertheless, theorists are having great fun exploring the potential of cosmic strings. Nature will give the final verdict on their work.

—*Alexander Vilenkin*

BIBLIOGRAPHIC NOTE

The classic paper that inspired many physicists to study cosmic strings and other symmetry defects was that by T.W.B. Kibble, "Topology of Cosmic Domains and Strings," *Journal of Physics, A9*:(1976), 1387–1398. That strings could be of importance for galaxy formation was suggested by Y.B. Zel'dovich, "Cosmological Fluctuations Produced Near a Singularity," *Monthly Notices of the Royal Astronomical Society, 192* (1980), 633–667, and by Alexander Vilenkin, "Cosmological Density Fluctuations Produced by Vacuum Strings," *Physical Review Letters, 46*:(1981), 1169–1172 (and a correction on p. 1496). For a more up-to-date review of the subject, see the collection of papers in G.W. Gibbons, S.W. Hawking, and T. Vachaspati, eds., *Formation and Evolution of Cosmic Strings* (Cambridge: Cambridge University Press, 1990).

25. QUANTUM COSMOLOGY AND THE CREATION OF THE UNIVERSE

"Where did this all come from?" is a question we can hardly help but wonder on staring out into space on a clear night. Such an esoteric question, for centuries pondered by philosophers and theologians, may seem to lie far outside the traditional domain of respectable scientific investigation. In recent years, however, developments in the cosmology of the very early universe not only have led to the possibility of understanding the very beginning of the universe from within the laws of physics, but also have underscored the necessity of understanding it in order to have a complete description of the universe in which we live.

The precise field in which these developments have taken place is quantum cosmology, in which quantum mechanics—usually applied only to the subatomic scale—is applied to the entire universe. The field is not a new one. Much of the groundwork was carried out in the 1960s by Bruce DeWitt, now at the University of Texas at Austin, by Charles Misner of the University of Maryland, and by John Archibald Wheeler, then at Princeton. But new impetus came to the field in the early 1980s, when a number of cosmologists began to take quantum cosmology very seriously as a realistic framework in which to address the issue of cosmological initial conditions. In particular, quite definite, testable *laws of initial conditions* within the context of quantum cosmology have been put forward, most notably by James Hartle of the University of California at Santa Barbara in collaboration with Stephen Hawking of the University of Cambridge, by Andrei Linde of the Lebedev Physical Institute in Moscow, and by Alexander Vilenkin of Tufts University. Adjoined with suitable laws governing the evolution of the universe, such laws of initial conditions could conceivably lead to a complete explanation of all cosmological observations. Although this goal is still a long way off, the initial steps in this direction show considerable promise.

The Hot Big Bang Model

Central to modern cosmology is the hot big bang model of the universe put forward by George Gamow in 1948. Based on general relativity (Einstein's theory of gravitation), together with some basic physics about the behavior of matter, this model envisages the universe beginning from a very small, hot, dense initial state some 15 billion years ago and expanding out to the large cold universe in which we live. The hot big bang model makes definite predictions about the universe we see today, and these predictions have been verified observationally. In particular, the model predicted the relative abundances of certain elements, as well as the existence and exact temperature of the microwave background—the glow of radiation permeating the universe left over from the initial explosion.

Although impressive in its predictions, the hot big bang model leaves many features of the universe unexplained. They include:

- *The horizon problem.* The universe that we see today consists of a vast number of regions that, in the hot big bang model, could never have been in causal contact at any stage in their entire history. This absence of contact makes it particularly difficult to account for the striking uniformity of the universe on very large scales.

- *The flatness problem.* This problem follows from the observational fact that the spatial geometry of the universe is extremely flat, rather than curved like the surface of a sphere. This flatness is rather surprising in the hot big bang model, because in this model the universe tends to become more curved as time evolves. Indeed, the universe could only be as flat as it is now if it started out almost exactly flat—flat to within one part in 10^{60}. The necessity to tune the initial conditions of the universe so finely if current observed conditions are to be reproduced from the hot big bang model is felt to be deeply unnatural.

- *The density fluctuation problem.* It has been known for a long time that large-scale structures in the universe (such as galaxies) can grow out of small fluctuations in the matter density at very early times in an otherwise homogeneous universe. The requisite form and magnitude of these fluctuations was worked out in the 1970s by Edward R. Harrison of the University of Massachusetts at Amherst and independently by the late Yakov B. Zel'dovich of the Institute of Physical Problems in Moscow. But the fundamental origin of these fluctuations remained completely unknown; the fluctuations had to be assumed as initial conditions.

This was the situation of thinking in modern cosmology around the beginning of the 1980s. These three features of the universe—its uniformity on very large scales, its lack of curvature, and the presence of inhomogeneities such as galaxies—plus a number of other related features could be *accommodated* by the hot big bang model, possibly by choosing very special initial conditions, but there was no sense in which the features were *explained* by the hot big bang model.

The Inflationary Universe

A crucial new element entered the game in 1983, when Alan H. Guth of the Massachusetts Institute of Technology proposed a compelling solution, known as the inflationary universe scenario, that could resolve the horizon problem, the flatness problem, and the density fluctuation problem (also called the smoothness problem). Like the hot big bang model, the inflationary universe scenario takes the gravitational field to be classical (described by Einstein's theory of general relativity). The matter content of the universe, however, is taken to be a type of matter called a scalar field, whereas in the hot big bang model the matter content is a uniformly distributed plasma or dust. Scalar fields arise naturally in many particle physics theories, and are generally the dominant matter content in the extreme conditions encountered in the very early universe. Guth showed that the presence of this scalar field may cause the universe to undergo a very brief but exceedingly rapid period of expansion, called *inflation*. This expansion is much more rapid than that in the hot big bang model— exponentially fast, in fact, like the rate of population increase in a population explosion. [See chapter 22, *The Inflationary Universe.*]

Inflation lasts for an almost inconceivably short time, some 10^{-30} second; but during this time the universe increases in size by an equally astounding factor of 10^{30}, from an initial size of about 10^{-28} centimeter to a size of about one meter. At the end of this period of inflation, the decay of the scalar field producing the expansion causes the initially cold universe to heat up to a very high temperature, and the subsequent evolution of the universe is exactly as in the hot big bang model. Inflation is, therefore, an incredibly brief glitch tacked onto the beginning of the hot big bang model, but during this time— a blink of the eye even on the subatomic scale—a whole host of cosmological problems are swept under the carpet.

- The horizon problem is solved because, due to the huge expansion, all of the observed universe emerges from a much smaller region than in the hot big bang model—sufficiently small that all of it can be in causal contact.

- The flatness problem is solved because the huge expansion wipes out all spatial curvatures; thus even if the universe really is spatially curved, it appears flat to us because it has been blown up to be so much larger than in the hot big bang model.

- The density fluctuation problem is also solved. The inflationary universe scenario offers a possible explanation of the origin of the density fluctuations necessary for the subsequent emergence of large-scale structures such as galaxies. As mentioned above, the dominant matter field present during the inflationary era is a scalar field. This field is taken to be largely homogeneous, but it may have small inhomogeneous parts. According to Heisenberg's uncertainty principle of quantum mechanics (discussed at greater length below), these inhomogeneous parts of the field cannot be exactly zero but must be subject to small quantum fluctuations. All types of matter are subject to such fluctuations, and for most purposes the fluctuations are so small that they are totally insignificant. The point of particular interest here, however, is that the huge expansion that the universe goes through during inflation may magnify initially insignificant microscopic fluctuations into macroscopic fluctuations in density. Indeed, detailed calculations show that, subject to certain natural assumptions about the scalar field, the resultant density fluctuations are of the type suggested by Harrison and Zel'dovich.

Thus in a single stroke the idea of inflation appears to solve three important cosmological problems.

However, the role of initial conditions in this "solution" to important cosmological problems should be understood. The hot big bang model can explain the observed features of the universe only with extremely unnatural fine-tuning of initial conditions; finding the present universe would be as unlikely as finding a pencil balanced on its point after an earthquake. The inflationary universe scenario relieves the universe of such extreme dependence on initial conditions by allowing the presently observed state of the universe to arise from a much broader, far more plausible set of initial conditions. Nevertheless, the presently observed state of the universe is not relieved from *all* dependence on initial conditions (figure 1).

Take, for example, the problem of density fluctuations. Certainly it is one of the successes of the inflationary universe scenario that it explains the *origin* of density fluctuations. But more detailed calculation of these fluctuations involves certain assumptions about the initial state of the inhomogeneous parts of the scalar field before inflation. Admittedly these assumptions are very natural, but they are,

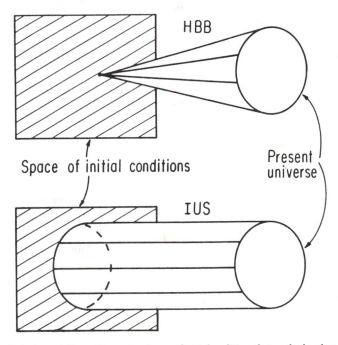

Figure 1. *A schematic figure illustrating the sets of initial conditions that can lead to the present universe in the hot big bang model and in the inflationary universe scenario. In the hot big bang model the present universe can arise, but only from an exceedingly small set of initial conditions. The inflationary universe scenario permits the present universe to arise from a much larger, far more plausible set of initial conditions, but it does not mean that the present universe could arise from* any *initial state.*

nonetheless, assumptions, and the correct density fluctuation spectrum cannot be obtained without them.

More important, the occurrence of inflation itself depends on initial conditions—it occurs only if the scalar field begins with a large, approximately constant energy density. Like it or not, therefore, the success of the inflationary universe scenario is contingent upon certain assumptions about initial conditions.

Where do these assumptions come from? Obviously one can go on asking an almost infinite sequence of such questions, like an overbearingly curious child during the "why" stage. But the cosmologist seeking a complete explanation of the observed state of the universe is ultimately compelled to ask: "What happened before inflation? How did the universe actually *begin*?"

Initial Singularities

The present successful description of the universe with the inflationary universe scenario followed by the hot big bang model involves an expanding universe. Following this expansion backward in time to the pre-inflationary era, we find that the size of the universe tends toward zero while the strength of the gravitational field and the energy density of matter tend toward infinity; that is, the universe appears to have emerged from an *initial singularity*, a region of infinite curvature and energy density at which the known laws of physics break down. Such singularities in cosmological models are not artifacts of unfortunate or inaccurate modeling of the universe, but are *generic*. This is a consequence of the famous "singularity theorems," proved in the 1960s by Hawking and by Roger Penrose of Oxford University. What these theorems show is that, under very reasonable assumptions, *any* model of the expanding universe extrapolated backwards in time will encounter an initial singularity (figure 2).

The physical implication of the singularity theorems is not, however, that a singularity actually occurs in the real universe, but rather that the theory predicting them—classical general relativity—breaks down at very high curvatures and must be superseded by some bigger, better, more powerful theory. What could this more powerful theory be?

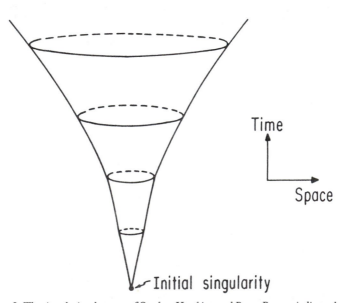

Figure 2. The singularity theorems of Stephen Hawking and Roger Penrose indicate that any reasonable model of the expanding universe will generally encounter a singularity when evolved backwards in time. At such a point, the curvature of the space-time becomes infinite, the known laws of physics break down, and predictability is completely lost.

A line of reasoning known as dimensional analysis yields some clues. Physics has three fundamental constants that control the *scale* of physical phenomena. They are:

- The velocity of light, c, which sets the scale of relativistic effects.
- Newton's constant, G, which is a measure of the strength of gravitational effects.
- Planck's constant, h, which controls the scale of quantum effects.

Around the beginning of this century, Max Planck, the forefather of quantum theory, showed that these three constants may be combined to produce fundamental units of length, time, and mass. They are known, respectively, as the Planck length, denoted l_p, with a value of approximately 10^{-33} centimeters; the Planck time, t_p, approximately 10^{-45} seconds; and the Planck mass, m_p, approximately 10^{-5} grams.

The Planck length and time are almost unbelievably small—smaller in relation to the atomic scale than the atomic scale is to the laboratory scale. The Planck mass may seem an unremarkable number (it is the approximate mass of a cell), but it must be compared to the typical mass scales of elementary particle physics: it is some 10^{19} times greater. The scales indicated by Planck's units are therefore very extreme, indeed. Their significance is that they are the length, time, and mass scales at which relativistic, gravitational, and quantum effects become simultaneously comparable.

Such scales could never be achieved in any laboratory situation, not even in the most powerful particle physics accelerator. These scales *are* approached, however, in the neighborhood of the initial singularity in the universe. This suggests that physics in the neighborhood of the initial singularity is best described using a theory in which relativistic, gravitational, and quantum effects are combined. General relativity is a theory in which relativistic and gravitational effects are already combined. What is needed, therefore, is a quantized version of general relativity—a *quantum theory of gravity*.

Quantum Mechanics

Quantum theory was originally developed in an attempt to account for phenomena whose description appeared to lie beyond the scope of classical mechanics. Classical physics failed to account, for example, for the structure of the atom. Experiments suggested that the atom consisted of a central nucleus with the electrons somehow in orbit about it, like planetary orbits about the sun. Attempts to describe this model mathematically using the classical theories of Newtonian mechanics and Maxwell's electrodynamics led to failure;

there was nothing to hold the electrons in orbit, and they would radiate their energy as electromagnetic radiation, spiraling into the nucleus in a fraction of a second.

With this sort of difficulty in mind, quantum mechanics was developed first by Niels Bohr in 1912 and subsequently by Erwin Schrödinger, Werner Heisenberg, Paul Dirac, and others in the 1920s. In quantum mechanics, motion is not deterministic, as in classical mechanics, but is *probabilistic*. The dynamical variables used to describe a system in classical mechanics, such as position and momentum, cannot in general be ascribed definite values in quantum mechanics. Instead, the fundamental notion in quantum mechanics is that all systems, such as point particles, are fundamentally wave-like in nature. Mathematically, they are taken to be described by a quantity called a *wave function*, into which is encoded probabilistic information about positions, momenta, energies, etc. The wave function for a particular system is found by solving an equation called the Schrödinger equation.

For the case of a single point particle, the wave function may be thought of as an oscillating field spread throughout physical space. At each point in space it has an amplitude and a wavelength. The square of the amplitude is proportional to the probability of finding the particle at that position; the wavelength, for constant amplitude wave functions, is related to the momentum of the particle. The particle will therefore have a definite position if the wave function is tightly bunched about a particular point in space; and it will have definite momentum if the wavelength and amplitude of the wave function are uniform throughout all of space (figure 3). Typical wave functions for a system will not, however, be of either of these types, and there will be a certain amount of indefiniteness, or *uncertainty*, in both position and momentum. In particular, because of the mutually exclusive types of wave function required for definite position and definite momentum, position and momentum cannot be definite simultaneously. This is Heisenberg's uncertainty principle, and it is an elementary consequence of the wavelike nature of particles.

Quantum mechanics exhibits many phenomena that are qualitatively very different to those exhibited by classical mechanics. Two particular quantum phenomena are relevant to this account. The first is the fact that in quantum mechanics a system can never have an energy of exactly zero. The total energy is generally a sum of two positive parts: the first, the kinetic energy, depends on the momentum; the second, the potential energy, depends on position. Since the uncertainty principle forbids both momentum and position to be definite simultaneously, it is impossible to say that both the kinetic

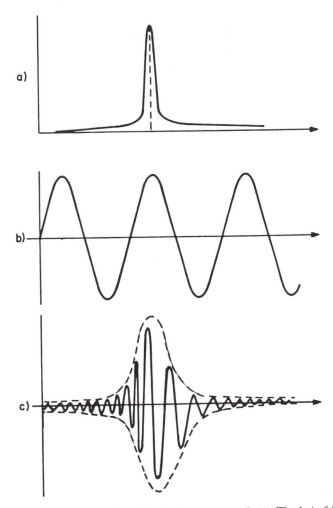

Figure 3. Plots of different types of wave functions in quantum mechanics. The plot is of the wave function against position in real physical space.

- 3A shows the wave function for a particle in a state of definite position. The wave function is sharply peaked about a particular point in space. The uncertainty in momentum, however, is essentially infinite, in that there is equal probability of finding it with any wavelength.
- 3B shows the wave function for a particle in a state of definite momentum. The wave function has a definite wavelength and constant amplitude throughout all of space. The particle position, however, is completely uncertain, in that there is equal probability of finding it to be anywhere.
- 3C shows a wave function for a particle in a state referred to as a "coherent" state, which is the perfect compromise between definite position and definite momentum. There is uncertainty in both position and momentum, but the uncertainty is as small as it can be consistent with Heisenberg's uncertainty principle.

and the potential energy are zero exactly. The system always has a state, called the *ground state*, in which the energy is as small as it can be, consistent with the uncertainty principle. These ground state fluctuations are important for galaxy formation in the inflationary universe scenario, as mentioned above.

The second particular quantum phenomenon relevant to this account is that of "tunneling." In classical mechanics, a particle traveling with fixed energy cannot get through an energy barrier. For example, a ball at rest in a bowl will never be able to get out. In quantum mechanics, in contrast, because position is generally not a sharply defined quantity but has a spread over a (typically infinite) range, there is a finite probability that the particle will be found on the other side of the barrier. We say that the particle may "tunnel" through the barrier. This purely quantum mechanical effect is responsible for alpha-decay in radioactive atoms; the alpha particle is trapped by an energy barrier inside the nucleus but manages to escape by tunneling. The tunneling process should not, however, be thought of as actually occurring in real time. In fact, in a certain well-defined mathematical sense, the particle is conveniently thought of as penetrating the barrier not in real time, but in "imaginary" time (time multiplied by the square root of minus one).

These distinctly quantum-mechanical effects—*ground state fluctuations* and *tunneling*—should not be thought of as being in contradiction with classical physics. Rather, the point of view typically taken is that quantum mechanics is a *broader* theory than classical mechanics and supersedes it as the correct description of nature. On macroscopic scales, it is usually argued that the wave-like nature of particles is highly suppressed and quantum mechanics reproduces the effects of classical mechanics to a very high degree of precision (although the exact manner in which this "quantum to classical transition" comes about is still a matter of current research). On microscopic scales, however, the predictions of quantum mechanics depart quite radically from those of classical mechanics. For example, the orbiting electron model of the atom discussed earlier is saved from the rather singular behavior predicted by classical mechanics by virtue of the phenomenon of ground state fluctuations: the electrons have an orbit of minimum energy from which they cannot emit radiation and fall into the nucleus without violating the uncertainty principle.

The important point to take away from this discussion of quantum mechanics is that in situations in which classical physics predicts singular behavior, the broader viewpoint of quantum mechanics may indicate that classical mechanics is invalid in such situations. A quantum treatment may replace the singular behavior with regular behavior, describable well within the laws of quantum physics. It is this observation that suggests that singularities in classical cosmology could be alleviated by quantum theory.

Quantum Cosmology

A modest approach to the issue of singular behavior might involve allowing the gravitational and matter fields to be classical for much of the evolution of the universe and investigating quantum effects only in the immediate vicinity of the initial singularity. A far more comprehensive and complete approach would be to apply quantum theory to the entire universe at all times and to everything it contains, i.e., to all the gravitational and matter fields describing the universe. To do this is to do quantum cosmology.

In quantum cosmology, as for a simple system such as an atom in ordinary quantum mechanics, the fundamental description of the system—the entire universe in this case—is in terms of a wave function, "the wave function of the universe." This wave function is found by solving an equation that is the cosmological analog of the Schrödinger equation, and is known as the Wheeler-DeWitt equation. Given such a wave function, we can, in principle, extract probabilistic predictions concerning the outcome of observation. In the simplest models of the universe, the analog of position is the spatial size of the universe and the analog of momentum is the rate of expansion of the universe.

Quantum Gravity

Many conceptual and technical difficulties arise in quantum cosmology, above and beyond those in quantum mechanics. The first and most serious difficulty is the lack of a complete and manageable quantum theory of gravity directly applicable to cosmology. Three of the four fundamental forces of nature—electromagnetic, weak nuclear, and strong nuclear forces—have been made consistent with quantum theory, i.e., *quantized*, in a satisfactory fashion. In contrast, all attempts to quantize Einstein's general relativity directly have met with failure.

There exists a theory, called the theory of "superstrings," which is claimed to be a consistent, unified quantum theory of all four forces of nature and thus is, or at least contains, a quantum theory of gravity. Final judgment on superstring theory is yet to be made, however, and although it may well be the sought-after Holy Grail of a quantum theory of gravity, it is a long way off from being a manageable theory directly applicable to cosmology.

From the very beginning, therefore, quantum cosmology rests on shaky foundations. It is possible to develop a formalism and to perform some rather crude calculations, and this in practice is what we actually do in quantum cosmology. But unlike, for example, the quantized theory of the electromagnetic field, there is no systematic scheme for calculating the quantities of interest to arbitrary precision, and thus it is difficult to know whether our calculations have anything to do with a consistent quantum gravity theory (should it exist), let alone with observation.

Quantum Mechanics Applied to Cosmology

Not least of all for these reasons, much of the work carried out in quantum cosmology has focused on *issues* rather than on detailed technicalities. Undoubtedly the most difficult, interesting, and important issues concern the application of quantum mechanics to cosmology.

First, there is the issue of the *applicability* of quantum mechanics to the entire universe. Quantum mechanics was developed to describe atomic scale phenomena that were totally inexplicable from the perspective of classical mechanics. The beautiful agreement of quantum mechanics with experiment is one of the great triumphs of modern physics, and no physicist in his or her right mind is in any doubt as to its correctness on the atomic scale. We may, however, find a few dissenters if we suggest that quantum mechanics is equally applicable to the everyday macroscopic scale—for example, to tables and chairs. It is very hard to dispute this suggestion because, as we have already seen, the predictions of quantum mechanics coincide very closely with those of classical mechanics on this scale, and genuinely macroscopic quantum mechanical effects are extremely difficult to detect experimentally. Even more contentious is the most extravagant extrapolation we could conceivably make: that quantum mechanics applies to the entire universe at all times and to everything in it. Acceptable or not, this is the fundamental assertion of quantum cosmology.

Second, there is the issue of the *interpretation* of quantum mechanics as applied to cosmology. Because quantum mechanics generally prohibits the familiar variables of classical mechanics from taking definite values, it was found necessary in the development of the subject to supplement the formalism with an extra structure to translate the mathematical formalism into statements about what we would actually observe on making a measurement. The foundations of this structure, known as "quantum measurement theory," were laid primarily by Bohr in the 1920s and 1930s. Central to Bohr's scheme was the notion that the observer plays a crucial role in the act of measurement. Bohr assumed that the world may be divided into two parts:

microscopic systems (such as atoms) governed purely by quantum mechanics, and external macroscopic systems (such as observers and their measuring apparatus) governed by classical mechanics. A measurement is an interaction between the observer (or his or her measuring device) and the microscopic system leading to a permanent recording of the event. During this interaction, the wave function describing the microscopic system undergoes a discontinuous change from whatever initial state it was in to a final state in which it is definite in the quantity that is being measured. This discontinuous change is referred to, rather dramatically, as "the collapse of the wave function." For example, the wave function could start out in a state of definite momentum, but if position is being measured it "collapses" into a state of definite position.

This scheme, known as the "Copenhagen interpretation" of quantum mechanics, is felt by many to be deeply unsatisfactory from a philosophical point of view. It cannot be denied, however, that the Copenhagen interpretation has been thoroughly successful in allowing predictions to be extracted from the theory—predictions that agree with experiment. It is perhaps for this reason that the Copenhagen interpretation stood largely unchallenged for almost half a century. Indeed, physicists' reticence to abandon the Copenhagen view of the world has been aptly summarized by Murray Gell-Mann of the California Institute of Technology: "Most physicists don't believe there is a problem with quantum mechanics because Bohr brainwashed a whole generation of physicists into thinking so fifty years ago."

In attempting to apply quantum mechanics to the entire universe, however, we meet with acute difficulties that cannot be brushed off as philosophical niceties. In a theory of the entire universe, of which the observer is part, there should be no fundamental division between observer and observed. Moreover, although it is almost acceptable to believe that it happens at the atomic scale, most physicists feel very uncomfortable at the thought of the wave function of the entire universe collapsing when an observation of the universe is made. There is also the question of probabilistic predictions. What do these mean in quantum cosmology? In ordinary quantum mechanics, we envisage an ensemble of many identical systems (such as atoms), and probabilistic predictions may be tested by making a large number of measurements. But what do probabilities mean when we have just one system which we measure only once?

Many Worlds

One of the first physicists to take seriously the notion that we should apply quantum mechanics to the entire universe was the late Hugh Everett III of Princeton. In a 1957 paper he presented a framework for the interpretation of quantum mechanics particularly suited to the special needs of cosmology. Everett asserted that the universal wave function should describe macroscopic observers and microscopic systems alike, and that there should be no fundamental division between them. He further asserted that there is no discontinuous change brought about by the collapse of the wave function during the measurement process, but only smooth evolution described by the Schrödinger equation. A measurement is just an interaction between different parts of the entire universe, and the wave function should predict what one part of the system "sees" when it observes the other part. But in modeling the measurement process, Everett discovered a truly remarkable thing: the measurement appears to cause the universe to "split" into sufficiently many copies of itself to take account of all possible outcomes of the measurement. [See chapter 28, *Multiple Universes*, in section VIII, *Cosmology and Philosophy*.]

The reality of these multiple copies of the universe in this severely uneconomical aspect of Everett's "many worlds" interpretation of quantum mechanics has been hotly debated. Modern versions of this interpretation, notably by Gell-Mann and Hartle, play down the "many worlds" aspect of the theory and talk instead about "decoherent histories." These are possible histories for the universe to which probabilities may be assigned. For practical purposes, it does not matter whether we think of all or just one of them as actually happening. Gell-Mann and Hartle also address the problem of understanding probabilistic predictions for the entire universe. They insist that the only probabilities that have any meaning in quantum cosmology are *a priori* probabilities. These are probabilities that are very close to one or zero—definite yes-no predictions. Although most probabilistic predictions are not of this type, they can often be made so by suitable modification of the question we are interested in. Unlike quantum mechanics, therefore, in which the goal is to determine probabilities for the possible outcomes of given observations, the goal in quantum cosmology is to determine those observations for which the theory gives probabilities close to zero or to one.

Research on this fascinating topic is still in progress, but for the purposes of practical quantum cosmology, the following understanding has emerged: in certain regions, typically but not always when the universe is large, the wave function for the universe indicates that the universe behaves classically to a very high degree of precision. In this case we would say that classical space-time is a prediction of the theory. Under these conditions, moreover, the wave function provides probabilities for the set of possible classical behaviors of the universe. There will, on the other hand, be other regions in which no such prediction is made. In such regions the notions of space and time quite simply do not exist; there is just a "quantum fuzz," still describable by known laws of quantum physics but not by *classical* laws (figure 4). Such regions, as we will discuss below, may surround singularities.

Laws of Initial Conditions

Armed with a quantum theory of cosmology and a means of interpreting it, we may begin to ask about the beginning of the universe. It should, however, be emphasized that the problem of initial conditions is not actually *solved* by going to quantum cosmology. Going to quantum cosmology certainly puts us in a much better position to address the issue than were we to remain in classical cosmology, because there is no longer any worry that we might be trying to impose classical initial conditions in a region in which classical physics is not valid, such as near the initial singularity. But the question of classical initial conditions becomes one of quantum initial

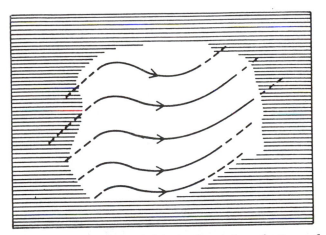

Figure 4. A schematic picture showing the behavior of a typical wave function as a function of, for example, the size of the universe, the energy density of matter, etc. In certain regions the wave function indicates that the notion of classical space-time is an appropriate one. Precisely, the wave function becomes strongly peaked about sets of classical histories for the universe. These are denoted by the bold lines. There will, however, be other regions in which classical space-time is not indicated to be a valid notion. These regions of "quantum fuzz" are denoted by the shaded region. They are still describable by quantum laws of physics, but not by classical laws.

conditions. It has become the following question: of the many possible wave functions permitted by the dynamics (i.e., of the many possible solutions to the Wheeler-DeWitt equation), how is just one singled out?

The cosmological situation should be contrasted with the laboratory situation, to which most of physics is directed. There, we have a physical system whose temporal and spatial boundaries are generally quite clearly defined. At those boundaries the experimenter may control, or at least observe, the physical conditions. Given suitable laws of physics, these initial or boundary conditions may be evolved in space and time, thereby predicting the physical state inside the system under consideration. This may then be compared with experiment or observation.

In cosmology, on the other hand, the system under scrutiny is the entire universe. By definition, the system has no exterior, no outside world, no "rest of the universe" onto which specification of boundary or initial conditions may be passed off (figure 5). It is therefore the

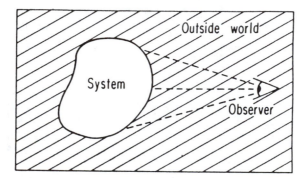

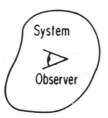

Figure 5. The difference between laboratory physics and cosmology. In laboratory physics the observer, who sits outside the system, may control (or at least observe) the external conditions, and these are used as the boundary conditions when determining what is going on inside the system. In cosmology, on the other hand, not only is the observer on the inside, but there is no outside world into which the specification of boundary or initial conditions can be passed off.

almost inescapable task of the quantum cosmologist to *propose* laws of initial or boundary conditions for the universe, in much the same way that it is the task of the theoretical physicist to propose laws of physics to govern the evolution of physical systems. Like such laws of evolution, the ultimate test of any law of initial conditions will be whether its predictions agree with observation.

There is a possible escape clause relieving the quantum cosmologist of his or her onerous task, which is why the word "almost" appears in the above paragraph. In his seminal 1967 paper, Bruce DeWitt expressed the hope that mathematical consistency alone would lead to a unique solution to the Wheeler-DeWitt equation, to a unique wave function for the universe. Such a possibility is entirely without precedent in theoretical physics and there is, unfortunately, little indication that DeWitt's hope will ever be fulfilled.

It was the realization by Hartle and Hawking, by Linde, and by Vilenkin that we must face up squarely to the issue of initial conditions in quantum cosmology that led to the revitalization of the subject in the early 1980s. These scientists made quite definite proposals that were intended to pick out a particular solution to the Wheeler-DeWitt equation, to single out a unique wave function for the universe.

The proposal made by Hartle and Hawking is known as the "no-boundary" proposal. It defines a particular wave function of the universe using a rather elegant formulation of quantum mechanics developed in the 1940s by the late Richard Feynman of Caltech, referred to as the "path integral" or "sum-over-histories" method. The proposal (made independently by Linde, by Vilenkin, and by others) is known as the "tunneling proposal," and is a particular way of picking out a solution to the Wheeler-DeWitt equation. Each of these proposals picks out a unique wave function for the universe (contingent, however, upon the resolution of a number of technical difficulties recently exposed by Hartle, Jorma Louko, and me). In both of these proposals, the wave function indicates that space-time is classical when the universe is reasonably large (typically larger than a few thousand Planck lengths), in agreement with observation. When the universe is very small, however, classical space-time is not indicated. In fact, both of these proposals view the beginning of the universe as being much like a quantum tunneling event; the universe "tunnels" to finite size, starting from zero size; it starts from "nothing" (figure 6). These proposals therefore have the desirable feature that the point that was an initial singularity in classical cosmology is surrounded by some kind of fuzzy quantum region.

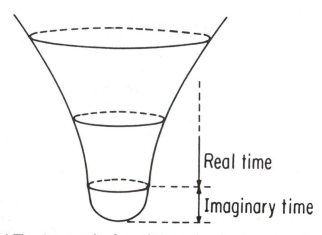

Real time

Imaginary time

Figure 6. The universe tunneling from nothing, as in the no-boundary and tunneling proposals for the wave function of the universe. The universe "starts" at zero size in a perfectly smooth way and evolves in imaginary time to a size of a few Planck lengths. Taking that small but finite size as its initial condition, the universe evolves thereafter in real, physical time. Although everything now appears to be smooth and nonsingular, the singularity theorems are not violated because the "smoothing off" (depicted in figure 2) is in imaginary time, while the singularity theorems refer to real time only.

Given these quantum theories of the creation of the universe, we may finally ask: "How did the universe actually begin?" The response of quantum cosmologists to this question may be something of a disappointment. Rather than answer the question, we would declare the question disqualified. In the neighborhood of singularities, such as the initial singularity, the wave functions given by the tunneling and no-boundary proposals indicate that classical general relativity is not valid and, furthermore, that notions of space and time are inappropriate. We cannot, therefore, ask questions involving notions of time or space. The picture that emerges is of the universe appearing from a quantum fuzz with non-zero size and finite (rather than infinite) energy density (figure 7).

In the subsequent classical evolution after quantum creation, the wave function provides a probability measure on different possible evolutions, and it is at this point that the no-boundary and tunneling wave functions differ. Workers in the field are not in complete agreement about this, but the current understanding appears to be that the tunneling wave function gives high probability for universes that un-

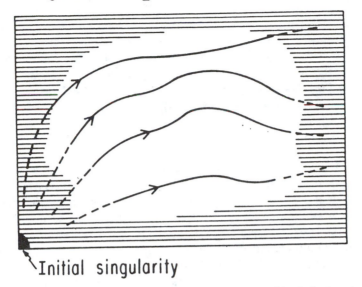

\Initial singularity

Figure 7. A schematic picture showing the possible classical (real time) histories for the universe indicated by the no-boundary and tunneling wave functions. The initial singularity is surrounded by a quantum fuzz, denoted by the shaded region (which actually corresponds to the imaginary time region in figure 6), where no notions of space and time exist. An observer in the present looking back in time would see the classical histories for the universe (denoted by the bold lines) emerging from this quantum fuzz at finite size in a nonsingular way.

dergo an initial period of inflation of sufficient duration to solve the horizon and flatness problems; the no-boundary proposal, on the other hand, does predict inflation, but does not appear to predict a *sufficient* amount of inflation.

The no-boundary and tunneling proposals also make predictions about the emergence of large-scale structure. Recall that in the inflationary universe scenario, although the origin of the initial density perturbation spectrum is explained, the exact form and magnitude depend on certain assumptions about the initial state of the scalar field. To be precise, the necessary assumption is essentially that the inhomogeneous parts of the scalar field start out in their quantum mechanical ground state—the lowest possible energy state consistent with the uncertainty principle. This is a natural assumption, but it is an assumption nonetheless. In a 1985 paper, Hawking and I demonstrated this assumption to be a *consequence* of the no-boundary proposal.

Conclusion

The achievements of quantum cosmology may be briefly summarized:

- First, quantum cosmology allows us to determine the conditions under which the notions of classical space and time are valid, and thus the conditions under which classical theories such as general relativity may be used.

- Second, quantum cosmology permits us to compress into a simple law of boundary conditions for the wave function of the universe the list of assumptions necessary for conventional cosmology consisting of inflation followed by a hot big bang.

How, though, can we test proposals of initial boundary conditions? This is an extremely difficult issue observationally. Much has happened in the universe since it began, and each stage of its evolution has to be modeled separately. While making observations now, it is difficult to distinguish between effects due to a particular set of initial conditions and effects due to a particular evolution of the universe. What is needed is an observation of some kind of effect produced at the very beginning of the universe and insensitive to the subsequent evolution of the universe. Leonid Grishchuk of the Sternberg Astronomy Institute in Moscow has argued that gravitational waves may be the sought-after probe of the very early universe. In quantum creation scenarios, gravitational waves of a calculable form and magnitude are produced, and they subsequently propagate largely without further hindrance, to be observed in the present day universe. Unfortunately, the detection of gravitational waves is extremely difficult, but we may hope for interesting results in the not-too-distant future.

It may be that the no-boundary proposal or the tunneling proposal are the correct boundary conditions for the wave function of the universe. But the severe problems of actually testing quantum cosmology observationally make this extremely difficult to check. It could be a very long time, therefore, before we can tell whether either of these proposals is an answer to the question at the beginning of this article: "Where did all this come from?" Nevertheless, through quantum cosmology we have at least been able to formulate and address the question in a meaningful way.

—Jonathan J. Halliwell

BIBLIOGRAPHIC NOTE

This chapter is an expanded version of a shorter article, "Quantum Cosmology and the Creation of the Universe," *Scientific American* (December 1991), pp. 76–85. One of the first papers published on quantum cosmology was B.S. DeWitt, "Quantum Theory of Gravity I: The Canonical Theory," *Physics Review*, *160* (1967), 1113–1148. Key papers that initiated more recent work on quantum cosmology include J.B. Hartle and S.W. Hawking, "The Wave Function of the Universe," *Physics Review, D28* (1983), 2960–2975; A.D. Linde, "Quantum Creation of the Inflationary Universe," *Lettere al Nuovo Cimento, 39* (1984), 401–405; and A. Vilenkin, "Quantum Creation of Universes," *Physical Review, D30* (1984), 509–511. A recent review of quantum cosmology is by J.J. Halliwell, "Introductory Lectures on Quantum Cosmology," in S. Coleman, J.B. Hartle, and T. Piran, eds., *Proceedings of the Seventh Jerusalem Winter School on Theoretical Physics: Quantum Cosmology and Baby Universes* (Singapore: World Scientific, 1991), pp. 159–243. This article contains an extensive guide to the literature on quantum cosmology. In the same volume is a review by Hartle of the decoherent histories approach, "The Quantum Mechanics of Cosmology," *ibid.*, pp. 65–157. A slightly earlier account of the decoherent histories approach is M. Gell-Mann and J.B. Hartle, "Quantum Mechanics in the Light of Quantum Cosmology," in W.H. Zurek, ed., *Complexity, Entropy, and the Physics of Information, Santa Fe Institute Studies in the Sciences of Complexity, vol. VIII* (Redwood City, California: Addison-Wesley, 1990). The many-worlds view of quantum mechanics is described in a collection of articles by Hugh Everett and others in B.S. DeWitt and N. Graham, eds., *The Many Worlds Interpretation of Quantum Mechanics* (Princeton: Princeton University Press, 1973).

VIII. Cosmology and Philosophy

Introduction

THINKING ABOUT THE UNIVERSE

e pense, donc je suis. With these words the seventeenth-century philosopher René Descartes asserted that he thought, and thus he existed. *I think, therefore I am* has become a widely acknowledged starting point of modern philosophy. And philosophers have not been lax when it comes to thinking about cosmological matters. Descartes, himself, was one of the first modern cosmologists, attempting to bring a single set of principles to bear on terrestrial and celestial phenomena. Though now overshadowed by Newton's science of motion, Descartes' coherent and plausible cosmology both destroyed ancient doctrine and held a commanding position of its own in natural philosophy for nearly a century. (Indeed, readers might well complain that there is not an essay on Descartes and his cosmology in this volume. There is just such an essay, however, in the *Encyclopedia of Cosmology: Historical, Philosophical, and Scientific Foundations of Modern Cosmology*.)

For some contemporary philosophers, Descartes's *Je pense, donc je suis* might be expanded to *I think, therefore I am, and therefore the universe is fit for human survival*. Obviously this is so, or we would not be here thinking about it. Recent evidence, however, suggests that the universe is just barely fit for human survival; if several of the basic physical features of the universe had been even slightly different, human life could never have arisen. Inevitably on a planet populated with thinking beings, there have followed attempts to account for the fact that the universe seems "finely tuned" for human life, attempts to explore the relationship between the presence of human life in the universe and the physical conditions that make such life possible in the first place. From all this thinking has risen the assertion that the presence of human life somehow constrains physical conditions and, in some sense, guarantees that the universe is habitable by human beings. Philosophers have dignified the idea that life in *some* way

501

constrains the fundamental features of the universe with a label, the "anthropic principle."

The anthropic principle hints at the existence of deep connections between the universe and humanity, makes striking claims about mankind's place in the universe, and evokes classical arguments for a divine cosmic designer. The principle involves additional philosophical and religious themes as well: the significance and value of human life, the goal-directedness of nature, and the degree to which the universe is anthropocentric. Although defended by prominent scientists, philosophers, and theologians, the anthropic principle is controversial and is subject to strong criticism.

Another highly controversial cosmological-philosophical idea is the plurality of worlds, the concept of other worlds beyond the earth. In the seventeenth century, debate centered on the possibility of earth-like planets complete with intelligent inhabitants. In the twentieth century, the plurality of worlds concept has bloomed into the extraterrestrial life debate, labelled *exobiology* by biologists and *bioastronomy* by astronomers. It is difficult to verify the existence of extraterrestrial life and other worlds, and the debate over a plurality of worlds thus incorporates a good measure of philosophy and a correspondingly smaller proportion of science. Critics have questioned whether the possible existence of extraterrestrial life is a scientific issue at all.

Religious controversy helped spread the idea of a plurality of worlds. [See chapter 31, *Cosmology and Religion*, the concluding chapter in the concluding section of this book.] So too did a number of social and political critiques appearing in the seventeenth century. A voyage to the moon increasingly was used as a literary convention, with the social and political organization of the lunar inhabitants becoming either the model of a perfect society or the reflection of all the vices of the earth's society.

Literary voyages to the moon helped to spread the new ideas and, not incidentally, furnished the critics some defense against outraged monarchs. Samuel Colville, in his 1682 *Whiggs Supplication*, described what one might see through the telescope:

> *If he once level at the moon,*
> *Either at midnight or at noon,*
> *He discovers* Rivers, Hills,
> Steeples, Castles, *and* Wind-Mills,
> Villages *and* fenced Towns,
> *With* Foussies, Bulwarks, *and* great Guns,
> *Cavaliers on horse-back prancing;*
> *Maids about a may-pole dancing;*
> *Men in Taverns wine carousing,*
> *Beggars by the high-way lowfing,*

Soldiers forging ale-house brawlings,
To be let go with their lawings . . .
Young wives old husbands horning,
Judges *drunk every morning,*
Augmenting law-fruits and divisions,
By Spanish *and by* French *decisions;*
Courtiers their aims missing,
Chaplains wido-ladies kissing;
Men to sell their lands itching,
To pay th' expences of their kitching,
Physicians cheating young and old,
making both buy death with gold . . .

Also helping publicize the concept of a plurality of worlds was the outburst of enthusiasm in England in response to persistent rumors that England intended to colonize the moon. Probably the intention was not to subjugate native Lunarians, but to send Englishmen to a previously uninhabited moon, because the rumors likely were begun by John Wilkins's quotation of Kepler's comment that as the art of flying were developed the successful nation would transplant a colony to the moon. None of the rumors were realized, but their publicity value was little diminished. The rumors also provided inspiration for Samuel Butler's satirical *The Elephant in the Moon*:

A virtuous, learn'd society, of late
The Pride and Glory of a foreign state,
Made an Agreement of a Summer's Night,
To search the Moon *at full, by her own Light;*
And draw Maps of her prop'rest situations
for Settling, and erecting new Plantations;
If ever the Society *should incline*
T' attempt so great, and glorious a Design;
A Task in vain, unless the German Kepler
had found out a Discovery to people her, . . .

The plurality of worlds, initially a doctrine more familiar to philosophy and literature than to science, was eventually accepted by astronomers as well. Paradoxically, it was in the eighteenth and nineteenth centuries, after the demise of the idea of an anthropocentric universe in the field of astronomy, that the idea of the Great Chain of Being received its widest circulation. The belief that there was a scale of nature, of divine design, and reading from the simplest animals at the bottom to man at the top, was a mind-set more appropriate to an earlier age before the idea of an anthropocentric universe had been rendered obsolete. The revolution in cosmological thought, though, seems not to have forestalled a later blooming of anthropocentric ideas in the biological sciences.

Curiously enough, however, the revolution in cosmological

thought that began in 1917 with Einstein's publication of his general theory of relativity rekindled thoughts of other worlds, although frequently of a different nature than earlier speculations. Like their predecessors, modern cosmologists have entertained concepts of universes other than our own, separated from our own in time, in space, or in other dimensions. But there their resemblance ends; totally divergent from the tradition of earlier other-worlds speculation is that which takes place within quantum theory today. Although this might well be expected, given the surprising originality of most quantum concepts, the degree of unfamiliarity that quantum other-world speculation presents can still prove distressing, especially to conservative physicists. Indeed, one of them referred to multiple universe theorizing as "comic book science." Many-worlds theorizing, however, is consistent with the philosophical foundations of quantum theory. Furthermore, the philosophical foundations of quantum theory both imply and are strengthened by many-worlds theorizing, at least according to some physicists.

In addition to the philosophical probings associated with many-worlds quantum theory, the twentieth-century revolution in cosmological thought has brought about challenges to the most general philosophical framework of physics itself. This framework was originally hammered into place through discussions, debates, and disputes among early modern physicists and philosophers, including Kepler, Descartes, Bacon, Locke, Leibniz, and Newton. It included consensus agreements about all the major philosophical issues, especially the three basic questions:

- What are acceptable kinds of objects to play the role of theoretical entities?

- What are acceptable sources and structures for scientific knowledge?

- What are acceptable methods and logic for constructing theories?

For almost three centuries, the received set of interrelated answers to these three questions guided scientific research, theory construction, and growing professionalization. But in the 1930s the proposal that the general theory of relativity, in conjunction with the expanding universe interpretation of galactic redshift observations, together implied that we live in a universe of curved and expanding four-dimensional space-time, almost immediately induced a fatal tension among the elements of the philosophy of science revered since the late seventeenth century. By the time a new consensus had formed around big bang cosmology, a new and oftentimes disconcerting philosophy of science had come into being.

26. THE ANTHROPIC PRINCIPLE

The universe is fit for human survival. Obviously this is so, or we would not be here discussing it. Recent evidence, however, suggests that the universe is just barely fit for human survival. If several of its basic physical features had been even slightly different, human life could never have arisen. The anthropic principle is an attempt to account for the fact that the universe seems "finely tuned" for human life by showing that, in some sense, the universe must be habitable by beings like us.

Discussion of the anthropic principle is considerably complicated by the fact that several principles commonly called anthropic are, though formulated in roughly the same way, of quite different status. All of them have to do with the relationship between the presence of human life in the universe and the physical conditions that make such life possible in the first place. They claim that the presence of human life somehow constrains these physical conditions and, in some sense, guarantees that the universe is habitable by human beings. I shall use the term "anthropic principle" in a general sense, to refer to this idea that life in *some* way constrains the fundamental features of the universe. Logical, methodological, and teleological versions of this general approach will be distinguished. I avoid the terms "strong" and "weak" anthropic principle, though they are common in the literature on the subject, because they do not accurately represent the logical differences between various anthropic ideas and because they have in practice become hopelessly equivocal.

Philosophical Aspects of the Anthropic Principle

Appealing to an anthropic principle in order to explain fundamental physical features of the universe is an attractive option for several reasons. Whereas standard scientific explanations fall short of providing satisfactory accounts of the ultimate origins and structure of the cosmos, the anthropic principle provides some hope of picking up where such explanations leave off. What makes the anthropic principle approach provocative is that it hints at the existence of deep connections between the universe and humanity, makes striking claims about mankind's place in the universe, and is reminiscent of classical arguments for a divine cosmic designer. The principle involves additional philosophical and religious themes as well: the significance and value of human life, the goal-directedness of nature, and the degree to which the universe is anthropocentric. Although it is defended by some prominent scientists, philosophers, and theologians, the principle, at least in some of its versions, is quite controversial and is subject to a number of telling criticisms.

The philosophical aspects and implications of the anthropic principle, which are the subject of this article, must be distinguished from the scientific questions associated with the anthropic principle. The scientific questions include:

- Claims about how the universe would have differed if various of its physical constants and initial conditions had been different.

- Alternative standard scientific explanations for the cosmic "fine tuning."

- Physical models involving multiple universes, along with theories concerning their relationships and consideration of the manner in which they arise and evolve.

- The physical circumstances necessary for the survival of various forms of life.

Leaving the above scientific issues for cosmologists and biologists, we shall be concerned with:

- Distinguishing the various versions of the anthropic principle with regard to their logical and scientific status.

- Evaluating the mode of explanation being provided by each version of the principle.

- Scrutinizing the principle's anthropocentric character.

- Investigating its scientific status and religious significance.

Versions of the Anthropic Principle

The anthropic principle is motivated by a desire to explain the fact that minor alterations of basic features of the world—conditions in the very early universe, the values of physical constants, the strengths of the fundamental forces, and the masses and charges of subatomic particles, for example—would have rendered the universe unsuitable for human habitation. The alterations in question are in some cases remarkably slight. For example, an increase in the early expansion rate of the universe by one part in a million would have meant that no galaxies would have formed. An increase in the strength of the gravitational force by a similar amount would have ensured that stellar lifetimes would be too short for the biological evolution of an intelligent species to occur. These scenarios, and many others, illustrate the claim that the physical requirements of human life are just barely met, and that the universe seems remarkably finely tuned to our needs—so remarkably, in fact, that the probability of getting a life-supporting universe out of all the possibilities is arguably zero.

- Why, of all the possible lifeless universes, is the actual universe so remarkably fit for life?
- What principle of selection is at work?
- What constraints on universe formation have taken place?
- Was this fitness somehow guaranteed, or was it merely a fortuitous occurrence?

Various versions of the anthropic principle attempt to answer these questions.

In the 1974 article in which he coined the term "anthropic principle," Brandon Carter describes a logical constraint on the observations of the universe that intelligent beings can make: such beings cannot (obviously) find themselves in spatial or temporal regions of the universe that are uninhabitable by intelligent beings. The universe, for example, must be old enough for stars and planetary systems to have formed, but not so old that all stars have cooled beyond the point of being able to support habitable planets. To generalize somewhat, all properties of the universe (not just those that may change with time or location), including its initial conditions and the values of its physical constants, must be consistent with the presence of intelligent beings, given that such beings are around to study the properties. This logical principle says no more than that one cannot observe what is inconsistent with the physically necessary prerequisites of one's existence. Carter's logical anthropic principle is a requirement of consistency, and is entirely uncontroversial.

Carter and others have also emphasized the methodological construal of this principle. Working scientists are advised to take account of the logical principle when assessing coincidences between seemingly unrelated phenomena or the significance of various observational data, especially when fine tuning for human existence is apparent. It may turn out that the "fine tuning" of the data is no more than the result of the logical constraints discussed above. For example, the remarkableness of the large-scale homogeneity of the universe is reduced once one realizes that humans could not exist if the universe were much different. Given that we exist, and that large-scale homogeneity of the universe is a necessary condition for our existence, then the universe is guaranteed, in a logical sense, to be homogeneous. It is important to realize this logical guarantee when contemplating the significance of any feature of the universe that happens to be a necessary condition for any actually existing beings. This methodological principle is no more controversial than its logical counterpart.

The way in which the presence of intelligent life guarantees the fulfillment of its own necessary conditions can also be interpreted teleologically. The teleological anthropic principle, suggested by John Barrow and Frank Tipler in a thorough recent defense of the principle, states that the universe is habitable precisely *so that* intelligent life can evolve in it. The principle does not argue from the presence of intelligent life to the fulfillment of the necessary conditions of such life; rather, it asserts that habitability is the goal or end of the universe, and for this reason the universe must be habitable. Thus, this principle differs sharply from the logical principle, and is not a stronger version of it, though it is certainly more controversial.

Several other ideas have been labeled anthropic principles, but because of their highly speculative nature they will not be discussed in detail here. The so-called participatory anthropic principle, suggested by John Wheeler, is an extension of the idea that the observer retroactively creates phenomena in delayed choice experiments. The claim is that the universe must be habitable because observers are necessary to make it real. Even Wheeler admits this is exceedingly speculative and notes that the observer in question need not be anything like a human being. Barrow and Tipler have proposed a "final" anthropic principle to the effect that life must arise in the universe, and then must never die out (otherwise it is hard to see why it must arise in the first place). Because the human species cannot survive into the late stages of the universe, as Barrow and Tipler readily admit, this principle is anthropic in only a weak sense.

Before turning to the several philosophical issues associated with the anthropic principle, a few words should be said about the common association of the logical version of the principle with observational selection effects. A selection effect is a systematic bias in observational data, commonly caused by the limitations of an instrument or by the restriction of what is to be observed. For example, a telescopic survey of stars in our galaxy will be biased in favor of bright stars because telescopes cannot detect very dim ones. The survey does not represent a random sample of stars in the galaxy, because certain types of stars have been "screened off" from observation. Similarly, bias may be present in telephone surveys because the views of people who do not own phones or who do not answer them will not be represented. Such surveys are not representative of all people's opinions, because they do not draw on a truly random sample of the population.

The logical anthropic principle shares some features, but not others, with the sources of such observational selection. The sort of universe we live in is not a random sample from among all possible universes, and so resembles the nonrandom samples mentioned above. At a minimum, we know that the universe must satisfy the necessary conditions for our existence; and it appears that such conditions are satisfied in only a small portion of possible universes. Because it is not even possible that we observe an uninhabitable universe, the fact that we inhabit the universe (and that the necessary conditions for our existence are fulfilled) in a sense ensures that a habitable universe has been "selected" from all the possible universes. The methodological anthropic principle is an exhortation to take into account the fact that certain features of the universe, such as its proximity to homogeneity, may be the result of this selection rather than of some undiscovered cause or of some sort of cosmic design.

There are significant differences, however, between the causes of selection effects and this logical principle. The screening off involved in the latter is of a logical, not observational, sort. No actually observable data or information is being screened off, as in the case of the telescope and the phone survey. All that is being screened off are conditions that, given our existence, cannot possibly obtain, and hence cannot be observed. Rather than being a bias in favor of what we choose to observe or one due to the properties of an instrument, the bias in this case is a bias in favor of the possible. No intervention or manipulation on the part of the observer or observing apparatus is involved. Unless one interprets "selection" in a very broad sense, this bias cannot be said to be of the same sort as that which often skews experimental results and gives rise to selection effects.

Explanatory Power of the Anthropic Principle

To what extent is the anthropic principle successful as an explanation? The logical form of the principle can be explanatory in only a very weak sense. Taking it into consideration at most reduces some of the initial puzzlement over certain unexplained physical phenomena. Even if reduction of puzzlement is taken as a step towards a satisfying explanation, the scientific claims about what the necessary conditions are for human existence are more central to the explanation than is the logical principle. Much puzzlement remains, because the logical principle makes no progress in explaining *why* the constants of nature and conditions in the early universe are such in the first place.

It might be possible to supplement the logical principle with a multiple-universe scenario and thus to arrive at a fuller explanation. If there were an ensemble of actual universes that together embody all physical possibilities, then this ensemble, together with the logical principle that we must find ourselves in a habitable member of the ensemble, would guarantee the actuality of a habitable universe like ours. But this guarantee would be obtained in virtue of the fact that all possible universes are actual, not in virtue of any anthropic principle. The ensemble, not the anthropic principle, does the explanatory work in such a scenario. There are significant problems with this version of an anthropic explanation:

- The ensemble is decidedly nonanthropic, because most of its members presumably are lifeless.

- Commitment to a vast multitude of physical entities above and beyond what can possibly be observed represents an exorbitant ontological price.

- It is incumbent on defenders of this view to provide a plausible account of how the multiple universes evolve and diversify.

A further complication of this line of thinking is that there may be relevant differences between ensembles in which the universes exist simultaneously, and those in which the universes come into being one after the other with no memory of previous universes being carried over into subsequent ones. Ian Hacking contends that cosmic fine tuning at most allows inference to a simultaneous ensemble. Any inference from the existence of a finely tuned universe to the existence of a cyclical ensemble is as fallacious as assuming that a good poker hand is evidence that a long series of poor hands has been previously dealt. It is what Hacking calls the "inverse gambler's fallacy." Thus, the strategy of adopting a cyclical ensemble to avoid the ontological penalties of a simultaneous one may have drawbacks of its own.

The teleological anthropic principle is certainly explanatory in intent, but turns out not to constitute a satisfactory teleological explanation. This conclusion can be reached without considering the thorny question of the appropriateness of teleological explanations in science. Such explanations appeal either to the explicit purposefulness of an agent desiring to achieve a goal, or to the immanent tendency of a process to achieve an end. Because defenders of the anthropic principle typically do not endorse the concept of a divine designer, they are appealing to the latter sense of teleology. The explanation for the habitability of the universe, they say, is that habitability is the goal of the universe. It is important to note that only the mere possibility, not the actual presence, of humans or other life forms is taken as the goal. (Assuming that the actual presence of humans is the goal would virtually commit one to a goal-directed evolutionary process.) Although the phrase "mankind is the goal of the universe" is often associated with the anthropic principle, the actual claim is the weaker one that the possibility of mankind evolving is the goal of the universe. But the mere possibility of something is hardly a robust enough goal to count as the goal of the entire universe. Furthermore, the evolutionary process that eventually yields intelligent life is emphatically not goal-directed. Because the goal posited is a mere possibility and because there is no readily identifiable goal-directed process leading to intelligent life, there are serious inadequacies in current construals of the anthropic principle as a teleological principle.

Anthropocentrism

Perhaps the most significant difficulty with the anthropic principle is that it is not really anthropic at all. The reasons for this are somewhat different for the different versions of the principle. The logical and methodological principles are not at all anthropocentric. For any X it is true that, given the existence of X, the necessary conditions for X must be fulfilled. The requirements for the existence of dogs, flowers, and planets are fulfilled just as much as are the requirements for human existence. The logical principle is better seen as a principle of consistency than as an anthropic principle. Its methodological version has a superficial affinity to human beings, because human scientists are the ones who should take into account the results of the logical principle. But, given that the logical principle applies equally to the necessary conditions for the existence of all things, there is no reason to assume that the methodological principle should place special emphasis on mankind. In assessing apparent coincidences and apparent fine tuning, one should realize that these phenomena may be the result of a logical "selection" by *any* existing thing, whether it be a human being, an insect, or a hydrogen atom. Indeed, intelligent insect-like beings would think the universe is finely tuned for *them* if they did not appreciate the generality of the principle.

The case of the teleological principle is more complex. The principle is clearly intended to be anthropocentric to some degree. But specifying the degree of anthropocentrism—which is not adequately done in the literature—proves to be quite problematic. Up to this point we have avoided this problem by assuming a human-oriented anthropic principle. The name of the principle suggests that it applies to human beings; but such a highly anthropocentric principle is indefensible, because it would draw an unjustifiable distinction between humans and hypothetical beings similar to us in all but their physical characteristics and physically necessary conditions. One could generalize the principle, as some do, to apply to (say) all types of intelligent life, in an attempt to reduce anthropocentrism but to retain the claim that the universe is finely tuned. The problem with this move is that it requires us to define life and intelligent life in such a way that we still know what its physical prerequisites are but have not made it too human-centered. Given that we are ignorant of the range of physical beings that can instantiate properties such as intelligence, such a task seems hopelessly beyond our reach. Yet the broader one makes the definition of intelligent life, perhaps in an attempt not to rule out hitherto unforeseen physical possibilities, the less one can rely on the original fine tuning arguments. For it may turn out to be the case that most of the fundamental features of the universe could be quite different without ruling out the possibility of *some* form of intelligent life.

Status as a Scientific, Heuristic, or Religious Principle

The status of the anthropic principle as a scientific principle depends largely on its success as an explanatory and predictive principle. The issue of the principle's explanatory power has been treated above. Unfortunately, none of the versions of the principle seems capable of giving rise to testable predictions. The logical principle is clearly not open to experimental confirmation or refutation. Likewise, it is hard to see how a prediction could be based on the claim that the possibility of intelligent life is the goal of the universe. No amount of fine tuning could show that habitability is the *goal* of the entire universe, nor could any conceivable observation. Whether the universe has a purpose is not a simple empirical matter to be decided by future observations.

Whether or not the anthropic principle has problems with explanatory adequacy or unjustified anthropocentrism, it may have heuristic value. The methodological principle is heuristically fruitful insofar as it leads us to look for features of the universe that are prerequisites for our existence and then to view them as less remarkable than they might otherwise appear. As we have shown, however, this principle is not really *anthropic*; and any heuristic value it has belongs to a much more general principle. The teleological principle is heuristically sterile: thinking that habitability is the goal of the universe can be no more fruitful than thinking that the universe must be habitable. The principle is perhaps even detrimental from a heuristic point of view, insofar as accepting it as explanation tends to stifle the search for standard types of scientific explanation. One may be tempted to give up the search for further explanation in the presence of an anthropic explanation for all the features of the universe that are necessary for our existence.

Although the anthropic principle was not proposed as a religious principle, the evidence motivating it resembles evidence used in classical design arguments, and its teleological form lends itself to being interpreted as a religious instead of a scientific principle. As a religious principle, the principle's lack of predictive ability is not a crucial problem. Its anthropocentric construal fits in well with theistic views of the lofty place of mankind in the universe. And the teleological mode of explanation it employs makes sense if one understands the origin and evolution of the universe as an ongoing creative process that serves divine purposes. Looked at in this way, the anthropic principle seems more of a religious principle, and less of a scientific one, than its defenders are willing to admit.

Conclusion

The anthropic principle raises a host of interesting questions, challenges traditional modes of scientific explanation, and touches on theistic world views, but does not live up to its promise as some new and revolutionary scientific principle: it is either a rather trivial truth or an impotent teleological principle in need of a more robust and justifiable goal and of a mind to guide its purposefulness.

—*Patrick A. Wilson*

BIBLIOGRAPHIC NOTE

The best general introduction to the subject is George Gale, "The Anthropic Principle," *Scientific American*, 245:12 (1981), 154–171. John D. Barrow and Frank J. Tipler, *The Anthropic Cosmological Principle* (New York: Oxford University Press, 1986) is the definitive if highly speculative defense of anthropic reasoning, and includes extensive historical and scientific background material as well as a brief and accessible introductory chapter. Much of the scientific material appears also in P.C.W. Davies, *The Accidental Universe* (Cambridge: Cambridge University Press, 1982). Martin Gardner's stimulating and highly critical review of the Barrow and Tipler volume in *The New York Review of Books*, 33:8 (1986), 22–25, is well worth reading. The philosopher John Leslie has written numerous articles on the topic, representative of which is "Anthropic Principle, World Ensemble, Design," *American Philosophical Quarterly*, 19 (1982), 141–151. Other articles mentioned in the essay above are Brandon Carter, "Large Number Coincidences and the Anthropic Principle in Cosmology," in M.S. Longair, ed., *Confrontation of Cosmological Theories with Observational Data* (Dordrecht:

Reidel, 1974), pp. 291–298; and Ian Hacking, "The Inverse Gambler's Fallacy: The Argument from Design: The Anthropic Principle Applied to Wheeler Universes," *Mind*, *96* (1987), 331–340.

27. PLURALITY OF WORLDS

Plurality of worlds (*plures mundi, Mehrheit der Welten, pluralité des mondes*) is the term historically used by many cultures for the concept of other worlds beyond the earth. In ancient Greek times this meant a plurality of ordered world systems, referred to as *kosmoi*. Beginning in the seventeenth century the term came to mean earth-like planets, complete with intelligent inhabitants. In the twentieth century the tradition has become known as the extraterrestrial life debate, a pursuit that biologists have labelled *exobiology* and astronomers have labelled *bioastronomy*. Because it is intrinsically difficult to verify the existence of other worlds and extraterrestrial life, the debate has always incorporated a good measure of philosophy. Critics have called it "a science without a subject," and some even have questioned whether it is a scientific issue at all. Depending on the period under consideration, historians have concluded both that the debate has been conducted primarily in philosophical terms and that the debate has been conducted primarily in scientific terms. To an extent, these conclusions depend on how one defines *science*; any account must take into consideration the changing nature of science over the millennia during which the subject of multiple worlds has been discussed.

This article develops the view that the plurality of worlds tradition:

- Originated with, and was sustained by, cosmological world views through the middle of the eighteenth century.
- Was dominated for the following century by philosophical explorations.
- Received late in the nineteenth century its scientific foundations in modern terms.

Even so, the extraterrestrial-life debate in the twentieth century remains, at best, an example of science functioning at its limits. In this

very fact is found a large part of its interest for historians and philosophers of science.

The Cosmological Connection

The idea of other worlds already was present in the earliest of man's
cosmological world views, ancient atomism. Constructed in the fourth
and fifth centuries B.C. by Leucippus, Democritus, and Epicurus, this
system held that:

> *There are infinite worlds both like and unlike this world of ours. For the atoms*
> *being infinite in number, as was already proved, are borne on far out into space.*
> *For those atoms which are of such a nature that a world could have been created*
> *by them [. . .] have not been used up either on one world or a limited number of*
> *worlds [. . .] So that there nowhere exists an obstacle to the infinite number of*
> *worlds.*

The important point here is that the atomists directly tied their
cosmology to the physical principles of the atomist system; it was no
half-hearted afterthought, but an integral part of the theory. Moreover, it is remarkable that these infinite number of worlds (*aperoi
kosmoi*) existed completely beyond the human senses, for the entire
visible world by the Greek definition composed a single *kosmos*. It was
this nonempirical aspect that caused some seventeenth-century critics to complain that atomists should not posit infinite unseen worlds
when so little was understood about the world we did see.

The Roman poet Lucretius spread the atomist doctrine of an
infinite number of worlds, together with the rest of atomism, throughout Europe in his *De rerum natura* (*On the Nature of Things*). The
atomist system, however, was not destined to win the day; it was
almost two thousand years before it would be revived in the sixteenth
and seventeenth centuries with the birth of modern science. In the
meantime, a far more elaborate cosmology was constructed by
Aristotle, whose life overlapped that of Epicurus by two decades and
who gave new meaning to the word *kosmos*. Aristotle's cosmology
placed the earth at the center of a nested hierarchy of celestial spheres,
from the spheres of the moon and planets to the sphere of the fixed
stars. The earth in this system was more than a physical center; it was
also the center of motion. According to one of the basic tenets of
Aristotle's cosmology, the doctrine of natural motion and place, everything in the cosmos moved with respect to that single center: the
element earth moved naturally toward the earth, and the element fire
moved naturally away, while the elements air and water assumed
intermediate natural places. Aristotle's belief in the impossibility of
more than a single *kosmos* was directly tied to this basic tenet. In his

cosmological treatise *De caelo* (*On the Heavens*) he reasoned that if there were more than one world:

> *It must be natural therefore for the particles of earth in another world to move towards the center of this one also, and for the fire in that world to move toward the circumference of this. This is impossible, for if it were to happen the earth would have to move upwards in its own world and the fire to the center; and similarly earth from our own world would have to move naturally away from the center, as it made its way to the center of the other, owing to the assumed situation of the worlds relative to each other.*

The issue of a plurality of worlds thus was reduced to a confrontation with the most basic assumptions of Aristotle's system. Either he must reject his doctrine of natural motion and place (on which he had built his entire physics) and reject as well his belief in four elements (on which his theory of matter rested), or he must conclude that the world was unique. The choice was not a difficult one; indeed, he must have taken comfort in reaching a conclusion so diametrically opposed to the atomists, whose system differed from his in so many other ways. Aristotle also proposed auxiliary metaphysical arguments for a single world, but this physical argument was the central argument.

It was Aristotle's system that was transmitted to the Latin West, where it was commented upon again and again, but now in the context of the Christian system. The problem that Christianity had with the doctrine of a plurality of worlds was as follows: *Suppose God wished to create another world. How could He do so given the principles of Aristotle?* Either Aristotle was wrong, and by his own admission wrong in some very basic principles, or God's power was severely limited. This dilemma was handled in various ways by commentators in the Middle Ages. Thomas Aquinas found God's perfection and omnipotence in the unity of the world rather than in its plurality. By late in the thirteenth century several commentators at Paris and Oxford universities argued that the plurality of worlds was not theologically impossible because God can act beyond the Aristotelian laws of nature. In the fourteenth century Jean Buridan and William of Ockham argued that the elements in each world would return to the natural place within their own world, either supernaturally or naturally. By 1377 the Paris master Nicole Oresme had completely reformulated the doctrine of natural place to state in no uncertain terms that other worlds were possible without any supernatural intervention.

The transition to the more modern plurality of worlds tradition, however, was not made through successive rebuttals to Aristotle's doctrine of a single world. Rather, it stemmed from the complete overthrow of Aristotle's geocentric universe and its replacement with the Copernican system of the world. By placing the sun in the center of the system of planets, and making the earth one of those planets, Copernicus gave birth to a new tradition where the term *world (mundus)* was now redefined to be an earth-like planet, and each of these earth-like planets took on the kinematic or motion-related functions of the single earth in the old geocentric system. Just as the kinematic implications of the decentralization of the earth led to the birth of a new physics, so the implications of that move for the physical nature of the planets led to the concept of inhabited worlds. All discussions of life on other worlds since the Copernican Revolution recall the argument set in motion by Copernicus:

- If the earth is a planet, then the planets may be earths.
- If the earth is not central, then neither is man.

Copernicus himself did not pursue the implications of his system for planetary physics, but the Italian philosopher Giordano Bruno, an avowed Copernican, showed just how far such implications might go. Although in his *De l' infinito universo e mondi* (*On the Infinite Universe and Worlds*) of 1584 Bruno pointed primarily to metaphysical ideas such as the unity and plenitude of nature as the source of his belief in an infinite number of worlds, this was the view toward which the Copernican system inexorably led. Even before the invention of the telescope, the young astronomer Johannes Kepler, already a convinced Copernican under the influence of his teacher Michael Maestlin, would ascribe inhabitants to the moon. The telescope accelerated this trend: Galileo in his *Siderius nuncius* (*The Sidereal* [or *Starry*] *Messenger*) of 1610 noted that the surface of the moon was "not unlike the face of the earth." Because of theological difficulties, Galileo himself sought to downplay the similarities between earth and moon, and he admitted in 1632 in his *Dialogue on the Two Chief World Systems* only that if there was lunar life, it would be "extremely diverse and far beyond all our imaginings." The Copernican tide, however, could not be stemmed. Six years later, in 1638, in the less repressive atmosphere of Anglican England, Bishop John Wilkins penned his *Discovery of a World in the Moone*, in which Galilean caution was thrown to the wind. Copernicanism was not synonymous with inhabited planets, but it did give theoretical underpinning to habitable planets. The proof or disproof of this implication remained a goal of astronomers until the Viking landers touched down on Mars late in the twentieth century.

All important cosmological world views of the seventeenth century and later incorporated the Copernican system as a basic truth. Such was the case with the first complete physical system proposed

Title page from John Wilkins' combined Discovery of a World in the Moone (1638) *and the* Discourse concerning a New Planet *(1640). At left Copernicus offers his heliocentric world view, while at right Galileo offers his telescope and Kepler wishes for wings so that he might visit the new world. The sun says "I give light, heat, and motion to all." [From the collection of the Library of Congress.]*

since Aristotle, that of the French philosopher René Descartes. His
Principia Philosophia (*Principles of Philosophy*) of 1644, greatly influenced by a revived atomism, offered a mechanical philosophy in which
atoms in motion once again formed the basis for a rational cosmology. For the plurality of worlds tradition it did even more, for it was
through the Cartesian cosmology that the quest for a biological universe was first carried to other solar systems, and in a fashion so
graphic that it has remained an ingrained concept to the present day.
Unlike the void space of his atomist predecessors (and his Newtonian
successors), Descartes proposed that the universe was a plenum, filled
with atoms in every nook and cranny. A consequence of this was that,
once set in motion by God, the particles of the plenum formed into
vortices (systems analogous to our solar system) centered around
every star. Though Descartes himself, again for religious reasons, was
careful not to specify that these vortices consisted of inhabited planets, his application of Cartesian laws to the entire universe and the
graphic vortex cosmology was plain for all to see.

Descartes's followers were not slow to realize the implications,
none more boldly than his countryman Bernard le Bovier de
Fontenelle. His *Entretriens sur la pluralité des mondes* of 1686 exploits
both the Copernican and Cartesian theories to shed light on the
question of life on other worlds. Fontenelle asked:

> *If the fix'd Stars are so many Suns, and our Sun the centre of a Vortex that
> turns round him, why may not every fix'd Star be the Centre of a Vortex that
> turns round the fix'd Star? Our Sun enlightens the Planets; why may not every
> fix'd Star have Planets to which they give light?*

In the same year, the Dutch astronomer Christiaan Huygens began
to formulate very similar ideas, published posthumously in his
Cosmotheoros in 1698. Although Cartesian vortices would be swept
away by the Newtonian system, the general idea of planetary systems
would not.

It is ironic that of all cosmological world views, the scientific
principles of the Newtonian world view entailed extraterrestrial life
least of all. Although a mechanical philosophy like that of Descartes,
Newton's atoms and void, with each body subject to universal gravitation according to fixed laws, did not necessarily imply other solar
systems. No mechanical necessity dictated the formation of solar
systems as in Descartes's world view; indeed, under Newtonian principles the whole question of other solar systems has proved to be one
of greatest complexity even to the present day. Newton himself de-
clined to expound any rational cosmogony that might shed light on
the question. He insisted only that the formation of ordered systems
was contingent upon God's will, contenting himself with the obser-

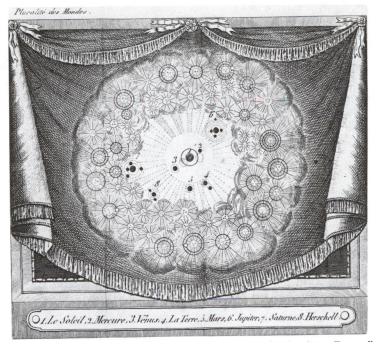

Plularité des Mondes.

1.Le Soleil,2.Mercure,3.Venus,4.La Terre,5.Mars,6: Jupiter,7. Saturne,8.Herschell.

Frontispiece for Fontenelle's Entretiens sur la pluralité des mondes. *It indicates Fontenelle's belief in the plurality of planetary systems and depicts planets circling other fixed stars. The depiction of the planet Uranus (discovered by William Herschel in 1781) in this 1821 French edition is an indication of how Fontenelle's work was often updated by new astronomical discoveries long after the vortex cosmology was abandoned. [From the collection of the Library of Congress.]*

vation in the second edition of the *Principia* in 1713 that "if the fixed stars are the centres of other like systems, these, being formed by the like wise counsel, must be all subject to the dominion of the One." The major effect of the Newtonian world view on the plurality of worlds tradition was to incorporate it into the tradition of natural theology, where it assumed the role of an important counterbalance in a system that had lessened the need for a Deity to keep the universe running. In one Newtonian treatise after another, the theological view of an inhabited universe was joined to the physical principles of Newton's system. Again and again, a universe full of inhabited solar systems was applauded as one "far more magnificent, worthy of, and becoming the infinite Creator, than any of the other narrower schemes." Once this argument had been made, overwhelming all Scriptural objections, other arguments such as teleology could be adduced in its favor.

This satisfying vision of the universe, operated by Newtonian laws and reflecting the power of the Deity by spreading intelligence through the universe, was passed on to the modern world. The proof of other solar systems by observation, and the proof of their likely formation by Newtonian principles, remained a desired goal in the centuries to follow. But the basic predisposition toward a universe of inhabited solar systems was set, almost within the lifetime of Newton himself. Philosopher-cosmologists such as Thomas Wright, Immanuel Kant, and Johann Lambert—committed Newtonians all—spread the vision of an inhabited universe in their eighteenth-century treatises. In the new Newtonian cosmological world view, the plurality of worlds tradition was joined to Christianity and natural theology. That conjunction would take a central place in the plurality of worlds debate in the nineteenth century.

Philosophical Explorations

Following the triumph of the Newtonian system in the middle of the eighteenth century, the extraterrestrial-life debate was waged not so much on a cosmological scale as on a scale of world views a level or more below the cosmological. Though sometimes discussed by the elaboration of Newtonian science (such as the Laplacian nebular hypothesis), more often the extraterrestrial-life debate fell into the domain of philosophical explorations, both secular and religious. If cosmological world views gave birth to the idea of extraterrestrial life, then philosophy and literature, in their traditional role of examining the human condition, explored the ramifications of the idea borne of that cosmological context.

In particular, much of the plurality of worlds debate late in the eighteenth century and into the nineteenth century—at least in the West—may be understood as a struggle with that widespread philosophical world view known as Christianity. If in the Newtonian system the plurality of worlds concept was reconciled with theism through natural theology, this was not equivalent to a reconciliation with Christianity; as Professor Michael Crowe succinctly states in his study of the nineteenth-century plurality of worlds tradition: "structures of insects or solar systems may evidence God's existence, but they are mute as to a Messiah."

Three choices were logically open to Christians who pondered the question of other worlds:

- Reject other worlds.
- Reject Christianity.
- Attempt to reconcile other worlds with Christianity.

Historically, all three of these possibilities came to pass in the eighteenth and nineteenth centuries.

Although the Scriptural and doctrinal problems of the issue had been widely discussed throughout the seventeenth century, only to be overwhelmed by natural theology, no one more forcefully expressed the continuing difficulties of the plurality of worlds doctrine for Christianity than did Thomas Paine. In 1793 in his influential *Age of Reason*, Paine bluntly stated that:

> To believe that God created a plurality of worlds at least as numerous as what we call stars, renders the Christian system of faith at once little and ridiculous and scatters it in the mind like feathers in the air. The two beliefs cannot be held together in the same mind; and he who thinks that he believes in both has thought but little of either.

Pointing to the Christian doctrines of Redemption and Incarnation, and to the absurdity of a planet-hopping Savior, Paine rejected Christianity.

Though few would reject Christianity because of Paine's argument, few would reject plurality of worlds either, a testimony to its entrenchment by the end of the eighteenth century. This left but one alternative: the two systems and all they implied would have to coexist. That other worlds could be incorporated into Christianity, despite Paine, was demonstrated by the Scottish theologian Thomas Chalmers. His 1817 *Astronomical Discourses* incorporated plurality of worlds into evangelical religion, and his countryman Thomas Dick made it a staple of Christianity in a number of works during the first half of the nineteenth century. Even astronomers such as John Herschel were strongly influenced by philosophical arguments in this issue.

Paine's objections, however, would not disappear. By mid-century the consonance of plurality of worlds with Christianity was once again called into serious question in one of the most interesting intellectual disputes of the nineteenth century. The instigator was William Whewell, philosopher, scientist, and Master of Trinity College, Cambridge. Influenced by Chalmers, Whewell was a pluralist from at least 1827. By 1850, as Professor Crowe has shown in his analysis of an unpublished Whewell manuscript, Whewell opposed pluralism. And in 1853 his treatise *Of the Plurality of Worlds: An Essay*, the most learned, radical, and influential anti-pluralist treatise of the century, appeared anonymously.

In his *Essay* Whewell confirmed that the existence of other earth-like planets and solar systems was commonly accepted. It is clear that his own Christian concerns were the source of Whewell's treatise, in particular the overwhelming reinforcement other worlds gave to the cry of the Psalmist "What is man, that thou art mindful of him?" and to the doctrines of Redemption and Incarnation. Before contemplating radical changes to Christianity, Whewell insisted, one should first examine the plurality of worlds doctrine. This was the purpose of the *Essay*, and the result was that Whewell would argue that it was pluralism, not Christianity, that should be rejected. Quickly disposing of one of the chief philosophical arguments of the pluralists, the teleological argument that the vast space must have some purpose, Whewell argued that confining intelligence to the "atom of space" that was the earth was no worse than confining humanity to the "atom of time" that geology revealed it had existed on the earth. On the more empirical side, Whewell argued that no proof existed of other solar systems, that the stars might not be exactly similar to our sun, and that in any case many of them were binary stars whose putative planets would therefore not have conditions conducive for life. In our own solar system, only Mars approached the conditions of earth, and it was just as likely as not that Mars was still in a condition of "pre-intelligence." Finally, Whewell cautioned against the unbridled use of the analogy argument in science. Altogether, Whewell's was a serious challenge to a doctrine that had come to be cherished by both science and natural theology.

Whewell's treatise generated a tremendous amount of debate, but in the end it did little to weaken support for a plurality of worlds among scientists or the religious. Professor Crowe documents twenty books and some fifty-four articles and reviews in response to Whewell; of these about two-thirds still favored pluralism despite Whewell's arguments. Treatises such as Sir David Brewster's *The Creed of the Philosopher and the Hope of the Christian* continued to be driven by an attachment to teleology and natural theology. Reconciliation with the doctrines of Incarnation and Redemption was never achieved. The claim that Christ's incarnation on earth was of great enough force to save extraterrestrials prevented a planet-hopping Christ, but strained credulity. The concept of a plurality of worlds even became a central doctrine for at least two nineteenth-century religions: the Mormons and the Seventh Day Adventists. Yet another religion, the Swedenborgians, had held it as one of their beliefs since the middle of the eighteenth century.

Thus Christianity holds the distinction of being the philosophical world view that most influenced the plurality of worlds doctrine in the nineteenth century, at least in the Western world. Secular philosophies also interacted with the concept of other worlds, though none so strongly or persistently as Christianity. Already in the seventeenth century the British empiricist John Locke had pointed out in his *Essay Concerning Human Understanding* in 1689 that human ideas are limited by the human senses, and that extraterrestrials might have no such limitations, or at least different ones. The German philosopher Gottfried Wilhelm Leibniz, well known for his belief that ours is the best of all possible worlds, may have been influenced in that belief by the possibility of actual worlds. This view was satirized by the most famous of the philosophes, Voltaire, who nevertheless also made use of extraterrestrials in his writings.

By contrast, many German philosophers of the nineteenth century were opposed to a plurality of worlds, not because of science or religion, but because of their anthropocentrism. G.W.F. Hegel held that the earth is the most excellent of all planets, and several of his students argued strongly against the pluralist position. Friedrich Schelling and his disciple Heinrich Steffens, though not themselves Hegelians, also opposed with anthropocentric arguments the idea of other worlds. The ever-pessimistic Arthur Schopenhauer, though he accepted the existence of extraterrestrials, also believed that man was at the pinnacle of creation.

Literature, in its role of emotional exploration of human purpose, played an important role in responding to the challenge of other worlds. While John Milton in seventeenth-century England had cautioned, "Dream not of other worlds; what creatures there live in what state, condition or degree," Alexander Pope's *Essay on Man* suggested that he who contemplated the "worlds on worlds" of the universe might "tell why Heaven has made us as we are." A century later Tennyson expressed similar sentiments, and the Romantic poets Byron, Shelley, and Coleridge used other worlds in a religious context. In prose no less than in poetry, the extraterrestrial perspective became entrenched. From the cosmic-voyage genre of the seventeenth century to the science fiction of H.G. Wells at the end of the nineteenth century, in which implications of extraterrestrials and their worlds attained the status of a classic theme, the implications began to be explored at an increasing pace in ever more detail.

By the last quarter of the nineteenth century, plurality of worlds clearly had won the day, but based more on philosophical than scientific arguments. The Copernican system remained in the background as a driving force, and science kept chipping away at the problem with its limited empirical tools, until two unexpected developments laid the scientific foundations for all subsequent discussion of life beyond the earth.

Scientific Foundations

As the Whewell debate reached its height, two fundamental developments in science profoundly affected the plurality of worlds debate: in 1859 Charles Darwin and A.R. Wallace published their theory of the origin of species and of evolution by natural selection, and in the early 1860s the new technique of spectroscopy began to be applied to astronomy. Although these and less sweeping developments in other fields did not effect an immediate and radical change in the character of the plurality of worlds debate, they did signal the beginnings of a long-term change that would bring the subject of other worlds increasingly under the purview of modern science. Natural selection not only provided the basis for a discussion of the evolution of life under differing conditions beyond the earth, but also gave impetus to the idea of the physical evolution of the universe. And spectroscopy provided a tool for studying the nature of the planets and stars in ever increasing detail, as well as a means of proving the truth of the evolutionary universe, prerequisites for determining the possibility of life.

Of the two developments, spectroscopy would have the more immediate and profound impact on the debate over other worlds. Although the arguments of analogy and uniformity of nature had for a long time given credence to the belief that the building blocks for matter—and life—were alike throughout the universe, now for the first time this great truth could be observationally proven. Many of the spectroscopic pioneers, including Sir William Huggins, did not fail to see the connection of their research to life in the universe.

Huggins and his collaborator William Miller wrote in 1864 that their work contributed toward an "experimental basis" that the stars were "energising centres of systems of worlds adapted to be the abode of living beings." Huggins's early attempts to probe planetary atmospheres spectroscopically was the first step toward yet another research program that would be increasingly central to the extraterrestrial-life debate in the future. Other pioneers in the new science (including Angelo Secchi, Heinrich Schellen, Jules Janssen, John and Henry Draper, and S.P. Langley) did not fail to make similar connections.

The Darwinian theory of evolution had a more gradual effect, but one eventually no less significant. Its earliest effect was in its general application to the idea of the physical evolution of the universe. This is evident already in one of the most prominent treatises on the plurality of worlds, Richard A. Proctor's 1870 *Other Worlds than Ours*, subtitled *The Plurality of Worlds Studied Under the Light of Recent Scientific Researches*. Like Proctor, his French counterpart Camille Flammarion professed to take a scientific approach to the problem of other worlds, though it is clear already from his 1862 *La pluralité des mondes habités* and subsequent works that far more than empirical science drove Flammarion to his belief in other worlds. (He had read pluralist authors, and also Jean Reynaud, who advocated the transmigration of souls from planet to planet, progressively improving at each stage.) By the end of the century it was not Flammarion's radical pluralism but Proctor's more limited version that became prevalent in the writings of British and American writers alike.

Other less encompassing theories born during this era also were destined to play an important, if delayed role in the plurality of worlds debate. G. Johnstone Stoney undertook the application of the kinetic theory of gases to planetary atmospheres as early as 1869. Laplace's nebular hypothesis was elaborated in more subtle form late in the nineteenth century, before being eclipsed for a while by the theory of close stellar encounters as a means of producing solar systems. Based on theories of solar system formation, the rarity or abundance of planetary systems would play an extremely important role in the debate over life.

Not all of these approaches achieved fruition in the nineteenth century, but as the old century transformed into the new, there was an increasing consensus that if there was to be an answer at all, it could only emerge from a scientific approach. At the turn of the century, in 1903, A.R. Wallace himself wrote a treatise, *Man's Place in the Universe: A Study of the Results of Scientific Research in Relation to the Unity or Plurality of Worlds*. The problem was that the intrinsic observational difficulties allowed scientific research to give no definitive answer; yet so great was the urge to resolve this age-old question that many scientists plunged ahead nevertheless. The result is a case study of how science and scientists function at their limits.

The twentieth-century extraterrestrial-life debate has quite naturally been dominated by the relatively nearby search for life in our solar system, but increasingly in the second half of the century it has been joined by three other scientific components: the search for other planetary systems, experiments on the origin of life as applied to life beyond the earth, and finally the Search for Extraterrestrial Intelligence (SETI) by means of radio telescopes. Each component of the debate has risen, in turn, to prominence, and in the 1960s these components began to converge toward a new scientific discipline, known first as exobiology and then as bioastronomy. Characterized by a coherent research program, federal funding, a tight-knit community of scientists, and formal institutionalization into the structure of scientific organizations (such as the International Astronomical Union), what began early in the century as a loose set of ideas ended as a protoscience with a broad-based constituency.

The search for life in the solar system has been focused on Mars ever since Percival Lowell founded the Lowell Observatory in 1894 especially to search for life on that tantalizing planet. The very next year—much too quickly, critics said—Lowell published his book *Mars*, in which he claimed to have mapped "canals," long straight markings crisscrossing the planet in a system that he argued had been engineered by intelligent Martians. Lowell was not the first to claim the existence of such markings on Mars—the Italian astronomer Giovanni Schiaparelli had mapped *canali* beginning in 1877, leaving their origin unexplained. But the connection that Lowell made between canals and intelligence began a controversy that peaked about 1910, and continued sporadically even after Lowell's death in 1916. The chief issues were observational, revolving around the ability of telescopes to detect fine detail on planetary surfaces and the ability of the human eye and brain to interpret this detail. Further observational issues centered around the very difficult determination of atmospheric and surface conditions on Mars. The resolution of these issues was complicated by the varying distance of Mars, which approached relatively close to the earth every fifteen years only to recede again. Nevertheless, during the close approach of 1909 the French astronomer E.M. Antoniadi, using the 33-inch telescope at the Meudon Observatory in France, resolved some canals into dark splotches, a feat that began the downfall of Lowell's theory. This achievement left unexplained, however, why Lick Observatory astronomers, who were critical of Lowell's observations, did not avert the whole episode by resolving canals during observations made in 1888 with their even larger 36-inch refractor. Even worse, modern spacecraft observations have shown little correlation between Martian surface markings and the canals that Lowell mapped, indicating that most of the objects of the debate, including Antoniadi's resolved canals, were illusory. The whole episode raises fundamental questions about observation and evidence in science. It had a profound effect on planetary astronomy during the rest of the century, causing some to leave the field of planetary astronomy because of the fierce debate and inspiring others to enter the field in order to discover the real nature of Mars.

After Lowell's death, with the close approach of Mars in 1924, attention focused on the possibility of Martian vegetation rather than intelligence. In one particularly important case this was still tied to the old visual method and canals; using the 36-inch refractor Lick astronomer Robert J. Trumpler concluded that the canals were the result of natural topography, but that vegetation caused the dark Martian areas and made the canals visible. But the mid-1920s marked a new era in Martian studies: physical methods of spectroscopy and infrared astronomy now came into widespread use in the attempt to determine temperature and atmospheric conditions. Respected scientists such as W.W. Coblentz of the National Bureau of Standards,

C.O. Lampland at the Lowell Observatory, and Edison Petit and Seth Nicholson at the Mount Wilson Observatory, pioneering in the field of infrared astronomy, determined that the temperature conditions on Mars were adequate for some form of Martian vegetation. Using spectroscopic techniques, others found evidence of oxygen and water vapor in the Martian atmosphere, but in increasingly minute amounts, now known to be spurious. Despite the desert conditions revealed by the new physical methods, by 1957 and the dawn of the Space Age the existence of hardy, perhaps lichen-like Martian vegetation was widely accepted, especially in the wake of William Sinton's claims in that year to have discovered infrared bands in the Martian spectrum that were unique to vegetation.

These hopes were partially dashed in the early 1960s when the Sinton bands were found to be due to deuterated water in the earth's own atmosphere, and the water content of the Martian atmosphere was lowered almost to the vanishing point. But hopes were completely dashed two decades into the Space Age, when the Viking landers in 1976 demonstrated not only the lack of vegetation on Mars but also the complete absence of any organic molecules. Although one of the three prime biology experimenters on the Viking project still maintains that his results were compatible with life, the consensus today is that life is absent on the planet Mars. Thus the twentieth century has seen the question of life on Mars progress from intelligence to vegetation to organic molecules—all having been disproven. With the discovery at mid-century that Venus is a victim of the greenhouse effect, with temperatures consequently at the 800 degrees Fahrenheit level, the Viking results left a solar system largely bereft of life beyond earth, though organics are still considered possible on some of the moons of the outer gaseous giant planets. But because Mars in particular had been widely viewed as a test case for life in the universe, the absence of life there was a correspondingly great blow to the concept of a universe filled with life.

Nevertheless, it is not surprising that long before the Viking results were in hand, attention had turned beyond the solar system to the possibility of the existence of other planetary systems—a prerequisite for life in the realm of the stars. Belief in such systems has been greatly affected throughout the century by theories of their origin. The nebular hypothesis of Laplace, whereby planetary systems originated from rotating gas clouds that formed the stars themselves, indicated that planets were a natural by-product of star formation and, therefore, very abundant. At the turn of the century, however, this theory was under heavy attack. In its place the geologist T.C. Chamberlin and the astronomer F.R. Moulton proposed that solar systems originated by the close encounters of stars, which resulted in the tidal ejection of matter which cooled to form small planetesimals, which in turn accreted to form planets. This "planetesimal hypothesis," elaborated and modified by the British astronomer James Jeans from 1916 almost until his death in 1946, implied that solar systems were extremely rare, since stellar collisions in the vastness of space were extremely rare. For this reason, during the 1920s and 1930s belief in extraterrestrial life was at a low point; it was difficult to conceive of life without planets.

But the fifteen years between 1943 and 1958 saw once again a complete turnabout in opinion. In 1943 two astronomers independently claimed they had observed the gravitational effects of planets orbiting the stars 61 Cygni and 70 Ophiuchi. Although these observations were proven spurious decades later, they filled a need at the time. Doubts expressed about Jeans's stellar encounter hypothesis by the American astronomer Henry Norris Russell in 1935 had grown to a crisis point by the early 1940s. Carl Friedrich von Weizsäcker began the revival of a modified nebular hypothesis in 1944, and the theoretical basis once again was laid for abundant planetary systems.

Although the new nebular hypothesis has been elaborated in ever more subtle form since that time, attempts to pin down the abundance of planetary systems have proven very difficult. Observationally, the search has been dominated by the astrometric method, whereby the proper motions of stars are studied for the gravitational effects of planetary systems. Since the 1960s several claims have been made by Peter van de Kamp and others for planetary systems around several stars. In the 1980s another method for determining planetary effects on stars—this time, on their radial velocities—came into use. As with the astrometric method, at the distances of even the nearest stars these effects are so small as to be at the limits of observation and, therefore, are still controversial. Theory has therefore still predominated in the debate. Aside from the general fact that, according to the nebular hypothesis, solar systems are a normal occurrence during stellar evolution, subsidiary arguments also have been important. Especially since the 1950s, the knowledge that stars of the F spectral type exhibit a greatly slowed rotation rate has been used as an argument that they may have lost their angular momentum to planetary systems, as is the case in our own solar system. Circumstellar material also has been observed that may represent solar systems in formation. Although such arguments lead astronomers to believe planetary systems are abundant, and despite numerous attempts to detect them observationally, no other solar systems have yet been unambiguously confirmed beyond our own.

As the idea of abundant planetary systems was being revived in the 1950s, work also was progressing on the biological question of the origins of life, a crucial factor in the question of extraterrestrial life. In particular, the work of Urey and Miller showed how organics could be produced under simulated primitive atmospheric conditions. However, since that time a better appreciation of the difficulties of the many steps in the origin of life has somewhat tempered optimism among biologists. Whereas astronomers focus on the enormous size of the universe and the likelihood of planets emerging from an abundance of stars, biologists concentrate on the extremely complex steps in the origin and evolution of life. Thus some dichotomy of opinion has developed between astronomers and biologists, further widened by the biologists' recognition that the evolution of life beyond the earth might lead to forms of life and intelligence very different from the humanoid form and alien to the human concept of intelligence.

Given the inherent difficulties in the search for planetary systems and the uncertainties in our knowledge of the origin and evolution of life, interest since 1959 has focused on the detection of intelligent signals of extraterrestrial origin, the so-called Search for Extraterrestrial Intelligence (SETI), a detection that would leapfrog many of these uncertain arguments. The physicists Giuseppe Cocconi and Philip Morrison proposed in that year a search in the radio region of the spectrum using the 21-centimeter hydrogen line, and the American radio astronomer Frank Drake independently undertook the first search for such signals at the National Radio Astronomy Observatory in 1960. It was in the context of a meeting in 1961 in the wake of this search that the so-called Drake equation was formulated. A general equation embodying the various factors of star and planet formation, the likelihood of the origin and evolution of life and intelligence, and the lifetimes of technical civilizations, it came to serve in the last third of the century as a paradigm for discussion of the issues. Although almost everyone acknowledges that the parameters of the equation are not well known—values range from one planet in our galaxy with intelligence (our own) to a hundred million or more planets with intelligence—this uncertainty has not prevented use of the equation as a basis for discussion of the prevalence of technological civilizations in the galaxy. Many radio searches have been undertaken worldwide since 1960, with the National Aeronautics and Space Administration sponsoring the largest program now underway.

Although no radio searches have been successful, the existence of extraterrestrial intelligence is widely accepted in the scientific community, as well as among the public. Leaving aside the motivations of the public, much of which is swayed by nonscientific considerations such as unidentified flying objects, the scientific acceptance is an interesting commentary on the methodology of scientists, many of whom have preferred not to reject a theory that seems plausible on general grounds, even as it has awaited empirical confirmation for centuries.

To return full circle to the cosmological connection with which we began, the twentieth-century view of a universe full of life may perhaps best be seen as a cosmology in its own right, a "biophysical cosmology" that asserts the importance of both the physical and biological components of the universe. Like all cosmologies, it makes a claim about the large-scale nature of the universe; its claim is that life is not only a possible implication, but also a basic property of the universe. Like all cosmologies, the biophysical cosmology redefines our place in the universe. And most important, like other cosmologies, the biophysical cosmology has become increasingly testable in the twentieth century; this is the role and the importance of modern SETI programs. Viewed in this light, the transition from the physical world to the biological universe is one of the great revolutions in Western thought, no less profound than the move from the closed world to the infinite universe described by the French historian of science Alexandre Koyré more than three decades ago.

—Steven J. Dick

BIBLIOGRAPHIC NOTE

The major historical works on the plurality of worlds tradition are Michael J. Crowe, *The Extraterrestrial Life Debate 1750–1900: The Idea of a Plurality of Worlds from Kant to Lowell* (Cambridge: Cambridge University Press, 1986); Steven J. Dick, *The Origins of the Extraterrestrial Life Debate from Democritus to Kant* (Cambridge: Cambridge University Press, 1982); and Karl S. Guthke, *The Last Frontier: Imagining Other Worlds, from the Copernican Revolution to Modern Science Fiction* (Ithaca, New York.: Cornell University Press, 1990). The latter emphasizes the plurality of worlds theme in the literary tradition. Completing the story up to very nearly the present time will be Steven J. Dick, *The Twentieth-Century Extraterrestrial Life Debate: A Study of Science at its Limits* (in preparation). The case for plurality of worlds as a cosmological world view is made in Steven J. Dick, "The Concept of Extraterrestrial Intelligence—An Emerging Cosmology?" in *The Planetary Report*, 9 (1989), 13–17, and "From the Physical World to the Biological Universe: Historical Developments Underlying SETI," in Jean Heidmann and Michael J. Klein, eds., *Bioastronomy—The Search for Extraterrestrial Intelligence: The Exploration Broadens* (Berlin and New York: Springer-Verlag, 1991), pp. 356–363. William G. Hoyt, *Lowell and Mars* (Tucson: University of Arizona Press, 1976) explores one of the more controversial aspects of the extraterrestrial life debate late in the nineteenth century and early in the twentieth century. Also an important contribution to our knowledge about the extraterrestrial life debate is Paolo Rossi, "Nobility of Man and Plurality of Worlds," in Allen Debus, ed., *Science, Medicine and Society in the Renaissance*, vol. 2 (New York: Science History Publications, 1972), pp. 131–162.

28. MULTIPLE UNIVERSES

People seem always to have speculated about worlds in addition to that world they have happened to find themselves in at the time. In the sixth century B.C., the Greek Anaximander (born in 610 B.C.), the first of the well-known Western natural philosophers, proposed a model of the cosmos cycling over time, with the old universe destroyed and a new universe produced at the end of each cycle and the start of the next. Today, some cosmologists speculate that the present phase of expansion of our universe may be followed by a contraction—the *Big Crunch*—and infinitely repeated cycles of expansion and contraction. Such worlds may be called *temporally-multiple universes*, a series of single universes followed one after the other over time.

Temporally-multiple universes are but one type of multiple universe that has gained speculative attention. Anaximander's successor, Anaximenes, discussed the possibility of *spatially-multiple universes*, worlds existing simultaneously. These might occupy separate parts of an infinite space (such as different suns, each with its own inhabited planetary system, inferred by the sixteenth-century A.D. philosopher Giordano Bruno) or worlds-within-worlds, the Gulliver theme, which fascinated philosophers inspired, in turn, by Antoni van Leeuwenhoek's recognition in 1674, using his newly improved microscope, of live micro-organisms in a drop of water.

In addition to temporally-multiple universes and spatially-multiple universes, a third type of universe, *other-dimensionally-multiple universes*, has attracted attention since the seventeenth century A.D. Then, the physicist-philosopher Gottfried Leibniz noted that the Christian doctrine that God chose to create our world implied the existence of other possible worlds that God could have considered creating instead of our own world. These sets of possible worlds do not exist in the same temporal dimension or in the same spatial dimension as does our universe. Although the idea of *parallel* worlds is today more common in science fiction than in science itself, the concept has risen in contemporary quantum theory.

Before examining each of the possible types of multiple universes—temporally-multiple, spatially-multiple, and other-dimensionally-multiple—it is useful to ask what constitutes a *universe*. Unfortunately, although the question is simple, its answer is not. The root of the difficulty is the complete lack of agreement among cosmologists, both ancient and modern, about what might constitute a *universe*.

Two aspects of the hypothesized worlds are of prime relevance to the issue here: (1) the *completeness* of the individual worlds in question and (2) their degree of *separation* from one another. For example, in Anaximander's theory, each cycle of creation and disassociation was complete in and of itself, since each phase apparently had all of the qualitative entities thought possible. Yet each phase was *not* totally separate from the others, again by virtue of the fact that the selfsame qualitative entities constituted the basic stuff of each phase. A review of most examples of many-world-theories shows clear evidence of the same bias, namely, the individual worlds are complete, but in one way or another separation of each from the other is not total. In the end, the essential requirement for separation—the requirement that, indeed, makes genuine any talk at all about *separation*—would appear to be that the worlds are causally or informationally separated from each other. In other words, the worlds cannot signal one another.

With this evidence in mind, it would seem that to the question "What counts as a *universe*?" a practical answer is "Any world or world-system which is complete and informationally separated from all others." In what follows, this criterion will be used to select among possible examples. An additional selection criterion also will be at work. Since, as was initially noted, producing many-world-theories models is an ancient and honorable pastime among natural philosophers, there is an incredible number of them for us to choose among. To optimize the information presented herein, only those efforts that best exemplify their type, or were relevantly influential on the field, will be examined. As will be evident, even criteria as strict as these select a large number of candidates for investigation.

Historical Types of Universes

Spatially-Multiple Universes

Spatially-multiple universes come in only two fundamental styles: the infinite universe model and the worlds-within-worlds model. Let us begin with the notion of an infinite universe. Infinite universe models trace their ancestry to the sixth century B.C., in the proposals made by Anaximander's successor, Anaximenes. It is not clear that Anaximenes's speculations involved multiple worlds, but it certainly is clear that his hypothesized world was spacious enough to contain as many worlds as he wished it to. Among those who followed the Ionians in their proposals that the universe was infinite in extent we find the pluralists Empedocles and Anaxagoras and, of course, atomists such as Democritus and Lucretius. Yet, although their worlds were infinite in extent, it is not clear that any of these thinkers explicitly espoused models that would satisfy our criteria for many-world-theories.

One of the most outspoken of the earlier natural philosophers who saw in an infinite space the possibilities for multiple worlds was Giordano Bruno (born 1548). Speaking of the one, infinite universal space, Bruno (from the mouth of his character Philotheo) claims that "innumerable celestial bodies, stars, globes, suns and earths may be sensibly perceived therein by us and an infinite number of them may be inferred by our own reason." Bruno carefully distinguishes between universes, worlds, and earths. There is only one universe, he claims, but there are infinite worlds within it, and perhaps even an infinite number of earths within the many worlds. Each world is complete, but it is not clear whether all of them satisfy our separation criterion. Certainly, those worlds that "may be sensibly perceived" are not separated from our own. But those worlds whose existence may only be "inferred by our own reason" most certainly are separated from us. Since there is an infinite number of these complete and separate worlds, Bruno's theory is certainly a many-world-theory.

Perhaps just as interesting as his theory's nature is Bruno's rationale for proposing it. Being and existence, claim Bruno, are better than nonbeing and nonexistence. Moreover, just as one is better than none, many are better than one, and infinity is more perfect than finitude. Thus it follows that the most perfect universe is that one that contains an infinity of worlds within its own infinitude.

Bruno's argument here is just one example of a whole family of arguments, all of which gain what force they have by explicit appeal to some sort of principle of value. Usually, as here, the value targeted is some sort of metaphysical-cum-ethical one to the effect that "the abolition and nonexistence of this world would be an evil." Sometimes the value appealed to is an aesthetic one, namely, the belief that multiple worlds are somehow more *fitting* than would be only one. In any case, the ultimate result of such arguments is what has been called the great chain of being, according to which all possible niches in reality are filled with actual existences. As we shall see, this sort of motivation pervades many-worlds theorizing.

A second type of spatially-multiple universes is the worlds-within-worlds model. Milic Capek has discussed the impulse toward this type of speculation and refers to it by the apt term *Gulliver's Theme*. Among proponents of this notion are many, if not most, of the well-known natural philosophers of the seventeenth and early eighteenth centuries. One of the most delightful expressions of this type of many-world-theory is found in the correspondence between Jean Bernoulli and Leibniz. Following a brief discussion of van Leeuwenhoek's discoveries, Bernoulli turns to a grain of pepper on his table and begins an inquiry into the possible experiences to be had by "pepperlings"—inhabitants of the "world-within-a-grain-of-pepper." Far from finding Bernoulli's fancies fanciful, Leibniz takes it as a simple matter of course that the pepperling world exists, is complete, and is separated from our own. Over the course of several exchanges, the two scientist-philosophers continue investigating the confines of this and any other of the infinite worlds-within-worlds, until finally their focus turns to another of their many mutual interests.

Several points here are worth pursuing. First, we must note a metaphysical presupposition shared by all those who espoused Gulliver's Theme. It is obvious that the whole plausibility of worlds-within-worlds rests on the plausibility of (1) the infinite divisibility of matter, and (2) linear reductions at all scales. Secondly, Bernoulli and Leibniz do not for an instant hesitate to proclaim the existence of life (indeed, *rational* life) in the many other worlds. This most likely is linked to the fact that it was the microscopic researches into "animalcules" in a drop of water that initially triggered the many-world speculation in the first place. Finally, it appears that the relationship that we call *containing* does not violate the separation criterion. Thus, for example, although Bernoulli and Leibniz's pepper-grain world is *contained* in our world, it apparently is not causally connected to it in any usual way. That we cannot communicate with the pepperlings, for example, would seem to be implied by the fact that Bernoulli and Leibniz make absolutely no effort whatsoever to discuss this possibility. It would seem plausible, of course, that we could *destroy* the pepper world by, for example, eating it in a sandwich; but this possibility doesn't seem to be part of the two philosopher-scientists' concerns about their other world.

A.N. Whitehead provided the exemplary statement of Gulliver's Theme many-world-theories:

> In this little mahogany stand may be civilizations as complex and diversified in scale as our own; and up there, the heavens, with all their vastness, may be only a minute strand of tissue in the body of a being in the scale of which all our universes are as a trifle.

As will later be shown, contemporary theories of spatially-multiple universes for the most part conform to the standards set up by the two styles just discussed, namely, the infinite space many-world-theories or the worlds-within-worlds many-world-theories.

Temporally-Multiple Universes

Anaximander's cyclical universe is typical of the temporally-multiple universe type. Some mechanism or force operates for an extended time upon the cosmos' material stuff thereby producing a given state of the universe. At that point either the initial mechanism ceases functioning or a countervailing force begins to override the initial one, thereby undoing all the latter's previous accomplishments. Opposed forces might include, for example, the metaphorically named Love and Strife; or conceived more mechanically, opposed forces might include condensation and rarefaction. Often, the cosmological cycles are associated with astronomical phenomena. Greek thought inevitably returned to the notion of the *Great Wheel* of the sky, a millenia-long cycle during which the major celestial observables wheeled about the pole star and returned finally to their original positional relationships.

Beginning with Anaximander, the phase transitions between cycles invoke a scrambling of all information from the previous cycle. Obviously, it is only in this manner that the succeeding cycle might in any sense be *new*, that is, in any sense be *separated* from the earlier cycle. On this notion the many-worlds of temporally-multiple universe theory correspond to the individual cycles undergone by the one universe. Modern temporally-multiple universe versions, for example those of Friedrich Nietzsche and Hans Reichenbach, speak of the phase-transition scramblings by reference to statistical-probabilistic notions. But one constant throughout all this scrambling is the eventual emergence of life, much as we saw earlier in the case of spatially-multiple universes. Perhaps it is one aspect of the completeness of any of the many worlds that it contains life.

In consonance with the statistical-probabilistic flavor of the intercycle scramblings, we find that, since Darwin, these notions are applied to the *mechanism* of the emergence of life as well. Herbert Spencer is quite explicit:

> And thus there is suggested the conception of a past, during which there have been successive Evolutions analogous to that which is now going on; and a future during which successive other such Evolutions may go on—ever the same in principle but never the same in concrete results.

Chief differences among contemporary temporally-multiple universes involve the mechanisms used to bring about the scramblings between cycles. Indeed, as we shall see, the very notion of the possibility of intercycle scrambling has been called into question. In all other essential aspects contemporary temporally-multiple universes are analogues of those discussed here.

Other-Dimensionally-Multiple Universes

Unlike spatially- and temporally-multiple universes, the other-dimensionally-multiple universe model is a relatively recent innovation. Leibniz created the notion with his hypothesis of "possible worlds." He saw that the Christian doctrine that God *chose* to create our world made no sense unless there were other worlds, other *possible* worlds, that God could have considered along with our own. Leibniz's vision is of an infinite set of worlds, each complete in itself and separated from the others by some infinitesimally small difference. For example, in one possible world Caesar didn't cross the Rubicon; in another possible world Judas didn't betray Christ; and so on.

One major notable feature of the possible worlds is that they—in some sense—exist together in a set. Although it is tempting to say that the worlds exist *simultaneously*, clearly such a temporal notion would be quite inappropriate. The dimension that the worlds occupy is neither a temporal nor a spatial one; rather, it is a modal dimension, the dimension of possibility. This means that, prior to God's choice, the possible worlds are *equally* possible, where the relation "equal" is analogous to the temporal relation "simultaneous."

Contemporary cosmological versions of Leibniz's other-dimensionally-multiple universe begin with Hugh Everett's quantum-mechanical, infinitely-branching ensemble of "parallel" worlds. But it must be noted that the very idea of a "parallel" world has been popular in science fiction literature since the 1930s. It can only be speculated that Everett's admittedly uncommon hypothesis in some sense derives from this literature. Such an influence, however, would not be surprising.

This completes our historical review of the various multiple-world-theories types. Let us now examine contemporary models.

Contemporary Examples of Many-World Theories

Spatially-Multiple Universes

All contemporary spatially-multiple universe theories rely on the concept, in one way or another, of infinite space. The reasoning is that if space is "large" enough, complete and separate universes will result. For the most part, these theorists practice under the aegis of the inflationary universe model. One exception is the work carried out by G.F.R. Ellis and a number of coworkers.

Ellis's models are minimal cases of many-worlds-theories. Indeed, he and his colleagues refer in their proposals to "*worlds* within just one infinite universe." Moreover, although we might consider Ellis's hypothetical worlds to be *separate* from one another, it certainly is not clear that each ought to be considered to be *complete*. Curiously enough, however, they are complete enough for Ellis and G.G. Brundrit in a recent work to consider the possibilities of life in their universe. Given the type of universe they postulate—low-density, homogeneous, open, and infinite—they claim that any set of conditions specifiable (e.g., earth-like conditions) will occur, and not just once but frequently: "It is highly probable that there exist infinitely many worlds on which there are 'duplicate' populations to that on our own world." Underlying their view is the belief that any set of finitely probable conditions may occur any number of times within an infinite volume of space. A mechanism to bring about such variation is not proposed by Ellis and his coworkers.

In contrast to Ellis and his colleagues, many-worlds-theories proposals based upon inflationary universe models provide not only a mechanism to bring about variation among complete universes, but also the *space* (or, perhaps, *superspace*) large enough to keep them separated. Alan Guth, working from a basis in particle physics, helped create the first genuine synthesis between cosmology and high energy physics. Based upon his investigations of grand unified theories, Guth proposed that the universe shortly after its initial explosion in the big bang went through a sudden intense expansion—the inflationary period. [See chapter 22, *Inflationary Cosmology*, in section VII, *Particle Physics and Cosmology*.] Although it was originally intended to solve other problems, Guth's hypothesis later came to be tied to many-worlds-theories through the work of J. Richard Gott.

Gott noticed that Guth's inflationary model consisted of *domains*—sub-spaces that eventually coalesced into a single, large Friedmann universe. Yet, after investigating the inflating domains (which he called *bubbles*) in detail, Gott came to the conclusion that the bubbles would remain distinct and separated from one another. Moreover, each bubble would constitute a universe with a $k = -1$ cosmology; in other words, each bubble would have a spacetime like our own. And there are a lot of bubbles, as Gott himself notes: "Guth cosmologies . . . can each produce in principle an infinite number of disjoint $k = -1$ (bubble) cosmologies."

K. Sato and his co-workers' efforts parallel and even extend that of Gott. They propose that separate bubbles of false vacuum would be trapped within the inflationary domain of true vacuum. These would eventually evaporate, but only after forming separate "daughter" universes. Moreover, within nucleated regions of the daughter universes would be formed a potentially infinite number of offspring: granddaughters, great-granddaughters, great-great-granddaughters, and so on and so on.

These examples show that producing spatially-multiple universe models is an active area within many-world theorizing. As A.D. Linde notes in an extensive review, inflationary universes are especially hospitable to multiple-worlds, in many of which will exist conditions favorable to life like our own.

Temporally-Multiple Universes

Today's temporally-multiple universe models resemble that of Anaximander: the universe itself cycles, and as each zero-node in a cycle is reached, the former phase is "scrambled" in such a way that the new cycle represents a new universe. Most contemporary research focuses sharply on what might happen during the scrambling. This focus was necessitated by one of the first modern analyses of the temporally-multiple universes scheme, an elegant, extended account produced by R.C. Tolman. His work was later furthered by P.T. Landsberg and D. Park. According to their results, entropy would not be canceled during the zero-phase, which implies that successive cycles would suffer increasing entropy, accompanied by physical enlargement. (Some philosophers, of course, would look askance at the notion that the universe's cycles could differ in "size"; "Relative to what?" would be the natural philosophical query.) Present observations suggest that only a hundred or so cycles might have preceded our present one. Although this number might sound large, it is quite likely that most many-worlds theorists would claim that it hardly counted as "many."

An interesting question regarding the separation of the successive cycles is also raised by Tolman's work. If it be true that the universe's state of entropy is preserved across the zero-phase, isn't it equally true that successive cycles communicate with one another? This question would be especially perplexing for the interpretation of entropy as information. Several thinkers have devoted special effort to solving the separability issue raised by Tolman's work. Most provocative have been the efforts of John Archibald Wheeler.

Wheeler is apparently responsible for naming the zero-phase the *Big Crunch*. In a most revealing metaphor, he speaks of the universe being "reprocessed" by gravitational collapse during the crunch. Since our theories simply are unable to describe what sorts of processes might go on during the collapse, Wheeler suggests that we adopt a "black box" approach to the problem of reprocessing: "Physics has long used the 'black box' to symbolize situations where one wishes to concentrate on what goes in and what goes out, disregarding what takes place in between." What goes out into the next cycle, Wheeler maintains, is completely different from what went in. Not only have all conservation laws (including those for charge, lepton number, baryon number, mass, and angular momentum conservation) collapsed, but Wheeler also believes that both the spectrum of particle masses and the physical constants "would seem most reasonably regarded as reprocessed from one cycle to another." In other words, gravitational collapse would indeed bring about total separation between the "worlds" represented by successive cycles of the universe.

Yet the question of the entropy remains. Although, in the face of such massive scrambling as that envisaged by Wheeler, it might seem odd to maintain still that entropy—in *some* form or other—persists between cycles, it has been so maintained, thereby reviving Tolman's original onus upon temporally-multiple universes. Recent work by Linde, however, flatly denies that entropy can persist through the crunch. His argument deserves some attention.

In the first place, Linde notes, "all particles are colored with respect to gravitational interaction." This means that unlike, say, the strong force or the weak force, which affect, respectively, only their own coupled quanta, the gravitational force affects *all* particles. Hence, in the instant of maximum collapse, all particles would be confined in the greatest of all possible black holes.

Second, if, as is expected, gravitation turns out to be a quantum-type interaction, then the normal rules of quantum confinement apply during the instant of maximum gravitational collapse. This implies that "in the gravitational confinement phase *any* particle would have infinite energy." Obviously, in such a regime "no real particle excitations can exist and the entropy of the universe vanishes." Linde, as Wheeler before him, believes that this scrambling of entropy allows new worlds to emerge in each cycle: "In our case all the entropy and all inhomogeneities of the contracting universe disappear in the purgatory of the de Sitter stage, and then are generated anew in each cycle." The emergence of new worlds—especially ones that differ, à la Wheeler, in such fundamentals as the constants of nature—would certainly argue that the successive cycles of a Wheeler-Linde temporally-multiple universe satisfy all requirements of completeness and separation; that is, each cycle genuinely qualifies as a different universe.

Although several other models of temporally-multiple universes have been suggested, most resemble the Wheeler-Linde version in relevant respects. Perhaps the only other candidate that should be mentioned is the model proposed by M.A. Markov. His mechanism to avoid Tolman's rising-entropy problem is to allow certain extra-dense gravitational regions to close in upon themselves during the universe's collapse phase, but prior to the zero-phase. Once closed, the entropy of these interstitial regions would be hidden from the "mother" universe: since gravitationally closed regions have no externally observable details beyond their mass, charge, and angular momentum, the "daughter" universe could evolve at will during successive cycles of expansion and contraction of the "mother" universe. This process of hiding daughter universes within gravitationally closed interstices of the mother universe is in principle extensible ad infinitum. Although some might object to Markov's mechanism as a sort of sleight of hand, he replies that it does not contradict the law of entropy increase; moreover, it does not contradict the law of entropy increase to propose that our own universe is one of the daughter universes hidden within another mother. Perhaps it is proper to say that Markov not only has a temporally-multiple universe here, but also has brought into existence yet another version of worlds-within-worlds!

It is clear from this review of temporally-multiple universes that this mode of multiple-world-theorizing remains an attractive option for speculating about universes other than our own.

Other-Dimensionally-Multiple Universes

Although the "parallel worlds" (i.e., many worlds in adjacent "dimensions") theme has long been a favorite in science fiction stories, it has not found much favor (or use) in science. Leibniz's notion of possible worlds, as we saw earlier, was the first such model. Among contemporary theorizing, only quantum mechanics has employed a version of such a structure. The theory developed as a possible solution to a stubborn problem in quantum measurement.

Quantum theory is inherently statistical; that is, its formalism inescapably entrains probability statements in its description of the world. For example, when used to predict where electrons emerging from a slit will strike a screen, the formalism couches its predictions in strictly probabilistic terms in a "wave function": there is, for instance, a 21% probability that the electron will impact at point A and a 28% probability it will impact at A', and so on. Yet, when the electron in fact impacts, it impacts at one point and at one point only, no matter how long the string of probabilities called out in the wave function. In this situation, what is one to make of the string of probabilities? What is their status, given that one and only one of the predicted probabilities can be accurate? Since measurements are inherently unique and quantum predictions are inherently statistical, how are we to conceive of the relationship between the world as measured and the world as quantum predicted?

A majority of physicists hold to the *Copenhagen Interpretation*, which proposes that the probabilities are accurate until the measurement is made, at which point, and with infinite speed, the probabilities collapse into the unique value measured, which, obviously, at that point, itself becomes accurate. Some physicists are uncomfortable with the Copenhagen interpretation: according to John Barrow and Frank Tipler, "The wave function collapse postulated by the Copenhagen Interpretation is dynamically ridiculous, and this interpretation is difficult if not impossible to apply in quantum cosmology." Doubts such as these prompted Everett (alone at first, and later with B.S. DeWitt) to develop an alternative interpretation of what happens during measurement on quantum mechanical systems.

According to the interpretation, which, for obvious reasons is called the *Many-Worlds Interpretation*, the wave function is interpreted to be literally true; that is, each of the probabilities calculated applies to some particular world in which, were the measurement to be carried out, the value would correspond to the probable value in the wave function. From this it follows that measurement does not "collapse" the probability function into a unique value; rather, it simply moves the observer along his or her world's own worldline among the infinitely many possible ones. All of the probable states exist together, each in its own world, all "parallel" in some sense to one another in the quantum phase-space dimension.

Certainly the various worlds are complete. Until any given two of the measuring systems diverge, their worlds are identical; hence, if any world is complete, all of them are complete.

Separation, however, is controversial. V.F. Mukhanov, for example, takes an extreme—even for a proponent of an Everett-type scheme—view of the real existence of the parallel quantum worlds. Not only do the many worlds exist and, through their existence, account for the "interference" (superposition) terms in the wave function, according to Mukhanov, but they also enable quantum theory itself: "One can interpret quantum mechanics as a theory the very *existence* of which is due to the existence of many worlds." It is unclear, however, whether or not the superposition "interference" between worlds might call the individual worlds' separateness into question. Certainly, beyond the superpositions themselves, no causal influences (e.g., signals) may be propagated between worlds.

One major cosmological benefit flows from Mukhanov's extension of the Everett-DeWitt ensemble of quantum worlds. Since Mukhanov takes the continuous string of probabilities to be descriptions of real worlds, he in effect has discovered a continuum of worlds, varying each from the other in the least specifiable degree. This means that the ensemble will exhibit all possible values of all possible parameters, including both constants *and* laws. Only probabilities will control which values appear: "This means that all the conceivable processes, if the probability of their realization is not exactly equal to zero, take place in reality." This extends, says Mukhanov, even to events of extremely small probability, "for example, heat overflow from a cold body to a hot one."

Although the many-worlds interpretation is still a definitely minority position among quantum theorists, it is a position that nonetheless refuses to go away. Barrow and Tipler believe that, in the end, the many-worlds interpretation will replace the Copenhagen interpretation, exactly as "the Copernican system replaced the Ptolemaic." Moreover, just insofar as the many-worlds interpretation survives (let alone flourishes) in quantum mechanics and quantum measurement theory, it also will provide a curious model for the close scrutiny of cosmologists.

Conclusion

Although the first speculations about other universes occurred long ago, it is clear that such theorizing remains a vigorous enterprise even in our own day. Viable multiple-universe models exist in all three of the categories examined above. Moreover, as we have seen, many-worlds-theories seem to be bound quite tightly to most of the presently acceptable cosmologies. This would seem to imply that we shall be just one of the many worlds of cosmology for a long time to come.

—*George Gale*

BIBLIOGRAPHIC NOTE

Recent discussions of multiple universes tend to focus on their place in quantum theory. One of the best discussions is chapter 7, "Quantum Mechanics and the Anthropic Principle," in John D. Barrow and Frank J. Tipler, *The Anthropic Cosmological Principle* (Oxford: Clarendon Press, 1986). A different perspective on the same topic is provided in several chapters of John Leslie's *Universes* (London: Routledge, 1989). Leslie also is the editor of an anthology, *Physical Cosmology and Philosophy* (New York: Macmillan, 1990), containing several discussions relevant to the the many-worlds theme. Finally, a more limited but extremely useful analysis of the history of discussions from Copernicus until today about the possibilities of other inhabited world systems in the universe is found in Karl S. Guthke, *The Last Frontier: Imagining Other Worlds, from the Copernican Revolution to Modern Science Fiction* (Ithaca, New York: Cornell University Press, 1990).

29. PHILOSOPHICAL ASPECTS OF COSMOLOGY

Cosmological speculation and natural philosophy were born in the same gesture: Thales and his Milesian colleagues' naturalistic hypotheses about the nature and structure of the universe around them. During the more than two and a half millennia since these sixth-century-B.C. beginnings, cosmological thought and natural philosophy have been inextricably interwoven, with admixtures, from time to time, of varying degrees of empirical and theoretical astronomy, coupled almost always to attempts to mathematicize the entire process. Three flourishings—ancient, classical, and modern—may be easily identified; equally identifiable at the beginning of each period of flourishing is the emergence of a recognizable philosophy of science, the result of vigorous discussion among the initiators of the new epoch of cosmologizing. In each era, both the details of the science and the nature of the science itself were shaped by the new consensus about the science's underlying philosophy.

Philosophy of Cosmology: Ancient and Classical

In the ancient world, Milesian and Pythagorean traditions reached the consensus that cosmological thought would blend analogical reasoning from everyday experience, logical deduction, and mathematical modeling. Although the results were not subject to what we today would call experimental or strict observational control, they were, nonetheless, taken to represent efforts to understand the actual structure of the cosmos.

Ancient efforts to propose realistic physical models of the cosmos culminated in the model propounded by Aristotle nearly three centuries after Thales. Aristotle's model envisioned the physical mechanism of homocentric crystalline spheres centered on the earth, with varying axes of rotation and rotational velocities functioning to explain the known astronomical details. Cosmological work continued within this tradition for several centuries. [See chapter 6, *Aristotle's Cosmology*, in section II, *The Greeks' Geometrical Cosmos*.]

In the end, the ancient period's philosophical and scientific consensus dissipated in the purely instrumentalistic proposals of epicycles, deferents, and equants given by Ptolemy. These efforts are remarkable for their strict disavowal of any and all claims regarding the physical reality of the mathematical hypotheses. According to the explicit doctrine of Ptolemy, his hypotheses, unlike Aristotle's physicalistic proposals, were exhibitions of pure theoretical mathematics and had nothing at all to say about the actual mechanisms of the cosmos. [See chapter 7, *Ptolemy*, in section II, *The Greeks' Geometrical Cosmos*.] Truth about the structural realities of the universe, Ptolemy thought, was unapproachable by human reasonings. It was not until the second flourishing of cosmological speculation that efforts were again made to discern the genuine physical structure of the world and to link it to known astronomical findings.

Copernicus's fame resides for the most part in his proposal for the cosmological scheme that bears his name. Yet, careful analysis reveals that his genuinely worthwhile innovation lies not there, in his scheme; after all, he himself appeals to Heraclides' authority in the matter of the earth's rotation. Where Copernicus truly evinced innovation and, indeed, a sort of courage was in his unswerving philosophical commitment to producing a realistic, physicalistic cosmology. [See chapter 8, *Copernicus*, in section II, *The Greeks' Geometrical Cosmos*.]

Copernicus's vigorous metaphysical stance unfolds immediately in his introductory indictment of his predecessors' inability to "agree on any one certain theory of the mechanism of the Universe." Even though Ossiander's notorious preface attempted to re-insert Copernicus in that instrumentalist tradition stretching unbroken back to Ptolemy, immediate and close successors such as Thomas Digges rejected Ossiander's deceitful claim that Copernicus delivered "these grounds of the Earth's mobility only as Mathematical principles, feigned and not as Philosophical truly averred."

One generation later it was Kepler himself who unmasked Ossiander as the villainous author of the spurious call to metaphysical anti-realism in cosmology. But in addition to seconding Copernicus's metaphysical realism, Kepler himself provided a philosophical innovation with his new methodology for cosmology. Mathematical hypotheses, couched in forms recognizably different from those of their predecessors, were to be confronted directly by the "irreducible, obstinate facts" of precise, accurate observations, those provided in this case by Tycho Brahe. Although Kepler's position resembles in some ways that of his ancient predecessors, its commitment to universal, strict, and precise comparison of theory with observation is totally novel. [See chapter 13, *Kepler*, in section IV, *The Scientific Revolution*.]

With this change in methodology, classical cosmology had a full-blown philosophy.

- First, it was to be a genuine branch of natural philosophy, that is, a discipline dependent solely on verifiable, empirical claims and proposals. No more angels and planetary souls.

- Second, it was to operate according to a mathematical-observationalist methodology, as exemplified by Kepler and, after him, by Newton. No more qualitative, discursive storytelling.

- Finally, the claims of cosmology were to be taken realistically, not instrumentally; that is, as being about actual, real mechanisms and not about fictitious movements and crystalline bodies. No more epicycles and deferents.

With later philosophical additions—such as Galileo's development of experimental observationalism, Bacon's call for a method based in inductive logic, the debates between Descartes, Boyle, and Locke regarding the role of theoretical reason, and the debates between Newton and Leibniz about the nature of fundamental scientific realities such as matter, force, space, and time—the classical philosophy of science came to be completed.

In short, led by cosmology, classical physics ultimately settled upon a set of philosophical canons that sufficed to lead it vigorously into the dawn of the contemporary era. But then, led again at least in part by cosmology, the classical physical synthesis abruptly began to break down. Once attempts began to reestablish science in the unsettled areas, new foundational-philosophical controversies engaged practitioners; in the end, these new disputes provided for an enriched, pluralistic, and indeed stronger cosmological science, one that continues to mature even unto today. The story of this most recent flourishing of the natural philosophy-cosmology interaction is perhaps the most fascinating of them all.

Historical Aspects of Modern Cosmology

Modern cosmology began in two separate places, as two distinct disciplines. On the empirical side of things, workers in the United States pushed their astronomical focus farther and farther out from the earth. America was a productive home for astronomers for two simple reasons: climate and telescopes, with excellent instances of both abounding. A heady era of great telescope-building marked the first decades of the twentieth century; accelerated observational schedules accompanied the new glass. Ultimately, by the mid-1920s, it was quite clear to most observers that the universe is an extremely large place, a place, moreover, quite clumped with huge collections of material objects—galaxies of stars. With this realization came as well the realization that observational astronomy would be able to assist in the efforts to understand the cosmos at large.

In contrast to this lush observational growth, theoretical cosmology failed to germinate on American soil. This was in part due to the typical American intellectual climate of skepticism toward theoretical activities in general and, in particular, toward those which, like cosmology, had not even a hope of eventual practical benefits. In addition, modern theoretical cosmology required a level of mathematical accomplishment generally unavailable at that time. As the English philosopher A.N. Whitehead noted with respect to Einstein's new cosmology, "The announcement that physicists would have in future to study the theory of tensors created a veritable panic among them." Because of these difficulties, theoretical cosmology—with the exception of work by R.C. Tolman, H.P. Robertson, and, to a certain extent, Edwin Hubble—got short shrift among U.S. astronomers and astrophysicists.

England was where modern theoretical cosmology first took hold and began to grow. This first phase of development depended intimately upon the role and status of the newly arrived general theory of relativity.

Possible connections between cosmology and the general theory of relativity were first displayed by Einstein in 1917. But his solution to the field equations, which envisaged a cosmos with uniform density of matter, was immediately criticized by the Dutch astronomer Wilhelm de Sitter, whose proposed alternative cosmos apparently was empty of all matter.

Both solutions were arrived at with the presupposition that there was no net "movement" or evolution of the universe as a whole. The universe, in this view, was static. Needless to say, at this point neither of the theoretical models aroused much interest among observational astronomers. On the one hand, the observed universe was clearly not an empty one, as de Sitter's interpretation required; on the other hand, Einstein's model was equally unacceptable for the opposite reason: it required too much matter. The universe as conceived by many leading empirical astronomers (such as the Dutch astronomer J.C. Kapteyn and the American astronomer Harlow Shapley) was nowhere near the size required by the large-scale distribution of matter called for by Einstein's interpretation.

Thus, in both cases, fitting the theoretical model to the observational situation required much awkward and often ad hoc modifications and special assumptions. This clash between observationalist conceptions and the general theory of relativity would continue until the widespread acceptance of the "island universe" model of extragalactic space following Hubble's 1924 observations. Such a universe most certainly provided enough room for Einstein's general theory of relativity solution, thus opening the way for pursuit of this model. Until this happened, however, the general theory of relativity offered little to the observationalists.

What the general theory of relativity *did* offer, however, was an intrinsic fascination for mathematicians, particularly for Arthur Stanley Eddington, the man probably best situated and best prepared to establish and nurture theoretical cosmology in England. Eddington presents an unusual case. He studied mathematical physics as a scholarship student at Cambridge University under E.T. Whittaker, became Senior Wrangler in 1904, and became a Fellow of Trinity College in 1907. That same year he joined the Greenwich Observatory as chief assistant. In 1914 he became director of the Cambridge Observatory; he already held the Plumian Chair at Cambridge. In 1915 Eddington presented his highly favorable *Report on the Relativity Theory of Gravitation* to the Physical Society, thereby introducing the general theory of relativity to the English-speaking world. The astrophysicist S. Chandrasekhar reports that by this time "Eddington must surely have been caught in its [relativity theory's] magic."

Because of his background in both astronomy and mathematical relativity theory, Eddington was the natural choice to head one of two expeditions to observe the solar eclipse of 29 May 1919, which would provide a noncosmological test of the general theory of relativity's prediction about the deflection of starlight. Both expeditions reported success.

Upon his return, Eddington gave enthusiastic and favorable reports on the success of the general theory of relativity to a wide variety of audiences: a packed joint meeting of the Royal Society and Royal Astronomical Society in November, essentially the entire Cambridge academic community on 2 December, and on 12 December the Royal Astronomical Society, which devoted its entire meeting to a discussion of the general theory of relativity, led by Eddington. Whitehead, who was present at the first of these meetings, commented that "the whole atmosphere of tense interest was exactly that of the Greek drama: we were the chorus commenting on the decree of destiny as disclosed in the development of a supreme incident."

These performances, plus the 1920 publication of his superb popularization of the general theory of relativity, *Space, Time and Gravitation*, established Eddington's reputation as one of the few scientists in the world who "really" understood relativity. An apocryphal story has Eddington, praised as one of only three persons in the whole world who understood relativity theory, pausing to think for a moment before asking "Who is the third?"

As Dirac later noted in no uncertain terms, in 1920 relativity was the place to be: "Everyone was talking about relativity, not only the scientists, but the philosophers and the writers of columns in the newspapers. I do not think there has been any other occasion in the history of science when an idea has so much caught the public interest as relativity did in those early days."

Eddington taught relativity at Cambridge University, and by the middle to late 1920s the subject was well established there and at Edinburgh, under the auspices of Eddington's former teacher, Whittaker. Students entering either university to study mathematics from that time on were routinely exposed to the formalisms of the general theory of relativity. Thus we find that W.H. McCrea (Cambridge 1923–26) was exposed to Eddington's lectures from time to time (although he admits not trying seriously to understand relativity until he became a colleague of Whittaker's in 1930), and that G.C. McVittie (Edinburgh 1922–28, Cambridge, 1928–30) learned relativity in the company of both Whittaker and Eddington.

It must be carefully noted, however, that the version of the general theory of relativity to which these students were exposed was conceived to be a strictly *mathematical* entity. This cannot be overemphasized.

- First, the theory was taught as a mathematical subject to mathematics students by mathematics instructors. Thus, even though Eddington and Whittaker were what we today would call physicists (or in the case of Eddington, an astronomer), it was as applied mathematicians that they presented the theory to their students.

- Second, the only two solutions—Einstein's and de Sitter's—that were widely known in the 1920s had been reached in classically mathematical fashion, with not the slightest admixture of observational input.

- Third, as noted earlier, both the Einstein solution and the de Sitter solution were awkward objects for observationally inclined astronomers.

Thus, throughout the period, the general theory of relativity remained for the most part a purely theoretical subject for study by mathematical physicists (or, perhaps more accurately from an Anglo-Scottish point of view, applied mathematicians).

Outside England, however, two men were working to find a way out of the impasse. But for reasons that remain cloudy even today, the efforts of Alexander Friedmann in Russia and Georges Lemaître in Belgium were without effect upon other workers in relativity theory, even though Friedmann's 1924 work had been inspected by Einstein and Lemaître's 1927 work was known to Eddington. Both of these leading relativists (Einstein and Eddington) failed utterly to see the cosmological significance of the other two men's solutions to the field equations. We might speculate that, because both Friedmann's and Lemaître's models envisaged a nonstatic, evolving universe, Einstein's and Eddington's predisposition toward a static universe blinded them to the significance of the new proposals. Another possibility is that, absent any alternative, the two static models remained just plausible enough not to be ruled out conclusively. In any case, the decade closed with the general theory of relativity still in its purely mathematical stage, with its community of students generally not oriented toward empirical application of the theory.

Hubble's 1929 announcement of the redshift-distance relation provided the push that ultimately would unite theoretical and observational elements into a modern cosmology—but not immediately. Even Hubble himself initially did not see the cosmological implications of his relationship. Rather, he proposed that the relation might be due to a purely local influence, which he termed "the de Sitter effect," the notion that the "displacements of the spectra arise from two sources, an apparent slowing down of atomic vibrations and a general tendency of material particles to scatter." [See chapter 19, *Hubble's Cosmology*, in section VI, *The Expanding Universe*.]

At nearly the same time, Eddington's student G.C. McVittie was working on a thesis that had, as part of its goal, finding a new route to an exact solution of the field equations. A few months later, in early 1930, Eddington suggested that McVittie interrupt his thesis work in order to concentrate on the theory of the redshifts of galaxies, which Eddington hoped would shed light on the stability or otherwise of the Einstein universe. At that moment, a letter came from Lemaître reminding Eddington that he (Lemaître) had already in his 1927 paper answered the question of the stability of Einstein's universe. Lemaître had shown that Einstein's universe was unstable and would tend to expand. With this suggestion, empirical cosmology and theoretical cosmology finally could come together.

If any date might be justly claimed as the birthday of modern cosmology, that date is 29 September 1931. On that day the British Association convened a special session of Section A devoted solely to the topic "The Evolution of the Universe." By the end of the session, all the major cosmological workers had reached consensus on two essential points:

- They had a science.
- The science was deployed around the general theory of relativity as its central model.

Thus was born relativistic cosmology. Present at the birth were many if not most of the participants who already had made contributions to the new science and who would, most of them, continue to dominate the field for the next two decades. The session was significant enough that the scientific journal *Nature* immediately devoted a supplementary issue to its coverage. Speakers included Sir James Jeans, Lemaître, de Sitter, Eddington, Oxford astrophysicist E.A. Milne, and Sir Oliver Lodge. McCrea and McVittie arrived together somewhat late and had to participate through a loudspeaker in a second hall hastily set up to handle the overflow crowd. By the end of the day, details of a consensus view about the nature and dynamics of the universe had become clear.

Eddington probably is the one most responsible for the development and promulgation of the core tenets of the consensus. As McCrea and McVittie reported Eddington's view just prior to the British Association meeting, it held that "our actual universe may probably be considered as expanding from an initial state something like 'Einstein's world' towards a limiting state like 'de Sitter's world'."

In this view, the crucial research problem became the discovery of the cause of the initial instability in the universe's Einstein state. It was this problem precisely that Eddington had set his student McVittie to work on in early 1930, with McVittie soon to follow McCrea's suggestion that they together investigate the possible role condensations might play in solving this problem, and finally, it was this problem that was to play the central research role in cosmology for the next eighteen months, until philosophical controversies deflected the community's interest toward more foundational issues raised by Milne. Until then, however, the consensus model held: it was affirmed initially by Eddington, de Sitter, Einstein, and Jeans, and seconded just a bit later by Tolman and Robertson.

Interestingly enough, Hubble at this point refrained from accepting the consensus relativistic model of an expanding universe; he would continue to do so for several more years, until at least 1937.

Unfortunately, the consensus view contained within itself philosophical and methodological strains, strains so severe as to almost immediately threaten the new structure itself. In particular, the delicate nature of scientific cosmologizing within the two-fold limitations of the general theory of relativity on the one hand, and observational perplexities on the other hand, strained the matrix of views providing the philosophical foundations laid down by the classical scientists and philosophers two hundred and fifty years earlier.

Philosophical Aspects of Modern Cosmology

It is a commonplace that new sciences emerge in bits and pieces, often in frightful disarrays of philosophical tenets, theoretical speculations, methodological arguments, and observational puzzles. At least a part of the effort of a science's founding parents must be spent in sorting out these disarrays and, in particular, working through the various philosophical issues essential to the consistent and well-founded basing of their science. Settling a new science's philosophical issues is an always energetic, sometimes downright loud activity—such was the case with relativistic cosmology.

The period immediately following the 1931 British Association meeting saw the initial philosophical salvos fired; the battle was to last until it was quieted down by World War II, and in the meantime it engaged nearly everyone even remotely concerned with cosmology. At one point, the exchange of volleys became so intense that *Nature* devoted an entire supplement to discussion of participants' philosophical views. It is instructive to examine the range and content of the elements taken from the philosophy of science's domain; they reveal much about relativistic cosmology's emergence.

Much of the dispute during the early days of relativistic cosmology derived from the fact that not only hadn't the basic issues underlying the question "What kind of a science ought this to be?" been settled, but also the nature and existence of the issues themselves had not yet been made distinct. In other, more mature, sciences (e.g., classical Newtonian physics) these issues had been settled for some time. But the turmoil involved in the discovery and development of both quantum physics and relativity theory had unsettled things once again, leaving classical physics and, perhaps more important, the *philosophy* of classical physics rather severely disrupted.

In particular, just to take related examples from the two new theories, philosophical questions about the existence-metaphysical reality of both quantum matter and curved space were raised alongside equally perplexing questions about the observational-epistemological status of either or both. Moreover, the two theories raised fundamental questions about what method (or perhaps, methods) was (were) appropriate for their scientific study. Insofar as the newly born cosmology was perceived to be linked to these unsettling developments in contemporary physics (relativity theory, in particular), its own basic issues were unsettled.

But there was more involved in cosmology's unsettled basic issues than just the connection to relativity.

- Cosmology really was a potentially new *kind* of science, that is, a potentially new intellectual domain involving both a novel theory and a new sort of beyond-the-limits style of observation.

- Also, cosmology had to separate itself *as a science* from its philosophical and theological ancestors.

- Finally, as an emerging new science, cosmology thereby brought into view the ultimate basic question that faces *all* new sciences: What ought to count as *being scientific* in this new domain?

The question of what ought to count as being scientific is the most primitive question faced by any science. Moreover, its form alone—the fact that it is an "ought" question—reveals it to be a philosophical question and not a scientific question. After all, if what is at issue is precisely *what is to be allowed to count as scientific*, then surely it cannot be that the issue is a scientific issue, because there cannot (by definition) be any scientific issues in a given field until it has been decided what sorts of things the term *scientific* is going to be allowed to apply to. In the end, a new science can emerge—as had been discovered by Galileo, Bacon, Descartes, Leibniz, and Newton during their own science's period of emergence—only after its philosophical foundations have been clearly and distinctly laid.

A science's philosophical foundation involves three essential stances: *metaphysical* (its stance toward what really exists), *epistemological* (its stance toward what the sources of knowledge are), and *methodological* (its stance toward how to construct scientific theories).

- **Metaphysical.** A metaphysical stance must be taken in regard to the question: *"What sorts of entities, among those referred to by the statements of the science, are to be understood as actual existents?"*

    Operationalism. Milne believed that only those entities that could be simply and directly observed were to count as real. This position is often referred to as *operationalism*, a term that derives from the fact that the entities involved are those that may be defined by some sort of operation with various physical instruments or equipment. Milne's colleague Herbert Dingle, an astrophysicist, then secretary of the Royal Astronomical Society, and a major, outspoken philosophical combatant, for some reason called all those who accept some sort of observability existence-criterion "idealists." Dingle was joined by Milne in referring to Einstein as the founder of the idealist position, and Dingle cites, apparently with favor, Rudolph Carnap's use of the operationalist criterion to demarcate "metaphysics" from science.

Explanatory Realism. Another metaphysical position, but one opposed to *operationalism*, might best be called *explanatory realism*. This position adopts as an existence criterion something like "If a theoretical entity *e* must be hypothesized in order to explain observable phenomenon *p*, then *e* exists." Eddington in his *Expanding Universe* adopts this position with respect to the cosmological constant.

During the classical period, Leibniz and Newton differed vigorously regarding the metaphysical status of space, time, matter, and force.

- **Epistemological.** An epistemological stance must be taken in answer to the question: *"What are the legitimate sources of scientific knowledge?"*

  Empiricism. When Robertson claimed that only information from observational and experimental results was acceptable as a source for cosmological theorizing, he was stating the *empiricist* position.

  Rationalism. When Milne asserted that it was perfectly all right for cosmologists to originate their theoretical efforts in very general ideas about what the universe should be like, he was describing a *rationalist* position.

Precisely the same disagreement as that between Robertson and Milne occurred two and a half centuries earlier when proponents of the views of the empiricist Locke and the rationalist Descartes clashed over what constituted the proper sources of scientific knowledge. This clash was among the first of many such disagreements between empiricists and rationalists during the early days of classical physics.

- **Methodological.** A methodological stance must be taken in answer to the question: *"What are the acceptable procedures for theory construction in this science?"*

  Hypothetico-Deductivism. Milne adopted a particular *methodological* position when he claimed that he generated hypotheses on the basis of his general ideas about the universe, and then went on to deduce the consequences of these hypotheses, hoping that at least some of the consequences would involve observational possibilities. This position is usually called *hypothetico-deductivism*.

Inductivism. Robertson argued that the scientist ought to proceed only by generalizations and extrapolations from solid observational and experimental information. This position is usually called *inductivism* because it relies upon inductive logic.

Three hundred years earlier, Francis Bacon had been the first classical author to argue explicitly that scientific thought ought to be grounded in inductivist methodology.

A fully articulated philosophy of science typically involves commitment to some position in each of the three categories—*metaphysical, epistemological,* and *methodological*. Indeed, such commitment may be taken as providing an individual's answer to the question "What is it to be scientific?"

A specific position in any one of the categories is, in the end, relatively independent of a specific position in one of the other categories. It is possible, for example, for someone to be a *hypothetico-deductivist* methodologically even while being an *empiricist* epistemologically and an *operationalist* metaphysically. This would, though, be a somewhat odd position, at least historically, since there haven't been many individuals with this precise cluster of philosophical loyalties. On the other hand, Milne's cluster of positions—*operationalism, rationalism,* and *hypothetico-deductivism*—has been relatively popular in several scientific fields. As third and fourth (but certainly not final) options, Dingle's position of *operationalism, empiricism,* and *inductivism* is not uncommon, although perhaps the most typical position of a physical scientist would be exemplified by McVittie's *realism, empiricism,* and *inductivism*.

At the time of the 1931 British Association meeting, when consensus was reached about the constitutional elements of relativistic cosmology, these philosophical questions and stances had not yet been explicitly and self-consciously brought forward. Of course, certain commitments already had been made, but only unconsciously, not explicitly. For example, in terms of metaphysical commitment, most cosmologists quite unquestioningly accepted the actual existence of a curved space which was, moreover, expanding. In the case of epistemology, Robertson, Eddington, and de Sitter almost always spoke as empiricists, yet it certainly cannot be claimed that they typically were either as explicit or as self-conscious about taking a philosophical stance as was Dingle (and frequently, McVittie) in adopting specific epistemological and methodological positions. Although the three indeed *held* particular views, they were not always aware at the time that they did. Explicit awareness came only later, *after* fuller debate with Milne.

Tolman was an exception. Not only was he completely clear on the differences between, say, a rationalist-deductive position on the one hand and an empiricist-inductive position on the other, and on the historical genesis of these two positions in Einstein and Galileo, respectively, but he also argued that *both* positions are of great merit in cosmology. In particular, Tolman asserted that "We must not let this just admiration [for Galileo's method of "immediate generalization of experimental findings"] blind us to the power and skill of those other theoretical physicists who obtain the suggestion for physical principles from the inner workings of the mind." It is unfortunate that the degree of philosophical sensitivity, not to mention astuteness, exhibited by Tolman was rare during the crucial initial period of the relativistic cosmology debate.

The debate began in a curious fashion. In late May, 1932, just seven months after the British Association meeting, Jeans became embroiled in a heated public debate about the reality of curved, expanding space. The debate was conducted chiefly in the "Letters" column of *The Times*. Against all comers, Jeans steadfastly maintained that relativity theory necessitated the existence of curved, expanding space, no matter what the cost to common sense and common conceptions.

Milne, well known for his work on stellar physics and for his spirited defense of his views in rigorous conflicts with Eddington, had never before evinced the slightest interest in cosmology; even during his participation in the British Association meeting he had said nothing at all about anything beyond the interior of stars. But now Milne reacted violently against Jeans's views. On 7 June, ten days after Jeans's final comment in *The Times*, Milne presented to a colloquium at Wadham College a novel theory of an expanding universe. Milne's theory was explicitly *anti*-relativistic cosmology.

Three weeks later, a two-page précis of the Wadham lecture appeared in *Nature*. The article, which argued vehemently against the supposed reality of curved and expanding space, staked its claim in a completely original theoretical model, with the model itself embedded in a thoroughgoing, explicitly formulated, but highly controversial minority-view of the philosophy of science. One or more elements of Milne's mixture of model and method would prove irresistibly fascinating to essentially all the other practitioners of relativistic cosmology.

Milne presented three things in his *Nature* article and its subsequent augmentations:

(1) *A Substantive Model.* The model was developed from a precise physical intuition, based on the notion of an observer located on a fundamental particle. Fundamental particles may be likened to the elements of a dense gas swarm, each of which has a particular velocity. Velocities were randomly distributed among particles in the swarm, ranging from very slow to nearly the speed of light. At some arbitrary time t = 0, the volume of the swarm was at a minimum. It follows necessarily from these parameters that, at some time sufficiently later than t = 0, the distribution of particles will be naturally arranged according to a linear distribution based on their velocities. Once the restriction that any observer on any particle will observe a situation identical to that observed by any other observer (i.e., the "cosmological principle") is imposed upon the system, the model exhibits the Hubble relation; that is, any observer will look out on the set of particles and see them distributed with velocities directly proportional to their distance from him or her. Interpreted in terms of the cosmological problem, an expanding universe is given.

(2) *A Metaphysical Program.* Elements of cosmological theories were to be linked to observational entities, and those entities alone; that is, no terms were to be allowed in cosmological theory which could not be associated directly with the results of observations made by either actual or "in principle" observers. Milne's criticism of the "curved space" of the relativistic cosmologists is typical:

> *If the curvature of space cannot be determined, if it is essentially unobservable, then it should be rejected. The time has come when we should remember William of Occam's maxim: "Entia non sunt multiplicanda præter necessitatem."* ["Do not multiply theoretical objects beyond what is necessary for an explanation."]

Milne adopted his metaphysical views early on. Several years prior to bursting into cosmology he had remarked that:

> *It may be necessary to introduce into our equations symbols corresponding to unobservables, but the physical content of the assertion which results from a piece of mathematico-physical reasoning must be a relation between observables only. It frequently appears, when this mode of thought is followed, that the same results are deduced as before, but they are now shown to be true independently of the introduction of concepts corresponding to*

unobservables. Theory is enriched by being pruned of unnecessary assumptions.

And, as far as Milne was concerned, relativistic cosmological theory *ought* to be pruned of its notions of space and time. He said so quite clearly in the *Nature* article:

> *We cannot observe space. We observe point events[. . . .] It seems best to avoid the phrase "the structure of space" or of "space-time" and consider simply the structure of the hyper-complex of world lines which can be reconstructed from our observations.*

Because of his strict adherence to the kind of *observationalist-operationalism* we see in these last two passages, Milne was forced to develop an account of space and time based on a strict application of observational techniques. His method ultimately reduced space to time, and time to signaling operations shared among observers. Moreover, a second of Milne's philosophical desires—a methodological one—was satisfied as well by this method of analyzing time. Reducing space to time and time to signaling allowed an enormous simplification in the axiomatization of cosmology, which Milne strongly desired. In the end, as he noted in a letter to his student and later colleague A.G. Walker, his philosophical goal was to base cosmology upon something "*deeper* than geometry," an end which was fully well served by his account of time.

(3) *An Epistemological-Methodological Program.* Linked to Milne's metaphysical minimalism was a strongly rationalist position on the origin of scientific knowledge. To wit, Milne believed that pure, speculative, *a priori* reasoning is a perfectly legitimate source for theoretical concepts. Coupled to this *positive* account of the rationalist origins of knowledge, Milne maintained a vigorous *negative* attitude against inductive accounts, agreeing with DeMorgan's devastating analysis of Bacon's inductive method of discovery as "*nonsense.*" Not surprisingly, Milne supported a hypothetico-deductive testing methodology for theories. What *is* a bit surprising, however, is the fact that Milne was an extremely strong proponent of theory axiomatization. Furthermore, the axiomatic desideratum is embedded in a particular theory of the evolution of physical sciences, according to which a given science progresses from a set of empirically founded generalizations to a set of free-standing axioms. Milne typically refers to a putative history of geometry as the paradigm of this type of evolution. His model of the origin and development of a physical science later came

to be the model accepted for relativistic cosmology by its practitioners.

Milne's three-pronged approach to cosmology—consisting of a substantive model, a metaphysical program, and an epistemological-methodological program—came to be called "kinematic relativity."

Milne's *Nature* article took the relativistic-cosmology community by storm. A philosophical *battle royale* began immediately. In the ensuing fray there was no such thing as a neutral, disinterested response to Milne's proposals. In the first place, everyone—simply everyone—thought Milne's central theoretical model a clever conception. McVittie's appraisal is a typical example: "The basic idea of this theory is both simple and elegant." Secondly, although some cosmologists had serious doubts as to exactly what the nature of Milne's theory was, with Eddington doubting "whether it differs from current relativity cosmology," others were rather more certain, with Whittaker asserting that "the cleavage between relativistic cosmology and kinematical relativity is fundamental."

Milne's first extensive treatment of his new view did not appear until almost a year later, in the *Zeitschrift für Astrophysik*. The initial serious challenges—both scientific *and* philosophical—came from Robertson's and Dingle's articles published in tandem with this first, thorough version of the new "kinematical relativity."

Robertson's comments on Milne's theory for the most part concentrated on showing that, substantively, the theory was either "the kinematical preliminary" to general relativity's dynamic, or "a special case of the alternate theory." Robertson was never to swerve from this interpretation of the physical substance of Milne's theory.

On other, more philosophical points, however, Robertson's interest was piqued from the first moment. For example, concerning what he (and Milne) would later call the "cosmological principle," Robertson called specific attention to its "metaphysical status": it was an "a priori principle." Milne would later attempt to deflect the negative effect of this characterization.

Oddly enough, Robertson did not initially respond in the least either to Milne's attack on the metaphysical status of curved, expanding space (i.e., its purely "conceptual character") or to Milne's call for an operational metaphysical-methodological criterion. In contrast to his initial lack of interest in these aspects, however, Robertson's later interactions with Milne's work, interactions that would secure Robertson's place in relativistic cosmology's pantheon, would focus almost exclusively on these philosophical elements.

Dingle's response emphasized epistemological and methodological "deviations" in Milne's work. Dingle argued, for example, that
"the spirit of relativity is simply a re-affirmation of Newton's principle of induction from phenomena. . . . Milne approaches the problems of physics in precisely the opposite way." Dingle's conclusion
could not have been stated more pungently: "It would seem that the
general course of Milne's theory is at variance with the fundamental
principles of scientific method." Over the next few years, Dingle's
hostility only grew from this low point.

The next several years continued to exhibit the same wide spectrum of responses seen here, ranging all the way from the strictly
substantive to the purely philosophical. During the course of the
debate, one of Milne's most significant moves was to present his
philosophical views explicitly and clearly in an article in the journal
Philosophy.

In this article Milne laid out the foundational elements of his
position and defended them against the attacks—both scientific and
philosophical—of his mainstream critics. But Milne's exposition
seemed only to exacerbate the debate. Other cosmologists energeti-
cally responded, each in his own way, to every one of the various
aspects of Milne's position. Some of their efforts were focussed quite
narrowly upon understanding and clarifying the *physical* nature and
character of Milne's theory, which remained controversial in spite of
everyone's efforts (including papers by W.O. Kermack and McCrea
in 1933, by McCrea and Milne in 1934, by McCrea in 1935, by
Walker in 1934 and 1935, and by J. Narlikar in 1935). Other efforts
instead raised issues regarding various of the *philosophical* aspects of
Milne's views (including papers by de Sitter in 1934, by McVittie in
1935, by Eddington in 1935, by Whittaker in 1935, by Walker in
1936, by Dingle in 1933 and 1936, and by McCrea in 1939). Even
cursory examination of the exchanges exhibits the fact that Milne set
the cosmological agenda—either positively or negatively—during the
period.

The zenith—or perhaps the nadir—of the philosophical contre-
temps occurred in 1937 with an article by Herbert Dingle, by then
recently past secretary of the Royal Astronomical Society. Dingle
entitled his piece "Modern Aristotelianism." The article is remarkable both for its style and for its content. Dingle's style in the article
is vituperative; he is obviously engaged in a polemical attack upon his
adversaries, Milne most especially. Emotionally loaded terms such as
"paralysis of reason," "intoxication of the fancy," "'Universe' mania,"
and the like frequently appear, to be topped only by references to

"something so rotten," "delusions," "traitors," and ultimately, "treachery."

Such intemperate language is rare in public discourse between physical scientists. Typically, there are just two goads which can drive a physical scientist to such strong language: deep philosophical disagreements and priority disputes. Some cases, such as the violent two-century-long battle between Leibnizians and Newtonians, involve both, which explains their extreme hostility. But in this case, Dingle was mostly interested in the philosophical aspects of the case.

Above and beyond his extreme language, Dingle makes certain substantive claims bearing directly on the central philosophical questions. The issue, as he sees it, is nothing more than the question "Whether the *foundation* of science shall be observation or invention."

As always, talk about "foundations" is philosophical talk. The two opposing positions Dingle here calls "foundational" involve questions of both method and epistemology, suitably tangled together. (It should be remembered that Dingle and Milne are pretty much in agreement about their commitment to operationalist metaphysics.)

Dingle delineates the opposed alternatives as follows. The way of true science, he claims, shows that "the first step in the study of Nature should be sense observation, no general principles being admitted which are not derived by induction therefrom." Restated in philosophical terminology, Dingle here argues that authentic science is *empiricist* in epistemology (scientific knowledge is founded in sensory observation), and *inductivist* in method (general principles are reached via inductive logic).

Opposed to this view, he argues, is "the doctrine that Nature is the visible working-out of general principles known to the human mind apart from sense perception." Dingle cites Milne as representative of this latter view, and refers in particular to Milne's claim that "it is, in fact, possible to *derive* the laws of dynamics rationally . . . without recourse to experience." To describe the situation in philosophical terminology, Dingle is here arguing against the view (Milne's view) that authentic science may be *rationalist* in epistemology (scientific knowledge is founded in pure theoretical reasoning) and *hypothetico-deductive* in method (general principles are justified by deductively implying correct observations).

Along with Milne, Dingle indicts Eddington and, by implication, Dirac. Dingle believes that all three of them are guilty of inventing scientific hypotheses by free mental imaginings rather than by strict immersion in observations and observational data.

Dingle's stentorian piece did not go unnoticed. Within several weeks a sizable proportion of Great Britain's scientific community had responded to the *Nature* article. Indeed, the philosophical traffic became so heavy that the editors of *Nature* decided to devote a supplementary issue to the topic. Milne, as the principal target of the attack, led off. Dingle, as the attacker, brought up the tail end. In between was arranged a veritable bestiary of British science. The upshot of the whole debate was the gradual but observable establishment of the view argued so plausibly (and generously) by Tolman several years earlier: there is room within cosmology for *both* methods: the inductive empiricism propounded originally by Bacon, Locke, Hume, and defended by Dingle, as well as the deductive rationalism favored by Descartes, Leibniz, and so ably illustrated by Milne.

Yet even while the public philosophical debate flared so flagrantly, less noticed but, in the end, more significant developments were occurring. In this regard, Robertson's contributions would seem compelling. From the very first, the talented American cosmologist had adopted the view that kinematical relativity and relativistic cosmology were substantively equivalent. Yet even given this view of the *physical* contents of Milne's theory, Robertson clearly felt challenged by the more *philosophical* aspects of Milne's position.

- First, Milne's attack on the metaphysical "irrelevancies" of curved and expanding space exhibited the minimalism of his operationalist techniques in what Robertson evidently found to be an inviting fashion, inviting enough, at least, to cause him to attempt to emulate Milne's procedures.

- Second, Milne's demand for deductive—indeed, axiomatic— development of cosmological theory provided a worthy target for theory development.

In an astonishing series of three papers, beginning with his famous "Kinematics and World-Structure," Robertson produced a new, generalized cosmological model à la mode Milne, that is (according to Robertson himself), a cosmological model "analyzed from the standpoint of the operational methodology." These three works constitute the grounds for the metric which today bears Robertson's name, in part.

Perhaps most telling in this episode is the change evident in Robertson's work *after* he encountered Milne. Robertson had published in 1933 an absolutely masterful review of the entire field of relativistic cosmology; this paper immediately became a classic. But in this paper, Robertson had not adopted any particular philosophical viewpoint with respect to an appropriate analytical perspective on the subject. Moreover, his paper exhibited straightforward unexamined attitudes about, for example, the reality of space and time and their status as objects of scientific knowledge. Three years later, however, all this had changed.

- First, Robertson's perspective is now inexorably operational, focused unwaveringly on the observer and what it is that he or she might be expected to observe.

- Second, the order and significance of topics bears much more than a casual relationship to the corresponding development in Milne's recently published book (which, incidentally, Robertson had reviewed for the *Astrophysical Journal*).

This relationship was duly noted by McCrea, in an editorial published in *The Observatory*.

At precisely the same time, Milne's student Walker was also developing in his own way topics initially suggested by Milne. In the process, Walker produced the same metric as that discovered by Robertson. Both men admitted that their discoveries were independent, although, of course, the linkage to Milne in each case was evident. It can be clearly seen that one extremely significant result of the philosophical debates during relativistic cosmology's early days was the establishment of the Robertson-Walker metric which, without Milne's outspoken commitment to a minority philosophical view (i.e., operationalism), would never have come about.

A final accomplishment of the philosophical discussions was the acceptance of a consensus model of cosmology *qua* science. It is extremely important to note precisely what this accomplishment comes to. As remarked earlier, one of the most fundamental issues an emerging science can face involves the question "What kind of a science is this to be?" Until the participants in the new science (and to a lesser extent, their observers in the wider scientific and general intellectual community) can decide what sort of a science they in fact are practicing, it can be effectively argued against them that they are practicing no science at all. Thus, the acceptance of a model of itself by a scientific community not only legitimates it in its own eyes, but also provides immediate access to the set of conventional goals, processes, and standards entrained by the model. In the case of relativistic cosmology, the model was provided by Milne.

That Milne ended up providing the model is not so strange as it might initially seem. Certainly he had from the start been the odd man out, philosophically speaking. His views on the metaphysics of space and time, on the appropriate source for cosmological hypotheses, and on the role of deduction and axiomatization bore the brunt of heavy and frequent attack. But it is precisely these sorts of conditions which elicit a tough, coherent, and well-reasoned philosophical-cum-historical response, which is exactly the type of response Milne provided.

Milne had long been interested in the role played by axiomatization in the physical sciences, an interest originally aroused by a close and approving study of Whitehead and Russell. From his study, Milne had come to a particular view about the evolutionary dynamics of a science. According to this view, fields such as number theory and geometry that began life as empirical discoveries or "rules of experience" ultimately became free-standing, fully rational, pure disciplines. In a later discussion of this notion, Milne stated his position clearly and forcefully:

> *A theorem in the theory of numbers may be discovered first empirically, and the weight of evidence for it may be such as to compel a belief in its general validity; but no mathematician accepts this as the last word on the subject. [. . .] The theorem cannot be held to be connected with the discipline which is the theory of numbers so long as the evidence for it is wholly or partly empirical. The mathematician must, at some stage, perform an act of renunciation and proceed without reference to empirical facts.*

Once the act of renunciation has occurred, the full-scale axiomatic development of the science may begin. As it progresses, the science in effect becomes more and more abstract and less and less empirical:

> *The more advanced a branch of science, the more it relies on inference and the fewer the independent appeals to experience it contains. As a science progresses, the most diverse phenomenal laws are seen to be deducible from a few general principles, and the number of these, in turn, tends to become smaller and smaller. We deduce the more and more from the less and less, or at least from the simpler and simpler. The roles of observation and experiment, fundamentally important as they are for discovery and verification, play a less important part in the structure of the finished science, and the role of deduction a more important part.*

This view, quintessentially Milne, survives the attacks of Dingle and the other empiricist traditionalists, in the end becoming the official story of the scientific nature of relativistic cosmology. The official endorsements are delivered by McCrea and Whittaker.

McCrea provides a subtle and carefully argued account that subsumes Milne's methodology under a general aegis which covers, as well, Newton. Thus, an attack on Milne is equally an attack upon Newton. Moreover, Milne's account of development is seconded. Both the progressive axiomatization of a science and its entirely legitimate reliance upon hypothetico-deductive methodology are defended with strength and skill by McCrea. Whittaker, for his part, explicitly inserts himself into the fray with the expressed goal of trying to settle some of the thorny philosophical issues. As his general model for what a genuine science looks and acts like, he presents a model whose nature and evolution are essentially indistinguishable from that presented by Milne.

Conclusion

Cosmology has at this point come of age philosophically. It is now ready and able to do science within a concensus-formed community of practitioners.

—*George Gale and John R. Urani*

BIBLIOGRAPHIC NOTE

Although no thoroughgoing history of modern cosmology yet exists, several useful texts are available. Most recent is B. Bertotti, R. Balbinot, S. Bergia, and A. Messina, *Modern Cosmology in Retrospect* (Cambridge: Cambridge University Press, 1990), which contains memoirs of many of the participants in modern developments. A detailed, historical-philosophical appraisal is J.D. North's, *The Measure of the Universe* (Oxford: Clarendon Press, 1965); unfortunately, the focus is mainly on theoretical aspects and coverage ends in 1964. Similar coverage, but from a more scientific perspective, is in Jagjit Singh, *Great Ideas and Theories of Modern Cosmology* (New York: Dover, 1970). Compilations of original source material are in Jeremy Bernstein and Gerald Feinberg, *Cosmological Constants* (New York: Columbia University Press, 1986), and in M.K. Munitz, ed., *Theories of the Universe* (Glencoe, Illinois: Free Press, 1957); the latter provides coverage of more popular-style documents from the entire history of Western cosmology and, moreover, has recently become available in a new, cheaper, paperback edition. Finally, R.W. Smith, *The Expanding Universe* (Cambridge: Cambridge University Press, 1982) is an excellent source for the history of work leading to acceptance of the expanding-universe theory.

IX. Cosmology and Religion

Introduction

THE COPERNICAN REVOLUTION

imited changes envisioned by revolutionaries often are surpassed as a revolution set in motion runs its course. The Copernican Revolution, at its beginning little more than an attempt to develop a new geometrical model of the planetary system, quickly saw the demise of physics as it was then understood, the destruction of the Aristotelian world view, and a shattering of belief in an anthropocentric universe. All of this would be replaced with a new heliocentric model for the solar system, a new physics, the Newtonian mechanical world view, and a universe in which humans were neither necessarily at nor even near the center of the universe, nor necessarily its uniquely intelligent inhabitants.

An infinite universe was one possibility inherent in the Copernican world view. Soon it was made explicit and dramatically emphasized by Galileo's telescopic observations. The first printing of 500 copies of his book recounting his observations quickly sold out. Indeed, when Galileo sent a package to a friend in Florence, that person's neighbors demanded that the package be opened immediately, hoping to look through a telescope. When the package was found to contain Galileo's book, the neighbors insisted that the book be read aloud to them.

There were tensions between the teachings of the Catholic Church and the new cosmology, tensions which have led through various understandings and misunderstandings of historians to our modern stereotype of the relationship between science and religion. The stereotype is readily summed up in the title of Andrew Dickson White's 1897 book, in two volumes, on *A History of the Warfare of Science with Theology in Christendom*. White wrote the introduction while serving as American minister to Russia. In St. Petersburg, watching from his window workers chipping away in April at the ice barrier across the River Neva binding together the piers and the old fortress of the czars, White wrote that the work in his book was similar: "I simply try to aid in letting the light of historical truth into that decaying mass of outworn thought which attaches the modern world to medieval conceptions of Christianity, and which still lingers among us—a most serious barrier to religion and morals, and a menace to the whole normal evolution of society."

White stated that a flood of increased knowledge and new thought would sweep away outworn creeds and noxious dogmas. Certainly he had experienced his share and more of dogmatic opposition to new ideas in the course of steering through the New York State legislature, the enabling legislation in 1868 for Cornell University, which he subsequently guided and served as its first president. During the legislative hearings clergymen had warned against the atheism of the proposed university with an emphasis on science, and a Protestant bishop had proclaimed that all its professors should be in holy orders. The controversy bore in upon White a sense of antagonism reflected in his subsequent history of warfare between the science and theology—not between science and *religion*, it should be noted, but between science and *theology*, or dogma. White observed that "In all of modern history, interference with science in the supposed interest of religion no matter how conscientious such interference may have been, has resulted in the direst evils both to religion and to science, and invariably; and, on the other hand, all untrammeled scientific investigation, no matter how dangerous to religion some of its stages may have seemed for the time to be, has invariably resulted in the highest good both of religion and of science."

White's perspective on relationships between science and religion is now nearly a century old, and though inevitably enriched and modified by subsequent generations of scholars, is yet to be replaced in the minds of many. Religious values, however, could as easily as not be compatible with modern science, as indeed Robert Merton and others have argued was the case with Puritanism, which may thus have furthered, albeit largely unwittingly, the rise of modern science in England. Also, we could go back further in history, to the 1277 condemnation of 219 Aristotelian propositions by the Bishop of Paris. Whatever the bishop's intent, did his action harm science, or did the subversion of Aristotelian science help open the way for modern science?

It is all too easy to presuppose an inevitable battle between science and religion, and to view history through this lens. The primary opposition to Galileo and to the new cosmology, however, came from philosophers, particularly from Aristotelian professors of philosophy in Italian universities. Nonetheless, Galileo's relationship with Catholic authorities culminated in such a dramatic episode that it now lies at the base of a general stereotype concerning relationships between science and religion in general. Thus it is worth a separate discussion and receives such in this book, in the following essay on *Galileo and the Inquisition.*

Religious considerations provoked by the new cosmology were not entirely one-sided. As we have seen in *A Rational Order for the Cosmos*, the introduction to section V of this book, *Galaxies: from Speculation to Science*, a belief in the orderliness of the universe following from Newton's demonstration of the most beautiful system of sun, planets, and satellites as the creation of an omniscient and omnipotent God all moving in the same direction in the same plane stimulated further research in cosmology.

In Newton's cosmology, God's presence and activity pervaded infinite space and imposed a rational order on the cosmos. This theological cast to cosmology dominated much of astronomical speculation and science through the eighteenth century. William Whiston, Newton's successor at Cambridge University in 1703, argued that the system of the stars, the work of the Creator, had a beautiful proportion, even if frail man were ignorant of the order. Whiston was unable to propose an order for the Milky Way, a dense band of stars, but the self-taught English astronomer Thomas Wright did, in 1750. Immanuel Kant, inspired by an incorrect summary of Wright's book, later explained the Milky Way as a disk-shaped system viewed from the earth located in the plane of the disk. Thoroughly imbued with a belief in the order and beauty of God's work, Kant went on to suggest that nebulous patches of light in the Heavens are composed of stars and are other Milky Ways or island universes. Astro-theology, the determination of the rational order of the cosmos, had become an important theological, philosophical, and scientific endeavor.

Nor were religious considerations provoked by the new cosmology antagonistic to further study of the question of a plurality of worlds. See, for example, in the preceding section of this volume, *Cosmology and Philosophy*, the essay on *Plurality of Worlds*, which could as well have been placed in this section instead.

John Wilkins, the Master of Trinity College, Cambridge, for a brief period before the restoration of Charles II and also a major figure in the establishment of the Royal Society of London, serving as its first secretary, published in 1638 *The Discovery of a New World; or, a Discourse tending to prove, that it is probable there may be another habitable World in the Moon*, a book which was to go though four editions by 1684. Wilkins faced Scriptural objections, such as Moses telling of but one world and St. John speaking of the world, not worlds, as God's work. He answered that negative authority of Scripture is not prevalent in areas which are not fundamental to religion. Furthermore, while divine wisdom might construct only one universe (since they must be either the same or diverse, the former needless since one would have no more perfection than the other, and the latter impossible, since to be called a world, it needed to be of a similar perfection with other worlds), there could be many inferior worlds within it. Although there was no direct evidence of lunar inhabitants, Wilkins guessed that there were some inhabitants, because why else would Providence have furnished the moon with all the conveniences of habitation shared by the earth.

The plurality of worlds, initially a doctrine more familiar to philosophy and literature than to science, eventually was accepted by astronomers as well. Paradoxically, it was in the eighteenth and nineteenth centuries, after the demise of the idea of an anthropocentric universe in the field of astronomy, that the idea of the Great Chain of Being received its widest circulation. The belief that there was a scale of nature, of divine design, and reading from the simplest animals at the bottom to man at the top, was a mind set more appropriate to an earlier age before the idea of an anthropocentric universe had been rendered obsolete. The revolution in cosmological thought, though, seems not to have forestalled a later blooming of anthropocentric ideas in the biological sciences, nor a continuation of anthropocentric concepts in theology. Extensive as the Copernican Revolution has been, its theological implications—including the implausibility of sending an Adam and an Eve, and later a son of God, to every planet, or one of each to our particular planet—have yet to be drawn out.

In the later half of the twentieth century a specialized science of cosmology has developed no longer closely intertwined with religious beliefs and making no explicit mention of God. The new cosmology offers a special challenge to believers in a God of Creation. The history of interrelationships between cosmology and Western religion, and some hints as to the future of this relationship, are presented in the final chapter in this book.

30. GALILEO AND THE INQUISITION

Galileo's earliest conflicts with authority did not involve religion. Basic opposition to any new cosmology came from philosophers, particularly Aristotelian professors of philosophy in Italian universities. That the Catholic Church looked to them in all matters of a scientific nature is historically true, but is also a derivative rather than a primary fact.

The first known conflict between Galileo and authority concerned the question of equal speeds of fall by bodies differing in weight, as exemplified by the story of the Leaning Tower of Pisa, from about 1590. While many consider this story apocryphal, some historians, this author among them, believe that the story is more than a legend. Galileo's conflict in this matter was with professors of philosophy alone; they were the ones who taught physics.

A second grounds for conflict arose between 1604 and 1605 with a supernova, a bright star newly appearing in the heavens. As professor of mathematics in Padua, Galileo lectured publicly on the new star. He argued that the absence of a detectable parallax placed the new star at least as far away from the earth as were the outer planets (thus contradicting Aristotelian doctrine, in which all change in the heavens was limited to the region from the earth up to the moon). The ranking professor of philosophy, Cesare Cremonini, argued on Aristotelian principles that the new star was beneath the moon. The dispute nearly cost Galileo his professorship.

A third conflict occurred in 1611, shortly after Galileo had moved to Florence. At a court luncheon hosted by the Grand Duke, Galileo debated a professor of philosophy about floating bodies. Of the two cardinals present, one supported the philosopher and the other (later to become Pope Urban VIII!) took Galileo's side.

These three early conflicts foreshadowed the later division of opinion in the matter of the motion of the earth. Professors of philosophy unanimously opposed Galileo's science; high officials of the Catholic Church did not. Indeed, it was probably at Padua that Galileo heard Cesare Cardinal Baronius remark that the Bible tells us how to go to heaven, not how the heavens go. (Cardinals Baronius and Bellarmine visited Galileo's friend and patron G.V. Pinelli at Padua, where the question of the motion of the earth probably came under discussion.) Later Galileo would cite the remark, in his 1615 *Letter to Madame Christina of Lorraine, Grand Duchess of Tuscany, Concerning the Use of Biblical Quotations in Matters of Science.*

Not until 1613 did Galileo publish his support for the Copernican system. Observation of the phases of Venus at the end of 1610 had already shown Aristotle's cosmology and Ptolemaic astronomy to be untenable, and the discovery in 1612 of eclipses of Jupiter's satellites made Galileo certain that motion of the earth must be taken into account in calculating the occurrences of the eclipses. In his 1613 *Letters on Sunspots*, Galileo made this assertion.

Later in 1613 the first serious charge was made that belief in the motion of the earth was contrary to the Bible. This charge had previously been voiced by a Dominican priest, Niccolò Lorini, at Florence in a private discussion at his convent. In December 1613 the charge was asserted in Florence in the presence of Galileo's employers by a professor of philosophy from Pisa. Benedetto Castelli, previously Galileo's pupil, was asked by the Grand Duchess Christina to speak on the issue of the motion of the earth as a theologian (he was a Benedictine abbot). Castelli refuted the visiting professor and also wrote to Galileo, describing the incident. Galileo replied with a long letter on the place of scriptural passages in purely physical questions.

A year later from a pulpit at Florence a Dominican priest, Thomas Caccini, denounced the "Galileists" and all mathematicians as enemies of religion. Caccini was from the same convent as Lorini, who was absent when the denunciation took place. In Pisa, Lorini told Castelli that he was sorry that his young colleague Caccini had gone so far. Castelli showed Lorini Galileo's letter written on religion and science a year before, and Lorini copied part of the letter. On returning to Florence, Lorini discussed Galileo's letter with others of his convent, and then sent the partial copy of the letter to Rome for examination by the Inquisition. He did not make any charge against Galileo or his followers.

The letter was turned over to a qualified theologian, who reported that it contained good Catholic doctrine although there were passages that could offend pious ears. The inquisitors then requested the complete original letter, which Castelli had sent back to Galileo. They examined Caccini and two persons he named as corroborators, and then late in 1615 dropped the matter for lack of evidence against Galileo's sound Catholicism.

Meanwhile, Galileo greatly expanded his letter on religion and science, addressing it now to Christina, mother of the Grand Duke Cosimo II of Tuscany. The existence of a score of copies shows that the letter was circulated as an open letter. It was not printed, however, until 1636, after Galileo's condemnation.

Then the letter appeared with a Latin translation titled "New-old doctrine of the most holy Fathers, and theologians of probity, of the testimony of sacred Scripture in purely physical conclusions that can be evinced by sensate experiences and necessary demonstrations. . . ." The term "new-old" alluded to Galileo's appeal to the authority of St. Augustine and other Church fathers in support of Galileo's own thesis that no contradiction can exist between the Bible as God's word and nature as His will—when the Bible's words are interpreted correctly. Indeed, St. Augustine had cautioned against any astronomical doctrine ever being made an article of the Catholic faith, lest some heretic better informed in science use such a Church position on a matter of science to impugn the credibility of proper articles of faith.

Late in 1615 Galileo went to Rome to clear his own name and to persuade high Catholic officials of the advisability of taking no formal action on the Copernican issue when new evidence from the telescope was still coming in. He knew that professors of philosophy were trying to have Copernican books forbidden, and in his *Letter to Christina* Galileo charged them alone with responsibility for bringing the Bible into a question answerable by sensate experiences and necessary demonstrations. But in February 1616 Pope Paul V, whose background was legal and who disliked intellectual disputes that caused dissension within the Church, submitted the Copernican stability of the sun and motions of the earth to the eleven official qualifiers of disputed propositions.

Galileo had counted on the concept of metaphorical rather than literal interpretation of biblical passages, such as "the Sun goeth forth as a bridegroom from his chamber and strongly runneth his course." Evidently the qualifiers did not even consider the possibility of metaphorical interpretation, for they replied immediately that stability of the sun was false and absurd in philosophy, and formally heretical, while motion of the earth was also false and absurd in philosophy, and at least erroneous in the Catholic faith. Both decisions were unanimous.

The pope did not need theologians to tell him that the Copernican propositions were false and absurd in textbook philosophy, but he now had the legal basis he had wanted for action in the matter. Ignoring the opinion about formal heresy, which could be decreed only by a pope or a Church Council, Paul V directed Cardinal Bellarmine to advise Galileo of the finding and to instruct him not to hold or defend the forbidden propositions any longer. If Galileo resisted, the Commissary General of the Inquisition was to order Galileo—in the presence of a notary and witnesses—not to hold, defend, or teach the forbidden Copernican propositions in any way, lest he be imprisoned.

By their shift of responsibility for biblical interpretation to philosophers, the qualifiers had departed from ancient Church tradition. Disputed propositions had previously often been of a philosophical character and the censure rendered had been in a traditional form. But the Copernican propositions were of a factual nature, and the medieval phraseology of censure was thus inappropriate. In question was not agreement or disagreement with philosophy, but with nature. Yet because the pope wisely ignored the *obiter dictum* about heresy, no immediate damage was done.

Cardinal Bellarmine summoned Galileo and instructed him as the pope had ordered. Bellarmine so informed the pope at the next weekly meeting, and the Congregation of the Index was instructed to issue an edict regulating Copernican books. The edict was carefully worded, probably by Bellarmine himself, and forbade only two things:

- Reconciliation of Copernicanism with the Bible.
- Assertion of literal truth for the forbidden propositions.

One book by a theologian was suppressed outright, and another was suspended pending removal of a page. Copernicus's *De revolutionibus* was suspended pending removal of one passage about scriptural exegetes and some other passages in which the earth was called a "star" (implying it to move as a planet).

Before Galileo returned to Florence, he received in Rome two letters from friends at Pisa and Venice informing him that rumors were circulating to the effect that he had been compelled to abjure and do penance. Behind such rumors may have been an unauthorized intervention of the Commissary General at the interview between Galileo and Bellarmine. Only if Galileo had resisted Bellarmine's advice—which he did not—was the Commissary General supposed to order Galileo not to hold, defend, or teach the forbidden Copernican propositions in any way. But the Commissary General may have given this order anyway. Bellarmine did not report any such unauthorized intervention by the Commissary General, but it could have happened and it could have become known to malicious persons in Rome. Bellarmine appears to have told Galileo to treat the incident as never having happened, for it could only damage the Church if made public. Upon being shown by Galileo the letters from Pisa and Venice reporting the rumors that he had been compelled to abjure and do penance, Bellarmine wrote out an affidavit for Galileo stating that he was under no restriction beyond the 1616 edict applying to every Catholic.

Neither Galileo nor the philosophers had obtained from the 1616 edict what they had sought. Official action had been taken, but Copernican books were not suppressed outright. Galileo knew that science would not suffer, although Catholic scientists would come to appear overly cautions in treating as only hypothetical what other scientists would in time accept as fully established. For several years Galileo remained silent on cosmological topics.

In 1623 Maffeo Barberini, who admired Galileo, was elected pope. He took the name Urban VIII. The following year, Galileo visited Rome to pay homage to Urban, and was granted six audiences. By that time the 1616 edict was widely believed, especially in Germany, to forbid any discussion of Copernican astronomy by Catholics. Urban, himself an intellectual, wanted the support of other intellectuals everywhere, and Galileo proposed to write a book in which it would be made clear that science in Italy was unhampered by the edict, which had not been issued to interfere with anything except unauthorized interpretations of the Bible.

Between 1624 and 1630 Galileo intermittently worked on the proposed book. He intended to call it *Dialogue on the Tides*, to show a valid reason for fully discussing motions of the earth (as an hypothesis from which could be drawn an explanation of the tides). Galileo's proposed book was provisionally licensed for printing at Rome after censors had reviewed it and removed anything that might violate the edict. When this arrangement was reported to the pope, he ordered one more change: that the tides must not appear in the title of the book or as the subject of the book. The book's value to the Church was that it had official permission to be printed, and presumably Urban did not want anyone to suppose that in licensing a discussion of arguments for and against the Copernican system, the Church endorsed a theory of tides based on that hypothesis.

The book was published in 1632 under the simple title *Dialogue*. Since 1744 it has been reprinted and translated many times as *Dialogue Concerning the Two Chief World Systems, Ptolemaic and Copernican*. Clearly, though, the book could not have been licensed under any such title after the Copernican system had been ruled out by Church edict. The longer title under which the book has been read for the past two centuries is responsible for many misapprehensions about its author, about his science, and about his religion.

Soon after its publication, sale of the *Dialogue* was ordered to be stopped and Galileo was summoned to Rome for trial on the charge of "grave suspicion of heresy." The charge was based on the alleged intervention of the Commissary General in 1616. Galileo produced his affidavit from Cardinal Bellarmine, dated and signed. The allegation of intervention of the Commissary General was never signed by anyone. Legally, Galileo had won on the rule of best evidence, but in fact he was condemned by seven of the ten cardinals sitting as his judges.

Their action made even hypothetical discussion of Copernicanism *prima facie* evidence of heresy for Catholics. Such a conclusion had been carefully avoided by the more prudent action in 1616. As Pope John Paul II said in 1978, the *Letter to Christina* shows Galileo to have been sounder in theology than the judges who condemned him.

—*Stillman Drake*

BIBLIOGRAPHIC NOTE

Galileo's 1615 *Letter to Madame Christina of Lorraine, Grand Duchess of Tuscany, Concerning the Use of Biblical Quotations in Matters of Science* is conveniently available, with an historical introduction, in S. Drake, *Discoveries and Opinions of Galileo. Translated with an Introduction and Notes by Stillman Drake* (New York: Doubleday, 1957). Other books dealing with Galileo and the conflict between science and religion include: J. Broderick, *Galileo* (London: George Chapman, 1964); S. Drake, *Galileo* (Oxford: Oxford University Press, 1980); M.A. Finocchiaro, *The Galileo Affair* (Berkeley: University of California Press, 1989); K. von Gebler, G. Sturge, trans., *Galileo and the Roman Curia* (London: Kegan Paul, 1879); L. Geymonat, S. Drake, trans., *Galileo Galilei* (New York: McGraw-Hill, 1965); J.J. Langford, *Galileo, Science, and the Church* (Ann Arbor: University of Michigan Press, 1966); P. Redondi, R. Rosenthal, trans., *Galileo Heretic* (Princeton: Princeton University Press, 1987); and G. de Santillana, *The Crime of Galileo* (Chicago: University of Chicago Press, 1955).

31. COSMOLOGY AND RELIGION

Systematic attempts to understand the universe taken as a whole have in the past been closely intertwined with religious belief. Only within the last half-century or so has a specialized science of cosmology developed which makes no explicit mention of God, and in which human concerns appear to dwindle to insignificance in the scale of cosmic time and space. Before that, it often seemed that any serious investigation of the ordered universe led to a Mover or a Designer who could be assimilated to the Being whom men and women worshipped as God.

Introduction

In this brief essay we will trace the outlines of this story, focusing primarily on the religions of the West in which God came to be understood as the Creator of the universe, and the dealings of God with men and women were set down in a sacred book. [The corresponding story in the East takes a very different shape; see chapter 2, *Chinese Cosmologies*, in section I, *Cosmology and Culture*.] Long before these religions appeared, of course, people looked at the skies in awe and saw in the regularities of the movements of the celestial bodies a testimony of the gods, the powerful beings on whom human affairs depended. Why that early association of sun, moon, and planets with the gods? The dependence of human life on the sun's bounty was evident enough, as was the importance of the seasons which were most clearly marked by the phenomena of the skies. The immensity of the heavens, the invariability of the celestial motions, marked off the bright lights of sky from all the familiar things of earth. And among the celestial lights a few seemed to have a special role because their motions were more complex than the single daily circle that all the others followed.

For the Babylonians, the ancient people who developed the most detailed knowledge of the celestial movements (and who were fortunate in their choice of semi-permanent clay tablets for their records), cosmology and religion were very close. [See chapter 3, *Mesopotamian Cosmology*, in section I, *Cosmology and Culture*.] In omen-lists going back to almost 2,000 B.C., significant celestial appearances like eclipses and first and last appearances of planets over the horizon were correlated with such significant events as war, plague, or drought in a designated part of the kingdom. Gradually, records came to be kept of these appearances; by the fourth century B.C., a sophisticated mathematics enabled the significant sky-events to be accurately anticipated. What in the first place motivated this complex set of observational, archival, and mathematical practices was the belief that the intricate movements of the celestial bodies were significant for the affairs of man, that the skies served, as it were, as the message-board of the gods. The cosmic order thus had a direct religious significance; cosmology testified to religious belief, and religious belief prompted the construction of a detailed cosmology. Those who were charged with this construction, the astronomer-astrologers (the Magi of the Gospel story of the three wise men "from the East" who followed a star to Bethlehem), were also priests of the state religion, affiliated with the temple and serving as interpreters of the gods.

The tie between cosmology and religion was not always as specific and as direct as it was in ancient Babylonia. But as we range over what we know of the beliefs and practices of the ancient world, we commonly find an association of specific planets with specific gods and a conviction that the configuration of the night sky had a special religious significance. Furthermore, the seasonal order of equinox and solstice was universally celebrated, and indeed seems to have provided a framework for religious ritual in general. The discovery of an underlying order of season or of celestial appearance was itself a religious act. One function of religion was to reassure men and women of the reality of this order, on which human life so obviously depended.

At an even deeper level, there were myths of origins, of how the world first came about, of how seas and land separated off and the mountains were formed, of the first humans and what their role in the story was. All this was seen as the mighty work of the gods, of beings cast in a human image and yet recognized as being more than human. Order and disorder alike were of their making; they were worshipped and feared and placated. Always there was the attempt to understand their purposes, for only thus might the whole and the relations of humans to that whole perhaps be understood.

One of these religious traditions has a special interest for us. The Hebrews celebrated Yahweh as their special protector. To him they were bound in solemn covenant. It was Yahweh who had brought them out of Egypt and displaced other peoples to ready for them a new land, the land he had promised them. As they reflected on the powers of Yahweh, the conviction deepened that he was not just one god among others; he was the potent Creator of all that is—of earth, sun, moon, and stars—the power who had first separated land and water and fashioned all the kinds of living things. In some of the most evocative poetry ever written—in *Isaiah, Job,* and *Psalms*—a being gradually took shape who "tells the number of the stars and calls each by name" (*Psalms,* 147), who "stretches the heavens like a tent [and] fixed the earth on its foundations," who "made the moon to tell the seasons and [told] the sun when to set" (*Psalms,* 104). This is a being omnipotent and omniscient, on whom the entire universe depends and yet who cares deeply for human welfare. It is he who established the cosmic order, who perpetually holds back the forces of chaos, and who has communicated to his people a knowledge of how important a part of the whole is their own fragile history.

The universe is presented here as the work of a single being, who relates to men and women as a person would; knowledge of that universe comes either directly from Him by way of revelation or on the basis of human reflection on what His cosmic role may be. What is striking, in the context of our theme, is that there is no attempt to build a "scientific" cosmology like the Babylonian one, relying on systematic observation and mathematical analysis. The planets are not associated with particular gods; the heavens are not the continuing source of omens and portents that they were for the Babylonians. The gods do not have to be placated for the sun to begin its trip southward again after the winter solstice. Cosmology and religious belief are intertwined now in a rather different way. There is no strong religious incentive to pay attention to the details of the goings-on in the sky. The heavens display the glory of the Lord, but they carry no special messages beyond that. The cosmos is of God's making and we can rely on the regularities he has built into it; but reflecting on God's power and his message is a better means to a true cosmology than a study of planetary movements would be.

Greek Cosmology

Beginning in the Greek-speaking world of the Eastern Mediterranean around the sixth century B.C., a very different approach to a cosmology began to take shape. The first philosophers of ancient Greece were preoccupied, indeed, with cosmology, with questions about the origin and constitution of the universe. [See chapter 4, *The Presocratics,* in section I, *Cosmology and Culture.*] But they tended to disconnect them from the myths of making that characterized earlier religious traditions. They sought to answer these questions in terms of homely insights into natural process and analogies shrewdly chosen from the repertoire of human making. The unspoken presupposition was that human beings, relying on their own unaided resources, could in principle provide plausible answers to questions such as these. Notions of reason, evidence, cause, and proof were gradually developed; one could seek a reasoned understanding of any given subject matter.

The first recorded Greek philosophers suggested plausible cosmogonies. Perhaps the world had come from an initial simple watery state, in which earth is deposited (as it is in lake bottoms), and water vapor (air) is given off. Or perhaps there is more than one primary element. Perhaps the kinds of living things originated from random configurations of organs and limbs, only a few of which could survive and propagate their kind. And so on. The details are familiar. What is important to our theme is that religious beliefs, whether about the Orphic deities or the Olympic gods, were not invoked. The role of Mind (Spirit, Soul) was sometimes emphasized but it was not associated with any specific religious belief or practice. Of course, such an emphasis left open a role for a causality beyond the human. But the impersonal and abstract character of Mind as mover and maker contrasted starkly with the earthy stories of the gods of Olympia. Democritus and his followers, indeed, took the process a long stage further, and formally excluded the action of the gods from their conception of the universe as a whirl of atoms. And they decried the beliefs and practices of the popular religion as unworthy superstition.

The fate of Socrates bears eloquent testimony to the tensions that were beginning to destroy the ancient unity of cosmology and religion. Socrates, the tireless questioner of unsuspected presuppositions, was accused of impiety and of corrupting the youth of Athens. One (false) charge was that he had taught that the sun is no more than a "stone," an earth-like body, an offence for which Anaxagoras had earlier been exiled. What was at issue, in part, was philosophy itself and its proclivity for undermining traditions, religious and political. The growing resentment against the naturalism and rationalism of the "physicists" was finally visited on Socrates, himself ironically a critic of the very materialism of which his enemies so strongly disapproved. But it was not just a matter of doctrine; it was a matter of method. And Socrates was the exponent of a style of question-and-answer that would enable cosmologies to be constructed in a new manner, a manner that traditional religions would have either to recognize or oppose.

Plato and Aristotle, the two greatest Greek philosophers, each produced a distinctive cosmology of the new sort; these cosmologies affirmed the existence of a God, but a God quite unlike those of the pantheon sanctified by the popular culture of the day. In one of his last works, the *Timaeus*, Plato constructed a "likely story," as he calls it, of a Demiurge (Craftsman) who is responsible for the manifest evidences of intelligence we find everywhere in the physical world. [See chapter 5, *Plato's Cosmology*, in section II, *The Greeks' Geometrical Cosmos*.] Plato had earlier shown the importance of Form or Idea in explaining how knowledge itself is possible, and had argued that the world of sense participates only imperfectly in the intelligible world of Form. But imperfect though this participation may be, it is everywhere to be found; it is constitutive of the sense-world, particularly of its mathematically describable aspects. Plato suggests a speculative geometrical atomism as an account of the underlying constitution of sensible things, but is skeptical of observational astronomy as a means to genuine understanding. The Demiurge is not a creator in the full sense, but imposes form on a preexistent matter-space. The recalcitrance of that matter-space is responsible for the defects that are as evident in the sense-world as are its intelligible aspects. Although the world is a cosmos, an ordered whole, the ordering is not complete, nor can there be a strict science of it. A study of that order (i.e., a cosmology) should terminate in a plausible affirmation of the existence of a shaping Intelligence. (Plato elsewhere speaks of a "world-soul" as animating the whole; this would lead to a very different sort of cosmology.)

Aristotle adopted a different starting point for his cosmology. [See chapter 6, *Aristotle's Cosmology*, in section II, *The Greeks' Geometrical Cosmos*.] The final book of his massive work, the *Physics*, argues for the necessity of an Unmoved Mover if the motions of the physical universe are to continue. The fundamental premise of his physics is that whatever is in motion must be kept in motion by something other than itself. A rigorous application of this principle shows that all motions, celestial and terrestrial, must terminate in a Mover (or Movers) able to cause motion in others, although itself (or themselves) remains unmoved. This motion is imparted to the spheres carrying the planets and is transmitted downward to earth. Eudoxus, a contemporary of Aristotle, had proposed an ingenious system of circular motions to account for the complex movement of each planet, including the sun and the moon. [See chapter 5, *Plato's Cosmology*, in section II, *The Greeks' Geometrical Cosmos*.] Aristotle converted Eudoxus's invention into an extraordinarily complicated system of fifty-five interlocking spheres, the axis of rotation of each carried by the sphere next outside it, to which its poles were assumed to be attached. Each planet had a set of four spheres (three for the sun and the moon) and a set of "counteracting" spheres to cancel all the motions needed to explain the motion of that planet (save that of the sphere of the fixed stars), so that these "proper" motions were not transmitted to the next planet downward. The motion of each sphere was essentially the mechanical resultant of all the motions above it, plus an intrinsic motion proper to the particular sphere itself, for which the intelligence associated with that sphere was responsible.

Here was cosmology, with a vengeance! Aristotle attempted, not altogether consistently, to reduce all physical motions to a single order on the basis of systematic astronomical observations, on the one hand, and mechanical-teleological principle, on the other. The primary natural science, physics, was completed by the postulate of a First Mover itself unmoved. The primary virtue of the construction was its explanatory force: the solid spheres *explained* the motions of the planets in an intuitively satisfactory way, although the system was far too cumbersome to be of practical use for purposes of prediction.

The First Mover could also be regarded as God. But it was nothing like the gods of traditional religion. It was pure act, a thinking being reflecting on itself, totally self-enclosed, in no way concerned for, or even aware of, the changing fortunes of mortals. Hardly accessible to prayer or relevant to ritual, it was an abstract explanatory principle incapable of animating faith or devotion. Aristotle could use the term *theology* to describe his "science of God." But his cosmology would have had little interest, one suspects, for the religious believers of his day.

Christianity

The advent of Christianity was to have two very different implications for cosmology. On the one hand, the Genesis account of origins suggested a tidy scenario of a making that was spread over six days, when all the kinds of things were brought abruptly to be by God. Cosmology would then be largely derivative from the Bible. On the other hand, the philosophical elaboration of the Biblical notion of creation on the part of philosopher-theologians like Augustine and Aquinas gave an entirely new perspective on how cosmology and religion might be related at a deeper level.

Augustine developed, more fully than anyone before him had done, the consequences of taking God to be *fully* Creator, responsible not just for the movement of, or the forms imposed on, matter, but for the fact that there *is* any matter in the first place. The Creator not only brings the universe to be from no preexistent makings, but also holds it in being at each moment of its existence. As Augustine sees it, the Lord of heavens and earth evoked by the writers of Scripture brings time itself to be in the moment of creation. There is no time, he points out, at which the universe did not exist. The act of creation is a timeless act, an act outside time, since God is outside time. God's Providence is thus woven into that single Divine act from which the universe—past, present, and future—sprang.

Nature is whole and entire in its own right; the "seeds" of all natural kinds are implanted at the beginning—Augustine argues that the six days of the *Genesis* account have to be understood as metaphor—and the corresponding kinds appear when conditions are right. God's purposes in the natural order are brought about not by intervening (that is, by overriding natural causality), but by ensuring that the desired result comes about naturally. It is not easy to make a consistent story of all this, particularly when one has to go on to imagine how such a timeless God could enter time and become an active participant in the story of human salvation. There has to be room for miracle and grace, and above all for the Incarnation of God in the person of Christ. Augustine had to struggle to bring these diverse threads together, and was not always successful in doing so. One context in which he *did* succeed, however, was in making room (in principle, at least) for an autonomous cosmology based on "sense-observation and necessary demonstration," as Galileo would later put it. Such a cosmology, if pursued in the proper spirit, would recognize in the wonders of the natural world the signs of God's creative action.

The rediscovery of Aristotle's "natural works" in the Latin West in the early thirteenth century led to further developments, as it had already done in the Islamic world. [See chapter 9, *Medieval Cosmology*, in section III, *Medieval Cosmology and Literature*.] The naturalism of these works worried theologians: they seemed to present a self-contained world in no need of the Divine, except for the sustaining of motion. And their conception of science as rooted in necessary principles challenged the religious view of the Creator as radically free in his choice of world-kind. Aquinas set out to show that one could be both an Aristotelian and a Christian, that the cosmology of Aristotle could coexist with the theology of creation. He argued, even more strongly than Augustine had, for the existence of real causal relations between created things, and hence for the genuineness of a natural science based on such causal relations. Although God is the primary cause of the existence and action of each creature, He has endowed these creatures with natures and enabled them to act on one another in ways we can discover and understand.

Aristotle's proof of a First Mover is now subordinated to a larger scheme in which the Creator is responsible not just for the motions of the material world but, more fundamentally, for its continued existence. The supplying of motion is simply one index of the entire dependence of the cosmos on its Creator. There is a tension here, nevertheless, between two ways of inferring to God's existence. One situates God within the cosmology itself, as the supplier of motion required by mechanical principle. The other views God as a precondition to *any* cosmology, required by a metaphysical argument from the contingency of any material cosmos. If one accepts the first, as Aquinas did on the strength of Aristotle's physics, cosmology bears directly on religious belief, provided one can construe the god in whom the argument from motion terminates as the God who is central to theistic belief and practice. But, as we shall see, this direct link between cosmology and religion was not to endure.

Copernican Challenge

Aquinas's blending of Aristotelian science and Christian theology did not persuade everyone. Shortly after his death, an assembly of French bishops issued a condemnation of many of the features of Aristotelian thought that they found objectionable (1277). Yet, quite soon, opposition diminished and Aristotelian cosmology began to be comfortably linked with Christian belief. A new philosophical critique of Aristotelian thought was mounted by the nominalist followers of Ockham, but most Christian thinkers seemed to find Aquinas's synthesis satisfactory. People's imaginations were shaped by Aristotle's concentric model of the universe, with the earth at the center and the planets carried around it on nested spheres. There were some—like Nicole Oresme and Nicholas of Cusa, both prominent churchmen—who suggested the possible advantages of a sun-centered cosmology, but little attention was paid to what were no more than suggestions until Copernicus transformed suggestion into actual mathematical calculation and showed that a heliocentric model could explain several features of the planetary motions that were merely ad hoc postulates in the rival geocentric astronomy. [See chapter 8, *Copernicus*, in section II, *The Greeks' Geometrical Cosmos*.]

In his play *The Life of Galileo*, Bertolt Brecht has an elderly Cardinal fulminate against the Copernicans who would demote human beings from their rightful place at the center of the universe God had created around them. Brecht was echoing a standard modern reading of why the Copernican theory was so strongly opposed by the Church. But this is anachronism; there is nothing in the record to support it. The real issue was the interpretation of Scripture—and who should be the interpreter. Luther's challenge to Rome bore on this very issue. Should the authority to interpret rest in the individual believer, or should it rest in the Church, represented by the bishops? What role should Church tradition play? Both sides in the spreading dispute cited Scripture in support of their theological views. Not surprisingly, the stress on *literal* interpretation grew. Where an earlier generation of theologians had been comfortable with allegory and metaphor, now the suggestion that a particular passage or phrase of Scripture ought be taken nonliterally was likely to be greeted with suspicion, by Reformers and supporters of Rome alike. Copernicus's book simply came at the wrong time.

No wonder, then, that Copernicus's Lutheran friend, Osiander, charged with the publication of *De revolutionibus* as its author lay dying, added a brief *Foreword* in which the heliocentric astronomy of the book is passed off as a useful calculational device, of no relevance to the *real* motions of sun or earth. This instrumentalist interpretation of mathematical astronomy was, indeed, more or less standard; it had its roots in the physically uninterpretable epicycles and equants of the Ptolemaic tradition. But it ran quite counter to Copernicus's manifest intention to claim that the earth really is in motion. Osiander was clearly hoping to defuse the opposition he expected from literalist readers of the Old Testament who would recall passages in which the earth is said to be at rest or the sun in motion. And for a time he seems to have succeeded.

Kepler was the first, half a century later, to draw attention to the fact that the *Foreword* cannot have been written by Copernicus himself, as the reader would naturally have supposed it to have been, and that it in effect falsifies the author's intentions. [See chapter 13, *Kepler*, in section IV, *The Scientific Revolution*.] Kepler, himself a devout Lutheran, saw no difficulty in reconciling the new cosmology with the Scriptures. In his view, the disputed Biblical phrases testified only to popular usage at the time the works were written. There was absolutely no warrant for treating them as literal claims about which cosmic bodies were at rest and which were in motion.

But the battle was not really joined until Galileo made his famous discoveries with the telescope in 1609 and the years following. [See chapter 12, *Galileo's Cosmology*, in section IV, *The Scientific Revolution*.] These discoveries undermined Aristotle's cosmology of concentric spheres centered on the earth. Under siege, the defenders of Aristotle responded by calling on the authority of Scripture in their support. Galileo, outraged, penned an eloquent defense of Copernicanism against theological attack, the *Letter to the Grand Duchess Christina* (1615). [See chapter 30, *Galileo and the Inquisition*.] Galileo asks what is to be done when there is an apparent contradiction between a finding in the science of nature and the literal interpretation of a passage in Scripture, and draws skillfully on theological tradition (notably on Augustine and Aquinas) in his reply. He defends two rather different strategies, without perhaps realizing how different they were in their implications for his own defense of the Copernican cosmology. The first (traceable to Augustine) is that if a claim that is supported by "the senses and necessary demonstration" appears to contradict a Scriptural passage, then an alternative reading of Scripture must be found, since real contradiction is inadmissible. If, however, the claim falls short of demonstration, the literal reading of Scripture should be maintained. (The implication of this maxim is that he will have to find a *demonstrative* proof of the Copernican system.)

The second principle that Galileo enunciates, one that was not without precedent among theologians but was unlikely to gain favor in tense Counter-Reformation Rome, is that the Scriptures are simply not relevant to matters of natural science, that they were not written with deep truths about nature in mind, that they were accommodated to the capacities of the listeners, that their function was to teach people "how to go to heaven and not how the heavens go." If *this* position were to be accepted, Galileo would not have to produce a demonstration of the Copernican position in order to be heard. A degree of likelihood would suffice, since the Scriptural objection is ruled out from the beginning.

The arguments of the *Letter to the Grand Duchess Christina* fell on deaf ears in Rome. A few months later (1616), a committee of consultors to the Holy Office (popularly known as the Inquisition) declared that the claim that the sun was at rest in the center of the universe was heretical and the claim that the earth was in motion was close to heretical. (The stronger emphasis on the first of the two claims was due to the fact that the Biblical references to the sun's motion are more numerous than those to the earth's immobility.) The Holy Office banned the works of Copernicus "until corrections would be made" (i.e., until assertions about the "real" motions of earth or the "real" position of the sun should be removed), which was done a few years later. It was made clear that there was no objection to retaining Copernicus's work as a major resource in mathematical astronomy, in the purely instrumentalist sense given that discipline by philosophers and theologians of the day.

The decree of 1616 set the stage for the final scene, perhaps the most celebrated moment in the long history of interaction between cosmology and religion in the West. The accession of his friend

Urban VIII to the Papal throne encouraged Galileo to write the extended work in defense of Copernican cosmology that he had long been contemplating. The *Dialogue Concerning the Two Chief World Systems* appeared in 1632 and almost immediately caused a storm of controversy. Urban VIII was especially enraged, in part, perhaps, because an argument he had himself proposed against the realist Copernican claim had been put in the mouth of the hapless Simplicio, the representative of Aristotle in the *Dialogue*, who was on the losing side of every argument. Galileo was put on trial before the Inquisition (1633), forced to retract his defense of Copernicus's work as a cosmology, and sentenced to lifetime house arrest.

In retrospect, it seems clear that a clash of so dramatic a sort was in no way inevitable. Had the sensitivities over the interpretation of Scripture been less keen—the issue, after all, was primarily one about how the literal reading of the Scripture should be circumscribed— had the committee of consultors in 1616 taken the arguments of Galileo's *Letter* into account, had Galileo not antagonized so many potential supporters, and had he found a more convincing demonstration of the merits of the Copernican case (particularly over what was by then its *real* rival, the Tychonic cosmology), the outcome might have been different. But as it was, in Catholic Europe at least, the new cosmology was called into doubt, and a pattern of long-lasting distrust was established.

Newtonian Cosmology

When news of Galileo's condemnation reached Descartes, he abandoned work on an ambitious cosmological treatise, *Le monde*, in which he proposed to provide a mechanical basis for the heliocentric Copernican model. But in his *Discourse on Method* (1637) shortly after, he sketched what would turn out to be a more revolutionary idea, even, than Copernicus's reordering of sun and earth. What, Descartes asked, if a chaos of particles in motion were to have been created by God, simply obeying the laws of the mechanics Descartes believed he had discovered? Might the chaos of particles not in a perfectly natural way eventually form sun, planets, earth, and even on the earth the complex sorts of bodies that we know? He was convinced that a cosmology could be devised in which the origins of even the most complex kinds of organisms might ultimately be explained by means of mechanical laws only, with no particular specification of the initial universe-state (no "fine-tuning," as it would be described today) needed. This may be called the "Cartesian principle" in cosmology.

Here was the first hint of a potentially self-contained science-based cosmology, one that, if successful, would need no special Di-

vine intervention to explain the origins of even the most highly organized natural kind. It had as yet little evidence in its support, but it
had a plausible ring, for large material systems like the planets, at
least. But what of living things? Robert Boyle, John Ray, and other
students of nature argued that the adaptation of means to end everywhere found in the living world, particularly in the instinctual behaviors peculiar to each natural kind, could not be explained by the mere
operation of mechanical law. A Designer was needed, a being who
could shape the original structures and behaviors of each kind for the
benefit of that kind. Cosmology testified directly, therefore, to the
existence of a Designer, though no longer to a First Mover in Aristotle's
sense. The law of inertia, first formulated by Descartes and foundational to Newton's mechanics, obviated the need for a continuing
Mover, as a matter of mechanical principle, at least.

Newton was sensitive to the charge that the new science implicitly promoted atheism, and so he was at some pains to point out that
God would still be needed to maintain the stability of the planetary
system, for example. [See chapter 14, *Newtonian Cosmology*, in section
IV, *The Scientific Revolution*.] His own mechanics could not, he thought,
explain that stability. His own belief in an omnipotent Creator in no
way depended on such considerations as these. But in an age when
religious belief was under challenge, it was tempting to base an
apologetics as directly on the new science as possible. And so a new
sort of science-based natural theology appealing to adaptation as evidence of design in the living world as well as to various apparent gaps
in the new mechanical cosmology, became popular among Christian
believers, especially in Britain. Cosmology could still, it would seem,
cohere with, even sustain, religious belief.

But the Newtonian gaps were, to all appearances, gradually filled.
Laplace, for example, argued that the laws of mechanics *could* of themselves explain the stability of the solar system. And what had been
only a promise in Cartesian cosmology gradually took on concrete
shape as Newtonian mechanics was applied by Kant and others to the
problems of cosmogony. The formation of planetary systems, even
perhaps of the Milky Way galaxy itself, could be understood in mechanical terms as a gradual coalescence of particles acted on by gravitational forces. The geological complexities of the earth's surface
testified to aeons of gradual development under the action of erosion,
deposition on lake and sea bottoms, volcanic eruption, and the rise
and fall of land-surfaces—all potentially intelligible in physical terms.

As telescopic improvements continued, and astronomical knowledge became ever more detailed, this developmental approach to
cosmogony led more or less naturally to a belief in the vast number of

habitable planets. [See chapter 27, *Plurality of Worlds*, in section VIII, *Cosmology and Philosophy*.] It was a short step to supposing them actually inhabited, and the argument most often relied on in making this step was based on a religious premise: the Creator would not in vain have provided so many abodes for life. What had seemed little more than literary fancy when Fontenelle published his *Conversations on the Plurality of Worlds* in 1686 gradually became an almost universal belief, despite the lack of direct evidence in its support. Kant in his *Universal Natural History* (1755) was one of the most assured: the plenitude of God's creative power is such that it would be "sheer madness" to deny that most of the solar planets must be inhabited by thinking beings. Furthermore, since planets are generated in the process of stellar formation, and there are uncountable other systems of stars besides our own Milky Way, we should expect life to be found throughout the boundless universe. There is no suggestion that this might pose a difficulty for the Christian, perhaps in part because it was a specifically religious premise that, in those pre-Darwinian days, enabled Kant to infer that the innumerable planets would, in fact, be inhabited.

But not everyone saw it that way. Thomas Paine in *The Age of Reason* (1793) made use of the general belief in a plurality of worlds to argue *against* Christianity: "From whence, then, could arise the solitary and strange conceit that the Almighty, who had millions of worlds equally dependent on his protection, should quit the care of all the rest and come to die in our world because, they say, one man and one woman had eaten an apple!" If, on the other hand, a Redeemer were to be sent to each of those worlds, "the person who is irreverently called the Son of God, and sometimes God himself, would have nothing else to do than to travel from world to world." For a deist like Paine, cosmology had become a powerful argument against the particularity of any religion claiming a privileged place for God's dealings with earth. (Paine's critique has recently been revived by Roland Puccetti, who utilizes the optimistic Sagan-Drake estimates of the likelihood of extraterrestrial intelligence.)

The most celebrated response to this objection came from Thomas Chalmers, a Scottish divine, who, in a set of sermons later published as *Astronomical Discourses* (1817), acknowledged the insignificance of earth in purely physical terms but argued eloquently that the generosity of a Creator who cares for even the least organism on earth would not draw back from sending a Redeemer to the humblest of his provinces, even to a species as undeserving as the human one is. We know nothing of how the plan of redemption extends to the peoples of other planets; for all we know, they may not need redemp-

tion or, if they do, they may receive it through some form of participation in the redemptive action of Christ on earth. Chalmers labors
to transform the new cosmology from a challenge to an earth-centered religion into a celebration of God's magnificence and grandeur.

Later, a very different sort of response came from William
Whewell, the celebrated master of Trinity College, Cambridge, and
the ablest philosopher-historian of science of his generation. He disagreed with Chalmers and the many others who had made of the
plurality of worlds almost a Christian doctrine. Whewell saw the
force of the type of objection that Paine had leveled, but instead of
drawing the inference that Paine had, he preferred as a Christian to
argue that the premise of the plurality of worlds was unsound. The
scientific eminence of the author made the unfashionable thesis of
The Plurality of Worlds (1853) all the more unexpected to his contemporaries. How, he asks, can a Christian possibly assent to the view
that the earth, the scene of Christ's redemptive work, is merely one
among millions of planets inhabited by rational beings? And he goes
on to challenge at every step the astronomical arguments that purported to establish the plurality of worlds: the existence of galaxies
other than our own, the great multiplicity of sun-like stars in our own
galaxy, the likelihood of planetary formation around an average star,
and so on. His most effective objection was to the almost universal
supposition that the planets of the solar system are habitable. (One
author, Thomas Dick, inferred the total population of the solar system to be almost twenty thousand *billion*, based on an estimate of
surface areas and the assumption that the average population density
would be roughly that of England!) Whewell considers the planets
one by one, and argues that from what is known of each, only earth
could be the abode of complex life. His book was a *tour de force*, and
although it did not win the assent of many, it served indirectly to
underscore that the vastly expanded universe of the new cosmology
might raise more of a problem for specifically Christian belief than
Christians had up to that point acknowledged.

Evolutionary Cosmology

The publication of Darwin's *Origin of Species* in 1858 drastically altered the parameters of the debate about the implications of cosmology for religion. Darwin's use of the evidence of adaptation to support a theory of natural selection undermined the Design argument
on which (in Britain, at least) Christian apologetics had become so
dangerously dependent. Now, at last, it seemed, one could envision
an entirely self-contained cosmology, one in which neither Mover
nor Designer had any role. For many, including Darwin himself, the

increasing autonomy of cosmology meant a loss of the primary motive for belief in God. From having been a strong support for a religious world view, natural history had now become for many a strong disincentive.

Some of those who accepted an evolutionary origin of natural kinds remained unpersuaded, however, that this could be explained simply in terms of natural selection working on chance variation. Alfred Wallace, co-formulator with Darwin of the original hypothesis of natural selection, was one of the skeptics. Development of such complex organs as the human brain could not, he argued, be explained by a blind process of chance variation and differential reproduction; a more directive agency was required. And so he postulated an Intelligence acting to supplement the effects of natural selection in order to account for the orthogenetic aspects of evolutionary change. In his view, although Design in the older sense of an instantaneous creation of natural kinds was no longer needed, God's action in the world could still be discerned in evolutionary process itself.

By the turn of the century, a new and more comprehensive cosmology had begun to make its way. It was based on Darwinian theory rather than on physics. Although the term *cosmology* was not usually attached to it, it was in fact a cosmology, an attempt to give a speculative account of world-order generally. The evolutionary philosophy propounded by scientists like Herbert Spencer was explicitly cosmological in intent. Spencer extended the notion of evolution from the biological back into the physical and forward into the social realms. Everything was to be understood in terms of origins, and the path from these origins to the structures of the present day was governed entirely by scientific law. Newtonian mechanics had been able to explain the origin of some large-scale gravitationally bound structures, but not much more than that. The evolutionary mechanisms proposed by Darwin had far more potential for cosmology; in Spencer's view, nothing more would be needed.

Among the evolutionary philosophers, two distinct strains can be noted. On one side were those like Spencer and Ernst Haeckel, for whom a deductive Darwinist model of explanation could be used to support a broader materialism that excluded theism of any sort. They were especially critical of the major institutional religions of the West, which were, in their estimation, no better than superstitions serving mainly political ends. Their own mission was to propagate the new evolutionary cosmology, particularly by influencing the education of the young, and in this way lessen the influence of organized religion on its adherents. They were willing to recognize that religion had in the past provided support for desirable ethical values; this dimension of religion they hoped to maintain.

But there were others, on the contrary, who, like Wallace earlier, took evolution to testify rather to the direct action of God in the world and who built an entire cosmology around this intuition. Best known of these was Henri Bergson whose book *Creative Evolution* (1907) drew heavily on the French vitalist tradition in biology. He argued that many features of the evolutionary record could not possibly be explained by Darwinian mechanics only. The steady increase in organic complexity, the apparent coordination of changes occurring together in the organism, and the advent of the human required an agency capable of shaping outcomes in a way that mechanical energies alone could never bring about. An *élan vital*, or living impulse, was required, and evolution itself was the main testimony to the continued operation of this transformative force in all living things. Bergson originally doubted that evolution manifested an overall directional character, but in his later writings, as he more and more explicitly identified *élan vital* with God's guiding action, he sometimes spoke of the appearance on the scene of the human as the goal of evolution.

The Bergsonian inspiration is clearly recognizable in the speculative cosmology of Pierre Teilhard de Chardin, a Jesuit paleontologist, notably in his posthumous work *The Phenomenon of Man* (1955). Teilhard, like Bergson, saw a special nonmechanical agency (which he called "radial" or "psychic" energy) at work in the history of evolution, particularly in the growing "complexification" of the organic realm and in major transitions such as those between nonliving and living, nonsentient and sentient, nonreflective and reflective. For such transitions as these to occur, the higher properties must already have been present in a latent state from the beginning. Once the level of the human is reached, evolution shifts into the realm of the social, the "noosphere." As this latter develops toward a higher degree both of complexity and of unity, the universe progresses towards Omega Point, the end of history, where the complex becomes fully one and the distinction between material and spiritual is finally overcome. Evolution for Teilhard is "no longer an hypothesis but a light which illuminates all hypotheses." Someone who views the universe from the Christian perspective (he claimed) will see in radial energy the transformative power of Christ and will find all sorts of resonances in evolution for the doctrines of God's becoming man and of man's eternal destiny with God. Teilhard's vision was greeted with enthusiasm by those who sought for a closer unity between science and the spiritual, but it evoked rebuke from traditional theologians and angry hostility from defenders of the neo-Darwinian synthesis.

Alfred North Whitehead's system of thought was broader in basis and less specifically theological in inspiration. Beginning from mathematics and relativity theory, he developed a very general metaphysics, in which the world appears as a system of extended and interlocking events, and objects appear as recurrent patterns in this complex. In *Process and Reality: An Essay on Cosmology* (1929), he distinguishes between metaphysics (which deals with the formal character of all *possible* facts) and cosmology (which is generalized from the empirical sciences of a particular period and thus reflects the contingent character of one particular type of world-order). The fundamental metaphor of this cosmology is that of organism: all actual entities have a "subjective aim" toward which they strive. Even the molecule exhibits a form of sentience, though not of consciousness; the latter is for Whitehead an incidental, not a basic, property. Like Teilhard, he denies the emergence of strictly new levels of existence; all basic properties have been there from the beginning, in some sense. Evolution is one consequence of the striving on the part of all actual entities towards novelty and self-creation. The original ordering of the world from which this process takes its rise is identified as the "primordial nature of God." In the process of cosmic becoming, God also provides the other actual entities with the impetus to self-creation, and in that way acquires a "consequent nature" in which each entity is "objectified" or reflected. God is thus both the beginning and end of natural process, and evolution is ultimately God's own gradual self-realization.

Cosmology and religion in one sense could hardly come closer. Yet Whitehead's God is so abstract, so remote (it would seem) from the categories of traditional religion, that it is hard at first sight to see how to bring them together. But one school of theology, process theology, has labored to do just that. The traditional theology of creation stressed God's transcendence, his independence of the universe, even the possibility that the world might never have been. Evolutionary cosmo-theologies of the "process" type tend to place God in time and to describe God as "groping" through cosmic process towards an uncertain self-fulfillment. They lean to immanence; in traditional terms, they would be pantheistic rather than theistic. The best-known writer, perhaps, in this tradition is Charles Hartshorne, whose *A Natural Theology for our Time* (1967) defends a relationship between God and Universe (pantheism) that is neither pantheism nor theism of the traditional sort.

Recent cosmologies of biological, and specifically evolutionary, inspiration have obviously been far from neutral where religious belief is concerned. Jacob Bronowski's popular television series, *The Ascent of Man*, began from a broadly evolutionary account of human origins and went on to assert that as science has advanced, ignorance and superstition in the form of institutional religion have retreated. Writers like Jacques Monod (*Chance and Necessity*, 1971) and Eric Dawkins (*The Blind Watchmaker*, 1986) have argued that the basic explanatory concepts of the neo-Darwinian "modern synthesis" exclude design in any form, showing not just that it is not necessary, but more fundamentally, that it is not possible because of the crucial role that chance plays in evolutionary process. Not only is a natural theology blocked, then, but postulating a God is otiose, since there is nothing for such a being to do, and good reason furthermore to suppose that such a being could not exercise the sort of Providence traditionally assigned to the Creator.

Theologians, on the other hand, have criticized these naturalist objections as misplaced. Many theologians would agree that a natural theology of the traditional sort (i.e., one based on a specific feature of the world, like design, that science is supposedly unable to explain) no longer carries weight. But they would point out that cosmological arguments of this kind have only been in vogue since natural science attained authority in the seventeenth century, and that the argument most favored by theologians relies on a deeper existential contingency of the natural world as a whole, of the world that science simply presupposes as its object. Other theologians have insisted that religious belief is not motivated by an appeal to God's role as an explainer in the first place. And the objection regarding chance would be countered by the Augustinian response that a Creator who brings the universe to be in a single act is not dependent on the present for his knowledge of, or dominion over, the future.

Physical Cosmology

Evolutionary theory served to prompt cosmological speculation rather more than did Newtonian physics. True, Newton did postulate an absolute and unbounded framework of space and time as a necessary framework for his new way of conceptualizing the relation between force and acceleration. And his "Third Rule of Reasoning" allowed him to assume that the properties of the limited sample of the world where his mechanics had been shown to apply could be universalized to the world as a whole and at all levels of size. But there was no way to know how far outwards in space matter in the form of planets and stars extended, nor at what point in time these bodies had taken shape. Kant, indeed, argued that the attempt to characterize the universe as a whole inevitably led to antinomy. His own search for an adequate foundation for Newtonian science led him to seek this in mind rather than in an autonomous material world over against the knower. In later debates between materialists and their critics, the authority of Kant would often be invoked in support of some form of idealism. And idealism came to seem to many the most secure defense of religion against materialist attack, particularly in Germany, where Kant's influence was strongest.

In England in the period between the two world wars, this debate took on a distinctive form. The two leading astrophysicists of the day, James Jeans and Arthur Eddington, were best known to the public as the authors of immensely successful popular works on cosmology. Each defended his own brand of idealism, each linked it with the new scientific discoveries, and each also saw it as a natural bridge between science and religion. [See chapter 18, *Cosmology 1900–1931*, in section VI, *The Expanding Universe*.] In *The Mysterious Universe* (1930), Jeans proposed a view strongly reminiscent of Plato's *Timaeus*. As physics has progressed, it encompasses the world more and more in the symbols of pure mathematics. The universe is clearly built on a complex mathematical pattern that pervades all levels from the galaxy to the atom. The only way to explain this is to see it as the work of pure mind, specifically, of a Divine Mathematician. Science can thus testify directly to the existence of a Creator; it can furnish a natural theology sufficient to sustain religious belief.

In *The Nature of the Physical World* (1928), Eddington took a different approach. He argued, along Kantian lines, for the mind-dependence of the sense-world in general, stressing the various forms this dependence takes, in quantum theory in particular. He saw the influence of mind everywhere, not as an original creator but as actively and continuously moving within Nature. "I simply do not believe that the present order of things started off with a bang" (*NPW*, p. 84). "Philosophically, the notion of an abrupt beginning of the present order of Nature is repugnant to me" (*New Pathways in Science* [1933], p. 59). Abrupt beginnings suggested to him an unacceptable sort of deism, conveying that there had been "a single winding-up at some remote epoch" and that the work of the Creator was then done, whereas Eddington wanted to emphasize the continuing presence of a Universal Mind, a Mind that was in some way constitutive of Nature itself and whose role was effectively revealed in post-Newtonian science.

He opposed Jeans's project of a natural theology, since the Mind whose traces science everywhere discerns was too unspecific to serve as object of religious belief. Besides, according to Eddington, it is *our* mind whose traces we are finding, *our* mind that imposes mathematics upon the universe. All that he has himself tried to do (Eddington says) is to remove difficulties against religious belief that were rooted in the determinism and materialism of the Newtonian world view, and to give "grounds for an idealistic philosophy which is hospitable towards a spiritual religion" (*NPS*, p. 306). The most that science unaided could testify to would be a "colorless pantheism"; his own religion, by contrast, is a mystical one, based on a personal experience accepted in advance as fundamental.

The support offered to religion by Jeans and Eddington did not go unchallenged. In a sharply critical work, *Philosophy and the Physicists* (1937), Susan Stebbing likened them to "revivalist preachers" and accused them of "cheap emotionalism" and "serious mental confusion." She was much more critical of Jeans than of Eddington; in her view, Jeans's argument for a Creator was a tissue of fallacies from beginning to end; he had shown himself to be inexcusably unaware of the distinctions that philosophers had insisted on in arguments of the sort. She devoted more space to Eddington, but in the end concluded that his attempt to base a quasi-Kantian idealism on science also failed. Hence, his indirect mode of linking science and religion would not work. Stebbing's criticisms may have stung; in their later books, neither Jeans nor Eddington returned to the confident religious declarations of their earlier works.

There was by then a new and much more powerful cosmological synthesis in the making, and the term *cosmology* was coming to be appropriated almost exclusively by physicists as a convenient label for that part of physics that deals with the nature and mode of formation of the largest material structures. Einstein's general theory of relativity employed a non-Euclidean formalism that would permit one, under certain constraints, to represent the cosmos once again *as* a cosmos, as a single finite (though unbounded) whole. Further, the large galactic redshifts discovered by Hubble could best be understood, Lemaître argued, as indicators of a cosmic expansion beginning from a "primeval atom." [See chapter 19, *Hubble's Cosmology*, and chapter 20, *Big Bang Cosmology*, in section VI, *The Expanding Universe*.] Eddington gave the new cosmology wide currency in an immensely popular little book, *The Expanding Universe* (1932). The affinities between the "Big Bang" model, as it was later dubbed, and the Judaeo-Christian notion of creation did not go unnoticed. Neither Lemaître nor Eddington, however, would countenance any sort of inference from one to the other, Lemaître because (among other reasons) there was no way to show that the expansion had not been preceded by a contraction, and Eddington because, as we have seen, he found the notion of an abrupt beginning unsatisfactory on philosophical grounds and preferred to suppose that the expansion had begun from an instability in an enormously dense concentration of energy that had remained in an "embryo" state for an indefinitely long time.

There were problems with the big bang model; in particular, the calculated expansion appeared to be too rapid, since it gave an age for the universe (one to two billion years) that was too short to accommodate the findings of geology and paleontology. Bondi and Gold proposed an alternative in 1948, a steady-state model that avoided the suspect singularity, but at the cost of introducing a host of lesser singularities. Hydrogen atoms were supposed to appear *ex nihilo* at an imperceptibly low rate in order to maintain the steady state, as galaxies moved away from one another. [See chapter 21, *Steady State Theory*, in section VI, *The Expanding Universe*.] This model was adopted and extended by Fred Hoyle, who objected to the clear affinity he perceived between the big bang model and the traditional Christian view of creation. In a number of popular works, from *The Nature of the Universe* (1950) to *Ten Faces of the Universe* (1977), Hoyle combined an exposition of popular science with scathing attacks, after the manner of Haeckel, on organized religion. Reacting against the strong religious cast of the leading British cosmologists of the older generation, he pounced on anyone unwary enough to suggest (as Pope Pius XII, for example, did in an address to the Pontifical Academy of Sciences in 1951) that big bang cosmology gives aid and comfort to the Christian believer.

The identification of the ubiquitous microwave radiation, discovered by Penzias and Wilson in 1965, as having the "signature" of the 3 degree Kelvin radiation predicted by the big bang model, had two almost immediate effects. First, it gave this model what proved to be a decisive advantage over the steady state alternative, erasing the slight lead the latter had by then built up. Second, and even more significantly, it persuaded many of the skeptics that cosmology had at last become respectable. Critics, like the positivist philosopher of science Herbert Dingle, had dismissed relativistic cosmological models as irresponsible speculation based on idealized assumptions misleadingly described as "cosmological principles." So many assumptions, indeed, were involved even in the interpretation of the galactic redshift as expansion that it seemed as though almost any finding could be "explained" by an *ad hoc* adjustment of one kind or another. But now a precise unadjusted observational consequence of one of the models had been unexpectedly verified, traditionally the mark of a "mature" science.

Fine-Tuning the Universe

One last story remains to be told. The big bang model, unlike the steady state alternative, involved an initial moment when energy densities were extremely high and, hence, when quantum effects would be the controlling factor. Thus one could hope that the application of quantum theory to these extreme conditions could both help to elucidate the sequence of states as the universe cooled and also serve to test the current versions of quantum theory in a way that laboratory conditions on earth never could. These hopes have been amply borne out. But one major surprise was in store.

In the early 1970s, reconstruction of the likely sequence of events as, first, the various elementary particles and, then, hydrogen and helium atoms made their appearance made it seem that the sort of universe we have (long-lived, populated by galaxies of stars, and containing appreciable quantities of the heavier elements) required an extraordinarily delicate balance of initial conditions and of the laws relating the four fundamental forces. Had these been even slightly different, the universe might have been very short-lived, either collapsing on itself or expanding too rapidly for matter to form into stars and planets. Or it might have consisted of hydrogen alone or helium alone. In short, it seemed that the Cartesian principle had been refuted. Descartes, we recall, maintained that the application of a simple set of mechanical laws (themselves justified on *a priori* grounds) to an initial "chaos" would be sufficient to explain how the major structures of the universe originated. In the Cartesian system, no limitations had to be set on the initial cosmic conditions, whereas in quantum cosmology an extreme sort of "fine-tuning" seemed to be required both of the initial energy-density and of the contingent-seeming relative magnitudes of the fundamental forces. (*Fine-tuning* has something of the ambiguity of the term *creation*; if it be understood as an action, then the existence of a "fine-tuner" seems to follow. Perhaps a more neutral term would be better.) Among the *possible* universes (possible in terms of quantum cosmology), only a vanishingly small proportion would be of the sort we inhabit: long-lived, galactic, etc. Does the fact that the universe *is* of this sort require explanation? B.J. Carr, Martin Rees, and several other British cosmologists suggested that it does require an explanation and proposed an "anthropic principle" by way of explanation. [See chapter 26, *The Anthropic Principle*, in section VIII, *Cosmology and Philosophy*.] The so-called "principle" took a number of different forms, but in a general way it asserted that the presence of human beings somehow explains why the universe has the unlikely form it has. A universe of a fairly specific sort is *necessary*, they argued, in order for complex life-forms to appear.

But what follows from this? For there to be creatures of a broadly human sort, let us suppose that the universe must be of the specific kind described. But did there have to be creatures of this sort in the first place? This is where a religious answer suggested itself, to some, at least. According to the Biblical account of creation, God chose a world in which human beings would play an important role, and would thus have been committed to whatever else was necessary in order for this sort of universe to come about. If fine-tuning was needed, this would present no problem to the Creator. In this perspective, the significance of the human in God's plan could be said to explain why the universe is of the sort it is, if indeed this fact *needs* special explanation. This is a genuine explanation, *provided* that the premises are correct. From the Christian perspective, then, the appeal to a Creator in order to explain the supposed fine-tuning is attractive. The further step to a natural theology is, however, highly dubious. For this step to be permissible, the claim of fine-tuning would have to be validated *and* alternative explanations, or approaches to the issue of explanation, would have to be undermined. This would be hard, if not impossible, to do.

We are, after all, dealing with a highly speculative theory about a situation entirely remote from any of which we have direct experience. An important modification of the original big bang model was proposed by Alan Guth in 1981, involving a brief period of extraordinary inflation in the first fraction of a second of cosmic expansion. [See chapter 22, *The Inflationary Universe*, in section VII, *Particle Physics and Cosmology*.] This would help to solve the so-called horizon and flatness problems of the standard model, and would eliminate part of the need for "fine-tuning." An initial cosmic energy-density permitting a long-lived universe, for example, would cease to be so improbable. But inflation itself causes some new problems; the pros and cons of this imaginative modification are still under debate. The moral is, however, clear: what appears as "fine-tuned" today may find an explanation tomorrow in a new or modified model of that remote instant when the processes of expansion and cooling began. The force of the metaphor of fine-tuning depends on establishing a special sort of contingency or unlikelihood that requires explanation. As long as the theory is as speculative as it still is, what now appears contingent can easily turn out after all to be necessary. And some will argue that even if it remains contingent, any kind of scientific explanation must ultimately end with a state of affairs that is just *given*.

The theistic explanation is, of course, not a scientific one and depends for its force on a *prior* belief in a Creator who would be likely to assign privileged status to human nature in the work of Creation. A *scientific* alternative is offered by a variant of the "many-worlds hypothesis" postulated by Everett and Wheeler in quantum theory. [See chapter 28, *Multiple Universes*, in section VIII, *Cosmology and Philosophy*.] According to this variant, if there are immeasurably many *real* worlds constantly causally diverging from one another, we would, of course, find human beings only in those of the worlds in which the conditions for the development of advanced organic life were satisfied. And so the "because we are here" of Brandon Carter's original anthropic principle would become quasi-explanatory, in the sense that the apparent fine-tuning would vanish. The many-worlds approach is, however, faced with many difficulties in its own right and enjoys little support among scientists.

What made this episode in quantum cosmology so intriguing to religious believers was that, for the first time since the heyday of physico-theology, scientific theory seemed to point to a state of affairs that would itself require an explanation of a different order. And unlike all other such leads in the past, this was one that would not require an alteration of natural process on the part of the Creator, no intervention of a miraculous sort, only a creative setting of the original stage. Physicists like Freeman Dyson (*Disturbing the Universe*, 1979) and writers of popular works on cosmology like Paul Davies (*God and the New Physics*, 1983) have found this sort of resonance between cosmology and theistic belief appealing, even if in no sense constituting a natural theology, in no sense, that is, capable of yielding a persuasive argument for the existence of a Creator.

The popularity of Stephen Hawking's *A Brief History of Time* (1988) no doubt derived in part from his cryptic remarks therein about the negative implications of recent cosmology for belief in a Creator. Critics were quick to point out that he had improperly assumed that the Christian view of creation requires one to suppose that time had an abrupt beginning and that, therefore, his own novel conjecture (which dispenses with an abrupt temporal origin) would eliminate the need for a Creator. Aquinas, as we saw, faced with the Aristotelian thesis of a world without a temporal beginning, reminded his readers that even such a world would still need a reason why it should be rather than not be. Although Christians have always believed, on theological grounds, that the world did begin at a finite time in the past, this is not (Aquinas insisted) part of the content of the notion of creation itself, which merely signifies dependence on being, not necessarily on an abrupt origin in time.

Cosmologists are agreed that the question "Why should the universe have existed in the first place?" is not a scientific one. But they are divided on the further question of its legitimacy *as* a question. Some would insist on the traditional positivist tenet that only questions that are in principle capable of scientific answer are permissible in the first place. Carl Sagan gave this restrictive thesis memorable life in his widely watched TV series *Cosmos*. ("The cosmos—as known by science—is all there is, all there was, and all there will be.") The thesis is itself a philosophical one, of course, and has been put to work in the cosmological context rather more by philosophers of science (among whom Michael Scriven and Adolf Grünbaum have been particularly emphatic) than by cosmologists themselves. Indeed, the relatively few cosmologists who have been willing to venture into the published record on this undeniably nonscientific topic (among then Robert Jastrow, Allan Sandage, and Christopher Isham) have been on the whole more sympathetic to the legitimacy of the existence question than have their philosopher colleagues.

Conclusion

Cosmology and religion are not as intertwined as they once were. The naturalization of cosmology that began with Descartes, accelerated with Newton and Darwin, and now continues with Hawking and Dawkins has led religious believers to recognize that the motives for belief animating the three great religions of the Book (Christianity, Judaism, and Islam) were not in the first instance cosmological. Implicit in the notion of creation has always been the idea of a universe coming whole and entire from God's hands, without gaps or supplements that would give a handhold to those who seek "scientific" ways to assure themselves of God's existence. Does this make God an idle wheel? No, because God's role in salvation history has been direct and dramatic, even if not of the gap-filling sort that philosophers and scientists have debated. No, because one may still ask the basic question "If the universe that science presupposes as its object is (as it seems to be) contingent, why should it ever have existed in the first place?" It is not enough to answer in terms of vacuum fluctuations or the like. One would still want to know why there should have been a vacuum, a space-time alive with possibility.

Recent cosmology may not offer religious believers an easy argument in support of their faith, but it surely does present a universe that is a fit subject of wonder. The vaster heavens of the quantum cosmologist proclaim to believers the glory of a Creator far further beyond their comprehension than even the mystics could have dreamed. If the new cosmology offers a special challenge to the believer, it is not so much that the work of Creation is more hidden than it was, as that earth seems so limited an arena for a Creator who holds a billion galaxies in being.

—*Ernan McMullin*

BIBLIOGRAPHIC NOTE

There is an abundant literature on the topic of religion and cosmology, from Plato's *Timaeus* onward. Singling out just a few of the most recent works, one could begin with a collection of historical essays edited by David Lindberg and Ronald Numbers, *God and Nature* (Berkeley: University of California Press, 1986). Popular works by scientists underline the change in perspective that has occurred over the last half-century. See, for example, Paul Davies, *God and the New Physics* (London: Dent, 1983) and *The Mind of God* (New York: Simon and Schuster, 1992); Freeman Dyson, *Disturbing the Universe* (New York: Harper and Row, 1979); and Stephen Hawking, *A Brief History of Time* (New York: Bantam, 1988). Two rather more technical but quite readable collections can also be recommended: Arthur Peacocke, ed., *The Sciences and Theology in the Twentieth Century* (Notre Dame, Indiana: University of Notre Dame Press, 1981); and Robert Russell, William Stoeger, and George Coyne, eds., *Physics, Philosophy, and Theology* (Notre Dame, Indiana: University of Notre Dame Press, 1988).

INDEX

W

Waczenrode, Lucas 148

Walam Olum 11

Walker, A.G. 374, 392, 561, 563, 566

Walker, T. 448

Wallace, A.R. 525, 527, 595, 596

Wallenstein, Albrecht von 240

Warao Indians 22

Washington, D.C. 321

water and earth as the principle of all things 59

water as the principle of all things 57

weak interactions 425

weakly interacting massive particles 453, 461

Wealth of Nations, The 234

Wegner, G. 451

Weil der Stadt 239

Weinberg, Steven 388, 389, 426, 427, 430, 445, 461

Weizsäcker, Carl Friedrich von 377, 530

Welles, Orson *xi*

Wells, H.G. 525

West, M. 459

Western civilization 6

Western cosmology 6

Westernness 7

Westfall, Richard S. 199, 274

Wheeler, John Archibald 375, 477, 508, 540–541, 542, 604

Wheeler-DeWitt equation 492, 493

Wheeler-Linde temporally-multiple universe 541–542

When Stars Came Down to Earth: Cosmology of the Skidi Pawnee Indians of North America 23

Whewell, William 297, 523, 524, 525, 594

Whig history 6

Whiggs Supplication 502–503

Whiston, William 277, 323, 573

White, Andrew Dickson 571–572

Whitehead, A.N. 536, 550, 551, 567, 597

Whitford, Alfred 399

Whitrow, Gerald 385, 397–398, 404

Whittaker, Edmund 383, 551, 563, 567

Whorf, Benjamin 20

wide cosmological principle, *see perfect cosmological principle*

Wilhelm IV 169

Wilkins, John 503, 518, 519, 573

Wilkinson, D.T. 386

William of Occam, *see Ockham, William*

Wilson, Curtis 262

Wilson, R.E. 306

Wilson, Robert 386–387, 389, 417, 601

WIMPS, *see weakly interacting massive particles*

Winnebago Indians 19

Wirtz, Carl 316, 341

Witten, Edward 473–474

Wittich, P. 167

Wolf, Max 314, 316, 332

Wollaston, William 308

Wood, C. 224

world of ideas 81–84, 92, 95, 585

World War I 349

World War II 368, 393, 555

Wren, Christopher 266

Wright, Thomas 277, 278, 323, 522, 573

Wright, W.H. 306, 310, 316

Württemberg 239

X

X-ray stars 389

Xenophanes 59, 63

Xerxes I 80

Xuan Ye 5

Xuan Ye school 29–32

Y

Yahweh 582–583

Yerkes Observatory 306, 347

Yin and Yang 33–34

Ylem 377–378, 381, 384

Yoshimura, Motohiko 430

Young, C.A. 304

Z

Z Centauri, *see supernova of 1895*

Zeitschrift für Astrophysik 562